ESTUARINE MICROBIAL ECOLOGY

THE BELLE W. BARUCH LIBRARY IN MARINE SCIENCE, NUMBER 1

ESTUARINE MICROBIAL ECOLOGY

Edited by L. HAROLD STEVENSON *and* R. R. COLWELL

Published for the Belle W. Baruch Coastal Research Institute by the

UNIVERSITY OF SOUTH CAROLINA PRESS

COLUMBIA, SOUTH CAROLINA

Copyright © 1973 by the University of South Carolina Press

FIRST EDITION

Published in Columbia, South Carolina, by the
University of South Carolina Press, 1973

Manufactured in the United States of America by Kingsport Press, Inc.

Library of Congress Cataloging in Publication Data

Belle W. Baruch Symposium in Marine Sciences, 1st,
  University of South Carolina, 1971.
  Estuarine microbial ecology.

  (The Belle W. Baruch library in marine science, no. 1)
  Includes bibliographies.
  1. Estuarine ecology—Congresses. 2. Microbial
ecology—Congresses. I. Stevenson, L. Harold, 1940–
ed. II. Colwell, Rita R., 1934–   ed. III. Belle
W. Baruch Coastal Research Institute. IV. Title.
V. Series.
QH541.5.E8B4 1971   576'.15   73-9585
ISBN 0-87249-281-8

To CLAUDE E. ZOBELL in recognition
of his many years of leadership in
marine microbiology.

CLAUDE EPHRAIM ZOBELL was born in 1904 in Provo, Utah, of Mormon parentage and spent his childhood in the upper Snake River valley in Idaho. He began his career as a teacher and principal of an elementary school in Rigby, Idaho. He received his B.S. degree from Utah State Agricultural College in 1927, his M.S. degree from the University of California in 1929, and, as a Thompson scholar, he obtained his Ph.D. from the University of California in 1931. His early training was in the field of medical microbiology, and he specialized in the study of the *Brucella* group. Just prior to his baptism into marine microbiology at the Scripps Institution of Oceanography, he served as research associate at the Hooper Foundation for Medical Research. He joined the Scripps Institution of Oceanography as an assistant professor in 1932 and was promoted to associate professor and professor in 1936 and 1942, respectively. At Scripps, he was in charge of the biology program from 1936 to 1937, assistant director from 1937 to 1948, and chairman of the division of marine biology from 1957 to 1960. He was professor of marine microbiology until 1 July 1972 when he retired. He is now serving as professor emeritus.

Dr. ZoBell is the person mainly responsible for the development of modern marine microbiology. Through his untiring efforts, he pioneered research into the various areas of marine microbiology (e. g., petroleum microbiology, geomicrobiology, and microbial biofouling) so that it remains practically impossible to initiate a marine microbial research program without referring to his publications. His 1946 monograph, "Marine Microbiology," remains the bible and the reference cornerstone in the field. In addition to his research and publications, he nurtured marine microbiology through trying times so that it has become a recognized, integral part of oceanography. Much of the research he conducted came in an era which he refers to as the "BGG" (Before Government Grants). He served as President of the American Society for Limnology and Oceanography in 1949, as well as chairman or committee member for various organizations such as the Committee on Geomicrobiology (National Research Council), and the Polar Microbiology Committee (American Society for Microbiology). Many scholars (both from abroad and the United States) have studied under his supervision. His teachings, through them, will be reflected in marine microbiology for many years to come.

Words that I might write would not do justice to Dr. ZoBell's contributions. His published research will always outshine and endure longer than any words describing his accomplishments.

I would like to thank the editors of this book for giving me the opportunity to pay tribute to my mentor.

RICHARD Y. MORITA
Professor of Microbiology
and Oceanography
Oregon State University
Corvallis, Oregon

PREFACE

This volume contains a collection of papers presented at the first Belle W. Baruch Symposium in Marine Science. The meeting, held in Columbia, South Carolina, on June 7–9, 1971, was convened under the title "Estuarine Microbial Ecology." The broad objective of the symposium was to promote communication among individuals actively involved in the utilization of a variety of techniques in investigations of basic processes and diverse microorganisms within the estuarine environment.

Since the standard "journal style" was not followed, the papers vary in form and mode of presentation. Each contributor was asked, where appropriate, to present techniques in detail. Discussions following the oral presentations were recorded and are included in the "Comments" section of many of the papers. The volume has been divided into nine sections for the convenience of the reader. The introductory portion, Section One, includes a paper given as the keynote address by Dr. Claude E. ZoBell. Section Nine consists of impromptu summation statements made by several participants at the close of the meeting.

The editors wish to thank Dr. John D. Buck for his assistance with the program and the authors of each of the papers for their effort in preparing the manuscripts. We also extend thanks to those who helped in the preparation for and operation of the meeting. We would like to extend a special thanks to Robert B. Abel, Director of the National Sea Grant Program, for presenting an address at the banquet held in conjunction with the symposium.

The symposium was supported by funds from the University of South Carolina, the Belle W. Baruch Foundation, and grant number GA-28196 from the National Science Foundation.

The Editors

CONTENTS

Section Nine
SUMMATIONS   519

Participants and Contributors

Robert B. Abel, Director
National Sea Grant Program
United States Department of Commerce
National Oceanic and Atmospheric Administration
Rockville, Maryland   20852

Donald G. Ahearn
Department of Biology
Georgia State University
Atlanta, Georgia   30303

Lawrence J. Albright
Department of Biological Sciences
Simon Fraser University
Burnaby, British Columbia, Canada

James E. Alexander
New York Ocean Sciences Laboratory
P. O. Box 867
Montauk Point
Long Island, New York   11954

E. H. Anthony
Department of Zoology
University of Guelph
Guelph, Ontario, Canada

Ralph E. Boerner
Adelphi Institute of Marine Science
Oakdale, New York   11769

R. O. Brinkhurst
Department of Zoology
University of Toronto
Toronto, Ontario, Canada

Ellis R. Brockman
Biology Department
Central Michigan University
Mt. Pleasant, Michigan   48858

John D. Buck
Marine Research Laboratory
University of Connecticut
Noank, Connecticut   06340

J. A. Calder
Department of Oceanography
Florida State University
Tallahassee, Florida   32306

Joseph M. Cassin
Department of Biological Sciences
Fordham University
Bronx, New York   10458
Present Address: Adelphi University
Institute of Marine Science
Garden City, New York   11530

Peter K. Chen
Department of Biology
Georgetown University
Washington, D.C.   20007

K. E. Chua
Department of Zoology
University of Toronto
Toronto, Ontario, Canada

R. R. Colwell
Department of Biology
Georgetown University
Washington, D.C.   20007

David Cook
Gulf Coast Research Laboratory
Ocean Springs, Mississippi   39564

Bruce C. Coull
Department of Biology
Clark University
Worcester, Massachusetts   01610

G. T. Cowley
Department of Biology
Belle W. Baruch Coastal Research Institute
University of South Carolina
Columbia, South Carolina 29208

C. C. Crawford
Department of Zoology
North Carolina State University
Raleigh, North Carolina 27607

Sidney Crow
Department of Food Science
Louisiana State University
Baton Rouge, Louisiana 70803

John Dean
Belle W. Baruch Coastal Research Institute
University of South Carolina
Columbia, South Carolina 29208

Jack W. Fell
Rosenstiel School of Marine and Atmospheric Science
University of Miami
Miami, Florida 33149

Steven A. Fisher
New York Ocean Sciences Laboratory
Montauk Point, Long Island,
New York 11954

John W. Foerster
Department of Biological Sciences
Goucher College
Towson, Maryland 21204

Kenneth Frenke
Adelphi Institute of Marine Science
Oakdale, New York 11769

Edward R. Gonye, Jr.
Jackson Estuarine Laboratory
University of New Hampshire
Durham, New Hampshire 03824

Robert Grandjean
Department of Biology
Belle W. Baruch Coastal Research Institute
University of South Carolina
Columbia, South Carolina 29208

Robert P. Griffiths
Departments of Microbiology and Oceanography
Oregon State University
Corvallis, Oregon 97331

Robert D. Hamilton
Freshwater Institute
Fisheries Research Board of Canada
Winnipeg, Manitoba, Canada

Bruce H. Hebeler
Department of Biology
Belle W. Baruch Coastal Research Institute
University of South Carolina
Columbia, South Carolina 29208
Present address: Department of Microbiology
University of Rochester
Rochester, N.Y. 14620

J. E. Hobbie
Zoology Department
North Carolina State University
Raleigh, North Carolina 27607

Rudolph Hollman
New York Ocean Sciences Laboratory
Montauk Point, Long Island,
New York 11954

Osmund Holm-Hansen
Institute of Marine Resources
University of California, San Diego
La Jolla, California 92037

Bruce E. Hopper
Canada Department of Agriculture
Ottawa, Canada

Galen E. Jones
Jackson Estuarine Laboratory
University of New Hampshire
Durham, New Hampshire 03824

Larry P. Jones
Department of Microbiology
Oregon State University
Corvallis, Oregon 97331

Dale Kiefer
Institute of Marine Resources
P. O. Box 109
University of California, San Diego
La Jolla, California 92037

John A. Koburger
Department of Microbiology
University of Florida
Gainesville, Florida   32601

Richard Laddaga
Department of Biology
University of South Carolina
Columbia, South Carolina   29208

Paul A. LaRock
Department of Oceanography
Florida State University
Tallahassee, Florida   32306

John Liston
Institute for Food and Science Technology
College of Fisheries
University of Washington
Seattle, Washington   98195

Carol D. Litchfield
Department of Bacteriology
Rutgers University
New Brunswick, New Jersey   08901

J. W. Marek
Lander College
Greenwood, South Carolina   29646

I. M. Master
Rosenstiel School of Marine and Atmos-
   pheric Science
University of Miami
Miami, Florida   33146

Jack R. Matches
College of Fisheries
Institute for Food Science and Tech-
   nology
University of Washington
Seattle, Washington   98105

John J. A. McLaughlin
Department of Biological Sciences
Fordham University
Bronx, New York   10458

Samuel P. Meyers
Department of Food Science
Louisiana State University
Baton Rouge, Louisiana   70803

Charles E. Millwood
Department of Biology

Belle W. Baruch Coastal Research Insti-
   tute
University of South Carolina
Columbia, South Carolina   29208

Richard Y. Morita
Departments of Microbiology and Ocea-
   nography
Oregon State University
Corvallis, Oregon   97331

Steven Y. Newell
Rosenstiel School of Marine and Atmos-
   pheric Science
University of Miami
Miami, Florida   33149

Jim Oliver
Department of Biology
Georgetown University
Washington, D.C.   20007

W. J. Payne
Department of Microbiology
University of Georgia
Athens, Georgia   30601

E. Charles Pilcher
Marine Research Laboratory
University of Connecticut
Noank, Connecticut   06340

Darrell B. Pratt
Department of Microbiology
University of Maine
Orono, Maine   04473

Millicent L. Quammen
Department of Oceanography
Florida State University
Tallahassee, Florida   32306

John W. Reynolds
Department of Microbiology
University of Maine
Orono, Maine   04473

Merrily Severance
Department of Oceanography
Florida State University
Tallahassee, Florida   32306

Ronald K. Sizemore
Department of Biology
Belle W. Baruch Coastal Research Insti-
   tute

University of South Carolina
Columbia, South Carolina   29208
Present address: Department of Biology
Georgetown University
Washington, D.C.   20007

Kevin P. Smith
Adelphi Institute of Marine Science
Oakdale, New York   11769

Thomas E. Staley
Departments of Microbiology and Ocea-
  nography
Oregon State University
Corvallis, Oregon   97331

L. Harold Stevenson
Department of Biology
Belle W. Baruch Coastal Research Insti-
  tute
University of South Carolina
Columbia, South Carolina   29208

Gordon Thayer
Center for Estuarine Research
Beaufort, North Carolina   28516

Robert L. Todd
Institute of Ecology
University of Georgia
Athens, Georgia   30601

F. John Vernberg
Belle W. Baruch Coastal Research Insti-
  tute
University of South Carolina
Columbia, South Carolina   29208

G. Vilks
Atlantic Oceanographic Laboratory
Bedford Institute
Dartmouth, Nova Scotia, Canada

Stanley W. Watson
Woods Hole Oceanographic Institute
Woods Hole, Massachusetts   02543

K. L. Webb
Virginia Institute of Marine Science
Gloucester Point, Virginia   23062

Thomas C. Wicks
Department of Biology
Georgetown University
Washington, D.C.   20007

Lindsay W. Wood
Department of Zoology
University of Toronto
Toronto, Ontario, Canada

Richard T. Wright
Department of Biology
Gordon College
Wenham, Massachusetts   01984

Richard Zingmark
Department of Biology
Belle W. Baruch Coastal Research Insti-
  tute
University of South Carolina
Columbia, South Carolina   29208

Claude E. ZoBell
Scripps Institution of Oceanography
University of California, San Diego
La Jolla, California   92037

Section One
INTRODUCTION

Introduction.
The Estuarine Ecosystem[1]

F. John Vernberg

Welcome to the campus of the University of South Carolina and to the Belle W. Baruch Coastal Research Institute. We are particularly pleased to have this symposium as the first of what is planned to be a continuing series in marine science. This most important meeting has been made possible through the excellent planning of Dr. L. Harold Stevenson and the Arrangement Committee (Dr. R. R. Colwell and Dr. John D. Buck) and with financial support from the National Science Foundation, the Belle W. Baruch Foundation, and the University of South Carolina. Not only has this symposium on estuarine microbiology established very high goals for itself, but also its purposes fit into the broader picture of delineating the dynamics of estuarine systems. It is of significant interest to scientists and to society to know what is happening to our estuaries; to do this we must determine how they function and how best to manipulate and manage them. Knowledge of the multifaceted role of the microbial component in the estuarine system and its associated habitats is paramount to achieving this long-term objective of environmental control.

Estuaries are of particular importance in South Carolina. The coastline is relatively underdeveloped, but many economic forces are actively competing for this one resource. Historically, this type of competition has been resolved by economic-political forces, although in recent years environmental scientific considerations have taken on an ever-increasing importance. Although the ecological, scientific community has a voice, economic forces demand immediate answers to problems of utilizations of the coastal plain, and we, as ecologists, are expected to provide immediate answers to rather complex problems. Yet, as you well know, there is not enough existing scientific information to permit scientists to answer the important environmental questions. As a result, hastily executed survey studies are done to satisfy the public clamor for a scientific study. Long-range continuing environmental studies of coastal regions are imperative if we are to have the necessary information on hand to avert an environmental crisis, or if not avert it, at least to face it and make judgments based on fact, not hysteria. Models of estuaries need to be developed which can have predictive significance in the management of coastal regions. The biologist has a responsibility to help in developing these models by emphasizing the role of the biota in estuaries, for these areas are too important to be left entirely to political-economic

[1]Contribution No. 48 of the Belle W. Baruch Coastal Research Institute.

influences. A permanent, broadly based scientific effort must be forthcoming. This symposium will be able to address itself to some of these problems as they relate to one important biotic component of the estuarine ecosystem, the microbiota. Especially by reviewing where we were, discussing what we are now doing, and pointing the directions for future research, you will be able to influence greatly your colleagues, both those in attendance and those who will read the proceedings.

The proceedings of this symposium, published by the University of South Carolina Press, constitute Volume I in a continuing series to be known as the Belle W. Baruch Library in Marine Science. The second volume will be the proceedings of a symposium on symbiosis in the sea which was held 4–6 May 1972.

Since I have a captive audience, I want to present some information about the Belle W. Baruch Foundation of Georgetown, South Carolina, and the Belle W. Baruch Coastal Research Institute which is located on the campus of the University of South Carolina.

To appreciate the genesis of the Belle W. Baruch Coastal Research Institute it is necessary to know the history of the Belle W. Baruch Foundation. Belle Baruch's father, Bernard M. Baruch, was a native of South Carolina. Although he left the state as a child, he retained a life-long love for the State's coastal region, and early in the 1900s he purchased 17,500 acres bordering Winyah Bay, North Inlet, and the Atlantic Ocean near Georgetown. This land was called the Hobcaw Barony, for "Hobcaw" is an Indian word meaning "between the waters." Belle Baruch was the eldest of the three Baruch children, the one who most loved Hobcaw, and eventually Mr. Baruch gave all of Hobcaw Barony to her. Belle Baruch was well aware of the uniqueness of this unspoiled environment and the potential of the area for research and development of conservation practices. In time she drew up plans for a Foundation to foster the development of research programs in marine biology, forestry, and wildlife science by colleges and universities in South Carolina. Since Belle Baruch was so devoted to her land and South Carolina, it was only fitting that the Foundation should carry her name.

There are two significant components of this foundation: one is the 17,500-acre Hobcaw Plantation on the coast, which is perfectly situated for marine studies (see Fig. 1); the other consists of specific funds which have been set aside to support the Foundation. Of particular interest to South Carolinians is the fact that this foundation was established to further teaching and research in institutions of higher learning within the state, with particular emphasis on marine biology, forestry, and the conservation of wildlife. The Belle W. Baruch Coastal Research Institute was started in 1969, jointly funded by a grant from the Belle W. Baruch Foundation and the University of South Carolina. We now have more than twenty faculty associates, representing biology, geology, engineering, and law, and we have working arrangements with faculty members in chemistry, economics, political science, and international studies. At present, fifty-five graduate students are working on various marine-related projects.

The Baruch Plantation, called Hobcaw Barony, is located near Georgetown, south of the Grand Strand resort area which extends from Pawley's Island to the North Carolina line with Myrtle Beach near its midpoint. The plantation is

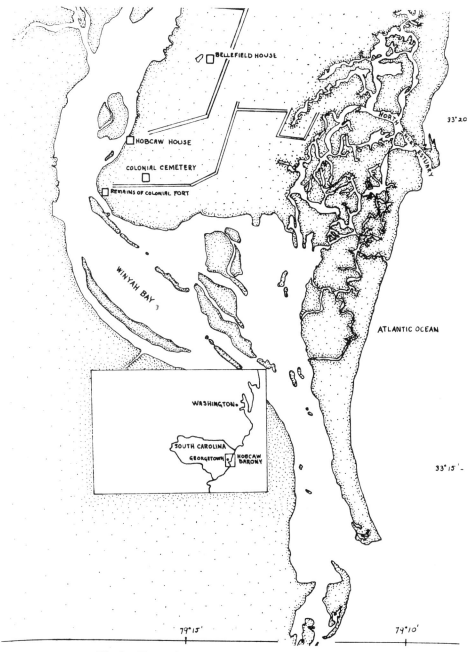

Fig. 1. **The region of Hobcaw Barony, South Carolina.**

approximately 60 miles north of Charleston. Winyah Bay, a low-salinity estuary, forms the western boundary, and the Atlantic Ocean, the eastern boundary. To the south lie 40,000 acres of undeveloped land. Thus, in this immediate vicinity there is a tremendous amount of land that is still unspoiled. On the Baruch property, and of particular interest to us, is North Inlet estuary (Figs. 2 and 3). Although it interconnects with Winyah Bay, there is little exchange between the two regions, and the North Inlet estuary tends to be an isolated one. This estuary is particularly suited for ecological study since there are no developments along

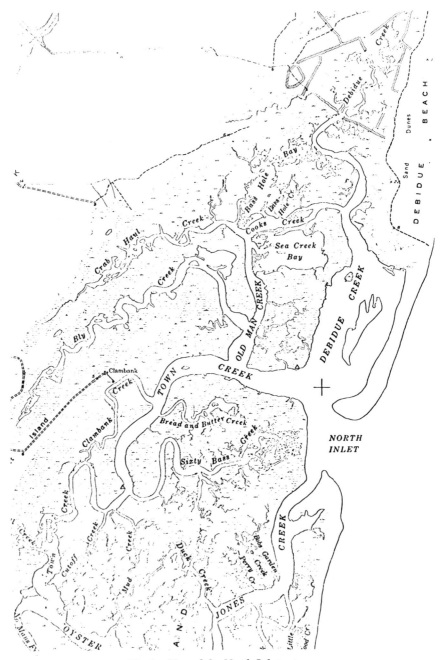

Fig. 2. **Map of the North Inlet estuary.**

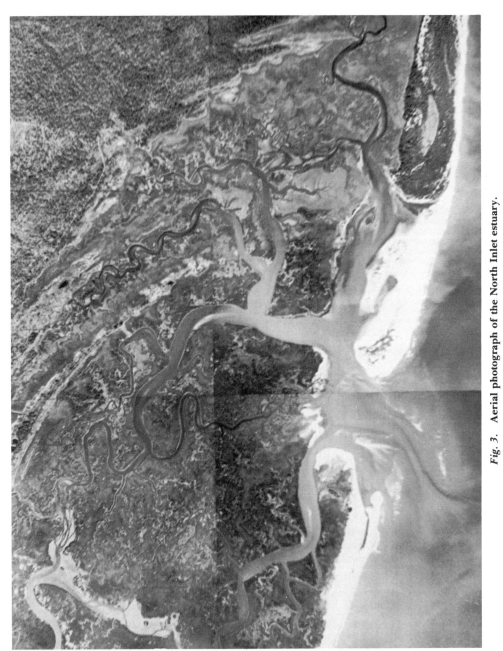

Fig. 3. Aerial photograph of the North Inlet estuary.

the shore, and the Belle W. Baruch Foundation plans to preserve this property for long-term estuarine studies. Thus, here is a relatively undisturbed estuary that can be studied on a long-term basis, a rarity along the East Coast of the United States.

Aside from the scientific value, great historic significance can be attached to the Barony. Presumably, an early Spanish settlement, San Miguel de Gualdape, was located here. Later, during colonial times, the King's Highway, which extended from Charleston north, bisected the property and remnants of a colonial fort and a colonial cemetery are in evidence. A twentieth-century addition is Hobcaw House. Here Mr. Baruch entertained many world leaders, including Sir Winston Churchill and President Franklin D. Roosevelt.

Associated with the marine waters are a rich variety of habitats. The approximately 6,500 acres include salt marshes, sandy bottoms, mud bottoms, sand-shell bottoms, protected intertidal zone, and beaches and open beaches; oyster reefs are abundant. At present our research facilities at the coast consist of various small boats, field equipment, a boathouse with a laboratory room, and limited laboratory space in Hobcaw House. Hobcaw House also has excellent meeting facilities for seminars and symposia.

One of our chief research objectives is to develop a model of an estuarine ecosystem based principally on the North Inlet estuary, since this unexploited estuary is available for extended studies. At present we have approximately 15 staff members, both here and at neighboring institutions, who are working cooperatively on various components of the estuary. This coordinated study includes primary productivity studies on both the aquatic and the terrestrial components. Secondary production studies of zooplankton, the meiofauna, macroinvertebrates, and vertebrates are under way. Geologists, chemists, and biologists are analyzing the physical environment. Also, systems analysts are working with us to develop a predictive model of estuarine processes. In addition to the overall objective of the analysis of an estuarine ecosystem, each staff member and his students have a number of individual projects with which they are involved. Of particular importance is our investigation of the physiological mechanisms which enable organisms to live where they do in a particular environment. Another area of work has been in aquaculture.

We hope your stay will be rewarding to you personally and scientifically. The results of this symposium will be of profound interest to microbiologists in particular, and marine scientists in general, for many years to come. Thank you for attending.

Microbial and Environmental Transitions in Estuaries

Claude E. ZoBell

The term *transition* is used in the sense of its dictionary definition: change in state, condition, or location; passage from one place or state to another. I can think of no better term to embrace all of the interacting biochemical, physicochemical, geochemical, and mechanical forces or processes and resultant conditions which characterize estuaries. In mod talk, this is "where the action is"—inalterably, irrevocably, and timelessly in estuaries. Let us hope for analogous actions, reactions, and interactions during this symposium on processes in estuarine microbial ecology.

Transition connotes actions, reactions, and interactions as well as processes. Rather than thinking of estuarine ecosystems as being static, which is implied by most histograms, curves, and tables of data descriptive of conditions observed in the field or in test tubes, estuarine ecology should be envisioned as a discipline that is highly dynamic in all dimensions: horizontal, vertical, diurnal, intertidal, distal, spatial, biochemical, physicochemical, and especially, biological. Numerous kinds of organisms, particularly microorganisms, cause countless transitions in estuaries. Estuarine organisms are undergoing protoplasmic or genetic transitions, either singly or in a multiplicity of ever-changing associations: symbiotic, commensal, antibiotic, and synergistic.

Besides bacteria, the microbial population in estuaries includes viruses, microalgae, certain fungi, yeasts, and microscopic animals (7, 58, 93, 98). The kinds of microorganisms present, their abundance, their growth, and their biochemical reactions are influenced by environmental conditions. In turn, the growth and biochemical reactions of microorganisms affect certain environmental conditions.

Environmental transitions include changes in temperature, salinity, osmotic pressure, pH, Eh or redox potential, surface and interfacial tension, partial pressure of oxygen and certain other gases, radiations, turbidity, suspended solids, turbulence or other movements of water, cations, anions, toxic substances, nutrients, and accessory growth factors. These and other environmental conditions are constantly changing from time to time and from place to place.

The microbiologist is primarily concerned with the effects of microorganisms on the environment and the effects of environmental conditions on microorganisms. But in order to assess these effects, the microbiologists must always be cognizant of the participation of higher organisms and physical forces such as water movements, sunlight, interactions between air and water, and sedimenta-

9

tion. The impact of man on estuarine ecology is of prime importance in many regions. People may pollute or they may bring about improvements in estuarine environments.

My keynote message is that many environmental and biological factors are involved in estuarine microbial ecology. Cognizance must be taken of all of these factors in order to obtain the complete picture.

Most urgently needed are quantitative data on the impact of microbial activities on the chemical components of estuaries, physicochemical properties of estuarine environments, and the well-being of higher organisms therein. More specifically, I mean the mechanisms by which, and the extent to which, microbial activities influence the health of higher plants and animals, including man, and the transformations or states of such labile elements as H, O, C, N, S, P, Fe, Mn, Ca, Pb, Hg, Se, Mo, Cu, Ni, Zn, I, and V. Currently, of utmost importance are the effects of microbial activities on pollutants such as sulfides or other O_2-consuming substances, toxic minerals, herbicides, insecticides, germicides, surfactants, and oil and its components.

Sampling procedures, analytical methods, and experimental models should be built around these concepts and objectives. To illustrate my "estuarine point of view," a few examples will be briefly reviewed.

Environmental Extremes in Estuaries

Salinity. The typical estuary where fresh water flows into the sea is usually characterized by transitions in salinity ranging from near nil to about 3.5%. (Note that salinity is expressed as percent, because, by definition there is a difference between S‰ and % of dissolved salts.) Low-density fresh water may flow over the denser saltwater for many miles seaward, sometimes hundreds of miles, as off the mouth of the Amazon River. With the flowing of the tide, particularly during the dry season or in places where fresh water has been impounded behind storage reservoir dams, saltwater may be found at all depths for long distances inland. Where mixing is more complete, for various reasons, the salinity gradient in the estuary may be negligible.

Perhaps more important than salinity per se or osmotic pressure resulting therefrom as an environmental factor affecting biota are other differences in the chemical composition of fresh water and saltwater. Of major significance are the kind and the concentration of organic matter, including living organisms normally indigenous to either fresh water or saltwater. Differences in the ratios of certain cations and anions may also be of great importance. The ratios of major ions in seawater differ greatly and in some cases are the reverse of such ratios in most rivers and freshwater lakes (Table 1). These ratios affect the properties of the water as well as the well-being of many kinds of organisms.

In the broadest sense of the term, estuaries include certain fjords, many lagoons, sloughs, ponds, saltwater marshes, mangrove swamps, Russian limans, other kinds of wet lands, and situations where rivers and streams flow into lakes *(60)*. In certain lagoons and other estuarine situations, the water may be only slightly brackish or the salt content may be double or triple that of normal seawater. For example, at certain times in the Laguna Madre of Texas the salinity

Table 1.   Ratios of major ions of normal seawater compared with mean values for many rivers and freshwater lakes, computed from data of Clark *(21)* and Hutchinson *(41)*

Ratio of	Normal Seawater	River Waters	Freshwater Lakes
CO_3^*:Cl	1:135	7.3:1.0	3.3:1.0
K:Na	1:28	1.0:4.6	1.0:2.2
Ca:Na	1:26	4.0:1.0	2.3:1.0
SO_4:Cl	1:7.1	1.6:1.0	2.1:1.0
Ca:Mg	1:3.2	3.6:1.0	1.3:1.0

* Including HCO_3.

may be as high as 8 to 10%, even higher in isolated tidepools. Several kinds of hypersaline lagoons or estuaries have been described *(25)*.

My first scientific observations in an estuary were made more than forty years ago in the Bear River estuary of Great Salt Lake. The salt content of this lake sometimes exceeds 25% *(22)*. Employing the submerged slide technique, supplemented by conventional culture work, living bacteria were demonstrated in water ranging in salt content from 0.02 to 25% *(82, 83)*.

As a class, bacteria and blue-green algae are more euryhaline than most kinds of higher aquatic organisms. In general, estuarine organisms of all kinds are more euryhaline than soil, freshwater, or open-ocean forms. Certain estuarine microbes grow almost equally well at salinities ranging from 0.05 to 2.0‰. Others grow throughout a range of 0.2 to 5‰. But a good many species are fairly stenohaline. None is known to be sufficiently euryhaline to grow throughout the range found in all estuarine situations, i.e., from near nil *(47)* to saturated salt solutions *(13, 42, 44, 59)*. Specific cations, notably Na^+, K^+, Ca^{++}, and Mg^{++} *(62, 63)*, and various heavy metals *(45)* may have more telling effects on microbial well-being than total salinity or osmotic pressure.

Temperature. Depending largely on latitude and season, the temperature in most estuaries ranges from about −2 to 40 C, usually less than 30 C. In highly saline situations as in the Great Salt Lake, for example, the water temperature may be as low as −18 C. Thermal pollution in localized areas may cause some warming, but it is not likely to exceed these limits. Within these limits, temperature will have a pronounced effect on the kinds of microorganisms which predominate and on their physiological reaction rates. Many psychrophilic species, though, grow well at the lowest temperature occurring in estuarine environments, exclusive of ice or air overlayering the water. There is unequivocal evidence for the growth of various microorganisms at sub-zero temperatures, such as −2 to −18 C *(24, 26, 65, 66)*. The highest temperatures at which microbial growth occurs are more than twice as high as the temperature occurring in any estuarine environment *(9)*.

Except for biochemical heating *(114)*, amounting to only a few degrees at the most in localized environments, microbial activity has little effect on the temperature of ecosystems. But the physiological reaction rates of microorganisms are temperature-dependent. Marked transitions in temperature, attributable mainly to water movements and insolation (diurnal and seasonal), occur from time to time and from place to place in many estuaries.

Hydrostatic Pressure. In inland seas or lakes located at high altitudes, the hydrostatic pressure may be less than 1 atm (Table 2). Many microbial species have been experimentally cultivated at pressures of less than 0.000,000,3 atm, the atmospheric pressure at an altitude of 100,000 m. Barophilic bacteria from the deep sea have been cultivated at pressures as high as 1,400 atm *(110)*.

Table 2. Environmental extremes in estuaries and the extremes at which microbial growth has been observed

Condition	Extremes in Estuaries[a]	Extremes for Microbial Growth
Salt content	<0.001–30%[b]	<0.001–30%
Temperature	−18[c]–40 C	−18–95 C
Pressure	<1–10 atm[d]	<1–1250 atm
pH	ca. 3–9	0.1–11.5
Eh	−450–+750 mv	−550–+950 mv
O_2 tension	0–14 mg/l	0–70 mg/l
Surface tension	55–75 dynes/cm	25–75 dynes/cm

[a] Including bottom deposits.
[b] Saturated NaCl solution.
[c] −18 C = freezing point of saturated NaCl solution.
[d] 10 atm = hydrostatic pressure at 100 m.

Taking 100 m as the greatest depth in estuaries, the greatest pressure would be about 10 atm; about 100 atm at a depth of 1,000 m. Such hydrostatic pressures have relatively little effect on the growth of microorganisms unless hyperbaric oxygenation or sudden changes in pressure are involved.

Hydrogen-Ion Concentration. Microorganisms catalyze many biochemical reactions which result in pH changes. Fortunately, most estuarine waters and sediments are fairly well buffered and there is generally a balance between pH-increasing and pH-decreasing reactions. Otherwise, ecosystems might become too acidic or too alkaline for optimal growth of aquatic organisms. Actually, this sometimes occurs in localized environments, thereby excluding all except specialized flora and fauna. Rarely, if ever, though, does the pH reach high enough or low enough levels under natural conditions to render the environment unfit as a habitat for all living organisms *(10)* (see Table 2).

Numerous species of Protista thrive at pH values lower than 2.5 or higher than 10 *(81)*. The minima for growth of acidophilous yeasts studied by van Uden *(92)* ranged from pH 1.1 to pH 1.8. *Thiobacillus thiooxidans* continued to grow and to produce sulfuric acid by oxidizing sulfur until the pH dropped to about 0.7 *(51)*. *Acontium velatum* and another fungus were reported to grow at pH 0.1 *(86)*. The method of measuring this extreme has been questioned.

At the other extreme, *Streptococcus durans* and five strains of *S. faecalis* were found to grow in nutrient media up to pH 10.5 *(17)*. *Bacillus circulans* grew in phosphate-buffered medium at pH 11.0, but not at pH 11.5 *(19)*. Three strains of *Agrobacterium tumefaciens* and 51 strains of *A. radiobacter* grew at pH 11.0, several at pH 11.5 *(40)*. The blue-green alga, *Plectonema nostocorum*, was found to grow on the surface of silica gel medium adjusted to pH 13 *(53)*. It was not reported whether CO_2 from the air lowered the pH.

Redox Potential. Redox potential is an often overlooked and frequently an important environmental parameter, particularly in microenvironments in or on particulate material such as bottom sediments or organic detritus. It is variously called oxidation, oxidation-reduction, O/R, or redox potential and is generally expressed in volts or millivolts as an Eh value *(21, 39)*. There is an extensive literature on the measurement of Eh values and the mechanisms whereby the redox potential influences the growth and biochemical activities of bacteria *(3, 14, 105)*.

The redox potentials of most natural ecosystems in contact with the atmosphere (aerated water, surface soils, etc.) range from Eh −50 to +750 mv at pH 2 to 11. Waterlogged soils, stagnant waters, bottom sediments, and other environments isolated from the atmosphere range roughly from Eh −450 to +100 mv. Organisms which grow at Eh values lower than −350 mv or higher than +650 are exceedingly rare. None has been reported growing in environments more reducing than Eh −600 mv or more oxidizing than Eh +950, within the range of pH 1.0 and 11.5. The pH is specified, because the Eh of any system is pH-dependent. The Eh of a system is also influenced by temperature and hydrostatic pressure.

Various biochemical activities of microorganisms bring about extensive transitions in the redox potential of ecosystems. Free or dissolved oxygen is the next most important factor affecting the redox potential.

Gas Tension. For their normal growth and metabolism, certain microorganisms utilize either free or dissolved oxygen, carbon dioxide, hydrogen, methane, carbon monoxide, nitrogen, or certain other gases. Most of these gases are produced by microbial activities. Free oxygen is required for the growth of all aerobic organisms, but it is produced only by algae and higher plants. Although photosynthetic bacteria bring about other important biochemical transitions in estuaries, they do not produce free oxygen.

At certain times or places, the oxygen tension in estuarine environments may range from nil to supersaturation, as during intense photosynthesis *(75)*. In muds and other reducing areas, there is often an oxygen debt *(76)*. The solubility of oxygen is a function of temperature and salinity and is influenced somewhat by other conditions. In closed systems, hydrostatic pressure influences the solubility of oxygen, but the O_2 content of natural bodies of water is influenced largely by the partial pressure of O_2 in the atmosphere.

The concentration of oxygen and its state (either molecular O_2 or in a great variety of inorganic and organic compounds) are highly transitional. Most generalizations concerning the effects of aerobic versus anaerobic conditions on microbial activities in situ are highly oversimplified. Redox potential and the presence of labile forms of fixed oxygen often influence the respiration of certain microorganisms.

The exchange of oxygen and other gases between water and the atmosphere is influenced by many static and dynamic conditions. Surface turbulence is affected mainly by wind velocity and to a lesser extent by the turbulent structure of the water (49). Turbulent structure is affected by the surface tension which, in turn, is affected by microbial activities in various ways.

Surface Tension. Surface tension or interfacial tension is the cohesiveness of molecules, a property which tends to bring the volume of a fluid into a form having the least superficial area. The intimacy of water and solutes therein with objects, including organisms and their exocrines, is affected by the surface tension, abbreviated S.T. The wetting, capillarity, and penetration of water are functions of its S.T. The S.T. of pure water is 73 dynes/cm at 20 C, about 62 dynes at 95 C. Some substances, notably calcium chloride and charcoal, raise the S.T. of water. Many substances, notably soap, bile salts, and detergents, reduce the S.T. The S.T. of conventional laboratory media, many body fluids, and cell saps ranges roughly from 45 to 65 dynes/cm. It is within this range that most microorganisms grow best (31), ostensibly due to the more intimate contact of water and solutes therein with the organisms or their exoenzymes. A reduced S.T. tends to increase the concentration of dissolved nutrients at the cell-water interface and to facilitate the excretion of waste products into the milieu. Reduced S.T. also increases the collision factor between substances in solution or suspension. Certain detergents irreparably disrupt cell membranes.

Although some microbial species grow feebly in nutrient media whose S.T. is less than 25 dynes/cm, an S.T. substantially lower than 40 dynes/cm is injurious to most cells. The low S.T. tends to (a) upset osmotic equilibria, (b) adversely affect permeability by damaging cell walls and membranes, and (c) promote leakage of essential metabolites. Thigmotactic forms, which normally grow attached on solid surfaces, are detached when the S.T. of the water is substantially reduced. Detergents and certain other substances in domestic sewage have a pronounced effect on the S.T. in certain estuaries. A good many bacterial species produce S.T. depressants but rarely enough to lower the S.T. below 45 dynes/cm.

During the last few years, numerous publications have reported observations on the adverse effects (on aquatic biota) of detergents used as oil-spill dispersants (16). Nontoxic dispersants, used in concentrations which do not reduce the S.T. below the threshold of tolerance of organisms, seem to promote the microbial degradation of polluting oils. Such applications appear to have other beneficial effects. But estuarine microbiologists are confronted by many unanswered questions concerning levels of tolerance and the biodegradability of various S.T. depressants and surfactants.

Probably the most far-reaching effects of S.T. depressants are on surface films, the topmost 50- to 100-μm film or layer of water, and the neuston therein. Samples of this submillimeter surface film, collected with special skimmers *(36)*, often contain from 2 to 100 times more detritus (mostly particulate organic matter), living bacteria, and phytoplankton than equal volumes of water collected by conventional methods from depths of 1 to 10 cm. This topmost 50- to 100-μm film appears to be a zone of exceptionally intense microbial activity, probably being second only to the slime layer at the mud-water interface in intensity of biochemical activity.

Higher Organisms and Environmental Factors. Many aquatic animals feed on microorganisms *(27, 79)*. Employing ^{14}C-compounds, Sorokin *(84)* demonstrated that appreciable amounts of carbon compounds assimilated by bacteria, micro-algae, and allied microorganisms eventually appear in animal biomass. Such ingestion results in the removal of extensive microbial populations *(46, 68)*. Animal and algal antibioses also adversely affect aquatic microorganisms *(2, 80)*.

On the other hand, higher plants and animals are beneficial to microbial populations in numerous ways. Providing nutriments and microenvironments is of greatest ecological and biochemical importance to microorganisms in soil and water. Slimes covering the surfaces of many kinds of plants and animals are veritable microbial gardens. The gut of most animals is an important habitat for bacteria. The ecological and biochemical interactions between microorganisms and macroorganisms merit much more attention than they generally get in microbial surveys.

Microenvironments

Although aquatic and soil microbiologists are admonished to "think big" by taking into account all of the physical, chemical, and biological factors which influence the distribution of microorganisms and their biochemical reaction rates, Brock *(11)* remonstrates, "When we think of microorganisms living in nature, we must also learn to think small." Adjacent microenvironments, little larger than a microorganism, may differ from each other in major or minor ways, including the microbial inhabitants. In certain situations, as in shallow bottom deposits, strict aerobes and strict anaerobes found in the same 1-g sample may seem to be living together, when actually they are millimeters apart in distinctive microenvironments.

Microelectrodes carefully inserted in undisturbed bottom sediments or other semisolid substances reveal substantial differences in pH and Eh within 1 or 2 millimeters. The submerged slide technique, first used in soil by Cholodny *(20)*, in freshwater lakes by Henrici *(38)*, and in marine environments by ZoBell and Allen *(112)*, provides much significant information on microbial growth in microenvironments. The capillary peloscope *(69)* and the modified Winogradsky column *(1)* are methods for dramatically demonstrating microbial activities in microenvironments. Autoradiography also has promising possibilities for exploring microenvironments (8).

Instead of shaking pooled samples 100 times in order to obtain uniform results, in many cases it might be more meaningful to make 100 determinations on the smallest manageable samples. This speculative suggestion applies more to sampling for numbers and kinds of microflora, pH, Eh, and solid substrates than to temperature, salinity, or dissolved substances. Careful sampling procedures have provided evidence for striking stratification of microbial populations in the topmost 100 m of water. Much more marked are microenvironments in bottom sediments and on or in suspended solids, animate as well as inanimate. The submerged slide technique has already been mentioned and there are many other ways of making measurements in situ.

This may be a good place to reflect on the philosophy of sampling and methods of making field observations. Is the prime purpose of collecting samples of water, mud, etc., to obtain as many species as possible for pure-culture study or for observations in a chemostat? Is the prime purpose to ascertain the total number of organisms present in a given sample regardless of whether they grow in the environment where found? Or is the prime purpose to ascertain quantitatively the effects of microbial activities on chemical, physicochemical, and biological conditions in estuarine environments?

Geochemical Effects of Microbes

The term *biochemical* connotes chemical composition or transformations in organisms. The term *geochemical* connotes transformations in the chemical composition or physicochemical conditions in the crust of the earth, of which more than two-thirds by area is water. Biogeochemistry is the science that deals with the effects of living organisms on chemicals primarily in the lithosphere and hydrosphere. There is continuous interaction between the lithosphere and hydrosphere, partly via the atmosphere, which currently is the reservoir for about 99.3% of the earth's free oxygen (111). By virtue of its great area and mass, the ocean is the site of the most extensive biogeochemical transformations (from 10^{10} to 10^{11} tons per year). The site of the most intensive biogeochemical transformations is believed to be at mud-water interfaces, particularly in the littoral zone and more specifically, in estuarine environments.

Besides contributing substantially to the reduction and formation of free oxygen, the biochemical activities of bacteria, micro-algae, and allied microorganisms have far-reaching effects on the state and transformation of C, N, H, S, P, Fe, Ca, and several other elements (56, 57, 102). These activities affect such environmental conditions as pH, redox potential, surface tension, electrostatic charges on surfaces, and gas tension. Considered from a viewpoint of the total masses annually transformed and their importance in the metabolism and chemical composition of organisms, carbon and oxygen rate the highest among the elements. From a viewpoint of the total number of atoms involved in biogeochemical transformations, hydrogen takes first place in the parade of elements in the biosphere.

Carbon. Carbon makes up only 0.02% of the total elementary composition of the crust of the earth, including the hydrosphere and atmosphere, as compared with 46.6% oxygen and 0.14% hydrogen *(55, 64)*. Unlike oxygen, most of the carbon is geochemically labile. Nearly 70% occurs in carbonates, a little in CO_2, about 25% in organic compounds, only 0.03% in fossil fuels, and barely 0.015% in biomass, i.e., living organisms and their undecayed remains. Limited space prohibits more than an outline of a few of the many ways in which microbial activities bring about the transformation of carbon and its compounds (Fig. 1).

It is questionable whether any organisms assimilate elementary carbon. Carbon monoxide (CO) is oxidized by certain bacteria, notably *Carboxydomonas oligocarbophilia, Methanosarcina barkerii, Methanobacterium formicum, Clostridium welchii,* and *Hydrogenomonas carboxydovorans.* The oxidation of CO by mixed cultures from marine mud, water, sewage, and soil has been reported by many workers *(43)*. The CO is oxidized to CO_2, with part of the carbon finding its way into organic compounds constituting bacterial biomass. The microbial oxidation of CO is of some significance, because CO occurs in many marine

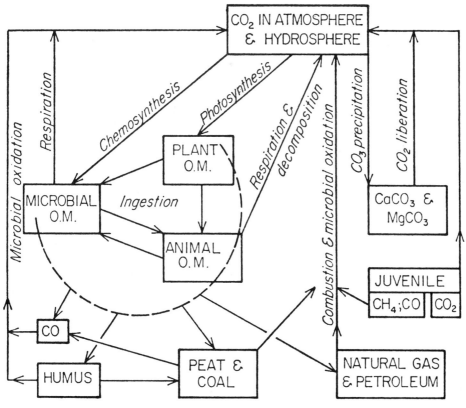

Fig. 1. **Transformation of carbon in nature, with emphasis on microbial reactions. The semicircular dashed line embraces organic materials (O.M.) that may contribute to the formation of other kinds of biomass, CO, CO_2, humus, peat, coal, petroleum, or other carbonaceous substances. (Reprinted from ZoBell *(107)* by permission of Pergamon Press Limited.)**

organisms *(61, 70)* and it is an important pollutant resulting from incomplete combustion of organic wastes and fossil fuels.

Carbon dioxide (CO_2) is assimilated by many kinds of microorganisms, including organotrophs as well as lithotrophs or autotrophs. The latter include photolithotrophs (micro-algae and photosynthetic bacteria) and chemolithotrophic bacteria, which obtain energy for the reduction of CO_2 by oxidizing such substrates as ammonium, nitrite, elementary sulfur, hydrogen sulfide, thiosulfate, molecular hydrogen, methane, and ferrous iron. Most of the carbon finds its way into biomass. The total amount fixed in the oceans by all kinds of photosynthetic organisms, mainly phytoplankton, is estimated to be about 1.5×10^{10} tons of carbon per year *(78)*. It is virtually impossible to estimate how much of this fixation is attributable to microorganisms or how much occurs in estuaries, because estuaries gradually merge with the sea and because there is no clear-cut demarcation between microorganisms and higher organisms. Chemolithotrophic bacteria might fix as much as 1.5×10^5 tons of carbon per year *(109)*.

A good many organotrophic bacteria also fix CO_2 *(91)*. CO_2 fixation by organotrophs is much more important biochemically, i.e., concerning what happens in living cells, than it is geochemically, because many organotrophs liberate more CO_2 in energy-yielding reactions than the amount fixed, as indicated by tagged atom experiments.

Virtually all kinds of organic compounds, including hydrocarbons *(113)*, are susceptible to microbial oxidation. Complete oxidation of biomass always results in the formation of CO_2 and usually various compounds of nitrogen, phosphorus, sulfur, hydrogen, etc. Under certain conditions of microbial decomposition of the organic remains of organisms, intermediate or end products may include alcohols, organic acids, ketones, hydrocarbons, humic acids, bituminoids, and lignites.

Often overlooked is the fact that a major primary product resulting from the microbial utilization of organic matter is microbial biomass or protoplasm. Depending on environmental conditions, the microflora, the organic compounds and their concentration, and other factors, from 5 to 85% of the carbon in assimilated organic matter may be converted into microbial biomass. Under ordinary conditions, an average of 30 to 40% of the carbon in organic matter assimilated by microbes in aquatic environments is converted into microbial biomass. Such biomass, probably amounting to millions of tons per year in the sea, has far-reaching geochemical significance. It may also constitute an important source of food for aquatic animals.

By producing or consuming CO_2 and by affecting the pH in various ways, both lithotrophic and organotrophic microorganisms play an important part in the CO_2-carbonate equilibrium in aquatic environments, including bottom sediments:

$$\text{Free } CO_2$$
$$\updownarrow$$
$$\text{Dissolved } CO_2 + H_2O \rightleftarrows H_2CO_3 \rightleftarrows H^+ + HCO_3^- \rightleftarrows CO_3^= + 2H^+$$
$$\updownarrow$$
$$CaCO_3 \rightleftarrows Ca^{++}$$

Oxygen. Oxygen is by far the most abundant element in the crust of the earth (46.5% by weight), including the hydrosphere *(64, 111)*. But unlike carbon, most of which is biochemically labile, more than 80% of the oxygen in the crust of the earth is quite stable, being bound or buried in silicates (nearly 50%), aluminates (about 12%), and various mineral oxides (20%) (see Table 3). By weight, only about 0.0001% of the total oxygen in the crust of the earth, including the hydrosphere, lithosphere, and atmosphere, is free or dissolved O_2. Biomass is the reservoir for 0.000,000,2% of the oxygen in the crust of the earth; about 5% occurs in phosphates, sulfates, and carbonates, including CO_2, 13% in H_2O, and 0.0006% in organic matter.

Table 3. Major reservoirs of oxygen in various regions or materials of the Earth

Region or Reservoir	Total Mass ($\times 10^{15}$ tons)	Total Oxygen Content		Free Oxygen, O_2 ($\times 10^{15}$ tons)
		(percent)	($\times 10^{15}$ tons)	
Whole Earth	5,975,000	27.2	1,625,200	1.1885
Earth's crust	24,000	46.6	11,185	$<1 \times 10^{-4}$
Hydrosphere	1,644	85.7	1,408.9	0.0084
Atmosphere	5.1	23.14	1.193[a]	1.18
Biomass[b]	0.145	*ca.* 30	0.0043	$<1 \times 10^{-5}$

[a] Besides free O_2, this includes about 0.013×10^{15} tons of oxygen in atmospheric H_2O, CO_2, CO, O_3, SO_2, NO_2, and N_2O.

[b] Living organisms and undecomposed organic matter.

The principal reservoir of labile oxygen is the hydrosphere which consists of 85.77% oxygen. Of this, only about 0.006% is free or dissolved O_2. Oxygen is liberated from H_2O by photosynthetic plants. Almost negligibly small amounts of free O_2 are liberated from water by various other processes such as photodissociation or electrolysis, apparently none by any known biochemical processes except photosynthesis *(5)*. The primary photosynthetic process, energized by light absorbed by chlorophyll, consists of the reduction of CO_2 with the formation of some simple carbohydratelike substances commonly designated (CH_2O) and the photolysis of H_2O to yield O_2 *(48)*:

$$nCO_2 + nH_2O \xrightarrow{\text{light}} (CH_2O)_n + nO_2$$

This oversimplified general equation fails to show the involvement of ATP, nucleotides, etc. in photosynthesis, nor does it show that the O_2 is derived from H_2O. Micro-algae and higher plants catalyze this reaction; bacterial photosynthesis yields no O_2 and usually very little carbohydrate.

Oxygen occurs in seawater of salinity 35‰ in forms and percentages as follows:

H_2O	99.787,5%
Sulfate ($SO_4^=$)	0.210,3

CO_2, HCO_3^-, and $CO_3^=$	0.001,1
Si(OH)$_4$, Si(OH)$_3$O, etc.	0.000,6
Borate ($BO_3^=$)	0.000,2
NO_2^-, NO_3^-, and $PO_4^=$	0.000,11
Dissolved oxygen (O_2)	0.000,09
Organic compounds	0.000,03
All other forms	<0.000,1

Carbonates and organic compounds are important reservoirs of oxygen, but not as direct sources of oxygen for oxidation reactions. The acidification of carbonates by microbial or other reactions results in the liberation of CO_2, which may be fixed by chemolithotrophic, photolithotrophic, or organotrophic organisms with the formation of organic matter. In ultimate oxidation processes, part of the oxygen in organic compounds may be liberated as water. Free O_2 could be released from this water by photosynthetic algae or higher plants.

All living cells, with the exception of a relatively few species of anaerobic organisms, consume O_2. In many bodies of shallow water such as estuaries, lagoons, and salt marshes, microorganisms appear to consume more oxygen than all higher organisms and abiotic processes combined. According to Kuznetsov (56), the basic processes whereby O_2 is consumed in natural waters fall into seven general categories: (1) microbial oxidation of organic matter, (2) bacterial respiration, which is usually greater than (3) respiration of higher organisms, including phytoplankton and zooplankton, (4) microbial oxidation of H_2 and CH_4 evolved in anaerobic bottom deposits, (5) microbial oxidation of H_2S, ferrous salts, ammonium, nitrite, thiosulfate, etc., (6) abiotic oxidation of inorganic substances, and (7) purely chemical oxidation of organic matter.

The biochemical oxygen demand (BOD) of carbon compounds ranges from 0.18 for oxalic acid to 3.99 for methane. The complete oxidation of most hydrocarbons requires from 3 to 3.5 mg O_2 per mg (113). For various reasons, the actual BOD of carbon compounds is often less than the theoretical or calculated values. The theoretical BOD of carbohydrates ranges from 1.0 to 1.2. The actual BOD may be considerably less when appreciable amounts of carbohydrate are converted into microbial biomass, or when CO_2 is fixed by the bacteria. The BOD of amino acids and proteins ranges from 0.64 to 2.58, depending in part on whether nitrogen is liberated as ammonia or oxidized to nitrate and whether sulfur is liberated as H_2S or oxidized to sulfate.

Of less importance than organotrophic organisms in consuming O_2 are chemolithotrophic bacteria which obtain energy from the oxidation of various inorganic substances. A few examples are given as general reactions. The theoretical free energy values are approximations based on standard conditions in aqueous solutions:

$$NH_4OH + 1\tfrac{1}{2}\, O_2 \rightarrow HNO_2 + 2H_2O + 76,000 \text{ cal}$$

$$HNO_2 + \tfrac{1}{2}\, O_2 \rightarrow HNO_3 + 21,000 \text{ cal}$$

$$H_2 + \tfrac{1}{2} O_2 \rightarrow H_2O + 56{,}000 \text{ cal}$$

$$H_2S + \tfrac{1}{2} O_2 \rightarrow H_2O + S + 41{,}000 \text{ cal}$$

$$S + 1\tfrac{1}{2} O_2 + H_2O \rightarrow H_2SO_4 + 118{,}000 \text{ cal}$$

$$2FeCO_3 + \tfrac{1}{2} O_2 + 3H_2O \rightarrow 2Fe(OH)_3 + 2CO_2 + 12{,}000 \text{ cal}$$

$$CO + \tfrac{1}{2} O_2 \rightarrow CO_2 + 74{,}000 \text{ cal}$$

Most of these reactions are coupled with other reactions in which CO_2 is converted into bacterial biomass whose oxygen content ranges from 30 to 45% (dry weight basis).

Sulfate is an important reservoir of biochemical oxygen. Sulfur- and H_2S-oxidizing bacteria are partly responsible for sulfate formation. Sulfate-reducing bacteria use the oxygen in sulfate to oxidize organic compounds or molecular hydrogen in accordance with the following reactions:

$$H_2SO_4 + 2(CH_2O) \rightarrow 2CO_2 + 2H_2O + H_2S$$

$$H_2SO_4 + 4H_2 \rightarrow 4H_2O + H_2S$$

Sulfate-reducing bacteria are widely distributed in bottom deposits and anoxic waters *(107)*. The H_2S produced by such bacteria promotes the depletion of O_2 in stagnant water basins, fjords, eutrophic lakes, lagoons, and the like *(76)*.

Although nitrate is only 0.05% as abundant on the average as sulfate in natural waters, it is an important reservoir of oxygen. The low concentration of nitrate in natural waters is a reflection of the ease with which it is reduced by so many microorganisms. A large percentage of aquatic and soil bacteria reduces nitrate. This means that the oxygen in nitrate is used to oxidize carbon compounds, hydrogen, etc. Nitrate is assimilated by virtually all micro-algae.

There are numerous other reservoirs of oxygen, meaning compounds which are produced or reduced by various kinds of microorganisms. For further details, see the review paper on the microbial biochemistry of oxygen *(111)*.

Hydrogen. By weight, hydrogen is the tenth most abundant element in the crust of the earth and the second most abundant in the hydrosphere. About 11% of H_2O is hydrogen. From 5 to 9% of the biomass of marine organisms is hydrogen (dry weight basis). Crude oils contain 11 to 14% hydrogen. Hydrogen ranks number one in atomic abundance, i.e., the number of atoms involved. It participates in all energy-yielding reactions in living cells. The hydrogen-ion concentration is an important physiological and environmental factor.

Free or molecular hydrogen is a minor component of most natural gases. Only 0.000,005% of the earth's atmosphere consists of H_2, most of which occurs at high elevations *(6)*. However, as much as 40 to 50% of the gas formed by bacteria during the initial stages of the anaerobic decomposition of certain organic compounds may be H_2 *(94, 106)*. Numerous microbial species, including anaerobes as well as aerobes, lithotrophs, and organotrophs, either produce or utilize H_2 *(30, 34, 106)*. H_2-producing and H_2-oxidizing bacteria are especially abundant in estuarine environments. Since so much H_2 is known to be produced, its low concentration or absence in many aquatic environments is believed to be a reflection of the widespread occurrence and avid activity of H_2-oxidizers.

Much more common than the formation or oxidation of molecular H_2 is the transfer of atomic hydrogen or hydrogen-ion from one component to another in biochemical reactions. Hydrogen prominently participates in the cycles of carbon, oxygen, sulfur, and nitrogen. It occurs in hydrocarbons and all other organic compounds, H_2O, H_2S, NH_4^+, etc. It commonly occurs as hydrogen-ion (H^+) or as hydroxyl-ion (OH^-).

Sulfur. By weight, sulfur is the thirteenth most abundant element in the crust of the earth. It is sixth in abundance in seawater, making up 1.56% of all dissolved solids. It is the tenth most abundant element in biomass. In various combinations with hydrogen, oxygen, and other elements, sulfur forms strong acids and/or reducing agents, which have far-reaching effects on physico-chemical conditions influencing geochemical and biochemical transformations *(4, 71, 90)*.

Free sulfur is deposited either intracellularly or extracellularly by certain bacteria which utilize H_2S. Extensive deposits of biogenic sulfur occur in geological formations. Sulfate-reducing bacteria are believed to have been contributory, but the mechanism whereby the free sulfur was formed is not known *(108)*. Sulfate-reducing bacteria are abundant and active in many soils, bottom sediments, and anoxic waters *(107)*. Much of the H_2S formed by sulfate-reducing bacteria finds its way into aerobic environments, where it tends to deplete free or dissolved O_2. H_2S also tends to precipitate iron and other heavy metals from solution.

Redfield *(73)* points out that one of the most important functions of sulfate-reducing bacteria in geochemical cycles is to extract the "fossil" oxygen from sulfate, which is used in the oxidation of organic matter or H_2. Thus the oxygen is transferred from sulfate to form CO_2 and H_2O. An average of 1.5 g of $SO_4^=$ is required to oxidize 1.0 g of carbohydrate:

$$2(CH_2O) + SO_4^= \rightarrow 2CO_2 + 2OH^- + H_2S$$

H_2S is produced by many microbial species during the decomposition of organic compounds which contain sulfur, such as cysteine, cystine, methionine, glutathione, thiozole, and thiophene. H_2S is oxidized by various colorless *(Thiobacillus, Thioploca, Thiothrix, Thiospirillum)* and colored *(Chromatium, Chlorobium, Rhodothiospirillum, Thiopedia)* bacteria *(71)*. Many of the colored sulfur bacteria fix nitrogen.

Figure 2 outlines some of the principal features of the sulfur cycle. There are several good review papers on this subject *(71, 85, 90, 96, 107, 108)*.

Nitrogen. Living organisms tend to concentrate nitrogen as well as carbon. Biomass contains from 9 to 15% nitrogen (dry weight basis) as compared with 0.002% in the crust of the earth and an average of only 0.0015% in the hydrosphere. The value for the crust of the earth includes the atmosphere, which is the principal reservoir of nitrogen. The atmosphere contains 755 g of nitrogen per cm² of the earth's surface *(33)*. More than 99.99% of the nitrogen in the

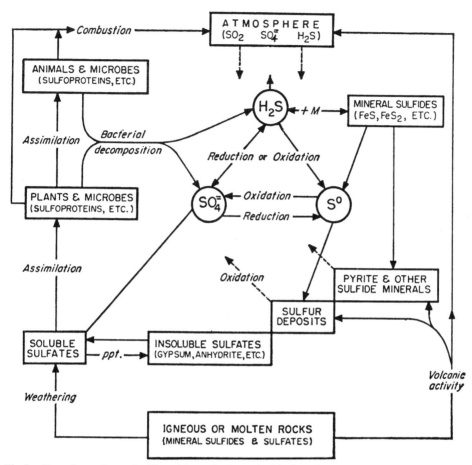

Fig. 2. Transformations of sulfur with emphasis on the biochemical activities of microorganisms. (Reprinted from ZoBell *(106)* by permission of Pergamon Press Limited.)

atmosphere is N_2, with only minute amounts of nitrogen oxides and ammonia. Seawater has a mean fixed nitrogen content of about 0.5 mg/l *(32)*, usually much more in estuarine waters. Approximately 60% of the nitrogen in the sea occurs as NO_3^-, 6% NO_2^-, 12% NH_4^+, 20% N_2, and about 2% as organic N. These values are subject to wide fluctuations in time and space, owing largely to biochemical transformations.

N_2 is the most stable form of nitrogen. Yet many species of bacteria and blue-green algae fix N_2 *(72, 87, 88, 89)*. See Burris *(15)* for an excellent review of N_2-fixing organisms, including the mechanisms whereby N_2 is reduced by hydrogen in the presence of reduced ferredoxin and ATP to form ammonia, which is incorporated into organic compounds. The use of acetylene reduction as an index of nitrogen fixation simplifies field surveys and laboratory tests for nitrogen fixers.

The liberation of N_2 from nitrate or other compounds is called denitrification. Relatively few microbial species are denitrifiers and these form N_2 only under rather special conditions. Much more common are bacteria and environmental conditions which bring about the reduction of nitrate to nitrite (52). For more detailed information on nitrate reduction, see the review by Payne in this volume.

Most microorganisms and plants obtain their nitrogen needs from either ammonium, nitrate, nitrite, or various carbon compounds which contain nitrogen. Animals depend largely on these carbon compounds. Deamination is a process (usually hydrolytic) in which amino-N is liberated from proteins and certain other compounds. The formation of ammonium by this and other processes is called ammonification. A high percentage of the bacteria in aquatic environments is able to form ammonium in one way or another (101).

Only a specialized few species oxidize ammonium to nitrite, but such nitrifiers seem to be widely distributed in the sea, particularly in the littoral zone (29, 50, 99, 100). The microbial oxidation of ammonium to nitrite is sometimes called nitrosification to distinguish this first stage of nitrification from the second stage:

$$NO_2^- + \frac{1}{2} O_2 \rightarrow NO_3^-$$

sometimes called nitration.

My cross-indexed card files contain more references (about 650) dealing with microbial transformation of nitrogen than any other element. But this voluminous literature contains relatively little quantitative data to warrant generalizations concerning rates, amounts, and places of nitrogen transitions in the hydrosphere. Urgently needed are new approaches to this problem.

Phosphorus. There is 5 to 10 times more nitrogen than phosphorus in marine organisms and in seawater. The vast majority of the phosphorus in water and bottom sediments is present as inorganic phosphates. Phosphorus is the eleventh most abundant element (0.105%) in the crust of the earth, including the hydrosphere. The bulk occurs in apatites, typified by the most abundant, fluorapatite, $Ca_{10}(PO_4)_6F_2$. Considerably smaller quantities occur in secondary phosphate minerals, orthophosphates, and organophosphates (33). Organophosphates and the relatively rare organophosphonates are the most labile and least abundant forms of organic phosphorus.

All living organisms contain phosphorus in amounts ranging from 0.05% in certain plants to as much as 6% of the dry weight of certain vertebrates. The mean phosphorus content of most microorganisms is between 0.5 and 1%. Most of it occurs in cells as organophosphates (nucleotides, nucleic acids, phospholipids, phosphoproteins, etc.) with various concentrations of ionic phosphate:

$$X-O-\overset{\displaystyle O}{\underset{\displaystyle O-X}{\overset{\|}{P}}}-O-\overset{\displaystyle H}{\underset{\displaystyle H}{C}}-X$$

where X represents either hydrogen, an inorganic cation, an organic moiety, or various combinations thereof.

Organophosphonates (phosphonolipids, aminophosphonates, etc.) have been demonstrated in a few species of bacteria, algae, protozoans, and higher organisms (54). Whereas in organophosphates phosphorus is bonded to carbon only through an oxygen bridge (P—O—C), phosphorus is bonded directly to carbon in organophosphonates (P—C):

$$\begin{array}{c} O \qquad H \\ \parallel \qquad | \\ X{-}O{-}P{-}\!\!\!-\!\!\!-C{-}X \\ | \qquad | \\ O{-}X \quad H \end{array}$$

The P—C bond is resistant to phosphatases, enzymes which release the PO_4 group from organophosphates. However, certain bacteria have specific phosphonatases.

Numerous species of bacteria, yeasts, fungi, and algae synthesize polyphosphates, commonly called poly-P. Tripolyphosphate has the following structural formula:

$$\begin{array}{c} O \qquad\quad O \qquad\quad O \\ \parallel \qquad\quad \parallel \qquad\quad \parallel \\ X{-}O{-}P{-}O{-}P{-}O{-}P{-}O{-}X \\ | \qquad\quad | \qquad\quad | \\ O{-}X \quad O{-}X \quad O{-}X \end{array}$$

Certain polyphosphates have up to a few hundred PO_4 groups linked by oxygen bridges (35, 102). Cyclic-condensed and branched chains as well as straight-chain polyphosphates are synthesized by certain organisms. Biochemically, polyphosphates are more labile than organophosphates or orthophosphates.

Most transitions of phosphorus take place intracellularly, especially in energy-yielding reactions and in the synthesis of protoplasm. Microbes and many animals decompose organophosphorus compounds in dead biomass with the liberation of PO_4. Such PO_4 is assimilated by most bacteria, yeasts, fungi, algae, and higher plants to satisfy their phosphorus requirements. The use of ^{32}P-labeled phosphate has contributed substantially to the study of phosphorus transitions in aquatic environments (37, 77, 95).

In certain situations in the euphotic zone, primary production by photosynthetic plants is restricted by a shortage of available phosphates. In other situations, termed "phosphate pollution" in popular parlance, an excess of phosphates may result in excessive growth of algae or higher plants. This triggers a chain of events involving imbalances in the transformation of carbon, oxygen, sulfur, and other biochemicals. Instead of being cycles, some of these transformations tend to become dead-end streets.

By their effects on pH, particularly in bottom sediments and microenvironments, microbial activities play an important part in the precipitation of phosphates (at higher pH values) or the solubilization of phosphates (at lower pH values). Orthophosphates are precipitated from solution in the sea largely as $Ca_3(PO_4)_2$, with some $FePO_4$. H_2S, produced by sulfate reducers, reacts with the $FePO_4$ to yield FeS_2. This leaves the PO_4 group free to combine with calcium or certain other cations which happen to be present.

Iron and Manganese. Iron is the fourth most abundant element (5.0%) in the crust of the earth. Manganese is the twelfth most abundant (0.095%). Only extremely low concentrations of iron occur in the hydrosphere or in biomass. Even less manganese occurs in biomass. Both elements are biogeochemically very interesting.

In the ferrous or ferric state, iron is easily oxidized or reduced. By accepting or donating electrons, iron functions in the oxidation or reduction of many elements and compounds, especially oxygen, sulfur, and carbon compounds. Many species of bacteria, including chemolithotrophs, facultative autotrophs, and organotrophs, participate in the oxidation of ferrous iron to the ferric state *(57, 67, 102)*. Kuznetsov's *(56)* scheme of iron and manganese cycles in lakes provides an elegant overview of the microbially catalyzed interactions of these metals with other elements.

A few iron-oxidizing bacteria, including certain *Leptothrix*, *Clonothrix*, and *Siderobacter* species, deposit oxidized manganese along with iron in their sheaths and capsules. These, like most other manganese oxidizers, are primarily organotrophs. It is questionable how much energy they derive from the reaction:

$$2MnO + O_2 \rightarrow 2MnO_2 + 40,000 \text{ cal}$$

Metallogenium personatum seemed to obtain part of its energy from the oxidation of bivalent manganese *(28)*. Zavarzin *(103)* found no evidence that the oxidation of bivalent manganese satisfied the energy requirements of *Metallogenium symbioticum*. In his comprehensive review of the evidence for the biological precipitation of manganese, Zavarzin *(104)* lists *Pedomicrobium podzolicum* as another bacterium which oxidizes both manganese and iron. Manganese but not iron was found to be oxidized by *Pedomicrobium manganicum*.

Probably much more iron and manganese in aquatic environments as well as in soil are changed in state by changes in redox potential caused by microorganisms than by the microbial oxidation of these elements in energy-yielding reactions. Additional aspects of the biogeochemistry of iron, manganese, and several other elements are discussed and documented elsewhere *(18, 22, 55, 97, 102)*.

Conclusions

Surveys on the numbers and kinds of microorganisms occurring in aquatic environments may yield significant information. More important, though, than accurate counts is information on whether the microbes are physiologically active. It is of utmost importance to determine microbial death rates, growth rates, and/or biochemical reaction rates.

Such rates depend in part on the nature of the microbial species or variety and more on the environmental conditions of the habitat. Many of the environmental conditions are in a state of constant flux or transition, being affected by each other and by microbial activities, often by the activities of higher organisms. Therefore, efforts should be made in microbial surveys to measure in the field and/or to control in the laboratory all of the limiting environmental parameters. Urgently needed are improved methods for observing conditions as well as transitions in situ. Such methods involve continuous sampling and measuring.

In this perspective overview, emphasis has been placed on the environmental and geochemical aspects. These aspects are of basic importance in seeking information on the solutions to the problems of pollution, biodegradation of pesticides, biofouling, biodeterioration of man-made structures, microbial effects on aquatic plants and animals, and the conservation or restoration of healthy estuarine ecology.

Literature Cited

1. Aaronson, S. 1970. *Experimental microbial ecology.* New York: Academic Press Inc.
2. Aubert, M. 1965. Le comportement des bactéries terrigènes en mer. Relations avec le phytoplancton. *Cah. C.E.R.B.O.M.* **19:**1–285.
3. Baas Becking, L. G. M., I. R. Kaplan, and D. Moore. 1960. Limits of the natural environment in terms of pH and oxidation-reduction potentials. *J. Geol.* **68:**243–84.
4. Baas Becking, L. G. M., and E. J. F. Wood. 1955. Biological processes in the estuarine environment. I and II. Ecology of the sulphur cycle. *Proc. Koniknl. Nederl. Akad. Wetenschap.* Ser. C **58:**160–81.
5. Berkner, L. V., and L. C. Marshall. 1964. The history of growth of oxygen in the earth's atmosphere. In *The origin and evolution of atmospheres and oceans,* ed. P. J. Brancazio and A. G. W. Cameron, pp. 102–26. New York: Wiley.
6. Brancazio, P. J., and A. G. W., Cameron, eds. 1964. *The origin and evolution of atmospheres and oceans.* New York: Wiley.
7. Brock, T. D. 1966. *Principles of microbial ecology.* Englewood Cliffs, N.J.: Prentice-Hall Inc.
8. Brock, T. D. 1967. Bacterial growth rate in the sea: direct analysis by thymidine autoradiography. *Science* **155:**81–83.
9. Brock, T. D. 1967. Life at high temperatures. *Science* **158:**1012–19.
10. Brock, T. D. 1969. Microbial growth under extreme conditions. In *Microbial growth,* ed. P. Meadow and S. J. Pirt, pp. 15–41. Soc. Gen. Microbiol., Cambridge: Cambridge University Press.
11. Brock, T. D. 1970. *Biology of microorganisms.* Englewood Cliffs, N.J.: Prentice-Hall Inc.
12. Brock, T. D. 1971. Microbial growth rates in nature. *Bacteriol. Rev.* **35:**39–58.
13. Brown, A. D. 1964. Aspects of bacterial response to the ionic environment. *Bacteriol. Rev.* **28:**296–329.
14. Bubenicek, L. 1964. L'oxydo-réduction en sédimentologie. *Bull. Bureau Rec. Géol. Minierès* No. **4:**1–36.
15. Burris, R. H. 1966. Biological nitrogen fixation. *Ann. Rev. Plant Physiol.* **18:**155–84.
16. Canevari, G. P. 1971. Oil spill dispersants—current status and future outlook, In *API/EPA/USCG Conference on Prevention and Control of Oil Spills,* pp. 263–270. Washington, D.C.: American Petroleum Institute. (Also see in same volume Dewling *et al.,* Dispersant use vs water quality, pp. 271–77, and ZoBell, Sources and biodegradation of carcinogenic hydrocarbons, pp. 441–51.)
17. Chesbro, W. R., and J. B. Evans. 1959. Factors affecting the growth of enterococci in highly alkaline media. *J. Bacteriol.* **78:**858–62.
18. Chester, R. 1965. Elementary geochemistry of marine sediments, In *Chemical oceanography,* ed. J. P. Riley and G. Skirrow, pp. 23–80. New York: Academic Press Inc.
19. Chislett, M. E., and D. J. Kushner. 1961. A strain of *Bacillus circulans* capable of growing under highly alkaline conditions. *J. Gen. Microbiol.* **24:**187–90.
20. Cholodny, N. 1930. Über eine neue Methode zur Untersuchung der Bodenmikroflora. *Arch. Mikrobiol.* **1:**620–52.
21. Clark, W. M. 1960. *Oxidation-reduction potentials of organic systems.* Baltimore: Williams & Wilkins Co.

22. Clarke, F. W. 1924. *The data of geochemistry.* U.S. Geol. Survey Bull. No. 770.
23. Ehrlich, H. L. 1964. Microbial transformations of minerals. In *Principles and applications in aquatic microbiology,* ed. H. Heukelekian and N. C. Dondero, pp. 43–60. New York: Wiley.
24. Elliott, R. P., and H. D. Michener. 1965. *Factors affecting the growth of psychrophilic micro-organisms in foods.* U.S. Dept. Agric. Tech. Bull. No. 1320. Washington, D.C.: U.S. Dept. Agric.
25. Emery, K. O., and R. E. Stevenson. 1957. Estuaries and lagoons. I. Physical and chemical characteristics. In *Treatise on marine ecology and paleoecology,* ed. J. W. Hedgpeth, vol. 1, pp. 673–750. Geol. Soc. Amer., Mem. 67.
26. Farrell, J., and A. H. Rose. 1965. Low-temperature microbiology. Adv. *Appl. Microbiol.* **7**:335–78.
27. Fenchel, T. 1968. The ecology of marine microbenthos. II. The food of marine benthic ciliates. *Ophelia* **5**:73–121.
28. Gabe, D. R., E. P. Troshanov, and E. E. Sherman, 1964. Formation of manganese-iron layers in mud as a biogenic process. In *The role of microorganisms in the formation of iron-manganese lake ores,* ed. M. S. Gurevich, pp. 95–117. Moscow: Nauka.
29. Genovese, S., G. Macri, and C. Scavuzzo. 1966. Contributo allo studio della nitrificazione nell'ambiente salmastro. *Boll. Pesca Piscicoltura Idrobiol.* **21**:273–98.
30. Gest, H. 1954. Oxidation and evolution of molecular hydrogen by microorganisms. *Bacteriol. Rev.* **18**:43–73.
31 Glassman, H. N. 1948. Surface active agents and their application in bacteriology. *Bacteriol. Rev.* **12**:105–48.
32. Goldberg, E. D. 1963. The oceans as a chemical system. In *The sea,* ed. M. N. Hill, vol. 2, pp. 3–25. New York: Interscience Publishers, Inc.
33. Goldschmidt, V. M. 1954. *Geochemistry.* Oxford: Clarendon Press.
34. Gray, C. T., and H. Gest. 1965. Biological formation of molecular hydrogen. *Science* **148**:186–92.
35. Harold, F. M. 1966. Inorganic polyphosphates in biology: structure, metabolism and function. *Bacteriol. Rev.* **30**:772–94.
36. Harvey, G. W. 1966. Microlayer collection from the sea surface: a new method and initial results. *Limnol. Oceanogr.* **11**:608–13.
37. Hayes, F. R., and J. E. Phillips. 1958. Lake water and sediment. IV. Radiophosphorus equilibrium with mud, plants, and bacteria under oxidized and reduced conditions. *Limnol. Oceanogr.* **3**:459–75.
38. Henrici, A. T. 1933. Studies of freshwater bacteria. I. A direct microscopic technique. *J. Bacteriol.* **25**:277–87.
39. Hewitt, L. F. 1950. *Oxidation-reduction potentials in bacteriology and biochemistry.* London: London County Council.
40. Hofer, A. W. 1941. A characterization of *Bacterium radiobacter. J. Bacteriol.* **41**:193–224.
41. Hutchinson, G. E. 1957. *A treatise on limnology.* New York: Wiley.
42. Ingram, M. 1957. Micro-organisms resisting high concentrations of sugars or salts. In *Microbial ecology,* ed. R. E. O. Williams and C. C. Spicer, pp. 90–133. Cambridge: Cambridge Univ. Press.
43. Inman, R. E., R. B. Ingersoll, and E. A. Levy. 1971. Soil: a natural sink for carbon monoxide. *Science* **172**:1229–31.
44. Jenkins, D. W., S. M. Siegel, and C. E. ZoBell. 1974. *The biological effects of environmental extremes (including laboratory simulation).* (In Press).
45. Jones, G. E. 1964. Effect of chelating agents on the growth of *Escherichia coli* in seawater. *J. Bacteriol.* **87**:483–99.
46. Jørgensen, C. B. 1961. The food of filter feeding organisms. *Rapp. Cons. Explor. Mer.* **153**:99–107.
47. Kalinenko, V. O. 1957. Multiplication of heterotrophic bacteria in distilled water. *Mikrobiologiya* **26**:148–53.

48. Kamen, M. D. 1963. *Primary processes in photosynthesis*. New York: Academic Press Inc.
49. Kanwisher, J. 1963. On the exchange of gases between the atmosphere and the sea. *Deep-Sea Res.* **10**:195–207.
50. Kawai, A., Y. Yoshida, and M. Kimata. 1968. Nitrifying bacteria in the coastal environment. *Bull. Misaki Mar. Biol. Inst., Kyoto Univ., No.* 12:181–194.
51. Kempner, E. S. 1966. Acid production by *Thiobacillus thiooxidans*. *J. Bacteriol.* **92**:1842–43.
52. Kimata, M., Y. Yoshida, and M. Taniguchi. 1968. Studies on the marine microorganisms utilizing inorganic nitrogen compounds. I. On the marine denitrifying bacteria. *Bull. Jap. Soc. Sci. Fisher.* **34**:1114–17.
53. Kingsbury, J. M. 1954. On the isolation, physiology and development of minute bluegreen algae. Ph.D. dissertation, Harvard University.
54. Kittredge, J. S., and E. Roberts. 1969. A carbon-phosphorus bond in nature. *Science* **164**:37–42.
55. Krauskopf, K. B. 1967. Introduction to geochemistry. New York: McGraw-Hill Book Co.
56. Kuznetsov, S. I. 1970. *Microflora of lakes and its geochemical activity*. Leningrad: Nauka.
57. Kuznetsov, S. I., M. V. Ivanov, and N. N. Lyalikova. 1963. *Introduction to geological microbiology*. New York: McGraw-Hill Book Co.
58. Lackey, J. B. 1967. The microbiota of estuaries and their roles, In *Estuaries*, ed. G. H. Lauff, pp. 291–302. Washington, D.C.: AAAS.
59. Larsen, H. 1967. Biochemical aspects of extreme halophilism. In *Advances in microbial physiology*, ed. A. R. Rose and J. F. Wilkinson, pp. 97–132. New York: Academic Press Inc.
60. Lauff, G. H., ed. 1967. *Estuaries*. Washington, D.C.: AAAS.
61. Loewus, M. W., and C. C. Delwiche. 1963. Carbon monoxide production by algae. *Plant Physiol.* **38**:371–74.
62. MacLeod, R. A. 1965. The question of the existence of specific marine bacteria. *Bacteriol. Rev.* **29**:9–23.
63. MacLeod, R. A. 1968. On the role of inorganic ions in the physiology of marine bacteria. In *Advances in microbiology of the sea*, ed. M. R. Droop and E. J. F. Wood, vol. 1, pp. 95–126. New York: Academic Press Inc.
64. Mason, B. 1966. *Principles of geochemistry*. New York: Wiley.
65. Michener, H. D., and R. P. Elliott. 1964. Minimum growth temperatures for food-poisoning, fecal-indicator, and psychrophilic microorganisms, In *Advances in food research*, vol. 10, pp. 349–396. New York: Academic Press Inc.
66. Morita, R. Y. 1966. Marine psychrophilic bacteria. *Oceanogr. Mar. Biol. Ann. Rev.* **4**:105–21.
67. Mulder, E. G. 1964. Iron bacteria, particularly those of the *Sphaerotilus-Leptothrix* group, and industrial problems. *J. Appl. Bacteriol.* **27**:151–73.
68. Paoletti, A. 1968. Organismes prédateurs dans l'autoépuration des eaux de mer polluées essais d'étude avec bactéries radioactives our autrement marquées. *Rev. Intern. Oceanogr. Med.* **10**:229–47.
69. Perfil'ev, B. V., and D. R. Gabe. 1969. *Capillary methods of investigating microorganisms*. Edinburgh: Oliver & Boyd, Ltd.
70. Pickwell, G. V. 1967. *Gas and bubble production by siphonophores*. San Diego: Naval Undersea Warfare Center. TP 8.
71. Postgate, J. R. 1969. The sulphur cycle. In *Inorganic sulphur chemistry*, ed. G. Nickless, London: Elsevier Publishing Co.
72. Pshenin, L. N. 1966. *The biology of marine nitrogen fixers*. Kiev: Naukova Dumka.
73. Redfield, A. C. 1958. The biological control of chemical factors in the environment. *Amer. Sci.* **46**:205–21.
74. Reid, G. K. 1961. *Ecology of inland waters and estuaries*. New York: Reinhold Publishing Corp.

75. Richards, F. A. 1957. Oxygen in the ocean. In *Treatise on marine ecology and pale-oecology*, ed. J. W. Hedgpeth, vol. 1, pp. 185–238. Geol. Soc. Amer., Mem. 67.
76. Richards, F. A. 1965. Anoxic basins and fjords. In *Chemical oceanography*, ed. J. P. Riley and G. Skirrow, pp. 611–645. New York: Academic Press Inc.
77. Rigler, F. H. 1956. A tracer study of the phosphorus cycle in lake water. *Ecology* **37**:550–62.
78. Ryther, J. H. 1969. Photosynthesis and fish production in the sea. *Science* **166**:72–76.
79. Seki, H., and O. D. Kennedy. 1969. Marine bacteria and other heterotrophs as food for zooplankton in the Strait of Georgia during winter. *J. Fish. Res. Bd. Can.* **26**:3165–73.
80. Sieburth, J. M. 1968. The influence of algal antibiosis on the ecology of marine microorganisms. In *Advances in microbiology of the sea*, ed. M. R. Droop and E. J. F. Wood, vol. 1, pp. 63–94. New York: Academic Press Inc.
81. Small, J. 1954. *Modern aspects of pH*. London: Baillière, Tindall and Cox.
82. Smith, W. W. 1936. Evidence of a bacterial flora indigenous to the Great Salt Lake in Utah. M.A. thesis, University of Utah.
83. Smith, W. W., and C. E. ZoBell. 1937. Direct microscopic evidence of an autochthonous bacterial flora in Great Salt Lake. *Ecology* **18**:453–58.
84. Sorokin, Ju. I. 1968. The use of ^{14}C in the study of nutrition of aquatic animals. *Mitt. Internat. Verein. Limnol.* No. **16**:1–48.
85. Starkey, R. L. 1964. Microbial transformations of some organic sulfur compounds. In *Principles and applications in aquatic microbiology*, ed. H. Heukelekian and N. C. Dondero, pp. 405–29. New York: Wiley.
86. Starkey, R. L., and S. A. Waksman. 1943. Fungi tolerant to extreme acidity and high concentrations of copper sulfate. *J. Bacteriol.* **45**:509–19.
87. Stewart, W. D. P. 1967. Nitrogen-fixing plants. *Science* **158**:1426–32.
88. Stewart, W. D. P. 1969. Biological and ecological aspects of nitrogen fixation by free-living microorganisms. *Proc. Roy. Soc.* (London) B **172**:367–88.
89. Stewart, W. D. P. 1970. Algal fixation of atmospheric nitrogen. *Plant Soil* **32**:555–88.
90. Trudinger, P. A. 1969. Assimilatory and dissimilatory metabolism of inorganic sulphur compounds by micro-organisms. *Adv. Microbiol. Physiol.* **3**:111–58.
91. Utter, M. F., and H. G. Wood. 1951. Mechanism of fixation of carbon dioxide by heterotrophs and autotrophs. *Adv. Enzymol.* **12**:41–151.
92. van Uden, N. 1962. Factors of host-yeast relationship. In *Proc. 8th Intern. Congr. Microbiol.*, pp. 635–643. Montreal: University of Toronto Press.
93. van Uden, N. 1967. Occurrence and origin of yeasts in estuaries. In *Estuaries*, ed. G. H. Lauff, pp. 306–10. Washington, D.C.: AAAS.
94. Warren, H. G. 1970. Feasibility study on the production of synthetic gas by subsurface disposal of urban refuse. *Report to U.S. Dept. Interior, Bureau of Mines.*
95. Watt, W. D., and F. R. Hayes. 1963. Tracer study of the phosphorus cycle in sea water. *Limnol, Oceanogr.* **8**:276–85.
96. Wiame, J. M. 1958. Le cycle du soufre dans la nature. In *Handbuch der Pflanzen-physiologie*, ed. W. Ruhland, pp. 102–120. Berlin: Springer Verlag.
97. Wolfe, R. S. 1964. Iron and manganese bacteria. In *Principles and applications in aquatic microbiology*, ed. H. Heukelekian and N. C. Dondero, pp. 82–97. New York: Wiley.
98. Wood, E. J. F. 1962. The microbiology of estuaries. In *The environment of marine sediments*, pp. 20–26. Occasional Paper No. 1. Kingston: University of Rhode Island.
99. Yoshida, Y. 1967. Studies on the marine nitrifying bacteria: with special reference to characteristics and nitrite formation of marine nitrite formers. *Bull. Misaki Mar. Biol. Inst., Kyoto Univ.*, No. **11**:1–58.
100. Yoshida, Y., A. Kawai, and M. Kimata. 1967. Studies on marine nitrifying bacteria (nitrite formers and nitrate formers). VI. Distribution of limnetic nitrifying bacteria in the coastal region. *Bull. Jap. Soc. Sci. Fish.* **33**:426–29.
101. Yoshida, Y., and M. Kimata. 1969. Studies on the marine microorganisms utilizing

inorganic nitrogen compounds. V. On the uptake and liberation of inorganic nitrogen compounds by the microorganisms as a whole in sea water. *Bull. Jap. Soc. Sci. Fish.* **35**:307–10.

102. Zajic, J. E. 1969. *Microbial biogeochemistry.* New York: Academic Press Inc.
103. Zavarzin, G. A. 1964. *Metallogenium symbioticum. Z. Allg. Mikrobiol.* **4**:390–95.
104. Zavarzin, G. A. 1968. Bacteria in relation to manganese metabolism. In *The ecology of soil bacteria,* ed. T. R. G. Gray and D. Parkinson, Montreal: University of Toronto Press.
105. ZoBell, C. E. 1946. Studies on redox potential of marine sediments. *Bull. Amer. Ass. Petrol. Geol.* **30**:477–513.
106. ZoBell, C. E. 1947. Microbial transformation of molecular hydrogen in marine sediments, with particular reference to petroleum. *Bull. Amer. Ass. Petrol. Geol.* **31**:1709–51.
107. ZoBell, C. E. 1958. Ecology of sulfate reducing bacteria. In *Symposium on sulfate reducing bacteria,* pp. 1–25. New York: St. Bonaventure University. (Reprinted in Producers Monthly **22**:12–29.)
108. ZoBell, C. E. 1963. Organic geochemistry of sulfur. In *Organic geochemistry,* ed. I. A. Breger, pp. 543–578. Oxford: Pergamon Press.
109. ZoBell, C. E. 1964. Geochemical aspects of the microbial modification of carbon compounds. In *Advances in organic geochemistry,* ed. U. Colombo and G. Hobson, pp. 339–56. Oxford: Pergamon Press.
110. ZoBell, C. E. 1970. Pressure effects on morphology and life processes of bacteria. In *Cellular Aspects of Hydrostatic Pressure,* ed. A. M. Zimmerman, pp. 85–130. New York: Academic Press Inc.
111. ZoBell, C. E. 1972. Microbial biogeochemistry of oxygen. Akad. Nauk SSSR Izvestia, Ser. Biol., No. 1: 23–42 (In Russian) Proceed. USSR Acad. Sci., Biol. Ser. No. 1: 23–42
112. ZoBell, C. E., and E. C. Allen. 1933. Attachment of marine bacteria to submerged slides. *Proc. Soc. Exp. Biol. Med.* **30**:1409–11.
113. ZoBell, C. E., and J. F. Prokop. 1966. Microbial oxidation of mineral oils in Barataria Bay bottom deposits. *Z. Allg. Mikrobiol.* **6**:143–62.
114. ZoBell, C. E., F. D. Sisler, and C. H. Oppenheimer. 1953. Evidence of biochemical heating in Lake Mead mud. *J. Sed. Petrol.* **23**:13–17.

Comments

STEVENSON: Would you comment on the relative productivity or activity of the estuarine region to that of the open ocean?

ZOBELL: I have been interested in this for many years experimentally and also in looking at very recent papers. Most of my time has been devoted to studying the conditions in the open ocean. It is my opinion that primary productivity per unit area is at least an order of magnitude higher in most estuarine waters than in the open ocean. More accurate estimates are difficult to make owing partly to the great variability in estuarine environmental conditions, especially plant nutrients, chemical composition, temperature, and other characteristics. In certain estuarine situations, primary productivity may be as much as a hundred times greater than the average for the open ocean.

ALEXANDER: If the parameters within one given estuary vary so much, how many samples would one have to collect to get a representative recovery?

ZOBELL: You know better than I what the statistical probabilities are. It depends in part upon what accuracy is desired and partly on the area and volume of the water body. In a large estuary like Chesapeake Bay, one sample per square mile probably would be adequate. In such a large body of water, microenvironments would not affect mean conditions very much. On the other hand, if you want to know what is happening in a small body of water, for instance, only one square mile, one sample is not enough. You may have to take several samples per acre. Moreover, it may be desirable to make determinations at different depths, which requires many more samples. One has to use judgment to determine how many samples. Conditions in the open ocean tend to be more uniform as contrasted with sharper diurnal, seasonal, distal, vertical, and other kinds of transitions in most estuaries. It seems almost axiomatic that the sharper the transition and the smaller the size, the more samples required per unit area.

ALEXANDER: What about the microenvironments? What norm would you choose to make samples with microsediments?

ZOBELL: There is no norm for all situations. The number, volume, and kinds of samples to be collected from various microenvironments depend upon the particular problems. By this, I mean whether you are concerned with determining microbial populations, microbial biomass, physiological activity rates, physiochemical effects of microbial activities, or something else. Suppose, as in the example discussed this morning, you are concerned with the vertical distribution and activities of microbes in bottom sediments, where strict aerobes and strict anaerobes appear to be growing together. In such a situation, several small samples and measurements must be taken, especially in the topmost few centimeters where transitions are greatest. If O_2 is present at all, most likely it will occur only at the mud-water interface. In material from the mud-water interface, the microbial population may be ten- to a thousandfold greater than in strata a few centimeters deeper. In such situations, there is no standard method for sampling or measurements. One must devise his own work model for each different situation. The point that I tried to make is that very different results can be expected, for example, from making replicate analyses of 1 kg of thoroughly mixed mud from various depths, as in a long core, than from examining 1,000 1-g samples carefully collected from various strata.

G. JONES: I would like to suggest that you might drop your lower pH to zero, based on work of Dr. Waksman in 1943, who worked on a fungus which grows at pH zero in saturated copper sulfate.

ZOBELL: Was that in an estuary?

G. JONES: No.

ZOBELL: Let us see that slide again. My slide should have shown that the range in most estuaries, except localized environments, is from about pH 3 to 9. The extremes at which microbial growth has been reported are from less than pH 0.1 to more than pH 11.5. Kingsbury at Harvard University reported the growth of blue-green algae at pH 13.5. This value has not been generally accepted because he did not state whether CO_2 from the atmosphere lowered the pH somewhat.

WATSON: Is there any place in estuaries where you have not found microorganisms?

ZOBELL: I don't know because I have not looked very extensively.

WATSON: How many places in the world can you, for example, find sterile environments? I have only found one and that is in the Red Sea. These were hot holes, where the temperature was 70 C, and the salinity was twice that of sea-water. Because of a combination of heat, salinity, and heavy metals, this whole area is completely sterile. As far as I know, this is one of the few completely sterile areas in the entire world. Even in the hot springs you can always find some organisms growing there regardless of the temperature.

ZOBELL: Your point is well taken. I would not attempt to say unequivocally that there is no place in the hydrosphere where you will not find living organisms if you look long enough. We have not made enough observations to say that there are no biological deserts. Personally, I have found none and know of none. The purpose of my environmental extremes chart was to show that most environmental parameters in estuaries are well within the range tolerated by certain microorganisms.

COULL: Dr. Gordon Riley, in his recent review of particulate organic matter in the sea, spent quite a bit of time on dissolved organics being put into particulate form via the bubbling process. There is some controversy, in that this may be the mechanism of sloppy technique or it may be the result of bacterial action. I would like your comments on it. What function do you feel bacteria play in the role of putting dissolved organics into particulate on the bubble?

ZOBELL: Unfortunately, I have not followed this work very carefully, and I have not done any work along this line myself. I do believe that bubble formation may play a part in the coalescence of soluble molecules to eventually form a dominance of particulate organic matter.

COULL: Do you then believe in the bubble formation hypothesis?

ZOBELL: I try to hedge on that question. I think that there is some evidence for the conglomeration of organic matter by the bubble phenomenon. This is the way I have interpreted the experiments. I have performed none of the experiments myself. I do not question the technique. I do question the significance of it as compared to other possible events that may be taking place. Perhaps more important is the microbial conversion of dissolved organic matter or small molecules to form microbial biomass which is particulate.

HOLM-HANSEN: I would like to add a comment to this discussion. You do not need to implicate bacteria to get particulate organic matter from the dissolved organic matter in seawater. This is distinctly outlined on the basis of filtration experiments in which you filter off all or nearly all of the particulate and let the filtrate sit very undisturbed in the lab. You will get the equilibrium formed again, thereby getting approximately the same concentration of particulate organic carbon and the same size distribution as determined by the atomic particle counting technique. This is not to say anything against the bubble or bacterial function in seawater, but you can also get particulate organic carbon from dissolved organic carbon without the intervention of bacteria.

ZOBELL: I would like to add that I agree with you and others who have worked on it. Moreover, there are invertebrates that can absorb and assimilate small molecules or dissolved organic compounds. An open question is the concentration at which animals can do this effectively. Can they effectively build biomass if the concentration of dissolved organic matter is only one-tenth part per million? Bacteria can.

HOLM-HANSEN: We are talking about two different things. You are concentrating on the actual biological use of assimilation of the dissolved organic carbon to form biomass. I was thinking of just the equilibrium between particulate organic carbon and dissolved organic carbon and there I think it is perfectly obvious that dissolved organic carbon can be used to a certain extent.

Section Two
TECHNIQUES

The Use of Selective and Differential Media in the Analysis of Marine and Estuarine Bacterial Populations

Darrell B. Pratt and John W. Reynolds

Methods are needed to enable microbiologists to make rapid analyses of the species composition of natural bacterial populations. If it is assumed that the composition of such populations is responsive to environmental changes of a chemical or physical nature and if the population changes could be rapidly detected, then these changes would have predictive value for subsequent changes in the macrobios. With such a tool, corrective measures could perhaps be taken to reverse undesirable ecological changes. The usefulness of such a procedure will ultimately depend on a considerable amount of correlative data relating the environmental conditions to population structure. It will be essential to discriminate between normal cyclic changes with a seasonal, tidal, or diurnal basis and abnormal changes caused by unnatural external factors.

Concepts and Approaches

Objectives. That selective and differential media might be employed for the analysis of natural bacterial populations is an extension of ideas long used in diagnostic microbiology. Ultimately, we hope to prepare a series of media which would allow the quantitative estimation of the various components of populations on initial cultivation on agar plates. The definition of what we mean by population component is now somewhat nebulous but our present goal is to recognize the principal genera of aerobic, heterotrophic bacteria capable of producing colonies. Quite obviously we probably will not achieve this objective but will have to compromise since it is unlikely that the separations produced by selective agents will coincide accurately with those produced by the techniques of modern taxonomy. Before continuing, I wish to say that we believe in the autochthonous flora and in obligate anaerobes but at the present time we feel the problem is sufficiently complicated without their inclusion in our considerations. Many selective and differential agents are available for use in preparing media for marine bacteria. The immediate task is to find which of these and what combinations will be most successful. For the purpose of initiating this study and developing our procedures, we have selected several dyes which have been widely used in the past with considerable success.

Limitations. The spread plate technique, as we usually employ it, serves nicely to estimate the viable count of estuarine and inshore waters. The use of 0.1 ml

inocula spread evenly over the surface of a plate is suitable if the counts are between 300 and 3,000 per ml. The procedure has limitations when it is used to
estimate components of the population. If a population of bacteria has as its
total viable count 1,000 per ml and represents 100 species with 10 representatives of each present, then 0.1 ml of the sample could be expected to contain 1
representative of each species. It becomes evident the presence or absence of a
particular representative in a sample will be governed by probability as predicted
by the Poisson distribution (4). Table 1 gives the probabilities of a given number
of colonies of a species being present as it is related to the expected number.
These probability values have been translated into numbers of plates in ten
showing a given number of colonies as related to the population density (Table
2). If ten representatives of a species are present in 1 ml, then four of ten plates
could be expected to be without representative colonies. Approximately forty
representative bacteria per ml are required to insure that every plate in ten will

Table 1. Probabilities calculated from the Poisson
distribution

m^a	P_0	P_1	P_2	P_3
0.01	.99	—	—	—
0.10	.90	.09	—	—
1.0	.37	.37	.19	.06
2.0	.13	.26	.26	.17
3.0	.05	.15	.23	.23
4.0	.02	.07	.14	.18

^{a}m represents the mean number per sample; P_0,
P_1, P_2, and P_3 are the probabilities that a given
sample will contain 0, 1, 2, or 3 representative
items.

Table 2. Calculated number of plates in a total of ten showing a given
number of colonies

Average Number in Sample^a	Plates in Ten with Indicated Number of Colonies			
	Zero	One	Two	Three
0.01	10	0	0	0
0.1	9	1	0	0
1.0	4	4	2	0,1
2.0	1,2	2,3	2,3	2
3.0	0,1	1,2	2	2
4.0	0	0,1	1,2	2

^{a}Sample = 0.1 ml (1.0 = 10 per ml).

contain a colony. Thus, it can be seen that if selective and differential media are used with the spread plate technique, the lower limits for usefulness will be a function of the population density of the species involved. Obviously, if the component being separated is, for example, the genus *Vibrio*, which may constitute 30 to 40% of a population, the spread plate may be useful. With selective media, it may be necessary to use membrane filters to provide a sufficient concentration of the sample. If sediments are being studied where the viable counts are high, the spread technique can be used.

Selective Agents. A selective agent separates a population by inhibiting the growth of some species but not others. For example a mixture of *Escherichia coli* and *Bacillus subtilis* may readily be separated by using media containing crystal violet *(1)*. The population would be clearly separated into a sensitive group and a more resistant group. Over a considerable range of dye concentrations the medium would give an accurate count of the resistant species and by difference the number of sensitive cells could be estimated.

In a natural population consisting of a number of species the action of a selective agent will be less precise; however, it seems possible that the very resistant and sensitive organisms present may represent homogeneous groups. In other words, the property of extreme sensitivity would be shared by only a few species and the same would apply to resistance (Fig. 1). We have found that with methylene blue as the selective agent, a rather large component of the natural population of marine bacteria was extremely sensitive to the dye and, indeed, more than 90% of the population was inhibited by higher concentrations (20 *μ*g/ml). An arithmetic plot of the data would show more graphically the sensitivity of these populations to concentrations as low as 0.2 *μ*g/ml of the dye (Fig. 2). In studies to be more fully described elsewhere, these sensitive

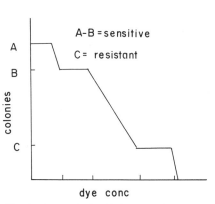

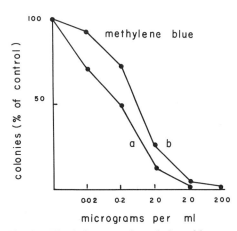

Fig. 1. **Hypothetical influence of an inhibitory dye on colony formation by a natural population. Two potentially homogeneous groups are projected: (A–B), the very sensitive, and (C), the very resistant.**

Fig. 2. **The influence of methylene blue on colony formation by populations in two samples of seawater (a and b).**

members of the population have been found to consist of gram-negative, motile, oxidase-positive, fermentative rods.

At the present time our approach to the general problem of the usefulness of a selective agent involves the use of both known cultures and natural populations. With the known cultures one establishes what may be potentially useful inhibitory agents by finding one which makes separations, for example, allowing only members of the genus *Vibrio* to grow. This agent should then be tested with natural populations to determine if randomly selected resistant organisms are always, usually, sometimes, or never *Vibrio*. Identification of the resistant strains will require the modern tools of taxonomy.

Results

Recently we have been studying the effects of dyes on populations of marine bacteria. Both known cultures and natural populations have been studied.

Dye Sensitivity. By means of a modified filter paper disc sensitivity test, we have observed the effects of twenty-four dyes on the growth of twenty test organisms. These marine bacteria were originally isolated from inshore waters on the Florida east coast (5). They have been maintained in stock with periodic transfer and repurification; they have retained their previously described requirement for salts. These organisms have been well characterized as to physiological and morphological properties. The dyes tested could be divided into three groups as follows: (a) those which inhibited none of the test cultures; (b) those which inhibited all of the test cultures; and (c) those which inhibited some of the test cultures. Congo red, neutral red, trypan blue, acid fuchsin, brom cresol purple, brom thymol blue, phenol red, and chlorphenol red were all without effect. Brilliant green, malachite green, para-nitrophenol, crystal violet, and acriflavin inhibited all twenty test cultures. The remaining dyes ranged from relatively ineffective to strongly effective (Table 3). In our opinion, those inhibiting ap-

Table 3. Dyes showing selective inhibition of twenty cultures of marine bacteria

Dye	Number[a] Inhibited	Dye	Number Inhibited
Eosin	2	Safranin 0	13
Erythrosin	2	Toluidine blue	15
Picric acid	5	Basic fuchsin	16
Dinitrophenol	10	Methylene blue	17
Resazurin	10	Ethyl violet	19
Rose bengal	13		

[a]Positive inhibition was indicated by the presence of a definite zone of inhibition in the lawn of bacterial growth around a filter paper disc which had been immersed in a 0.2% solution of the dye. The medium was Marine Agar (Difco).

proximately half of the test organisms might be more useful in developing selective media. The observation that cationic dyes such as crystal violet inhibited a broad range of marine bacteria may have ramifications which bear on the cationic requirements of the organisms. We did not anticipate that these gram-negative bacteria would be as susceptible as they were to basic dyes (1, 2).

Inhibition of Natural Populations by Dyes. Samples of seawater were planted on media containing one of several selected dyes in concentrations of 1, 2, 10, and 20 µg/ml. After incubation for 7 days at 15 C the colonies were counted and compared to the number on media without dye. Crystal violet was the most inhibitory of dyes tested in this manner while safranin and rose bengal were less toxic (Table 4). Dinitrophenol was of interest because the preliminary results suggested that an increase in colony numbers was induced by small amounts of this compound as compared to the control. The resistant colonies on media containing the higher amounts of safranin and rose bengal were rather uniform in morphology and may represent a homogeneous population. These now should be isolated and characterized to determine if, for example, 20 µg/ml of rose bengal per ml select a particular class of bacteria. Colonies growing in the lower concentrations of dye were stained by the dye in varying degrees and ways; this variation in color may have potential value for differentiating colonies. These preliminary experiments will require amplification.

Table 4. Influence of dye concentration on the number of colonies produced from natural populations[a]

Colonies Formed − 0.1 ml Sample				
Dye (µg/ml)	1	2	10	20
Crystal violet	24	10	4	0
Safranin	80	51	9	5
Rose bengal	80	33	11	4
2,4 Dinitrophenol	114	99	50	33
Resazurin	37	36	8	6

[a]Control 73 S.D. = 10.5. Samples were obtained at Schoodic Point, Maine. Incubation temperature was 15 C. Incubation time was 7 days. Basal medium was Casitone (Difco) and yeast extract in artificial seawater.

Differential Agents. Neutral red was not significantly inhibitory even at 20 µg/ml. In the marine agar without added carbohydrate, the colonies were either stained some shade of red, were colorless, or were yellow. This dye is red in an acid and yellow in an alkaline medium. The physiological basis of this may prove interesting but, in any event, the variations in colony coloration may well be of use in making colony separations.

Triphenyl tetrazolium chloride (TTC) is a dye which on reduction becomes insoluble and red. Its reduction is favored by basic environments and inhibited in acid. Bacterial colonies are stained red as the dye is reduced. Gram-negative bacteria such as *E. coli* produce colorless colonies in the presence of a ferment-able carbohydrate but are red if the carbohydrate is absent or not fermentable *(3, 6)*. When seawater samples were plated on media containing TTC and glucose, both red and white colonies were produced. In the absence of glucose all colonies were red (Table 5). The TTC showed some degree of colony in-hibition which possibly was reduced by the presence of glucose. From media containing carbohydrate, randomly selected red colonies were oxidative by the standard O/F test; the white colonies with one exception (1 in 30) were all fermentative. Colonies selected at random from plates without glucose or TTC were found to be *ca* 50% fermentative. Quite possibly this can be developed into a useful technique for estimating fermentative and nonfermentative bacteria in the population. The inhibitory action of tetrazolium may confound the results.

Table 5. Inhibitory and selective effects of TTC on Marine Bacteria[b,c]

Tetrazolium (μg/ml)	Without Glucose	With Glucose	
		Total	% White
0	1,090[a]	1,070	
2.5	860	1,030	28%
5.0	600	1,120	25%
10.0	440	840	37%
20.0	330	460	48%

[a]Bacteria per ml.
[b]Sample from Schoodic Point, Maine.
[c]Basal medium was 1% trypticase in artificial seawater.

Summary

It is difficult to bring these disjointed observations and speculations together into a unified report. They have been preliminary in nature, with the objective of developing the course for future research. Our final goal is to develop prac-tical, differential, and selective media for the analysis of marine and estuarine bacterial populations.

Literature Cited

1. Churchman, J. W., and L. Siegel. 1928. Cultural separation of bacteria on the basis of triphenyl methane coefficients. *Stain Technol.* **3**:73–80.
2. Dubos, R. 1946. *The bacterial cell.* Cambridge, Mass.: Harvard University Press.
3. Lederberg, J. 1948. Detection of fermentation variants with tetrazolium. *J. Bacteriol.* **56**:695.
4. Meynell, G. G., and E. Meynell. 1970. *Theory and practice in experimental bacteriology.* Cambridge, England: Cambridge University Press.

5. Tyler, M. E., M. C. Bielling, and D. B. Pratt. 1960. Mineral requirements and other characters of selected marine bacteria. *J. Gen. Microbiol.* **23**:153–61.
6. Zamenhof, S. 1961. Gene unstabilization induced by heat and nitrous acid. *J. Bacteriol.* **81**:111–17.

Comments

COLWELL: Do you think this technique might be used to isolate specific types of organisms like *Pseudomonous* and *Vibrio?*

PRATT: That is my intention. I want to separate the organisms in a dichotomous way. I am trying to develop some media we could isolate *Vibrio* on, some media for *Pseudomonous,* and some for *Flavobacter.* As I say, I am not that overly ambitious at this moment. It is hoped that we might do something like that.

WATSON: You mentioned the possibility of eventually using the Millipore filter technique in concentrating bacteria. Based on our work with nitrifying bacteria, I would like to offer a word of caution on this. Another investigator has developed a Millipore filter technique for use with these organisms. We decided it would be a very handy technique if it actually worked. But first, we wanted to determine if it would work on pure cultures. We put a known number of nitrifying bacteria on filters and measured CO_2 fixation. We found that as soon as the bacteria were placed on the filters, they lost about 99.99% activity. We then looked at urea-decomposing bacteria in the same manner, and again we found that more than 99% of the bacteria were killed with the technique. I do not know if this is a common phenomenon, but I think it certainly should be explored before anyone uses the Millipore filter technique for counting marine microorganisms.

PRATT: Do you think the actual process of bringing them down onto the filter is destructive?

WATSON: Yes, I think they get trapped. I do not know exactly what is killing them. I was talking to Dr. Wood and I think he has some evidence that would tend to substantiate what I have said.

HOLM-HANSEN: Did you wash or treat the filter before you performed the experiment?

WATSON: Yes. We always washed the filter before we started the experiment.

HOLM-HANSEN: We have been examining the Millipore filters recently and have found that they contain a variable content of strontium, nickel, copper, and other heavy metals. The ions may effect the survival or activity of the bacteria.

WATSON: The scanning microscope shows that there is a great variation in Millipore filters. We also tried to use Gelman and Teflon filters, but without much success. I have never seen this problem discussed elsewhere. Certainly Millipore filters are commonly used by people studying fresh water, and I wonder if we are doing something drastically wrong. Maybe this is a fairly common phenomenon.

May I offer another word of caution. Many of you probably use Niskin[1] samplers as I have. We have been using them since they were developed by Shale Niskin. We tested the toxicity of this plastic bag on the nitrifying bacteria

[1]Niskin, S. J. 1962. A water sampler for microbiological studies. Deep Sea Res. **9**:501–503.

and found it was not toxic. However, recently we started to measure changes like acetate oxidation and urea decomposition. We had planned a big cruise off the northwest coast of Africa, and planned to take samples every 5 m and measure acetate oxidation and urea decomposition. Unfortunately, we did not have a scintillation counter aboard, and had to make measurements when we returned. We found that none of our acetate samples showed oxidation. We became very curious as to why this should be and examined Woods Hole water. We took samples in an Erlenmeyer flask and placed the water in a Niskin bag for 5, 10, 15, and 30 min. We found that there was a toxic component in the Niskin bag inhibiting acetate oxidation by 80% and urea decomposition by 50%. Maximum effect was obtained when the water remained in the bags for a half an hour. I talked to Dr. Niskin about this, and we are certainly going to look into it. But again, a word of caution to people who are relying on the Niskin bag.

HOLM-HANSEN: Can you tell us how the killing of the bacteria varies with the pressure of the vacuum which you use to concentrate them with a Millipore filter?

WATSON: It did not seem to make any difference. Actually, we tried to do without any vacuum. We tried to let it drip through. But this just did not seem to be a the main problem at all. Bacteria simply seemed to get buried in the filter. I suppose the greater vacuum you have, the more killing effect you would have.

HOLM-HANSEN: How about the glass fiber filters?

WATSON: We have not tried that.

ZOBELL: I would like to comment with regard to this phenomenon of loss of activity on membrane filters. First of all, I would ask whether you used colony counts or activity as criteria. And then second, how many biochemical activities have you investigated? For example, have you analyzed the effects of membranes on exoenzymes, and the adsorption or retention of the exoenzymes by membrane filters, to the extent that they do activate or decrease the biochemical activity without killing the organisms?

WATSON: When working with nitrifying bacteria, you do not count colonies very often. We were actually looking at biological activities. The membrane may be inhibiting the activity for a short time without killing the bacteria. I do not know.

Isolation of Myxobacteria from Marine Habitats in the U.S. Virgin Islands

Ellis R. Brockman

Bacteria classified in the order Myxobacterales, the eighth order of the class Schizomycetes *(3)*, differ from other species of bacteria in several respects. Myxobacteria produce large quantities of extracellular slime, have a gliding type of motility, and, in the more complex forms, the formation of a fruiting body containing microcysts is a normal function of the life cycle. Although the most obvious difference between the fruiting and nonfruiting myxobacteria is the presence or absence of a fruiting body, another significant difference is the low GC% base ratio in the nonfruiting species as compared to a higher ratio in the fruiting forms *(10, 17)*.

Fruiting myxobacteria have not been reported to occur in seawater. These bacteria have been found, however, in beaches along the Atlantic coast of the United States *(19, 5)*.

The purposes of this investigation were to determine if fruiting myxobacteria were present in seawater, which species could be isolated, and whether the presence of these bacteria was associated with runoff from land to ocean.

A survey of the literature reveals that nonfruiting myxobacteria, *Cytophaga* and *Sporocytophaga* species, have been found in the marine environment on numerous occasions and in several geographical locations. Since the nonfruiting myxobacteria are currently classified in the same order as the fruiting species and because no review of the literature dealing with myxobacterial isolates from marine sources exists, the following brief review of the literature is presented.

Waksman, Carey, and Reuszer *(31)*, as cited by Kadota *(14)*, found *Cytophaga* species in seawater from the Gulf of Maine and Georges Bank. Stanier *(27)* isolated *Cytophaga diffluens* and *C. krzemieniewska* from the Pacific Ocean water. While investigating marine agar-digesting bacteria, Humm *(11)* found *C. sensitiva* off the Beaufort, North Carolina, coast. A halophilic myxobacterium that in chinook salmon caused a disease resembling columnaris disease was found in seawater by B. J. Earp and E. J. Ordal (unpublished data) as cited by Rucker, Earp, and Ordal *(24)*. Starr *(28)* reported the isolation of species of *Cytophaga* and *Sporocytophaga* from seawater and mud of Port Orchard Bay, Puget Sound, and strains of *Sporocytophaga* from the Beaufort Sea. Kadota *(12, 13)* described two species and one variety of cellulolytic *Cytophaga* that he isolated from the sea off the coast of Japan. *C. haloflava* and *C. haloflava* var. *nonreductans* were found on fishing nets immersed in Maizuru Bay (Kyoto) and Hiroshima Bay.

C. rosea was isolated from mud, immersed cotton cords, and water of Maizuru Bay. MacLeod, Onofrey, and Norris *(21)*, working with water collected near Vancouver, Canada, isolated an agar-digesting myxobacterium. Their isolate, B-9, was originally identified as *Flavobacterium* sp., but, as reported by MacLeod and Matula *(20)*, it was reclassified as a *Cytophaga* sp., on the suggestion of J. M. Shewan, Torry Research Station, Aberdeen, Scotland. Bachmann *(2)* described a facultatively anaerobic myxobacterium, *Cytophaga fermentans*, that she isolated from marine mud of the Pacific Ocean. Although Williams and Rittenberg *(32)* did not isolate myxobacteria from the ocean, they speculated concerning the possible close relationship between microcyst-forming, marine spirilla, and species of *Sporocytophaga*. As cited by Hayes *(9)*, Velankar *(29)* found a species of *Cytophaga* in the Gulf of Mannar (between western Ceylon and southern India) and Georgala *(8)* reported *Cytophaga* sp. on fish taken from the North Sea. Liston *(15)* also reported the isolation of *Cytophaga* strains from fish caught in Puget Sound. An organism provisionally classified as *Cytophaga* sp. was isolated from the North Sea by Spencer *(26)*. This isolate was later identified by Colwell, Citarella, and Chen *(6)* as *Cytophaga marinoflava*. Veldkamp *(30)* stated that three of the *Cytophaga* cultures he used were isolated from marine mud. Of these, *C. salmonicolor* did not digest agar; the two other isolates, *C. salmonicolor* var. *agarovorans* and *C. fermentans* var. *agarovorans*, were agarolytic. Isolates of *Cytophaga* have been reported in arctic marine and littoral waters and sediments by McDonald, Quadling, and Chambers *(18)* and by Quadling and Colwell *(22)*. During a study of filterable marine bacteria, Anderson and Heffernan *(1)* isolated a strain of *Cytophaga* (isolate I-12) from Narragansett Bay surface water. Lovelace, Tubiash, and Colwell *(16)* reported that strains of *Cytophaga* were found in Chesapeake Bay.

Materials and Methods

Surface water was collected from various locations in the U.S. Virgin Islands (Fig. 1) during December, 1970, and January, 1971. The samples were obtained in sterile, widemouthed plastic bottles, with care taken to avoid possible con-tamination from either the boat or the collector. The samples were held at ambient temperatures for approximately 2 hr before they were filtered. One liter of each sample was passed through a membrane filter (0.45 μm) held in a field moniter (Millipore Corp., Bedford, Mass.). The filter was aseptically transferred to the surface of a medium having the following composition: 0.6 g KNO_3, 0.8 g K_2HPO_4, 0.24 g $MgSO_4 \cdot 7H_2O$, 0.06 g $CaCl_2$, 0.004 g $FeCl_3 \cdot 6H_2O$, 15.0 g agar, 1 l distilled water.[1] After autoclaving at 121 C for 20 min, sterile cycloheximide (Actidione, Upjohn Co., Kalamazoo, Mich.) was added to the medium to give a final concentration of 0.0025% (w/v). The medium was poured into sterile plastic petri dishes and after solidification the dishes were stored at ambient temperatures in sealed plastic bags.

The plates containing the filters were incubated at ambient temperatures and

---

[1]Previous studies (unpublished) by the author, in which media were prepared with both aged and artificial seawater, as well as distilled water, demonstrated that fruiting myxobacteria could be isolated from seawater only with a distilled water medium.

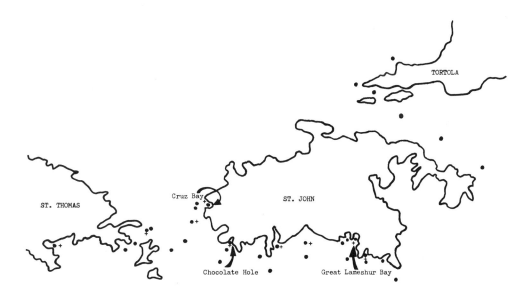

fruiting myxobacteria +

Fig. 1. **Sample locations. Multiple samples were collected in Cruz Bay, Chocolate Hole, and Great Lameshur Bay (scale 1:100,000).**

were observed with a stereomicroscope at weekly intervals for the presence of fruiting bodies. A filter showing no myxobacteria was discarded after eight weeks of incubation and observation. It should be noted that the sole carbon and energy sources in the plates consisted of cells and detritus on the filter. The fact that myxobacteria are capable of utilizing other microorganisms or their metabolic products as their sole nutrient support has been reviewed in detail by Brockman *(4)*, Dworkin *(7)*, and Quinlan and Raper *(23)*.

The myxobacteria growing on the filters were isolated by transferring fruiting bodies to freshly prepared plates of the previously mentioned medium that had been smeared with autoclaved bakers' yeast (one package of dry yeast mixed with 25 ml distilled water and sterilized at 121 C for 15 min) according to the method of Singh *(25)*. Pure cultures of the isolates were easily obtained by repeated subculture of the leading edge of the vegetative swarm to fresh yeast plates.

Dark-contrast, phase-microscopic observations of the pure cultures were made with the specimen mounted in lactophenol. On the basis of vegetative cell morphology and size, microcyst shape and size, and the pigmentation and shape of the fruiting body, the isolates were identified according to the procedures set forth in *Bergey's Manual (3)*.

Colony counts of the myxobacteria were not attempted because each bacterium does not form an individual colony on the culture medium employed in this study. When a species was isolated from a sample, its presence was simply noted. Furthermore, no attempt was made to count the number of fruiting bodies present on a filter since this count could not lead to an accurate correlation of numbers of cells present. In addition, it is not known whether these bacteria were originally present in the seawater as vegetative cells or microcysts.

Results and Discussion

Fruiting myxobacteria were isolated from 27 (41%) of the seawater samples collected in the U.S. Virgin Islands. The species obtained from this water, all members of the family Myxococcaceae, were *Myxococcus fulvus*, *M. xanthus*, *M. stipitatus*, and *Chondrococcus coralloides*. A close inspection of the area studied indicates that most of the myxobacteria were found in bays (Fig. 1 and Table 1). In addition, most of the isolates were obtained from one bay, Cruz Bay, St. John.

Table 1. Species of fruiting myxobacteria isolated from seawater

Species	Bay	Other	Total
M. fulvus	· 13	5	18
M. xanthus	12	5	17
M. stipitatus	2	0	2
Ch. coralloides	2	0	2

Cruz Bay,[2] the location of a small town of the same name, is the most densely populated area on St. John. It receives sewage and waste from the town of Cruz Bay, runoff from a small stream, and waste from a wide variety of watercraft ranging in size from small yachts to passenger ferry boats. The bay is partially protected from the adjacent sound (Pillsbury Sound) by a reef. The water of the bay was turbid and brown-green in color, indicative of nutrient enrichment and microbial growth. It is noteworthy that all samples from Cruz Bay contained fruiting myxobacteria (Fig. 2). It is also interesting that almost one-half of the

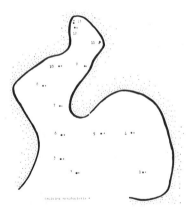

Fig. 2. **Location of samples in Cruz Bay, St. John.**

[2] A detailed description of Cruz Bay and Chocolate Hole, St. John, is found in: Caribbean Research Institute. 1970. *Report on Estuarine Environment at Cruz Bay, St. John. Phase II Report.* St. Thomas, U.S.V.I.: College of the Virgin Islands.

Table 2. Fruiting myxobacteria isolated from three St. John bays. (MF) *Myxococcus fulvus*, (MX) *M. xanthus*, (MS) *M. stipitatus*, (CC) *Chondrococcus coralloides*

Cruz Bay					Chocolate Hole					Great Lameshur Bay				
Sample	*MF*	*MX*	*MS*	*CC*	Sample	*MF*	*MX*	*MS*	*CC*	Sample	*MF*	*MX*	*MS*	*CC*
1		X			1					1	X			
2		X			2					2				
3		X			3	X				3	X			
4	X				4					4				
5		X			5					5				
6	X	X			6					6				
7	X	X			7					7	X			
8	X	X	X		8			X		8				
9	X	X		X						9				
10		X								10				
11	X	X		X						11	X			
12	X	X	X							12				
13	X									13				

samples contained two or three species of myxobacteria (Table 2). In tabulating the number of myxobacteria observed during the entire study, the twenty-three Cruz Bay isolates accounted for 59% of all myxobacteria obtained.

Multiple samples were also collected from Chocolate Hole and Great Lameshur Bay (Fig. 1). In comparison to Cruz Bay, these two bodies of water are relatively inactive in regard to human activity. A small amount of land around Chocolate Hole has been disturbed by home construction, but from the appearance of the water and the bottom of the bay, this activity has not caused any damage to the bay. A small pond connected to the northwest end of Chocolate Hole serves to trap a large amount of sediment from runoff that would otherwise pass into the bay. This bay is not protected by a reef. Myxobacteria were observed in only two of the eight samples collected in this bay and both samples were from along the shoreline (Fig. 3 and Table 2).

Great Lameshur Bay, like Chocolate Hole, shows little damage due to human activity. This bay was the location for Tektite I and II and is presently used extensively by the Virgin Islands Ecological Research Station. A mangrove swamp is adjacent to the northwest side of the bay. No streams flow into the bay and it is not protected from the Caribbean Sea by a reef. Thirteen samples were collected in this bay and only four were found to contain myxobacteria (Fig. 4 and Table 2).

The species of fruiting myxobacteria isolated from seawater during this study

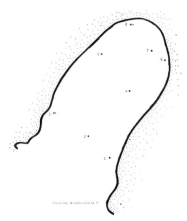

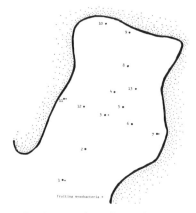

Fig. 4. **Location of samples in Great Lameshur Bay, St. John.**

Fig. 3. **Location of samples in Chocolate Hole, St. John.**

are also quite common to both the soil and fresh water of the Virgin Islands.[3] The relative absence of these bacteria in seawater collected outside the bays strongly suggests that they cannot be considered as normal inhabitants of the ocean. It is my belief that the fruiting myxobacteria are present in the seawater simply as a consequence of their being a normal part of the terrestrial microflora, and subsequently are transported to the ocean by runoff.

The survival time for fruiting myxobacteria in seawater is not known. It is reasonable to assume, however, that if these organisms are washed into the ocean in the microcyst stage, their survival rate would be greater than if they were present in the water as vegetative cells. With this assumption, it is quite possible that these bacteria might play a role in the future as biological indicators in studies of terrestrial pollution of the oceans.

Acknowledgments

I wish to express my sincere appreciation of the staff of the Virgin Island Ecological Research Station, Caribbean Research Institute, College of the Virgin Islands, Lameshur Bay, St. John, U.S. Virgin Islands. Without their help and the use of their facilities, this study would not have been possible.

Literature Cited

1. Anderson, J. I. W., and W. P. Hefferman. 1965. Isolation and characterization of filterable marine bacteria. *J. Bacteriol.* **90:**1713–18.
2. Bachmann, B. J. 1955. Studies on *Cytophaga fermentans*, n. sp., a facultatively anaerobic lower myxobacterium. *J. Gen. Microbiol.* **13:**541–51.
3. Breed, R. S., E. G. D. Murray, and N. R. Smith. 1957. *Bergey's manual of determinative bacteriology.* Baltimore: Williams and Wilkins Co.

[3]E. R. Brockman, unpublished data.

4. Brockman, E. R. 1964. A study of myxobacterial-eubacterial association. Ph.D. dissertation, University of Missouri.

5. Brockman, E. R. 1970. Isolation and distribution of fruiting myxobacteria in Atlantic coast beaches. *Bacteriol. Proc.*, p. 27.

6. Colwell, R. R., R. V. Citarella, and P. K. Chen. 1966. DNA base composition of *Cytophaga marinoflava* n. sp. determined by buoyant density measurements in cesium chloride. *Can. J. Microbiol.* **12**:1099–1103.

7. Dworkin, M. 1966. Biology of the myxobacteria. *Ann. Rev. Microbiol.* **20**:75–106.

8. Georgala, D. L. 1957. Quantitative and qualitative aspects of the skin flora of North Sea cod and the effect thereon of handling on ship and on shore. Ph.D. dissertation, Aberdeen University.

9. Hayes, P. R. 1963. Studies on marine flavobacteria. *J. Gen. Microbiol.* **30**:1–19.

10. Hill, L. R. 1966. An index to the DNA base composition of bacterial species. *J. Gen. Microbiol.* **44**:419–37.

11. Humm, H. J. 1946. Marine agar-digesting bacteria of the south Atlantic coast. *Bull. Duke Univ. Mar. Sta.*, No. 3, pp. 45–75.

12. Kadota, H. 1953. Microbiological studies on the weakening of fishing nets. III. The microbiological deterioration of fishing nets during storage. *Bull. Jap. Soc. Sci. Fish.* **19**:476–80.

13. Kadota, H. 1954. Microbiological studies on the weakening of fishing nets. V. A taxonomical study on marine *Cytophaga*. *Bull. Jap. Soc. Sci. Fish.* **20**:125–29.

14. Kadota, H. 1956. A study on the marine aerobic cellulose-decomposing bacteria. *Mem. Coll. Agric., Kyoto Univ.* **74**:1–128.

15. Liston, J. 1960. The bacterial flora of fish caught in the Pacific. *J. Appl. Bacteriol.* **23**:469–70.

16. Lovelace, T. E., H. Tubiash, and R. R. Colwell, 1967. Bacteriological studies of *Crassostrea virginica* in Chesapeake Bay. *Bacteriol. Proc.*, p. 46.

17. McCurdy, H. D., and S. Wolf. 1967. Deoxyribonucleic acid base composition of fruiting myxobacterales. *Can. J. Microbiol.* **13**:1707–08.

18. McDonald, I. J., C. Quadling, and A. K. Chambers. 1963. Proteolytic activity of some cold tolerant bacteria from Arctic sediments. *Can. J. Microbiol.* **9**:303–15.

19. McDonald, J. C. 1968. Fruiting myxobacteria from the North Carolina seacoast. *Ass. SE Biol. Bull.* **15**:45.

20. MacLeod, R. A., and T. I. Matula. 1962. Nutrition and metabolism of marine bacteria. XI. Some characteristics of the lytic phenomenon. *Can. J. Microbiol.* **8**:883–96.

22. Quadling, C., and R. R. Colwell. 1963. Taxonomic studies on gram-negative bacteria from the Arctic. *Bacteriol. Proc.*, p. 40.

23. Quinlan, S., and K. B. Raper, 1964. Development of the myxobacteria. *Encycl. Plant Physiol.* **15**:596–611.

24. Rucker, R. R., B. J. Earp, and E. J. Ordal. 1953. Infectious diseases of Pacific salmon. *Trans. Amer. Fish. Soc.* **83**:297–312.

25. Singh, B. N. 1947. Myxobacteria in soils and composts: their distribution, number and lytic action on bacteria. *J. Gen. Microbiol.* **1**:1–10.

26. Spencer, R. 1960. Indigenous marine bacteriophages. *J. Bacteriol.* **79**:614.

27. Stanier, R. Y. 1941. Studies on marine agar-digesting bacteria. *J. Bacteriol.* **42**:527–58.

28. Starr, T. J. 1953. A study of marine myxobacteria. Ph.D. dissertation, University of Washington.

29. Velankar, N. K. 1957. Bacteria isolated from sea-water and marine mud of Mandapan (Gulf of Mannar and Polk Bay). *Indian J. Fish.* **4**:208.

30. Veldkamp, H. 1961. A study of two marine agar-decomposing, facultatively anaerobic myxobacteria. *J. Gen. Microbiol.* **26**:331–42.

31. Waksman, S. A., C. L. Carey, and H. W. Reuszer. 1933. Marine bacteria and their role in the cycle of life in the sea. I. Decomposition of marine plant and animal residues by bacteria. *Biol. Bull.* **65**:57–79.

32. Williams, M. A., and S. C. Rittenberg. 1957. A taxonomic study of the genus *Spirillum* Ehrenberg. *Int. Bull. Bacteriol. Nomen. Taxonom.* **7**:49–111.

Comments

QUESTION: Do you believe there are cellulose-decomposing organisms in the Arctic marine environment?

BROCKMAN: Yes, I do. Although the myxobacteria from the Virgin Island marine samples I obtained were not cellulose decomposers, they have been reported from seawater. Kadota obtained strains of cellulolytic, nonfruiting myxobacteria from both Hiroshima Bay and Maizuru Bay. Sediment samples that are now being collected off the Virgin Islands will be examined for cellulolytic strains of the genera *Sorangium* and *Polyangium*.

Gas Chromatographic Analysis of Denitrification by Marine Organisms

W. J. Payne

Over the past three decades many of the questions formulated by microbial physiologists have stimulated a considerable part of the stir of activity we have seen in biochemistry and genetics. As a consequence, a closer association of scientists representing each of these diciplines resulted, and much has been learned of intermediary metabolism, macromolecular structure and function, and control mechanisms operative in microbial cells. Much is yet to be learned.

But I should like now to propose that a closer association between microbial physiologists and ecologists who work in the estuarine, marine, soil, and aquatic environments can be equally profitable. The physiologist, who studies intensely a single phenomenon, or at most a few activities, can provide insights that may not come to workers in multivariant systems; whereas, the ecologist is better prepared to evaluate an isolated phenomenon or set of activities in the context of their contribution to better understanding of complete, functioning systems.

Nitrate reduction may be taken as a representative phenomenon, because it has been studied in depth by physiologists *(1, 5, 6, 10, 12, 14, 26, 32, 35, 36, 37, 38, 52, 55, 56, 61)* and is coming increasingly to be assayed in complex natural environments *(4, 13, 19, 20, 21, 22, 64)*. This activity, therefore, may serve to exemplify my thesis. As a student of one aspect of nitrate reduction (denitrification) over the past decade *(2, 3, 43, 44, 54)* I am now persuaded that, before this activity can be fit into ecological studies, a fuller understanding of the various factors that influence both the synthesis and the operation of its components must be obtained. Appreciation of the interacting phenomena seems particularly to be desired at just this juncture when studies of interrelationships between respiration and nutrient conversion, maintenance, and synthesis are becoming more extensive *(23, 42, 58)*.

As a beginning, it should be noted that reduction of nitrate serves a dissimilatory as well as an assimilatory function *(38, 65, 68, 69)* for the microbial cell. When the reduction of nitrate is used as an index of the capacity of microorganisms in a complex system for incorporation of nitrate nitrogen, safeguards are needed. Awareness of various consequences of the reduction and the factors that influence them assumes particular significance. Let us examine these.

53

Various Types of Nitrate Reduction

Oxygen-Sensitive Respiratory Reduction to Nitrite. Although a number of bacteria, including the enterics, can grow fermentatively, many also synthesize a molybdenum-containing nitrate reductase which permits them to carry out cytochrome-linked respiration as well *(17, 30, 38, 52, 60, 62)*. The synthesis of respiratory nitrate reductase is repressed by oxygen, which additionally inhibits and inactivates the enzyme if it is exposed after synthesis *(9, 46, 51, 58, 65)*. A notable discovery of recent years is the observation that nitrate is not a necessary effector of derepression of synthesis of this type of nitrate reductase *(9, 46, 51, 59)*, although activity is greater if nitrate is present perhaps to provide for the generation of additional energy for increased enzyme synthesis. It seems apparent that a diminished supply of oxygen is an effector, for when enteric bacteria are incubated in the presence of very limited amounts of oxygen, even in the absence of nitrate, they are found to produce nitrate reductase in what appears to be the same membrane-bound form as that in cells grown in nitrate-containing medium *(59, 65)*. This enzyme is very closely associated with formate dehydrogenase in the membrane of coliform bacteria and *Proteus* species. Mutants that have lost the ability to produce one cannot produce the other *(8, 63, 65, 66)*. The presence of ammonium ion in the culture medium does not repress formation of the respiratory nitrate reductase system; and once it is produced, the enzyme-membrane complex resists disruption by sonic radiation *(65)*.

Studies of electron transport leading to respiratory nitrate reductase in cell-free extracts have revealed that flavin, *b*-type and *c*-type cytochromes *(12, 17, 38)* and ubiquinone *(12, 27)* are components of the transport chain. In *Bacillus stearothermophilus*, conversion from aerobic to nitrate respiration results in a rise in *c*-cytochromes and disappearance of cytochrome a_3 *(11)*.

In addition to the coliforms *(38, 65, 68)*, *Achromobacter (38)*, *Proteus mirabilis (8, 9)*, *Hemophilus parainfluenzae (60)*, *Rhizobium japonicum (30)* and *Spirillum itersonii (17)*, a number of bacilli *(67)*, and staphylococci *(38)* produce this type of respiratory nitrate reductase as well.

Oxygen-Durable Assimilatory Reduction of Nitrate. When incubated in a medium lacking either ammonium ion or an organic nitrogen source, several microbial species produce a different nitrate reductase activity. In many cases this occurs even in cultures that are fully aerated *(6, 23)*. Aeration does not suppress or inhibit this activity; it is not tightly bound to membrane; and sonic treatment is more deleterious to this than to the respiratory type of nitrate-reducing system *(65)*. It serves the cells, of course, by carrying out the first step in the generation from nitrate of an assimilable form of nitrogen. Still, both this type and respiratory reduction may be carried out by the same enzyme protein. When a number of factors that influence the activity of the assimilatory and dissimilatory enzymes were compared, no differences in metal requirement, pH optima, or temperature effects were noted between the two types. The oxygen-durable nitrate reductase protein has thus been postulated to be identical with the oxygen-sensitive reductase protein, but the two systems are clearly complexed differently within the cell where they are designed for such different ends *(65)*.

In bacteria producing the oxygen-durable assimilatory type of nitrate reductase, nitrite is but a transient product which is reduced to hydroxylamine (32) and then to ammonium ion. As ammonium ion, the nitrate nitrogen is in the form that is assimilated during amino acid synthesis and other anabolic events. Neurospora (39, 41) also produce this type of reductase.

Nitrate Reduction as the First Step in Denitrification. The nitrate reductase produced by denitrifiers is similar in some respects to its counterpart in the enteric bacteria. Its synthesis and operation are sensitive to oxygen (14, 45, 48, 49, 50, 54); its production is unaffected by the presence of ammonium ion; and the protein apparently contains molybdenum (16, 38). Nonetheless, there are notable differences which have been discerned principally by Pichinoty and his colleagues (49, 50). It has been found that the denitrifying as well as some other types of bacteria produce a characteristic enzyme, designated reductase A. In cell-free extracts, the enzyme has been found to reduce chlorate and to trans-

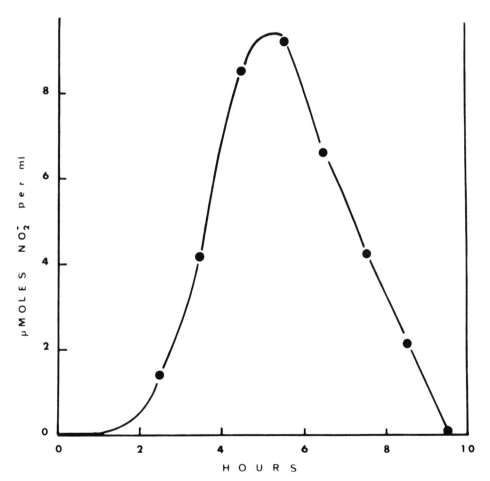

Fig. 1. Accumulation of nitrite in cultures of *Pseudomonas perfectomarinus* during initial period of anaerobiosis before the onset of gas release.

fer electrons from reduced nicotinamide adenine dinucleotide (NADH) to nitrate. On the other hand, the enzyme designated B, which is produced by the enteric and other bacteria, is inhibited by chlorate and can be shown to transfer electrons to nitrate from the reduced viologen dyes but not from NADH. Most bacteria that reduce nitrate produce one or the other type of nitrate reductase *(47)*, but certain of them produce membrane-bound enzyme A as well as some soluble B *(48, 49, 50)*. Quite recently two forms of enzyme B have been demonstrated. The distinction is that enzyme $B\alpha$ is activated by Na^+, K^+, and Cs^+, whereas enzyme $B\beta$ is not *(48)*.

The nitrate reductase A* produced by *Pseudomonas perfectomarinus (70)*, the marine bacterium we have studied, appears as in other bacteria to dominate denitrification. This conclusion is based on the observation that the nitrite produced as an intermediate accumulates in the medium and is rapidly reduced only after all the nitrate has been reduced (Fig. 1). Nitrite accumulation has been noted by other workers as well *(33, 54)*. Nitric oxide and nitrous oxide are the intermediates next produced by bacteria *(2, 10, 15, 31, 33, 34, 35, 37, 43, 44, 55)*, but nitrite first accumulates before these products are reduced to nitrogen. In one bacterium, *Corynebacterium nephridii*, it is known that nitrous oxide rather than nitrogen is the terminal product. It arises specifically from reduction of nitric oxide *(53)*; but otherwise, a variety of pseudomonads from soil and from both fresh and marine waters, *Micrococcus denitrificans*, and certain *Thiobacillus* species — all carry out complete reduction of nitrate to elemental nitrogen *(2, 18, 25, 54, 61)*.

Summary Commentary on Nitrate Reductase

Obviously, care must be taken in interpreting the significance of measurement of nitrate reduction by microorganisms in a complex microbial system such as marine or estuarine water or sediment. If studies are long-term, the influences of the presence or absence of oxygen and ammonium ion on the rate and extent of reduction must be weighed. If disappearance of nitrate is the parameter assayed, it must be realized that as many as three metabolic pathways may be operative — and not all will lead to assimilation of nitrate nitrogen by the cells. The fate of the nitrite that results from nitrate reduction must be carefully determined to differentiate (1) the assimilation of nitrate nitrogen as ammonium ion by the cells from (2) denitrification, the return of elemental nitrogen to the atmosphere.

Denitrification

Preliminary Enzymatic Events. Microbial reduction of nitrate to elemental nitrogen was recognized even during the last century *(18, 56)* and was characterized as respiration early in the progress of investigation *(25)*. The phenomenon occurs concomitantly with the oxidation of a variety of substrates. Thiobacilli may oxidize sulfur and micrococci oxidize hydrogen during denitrification *(38)*.

*Pichinoty, personal communication.

Table 1. Electron donors for denitrification by *Pseudomonas perfectomarinus*[a]

Original Substrate	Added Substrate		
	None	Glucose, 20 μmoles	Citrate, 20 μmoles
Acetate	−	−	−
α-Ketoglutarate	−	−	−
Citrate	−	−	−
Pyruvate	−	+	+++
Succinate	−	−	−
D-Arabinose	−	+	++
L-Arabinose	−	++	++++
Xylose	−	−	−
Glucose	−	−	−
Mannitol	−	−	−
Arginine	−	++	+++
D,L-Alanine	−	+	+
Aspartate	−	−	−
Glutamate	−	++	+++
Glycine	−	+	+++
Histidine	−	−	−
Leucine	−	++	+
Lysine	−	−	−
Proline	−	+	+++
Threonine	−	+	+
Tryptophan	−	−	−
Tyrosine	−	−	−
Urea	−	+	++
Peptone	++++[b]		
Vitamin-free casamino acids	++++		
Asparagine	++++		

[a] Reprinted from Rhodes, Best, and Payne *(54)* by permission of the National Research Council of Canada.

[b] Gas production in peptone-supplemented medium was taken as ++++. Fewer + signs indicate less gas released, and the − signs indicate no gas released. There was some growth in each of the tubes, varying from dense to barest traces (cf. ref. 54).

In addition, the usual array of sugars, organic acids, and polyhydric alcohols can serve the soil bacilli and pseudomonads as electron donors for denitrifying respiration *(1, 14, 55, 67)*. But, two Indian workers found that several marine isolates required a mixture of substrates to support denitrification in minimal media *(61)*. Thus, we began our studies of denitrification by a marine bacterium with a similar demonstration that mixtures of a variety of substrates with glucose and citrate supported denitrification in minimal seawater medium (Table 1). We observed, in addition, that asparagine was the only simple substrate assayed that served as an electron donor for denitrification *(54)* by this pseudomonad. This result should not have been unexpected since French workers had reported denitrification at the expense of asparagine in 1886 *(18)*. It would thus seem appropriate that asparagine should be included as a prime substrate in studies of nitrate reduction and denitrification in estuarine and marine systems. Provision of a mixture of substrates might also be advisable.

Continuing, we found by the use of manometric techniques (Table 2) that we could demonstrate denitrification at the expense of asparagine and citrate in

Table 2.   Effectiveness of various electron donors and cofactors in denitrification with cell-free extracts of *Pseudomonas perfectomarinus*[a]

Composition of Assay Mixture[b]	Specific Activity with Various Substrates (mμmoles N$_2$/mg protein per hr)	
	0 μmoles	20 μmoles
1. Extract	<1.0	
2. Extract + NO$_3^-$	<1.0	
3. Extract + NO$_3^-$ + substrate		
(a) Citrate		33.4
(b) Asparagine		47.8
4. Extract + NO$_3^-$ + FMN	41.2	
5. Extract + NO$_3^-$ + FMN + substrate		
(a) Citrate		85.3
(b) Asparagine		174.1
6. Extract + NO$_3^-$ + FAD	26.8	
7. Extract + NO$_3^-$ + FAD + substrate		
(a) Citrate		85.3
(b) Asparagine		159.7

[a] Reprinted from Rhodes, Best, and Payne *(54)* by permission of the National Research Council of Canada.

[b] Extract, 1.0 ml; 0.067 M phosphate buffer, pH 7.0, with or without 0.5 μmoles FMN or FAD, 1.0 ml; substrate, 0.5 ml; 10 μmoles potassium nitrate, 0.5 ml; 15% KOH in center well, 0.2 ml. Helium atmosphere. 30 C (cf. ref. 54).

cell-free extracts of *P. perfectomarinus (54)*. Addition of flavins increased activity, as workers who had studied denitrification in other pseudomonads had previously observed *(5, 35, 37)*. But, our assumption (Table 2) that the gas released was all N_2 was probably erroneous, in light of subsequent findings *(2, 43, 44)*; use of manometry did not offer much hope of greater analytical precision. We thus decided that determination of the preliminary enzymatic events leading to denitrification was a useful interim pursuit until better means of analysis were available, and studies of the intermediary reactions involved were undertaken.

It was necessary first to realize that asparagine is very unlikely to be an oxidizable species in its own right; but the carbon skeleton is transformable to one of several of the intermediates of the tricarboxylic acid cycle. It was not surprising, therefore, to find that cell-free extracts of *P. perfectomarinus* deaminated and deamidated asparagine (Fig. 2). Curve 1 represents the rate of release from asparagine, by asparaginase and aspartate deaminase, of both amide and amino nitrogen as ammonia. Curve 2 represents the rate of amino nitrogen released as ammonia from aspartate *(3)*. Fumarate was discerned, and a fumarase was also demonstrated in the extracts. By chemical analysis, malic acid was then shown to be the oxidizable end product from which electrons could be transferred to initiate denitrification. It has since been found that asparagine is taken up quite effectively by whole cells of *P. perfectomarinus*, whereas malic acid is not. Once the cells were disrupted, there was a lag in the apparent oxidation of asparagine; but malate was oxidized in the cell-free extracts (Fig. 3) with no lag by an NADP-linked malic decarboxylase *(3)*. A transhydrogenase was then ob-

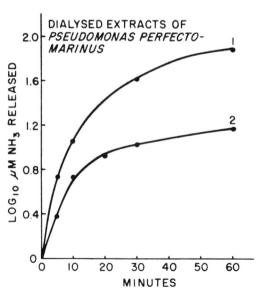

Fig. 2. Release of ammonia from asparagine and aspartate by cell-free extracts of *Pseudomonas perfectomarinus* (reprinted from Best and Payne *(3)* by permission of the American Society for Microbiology).

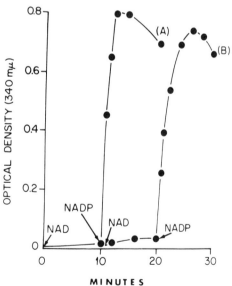

Fig. 3. NADP-linked malate oxidative decarboxylase activity in cell-free extracts of *Pseudomonas perfectomarinus* (reprinted from Best and Payne *(3)* by permission of the American Society for Microbiology).

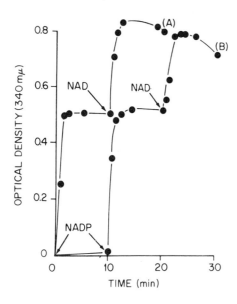

Fig. 4. **Pyridine nucleotide transhydrogenase in cell-free extracts of *Pseudomonas perfectomarinus* (reprinted from Best and Payne *(3)* by permission of the American Society for Microbiology).**

served to reduce NAD with electrons donated by reduced NADP (Fig. 4). Enzyme that reduced flavine adenine dinucleotide (FAD) at the expense of NADH was also demonstrated in cell-free extracts.

Thus, at the completion of this phase of the studies, our estimate of the sequence of events characterizing denitrification carried out by *P. perfectomarinus* was only partially documented (Fig. 5). The question remaining unanswered was: how are the cytochrome-linked *(11, 28, 29, 40, 57, 58)*, subterminal, and terminal events to be investigated?

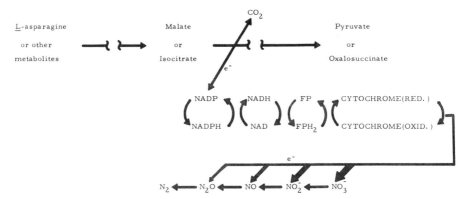

Fig. 5. **Schematic representation of electron transport in denitrification by *Pseudomonas perfectomarinus.***

Use of Gas Chromatography. Fortunately, at this point, it was learned that the gases of interest to us could be chromatographed on a resin called Porapak Q *(7, 24)*. We were able to duplicate the chromatographic separation (Fig. 6) of the components by using samples from control mixtures of the gases *(2)*. We employed an F & M Model 700 chromatograph with a thermal conductivity detector set at 250 C. The samples were injected at 28 C. Helium served as the carrier gas. By injecting several measured quantities of each gas separately and integrating the area under the response peaks traced by the recorder, we were able to establish reference curves that were used subsequently to quantitate the gases in the experimental samples.

To apply chromatography to analysis of the gases released by *P. perfectomarinus* during denitrifying growth in a complex medium, dehumidified samples swept from the cultures with helium were injected into the analyzer *(2, 7)*. It was found that CO_2 was released early in the growth phase (Fig. 7A) and N_2 later. The N_2 response represented in Figure 7B is actually very much larger than it appears.

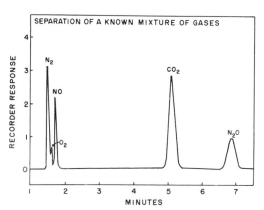

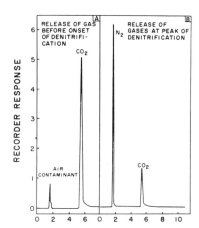

Fig. 6.   Chromatography of gases on Poropak Q (reprinted from Barbaree and Payne *(2)* by permission of Springer-Verlag, Berlin and Heidelberg).

Fig. 7.   Gases swept from denitrifying cultures of *Pseudomonas perfectomarinus* (reprinted from Barbaree and Payne *(2)* by permission of Springer-Verlag, Berlin and Heidelberg).

The response was attenuated 16X to keep the peaks relatively near the same size in 7A and B. Similar assays of the sweep gas obtained at intervals over several hours revealed that CO_2 production peaked at 16 hr and N_2 production at 30 hr (Fig. 8). It was then quickly found that denitrification by cell-free extracts easily determined by gas chromatography as well *(43)*.

Commentary on the Usefulness of Gas Chromatography

Concern with estimates of capacity for denitrification and with the factors that control it is assuming greater importance in studies of marine and other aquatic environments *(4, 19, 20)*. The phenomenon is not an artifact of pure culture but does occur in nature in a variety of climates *(21, 22, 64)*. Goering and Dugdale determined that denitrification takes place in marine water *(21)* by demonstration by mass spectrometry that microorganisms in samples of water from a Pacific bay released $^{15}N_2$ from labeled nitrate. Now that the validity of the phenomenon has been firmly established by the use of isotopic tracers, it seems likely that the precision and flexibility of gas chromatography highly recommend this

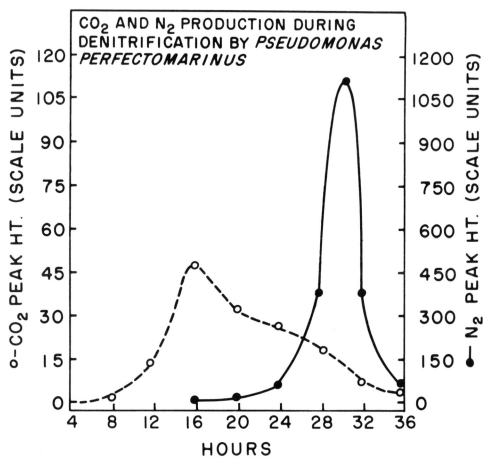

Fig. 8. Time course of release of gases by dentrifying cultures of *Pseudomonas perfectomarinus* (reprinted from Barbaree and Payne *(2)* by permission of Springer-Verlag, Berlin and Heidelberg).

technique for use in future studies of the activity in complex, multivariant systems as well as enzymatic reaction mixtures.

Conversion of P. perfectomarinus from Aerobic to Denitrifying Metabolism

Simultaneous Synthesis of Denitrifying Enzymes in the Presence of Nitrate. When aerobic, nitrate-deadapted populations of this marine bacterium were incubated anaerobically with nitrate, an average of 82% of the cells produced colonies *(44)*. Because we have seen that nitrite accumulates in cultures before gas release begins, we were curious about the sequence of synthesis of the denitrifying enzymes during this large-scale conversion. We reasoned that either (1) nitrite, nitric oxide, and nitrous oxide reductases were not formed until after nitrate was depleted, or (2) all the enzymes were produced simultaneously, but all except nitrate reductase remained inactive until the nitrate present in the culture was reduced. To test these hypotheses, deadapted cells were incubated anaerobically in complex growth medium with nitrate. Samples were removed at intervals. The cells in these were treated with 100 μg/ml of chloramphenicol to stop further protein synthesis, washed with 0.05 M $MgCl_2$ containing chloramphenicol, and assayed in a chloramphenicol-containing suspending medium for the ability to reduce nitrite and nitrous oxide at the expense of asparagine during 20-min incubation periods. (Whole cells do not reduce nitric oxide.) It should be noted that in control experiments, the presence of chloramphenicol did not influence the rate of reduction of either nitrite or nitrous oxide. Moreover, control cells incubated in medium containing chloramphenicol from zero time did not produce the reductases. Finally, it was found that within 40 min of anaerobic incubation with nitrate, the capacity to reduce both nitrite and nitrous oxide began to increase (Fig. 9). The activity reached its maximum rate at 60 min (43).

It was thus apparent that the activity of the denitrifying enzymes was influenced at some step(s) between nitrite reduction and release of nitrogen. Considering the various components of the reaction mixtures, either nitrate or nitrite, or a combination of them, was suspected as the most likely effector. In order to determine where denitrification was slowed or stopped, as well as to identify the effector(s), crude cell-free extract from denitrifying cells was divided into three portions. Each was supplemented with NADH, and each separately incubated anaerobically—one with nitrite, another with nitric oxide, and the last with nitrous oxide. Each reaction mixture was charged with 10 μmoles of nitrate/ml. Upon incubation and assay, it was found that only the rate of reduction of nitric oxide to nitrous oxide was suppressed by the presence of

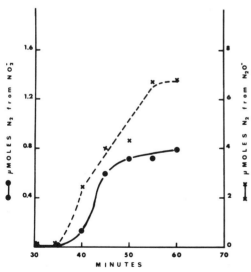

Fig. 9. Onset of synthesis of denitrifying enzymes by *Pseudomonas perfectomarinus* following initiation of anaerobiosis in nitrate-containing medium (reprinted from Payne and Riley (43) by permission of the Society for Experimental Biology and Medicine).

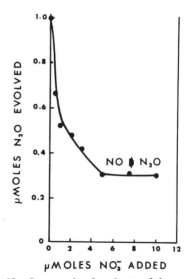

Fig. 10. Suppression by nitrate of the rate of reduction of nitric oxide by a fraction from a cell-free extract of *Pseudomonas perfectomarinus* (reprinted from Payne and Riley (43) by permission of the Society for Experimental Biology and Medicine).

nitrate *(43)*. Activity was diminished by 60%, whereas rates of reduction of nitrite to nitric oxide and nitrous oxide to nitrogen were not affected by the presence of nitrate.

The question then became: Is this effect specific for nitric oxide reductase, or is it a result of complicated reactions in crude, cell-free extract? At this time we were fortunate to find that a complex fraction, which contained nitric oxide reductase and released nitrous oxide at the expense of NADH, could be separated from the other reductases by DEAE-cellulose chromatography *(43)*. This fraction did not reduce nitrate, nitrite, or nitrous oxide. But most significantly, when assayed in media containing nitrate, its ability to reduce nitric oxide to nitrous oxide was suppressed by 60% by 1 μmole/ml of nitrate (Fig. 10). Since nitrate was not reduced in these assay mixtures, we concluded that nitrate was the suppressive agent. Additional nitrate did not increase the degree of suppression beyond 60%. When an identical experiment was performed with nitrite supplied as the potentially suppressive agent, four to five times as much nitrite as nitrate were required for the same degree of suppression (Fig. 11). It thus appeared that nitrate and nitrite exert a fine, and a less fine, controlling effect, respectively, impeding but not preventing reduction of the gaseous oxides until the ionic oxides have been used up.

Anaerobiosis as an Effector of Synthesis of Denitrifying Enzymes. Because denitrifying pseudomonads do not grow fermentatively, but must respire either oxygen or nitrate, the question of whether the presence of nitrate as well as anaerobic conditions would be required to initiate synthesis of the denitrifying enzymes had not been tested. We can now report that the presence of nitrate is not required. It was found *(44)* that fully aerobic, deadapted populations of *P. perfectomarinus* began to synthesize the enzymes within a short time after being deprived of oxygen during log-phase growth in asparagine-minimal medium in the absence of nitrate (Fig. 12). Apparent oxygen saturation decreased from 100% to zero within 10 min in these actively growing cultures when sparging was stopped. The result was that cells that were incubated anaerobically for 3 hr, harvested, and then treated with chloramphenicol were shown to have gained the capacity to reduce nitrite to nitrogen. Synthesis of the denitrifying enzymes during the assay period was obviated by carrying out the assays in media containing chloramphenicol. Curve 1 represents the control activity of cells grown on nitrate stoichiometrically releasing 15 μmoles of N_2 from 30 μmoles of nitrite. Chloramphenicol did not affect the activity of these cells. Curve 4 represents the opposite, the inactivity of aerobic control cells, which released no N_2. Curve 3 represents the activity of cells assayed after 1 hr of anaerobic incubation, and curve 2 the activity of cells incubated anaerobically for 3 hr. These latter cells had gained 40% of the activity displayed by the control bacteria grown anaerobically on nitrate.

Anaerobic incubation resulted in an increased capacity for reduction of nitrous oxide as well (Fig. 13). Although they displayed little or none on short-term incubation, prolonged incubation revealed that fully aerobic cells still had some nitrous oxide reductase activity, as curve 3 indicates. After anaerobic incubation for 1 hr, the activity rose to that represented by curve 2. And, after 3 hr,

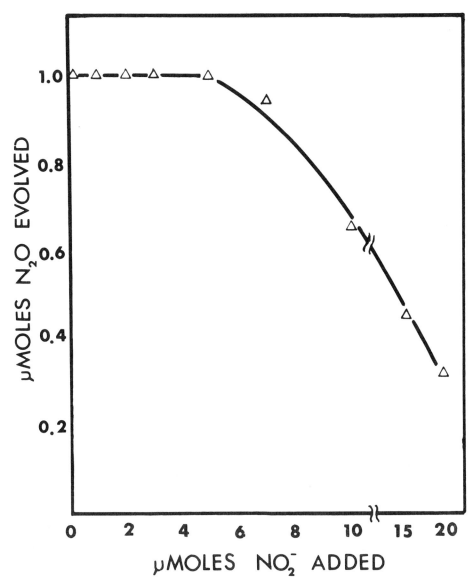

Fig. 11. **Suppression by nitrite of the rate of reduction of nitric oxide by a cell-free extract of** *Pseudomonas perfectomarinus.*

the nitrous oxide reductase activity was equal to that of cells grown on nitrate. Curve 1 represents the activity of both nitrate-grown cells and those incubated anaerobically for 3 hr after the cultures were deprived of air.

These activities were demonstrated in cell-free extracts as well *(44)*. As indicated by curve 1 (Fig. 14A), nitrous oxide was reduced at the expense of NADH by enzymes in extracts of cells incubated for 3 hr after the cultures were deprived of air. Curve 2 indicates the nitrite-reducing activity, and curve 3 either the lack of activity in controls when neither nitrite nor nitrous oxide was pro-

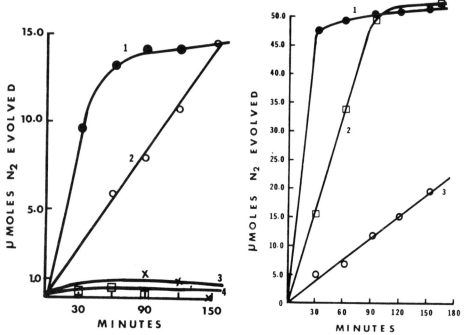

Fig. 12. Synthesis of nitrite-reducing enzyme by *Pseudomonas perfectomarinus* following initiation of anaerobiosis in asparagine-minimal medium in the absence of nitrate, nitrite, nitric or nitrous oxide (reprinted from Payne, Riley, and Cox *(44)* by permission of the American Society for Microbiology).

Fig. 13. Synthesis of nitrous oxide-reducing enzyme by *Pseudomonas perfectomarinus* following initiation of anaerobiosis in asparagine-minimal medium in the absence of nitrate, nitrite, nitric or nitrous oxide (reprinted from Payne, Riley, and Cox *(44)* by permission of the American Society for Microbiology).

vided or the lack in extracts of aerobically grown cells. In Figure 14B, curve 1 represents the reduction of nitric oxide by enzymes in extracts of cells incubated for 3 hr and curve 2 in the minimal and belated activity of extracts from aerobic cells. Nitrous oxide accumulated in these reaction mixtures during the hour before N_2 began to appear, suggesting an additional control by excess nitric oxide of the final reductive step. We have not examined this possibility further.

Summary Commentary on the New Approach to Studies of Denitrification

Introduction of the use of gas chromatography *(2, 43, 44, 53)* has enabled us to separate the individual steps in the denitrifying process and to determine the influences of environmental factors on each of them. Our most significant findings would appear to be (1) the discernment of the derepressive influence of anaerobiosis on synthesis of the denitrifying enzymes, and (2) control of their operation by nitrate and nitrite. Finally, separation of the three complex fractions, each with a single reducing capacity, holds great promise for usefulness. It may be possible now to determine the nature of the cytochromes involved separately in nitrite, nitric oxide, and nitrous oxide reduction.

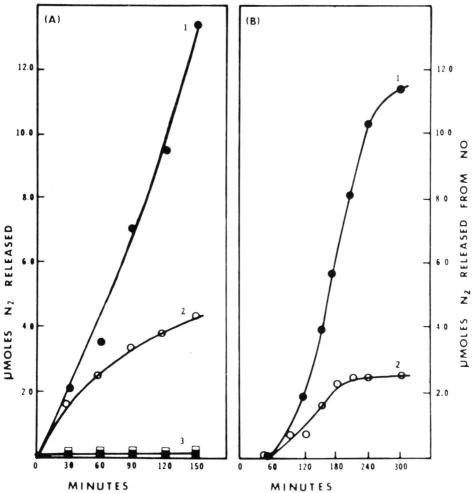

Fig. 14. Demonstration of denitrifying enzymes in cell-free extracts of *Pseudomonas perfectomarinus* after anaerobic incubation of cells in the absence of nitrate, nitrite, nitric or nitrous oxide (reprinted from Payne, Riley, and Cox *(44)* by permission of the American Society for Microbiology).

Acknowledgment

Several of the studies reported here were supported by grants from the Office of Naval Research, by stipends provided by training grants from the National Institute of General Medical Sciences and by grant no. GB-23347 from the National Science Foundation.

Literature Cited

1. Allen, M. G., and C. B. vanNiel. 1952. Experiments on bacterial denitrification. *J. Bacteriol.* **64**:397–412.

2. Barbaree, J. M., and W. J. Payne. 1967. Products of denitrification by a marine bacterium as revealed by gas chromatography. *Mar. Biol.* **1**:136–39.

3. Best, A. N., and W. J. Payne. 1965. Preliminary enzymatic events in asparagine-dependent denitrification by *Pseudomonas perfectomarinus. J. Bacteriol.* **89**:1051–54.

4. Bollag, J. M., M. L. Orcutt, and B. Bollag. 1970. Denitrification by isolated soil bacteria under various environmental conditions. *Proc. Soil Sci. Soc. Amer.* **34**:875–79.

5. Chung, C. W., and V. A. Najjar. 1956. Cofactor requirements for enzymatic denitrification. II. Nitric oxide reductase. *J. Biol. Chem.* **218**:627–32.

6. Cove, D. J. 1967. Kinetic studies of the induction of nitrate reductase and cytochrome *c* reductase in the fungus *Aspergillus nidulans. Biochem. J.* **104**:1033–39.

7. Dal Nogare, S., and R. S. Juvet. 1962. Gas-liquid chromatography, theory and practice. New York: Interscience Publishers.

8. deGroot, G. N., and A. H. Stouthamer. 1969. Regulation of reductase formation in *Proteus mirabilis.* I. Formation of reductases and enzymes of the formic hydrogenlyase complex in the wild type and in chlorate-resistant mutants. *Arch. Mikrobiol.* **66**:220–33.

9. deGroot, G. N., and A. II. Stouthamer. 1969. Regulation of reductase formation in *Proteus mirabilis.* II. Influence of growth with azide and of haem deficiency on nitrate reductase formation. *Biochim. Biophys. Acta* **208**:414–27.

10. Delwiche, C. C. 1959. Production and utilization of nitrous oxide by *Pseudomonas denitrificans. J. Bacteriol.* **77**:55–59.

11. Downey, R. J., D. F. Kiszkiss, and J. H. Nuner. 1969. Influence of oxygen on development of nitrate respiration in *Bacillus stearothermophilus. J. Bacteriol.* **98**:1056–62.

12. Egami, F. M., M. Ishimata, and S. Taniguchi. 1961. The electron transfer from cytochromes to terminal electron acceptors in nitrate respiration and sulfate respiration. In *Haematin enzymes,* ed. J. E. Falk, R. Lembery, and R. K. Morton, pp. 392–401. Oxford: Pergamon Press.

13. Eppley, R. W., T. T. Packard, and J. J. MacIsaac. 1970. Nitrate reductase in Peru current phytoplankton. *Mar. Biol.* **6**:195–99.

14. Federova, M. V., and R. Sergeeva. 1957. The effect of oxidation-reduction conditions of the medium on the rate of nitrate reduction by denitrifying bacteria. *Mikrobiologiya* **26**:151–59.

15. Fewson, C. A., and D. J. D. Nicholas. 1961. Nitric oxide reductase from *Pseudomonas aeruginosa. Biochem. J.* **78**:9.

16. Fewson, C. A., and D. J. D. Nicholas. 1961. Nitrate reductase from *Pseudomonas aeruginosa. Biochim. Biophys. Acta* **49**:335–49.

17. Gauthier, D. K., G. D. Clark-Walker, W. T. Garrard, and J. Lascelles. 1970. Nitrate reductase and soluble cytochrome *c* in *Spirillum itersonii. J. Bacteriol.* **102**:797–803.

18. Gayon, U., and G. Dupetit. 1886. Recherches sur la transformation des nitrates par les infiniments petits. *Mem. Soc. Sci. Phys. Nat. Bordeaux,* Ser. 3, **3**:201–307.

19. Goering, J. J. 1968. Denitrification in the oxygen minimum layer of the eastern tropical Pacific Ocean. *Deep-Sea Res.* **15**:157–64.

20. Goering, J. J., and J. D. Cline. 1970. A note on denitrification in seawater. *Limnol. Oceanog.* **15**:306–309.

21. Goering, J. J., and R. C. Dugdale. 1966. Denitrification rates in an island bay in the equatorial Pacific Ocean. *Science* **154**:505–506.

22. Goering, J. J., and V. A. Dugdale. 1966. Estimates of the rates of denitrification in a subarctic lake. *Limnol. Oceanog.* **11**:113–17.

23. Hadjipetrou, L. P., and A. H. Stouthamer. 1965. Energy production during nitrate respiration by *Aerobacter aerogenes. J. Gen. Microbiol.* **38**:29–34.

24. Hollis, O. L. 1966. Separation of gaseous mixtures using porous polyaromatic beads. *Anal. Chem.* **38**:309–16.

25. Jensen, H. 1904. Denitrifikation und Stickstoffenbindung. *Lofars Handbuch der Techmachun Mykologie* **3**:182–92.

26. Kluyver, A. J., and W. Verhoeven. 1954. Studies on true dissimilatory nitrate reduction. II. The mechanisms of denitrification. *Antonie van Leeuwenhoek J. Microbiol. Serol.* **20**:241–337.

27. Knook, D. L., and R. J. Planta. 1971. Function of ubiquinone in electron transport from NADH to nitrate and oxygen in *Aerobacter aerogenes. J. Bacteriol.* **105**:483–88.

28. Kodama, T., and T. Mori. 1969. A double peak *c*-type cytochrome, cytochrome *c*-552, 558, of a denitrifying bacterium, *Pseudomonas stutzeri. J. Biochem.* (Tokyo) **65**:621–28.

29. Lam, Y., and D. J. D. Nicholas. 1969. A nitrite reductase with cytochrome oxidase activity from *Micrococcus denitrificans. Biochim. Biophys. Acta* **180**:459–72.

30. Lowe, R. H., and H. J. Evans. 1964. Preparation and some properties of a soluble nitrate reductase from *Rhizobium japonicum. Biochim. Biophys. Acta* **85**:377–89.

31. Matsubara, T., and T. Mori. 1968. Studies on denitrification. IX. Nitrous oxide, its production and reduction to nitrogen. *J. Biochem.* (Tokyo) **64**:863–71.

32. McNall, E. G., and D. E. Atkinson. 1957. Nitrate reduction. II. Utilization of possible intermediates as nitrogen sources and as electron acceptors. *J. Bacteriol.* **74**:60–66.

33. Miyata, M., and T. Mori. 1968. Studies on denitrification. VIII. Production of nitric oxide by denitrifying reaction in the presence of tetramethyl-*p*-phenylene diamine. *J. Biochem.* (Tokyo) **64**:849–61.

34. Miyata, M., T. Matsubara, and T. Mori. 1969. Studies on denitrification. XI. Some properties of nitric oxide reductase. *J. Biochem.* (Tokyo) **66**:759–65.

35. Najjar, V. A., and C. W. Chung. 1956. Enzymatic steps in denitrification. In *Symposium on inorganic nitrogen metabolism*, ed. W. D. McElroy and B. Glass, pp. 267–78. Baltimore: Johns Hopkins Press.

36. Najjar, V. A., and M. B. Allen. 1953. Production of nitrogen from nitrite by cell-free extracts of *Pseudomonas stutzeri* and *Bacillus subtilis. Fed. Proc.* **12**:250.

37. Najjar, V. A., and M. B. Allen. 1954. Formation of nitrogen, nitrous oxide and nitric oxide by extracts of denitrifying bacteria. *J. Biol. Chem.* **206**:209–14.

38. Nason, A. 1962. Symposium on metabolism of inorganic compounds. II. The enzymatic pathways of nitrate, nitrite and hydroxylamine metabolism. *Bacteriol. Rev.* **26**:16–41.

39. Nason, A., and N. J. Evans. 1953. TPN-nitrate reductase in *Neurospora. J. Biol. Chem.* **202**:655–73.

40. Newton, N. 1969. The two-haem nitrite reductase of *Micrococcus denitrificans. Biochim. Biophys. Acta* **185**:316–31.

41. Nicholas, D. J. D., and P. J. Walker. 1964. A dissimilatory nitrate reductase from *Neurospora crassa. Biochim. Biophys. Acta* **86**:466–76.

42. Odum, E. P. 1971. *Fundamentals of ecology.* Philadelphia: W. B. Saunders.

43. Payne, W. J., and P. S. Riley. 1969. Suppression by nitrate of enzymatic reduction of nitric oxide. *Proc. Soc. Exp. Biol. Med.* **132**:258–60.

44. Payne, W. J., P. S. Riley, and C. D. Cox, Jr. 1971. Separate nitrite, nitric oxide, and nitrous oxide reducing fractions from *Pseudomonas perfectomarinus. J. Bacteriol.* **106**:356–61.

45. Pichinoty, F. 1964. A propos des nitrate reductases d'une bacterie denitrificante. *Biochim. Biophys. Acta* **89**:378–381.

46. Pichinoty, F. 1965. Regulation chez les micro-organismes. L'effect oxygene et la biosynthese des enzymes d'oxydo-reduction bacteriens. *Coll. Intern. du C.N.R.S.* **124**:507.

47. Pichinoty, F. 1969. Les nitrate-reductases bacteriennes. III. Propriétiés de l'enzyme B. *Bull. Soc. Chim. Biol.* **51**:875–90.

48. Pichinoty, F. 1971. Les nitrate-reductases bacteriennes. VIII. Etude preliminaire de l'enzyme de *Micrococcus halodenitrificans. Arch. Mikrobiol.* **76**:83–90.

49. Pichinoty, F., A. Mucchielli, and C. Pelatin. 1971. Les nitrate-reductases bacteriennes. VII. Mesure de l'activité des enzymes A et B par une méthode colorimetrique. *Arch. Mikrobiol.* **75**:353–59.

50. Pichinoty, F., E. Azoulay, P. Couchoud-Beaumont, L. LeMinor, C. Rigano, J. Bigliardi-Rouvier, and M. Piechaud. 1969. Recherche des nitrate-reductases bacteriennes A et B: resultats. *Ann. Inst. Pasteur* (Paris) **116:**27–42.

51. Pichinoty, F., and L. D. Orano. 1961. Inhibition by oxygen of biosynthesis and activity of nitrate reductase in *Aerobacter aerogenes*. *Nature* (London) **191:**879–81.

52. Quastel, J. H., M. Stephenson, and M. D. Weltham. 1925. Some reactions to resting bacteria in relation to anaerobic growth. *Biochem. J.* **19:**304–17.

53. Renner, E. D., and G. E. Becker. 1970. Production of nitric oxide and nitrous oxide during denitrification by *Corynebacterium nephridii*. *J. Bacteriol.* **101:**821–26.

54. Rhodes, M. E., A. N. Best, and W. J. Payne. 1963. Electron donors and cofactors for denitrification by *Pseudomonas perfectomarinus*. *Can. J. Microbiol.* **9:**799–807.

55. Sacks, L. E., and H. A. Barker. 1952. Substrate oxidation and nitrous oxide utilization in denitrification. *J. Bacteriol.* **64:**247–52.

56. Schloesing, T. 1868. Sur la decomposition des nitrates pendent les fermentations. *Compt. Rend. Acad. Sci.* **66:**237–39.

57. Scholes, P. B., and L. Smith. 1968. Composition and properties of the membrane-bound respiratory chain system of *Micrococcus denitrificans*. *Biochim. Biophys. Acta* **153:**363–75.

58. Schulp, J. A., and A. H. Stouthamer. 1970. The influence of oxygen, glucose and nitrate upon the formation of nitrate reductase and the respiratory system in *Bacillus licheniformis*. *J. Gen. Microbiol.* **64:**195–203.

59. Showe, M., and J. A. DeMoss. 1968. Localization and regulation of synthesis of nitrate reductase in *E. coli*. *J. Bacteriol.* **95:**1305–13.

60. Sinclair, P. R., and D. C. White. 1970. Effect of nitrate, fumarate and oxygen on the formation of the membrane-bound electron transport system of *Haemophilus parainfluenzae*. *J. Bacteriol.* **101:**365–72.

61. Sreenivasan, A., and R. Venkataraman. 1956. Marine denitrifying bacteria from South India. *J. Gen. Microbiol.* **15:**241–47.

62. Stickland, L. H. 1931. The reduction of nitrates by *Bacterium coli*. *Biochem. J.* **25:** 1543–44.

63. Stouthamer, A. H., C. Bettenhaussen, J. van Hartingsveldt, J. Van't Riet, and R. J. Planta. 1967. Nitrate reduction in *Aerobacter aerogenes*. III. Nitrate reduction, chlorate resistance and formate metabolism in mutant strains. *Arch. Mikrobiol.* **58:**228–47.

64. Thomas, W. H. 1966. On denitrification in the northeastern tropical Pacific Ocean. *Deep-Sea Res.* **13:**1109–14.

65. Van't Riet, J., A. H. Stouthamer, and R. J. Planta. 1968. Regulation of nitrate assimilation and nitrate respiration in *Aerobacter aerogenes*. *J. Bacteriol.* **96:**1455–64.

66. Venables, W. A., J. W. T. Wimpenny, and J. A. Cole. 1968. Enzymic properties of a mutant of *Escherichia coli* K12 lacking nitrate reductase. *Arch. Mikrobiol.* **63:**117–21.

67. Verhoeven, W. 1956. Some remarks on nitrate and nitrite metabolism in microorganisms. In *Symposium on inorganic nitrogen metabolism*, ed. W. D. McElroy and H. B. Glass, p. 67. Baltimore: Johns Hopkins Press.

68. Wimpenny, J. W. T., and J. A. Cole. 1967. The regulation of metabolism in faculative bacteria. III. The effect of nitrate. *Biochim. Biophys. Acta* **148:**233–42.

69. Woods, D. D. 1938. The reduction of nitrate to ammonia by *Clostridium welchii*. *Biochem. J.* **32:**2000–12.

70. ZoBell, C. E., and H. C. Upham. 1944. A list of marine bacteria including descriptions of sixty new species. *Bull. Scripps Inst. Oceanogr.* **5:**239–92.

Comments

ZOBELL: I think the understatement of the year was yours that the denitrification is of some importance in the sea. I think it is of tremendous importance. But I have been wondering if you have data on the conditions under which it occurs. Particularly, do you have information on the concentration of the source of energy (whether it be organic matter, sulfur, or hydrogen) required for denitrification to occur?

PAYNE: I do not and I do not know anyone who does, principally because it has been a difficult phenomenon to measure, up to now. I hope to spread the notion that gas chromatography can be used now to specify not only that denitrification is taking place, but what phase of it is going on. Just how can this sort of data best be gathered? I do not know of anyone working in the environment with this phenomenon, the entire phenomenon, do you? I know Eppley at Scripps is working on nitrate reducers. Dugdale and his colleagues may be studying the entire phenomenon. I am inclined to believe what you said earlier. The ecologists from the University of Georgia who went on the Antarctic cruise found anomalies in apparent respiration in oxygen-poor areas of the sea which I think might be accounted for by a switch from aerobic to nitrate respiration. At least I would like to try to make that connection someday.

WATSON: I also think you underplayed the importance of this process. Man is now fixing, by commercial means, 30 million metric tons of nitrogen per year. This is equivalent to all the biological nitrogen fixation in the world. Places where denitrification occurs, as you point out, are in anaerobic areas of low oxygen concentrations. On land we have destroyed one quarter of all the low-lands in the United States, thus decreasing the power of bacteria to reduce all this nitrate which is being produced back to gaseous nitrogen to keep the nitrogen cycle in balance. Thus, we have to rely primarily on the oceans, I think, for the denitrification that must occur if we are going to balance all this new nitrogen which is being fixed. Delwiche estimates something like 30 million tons of nitrate must be reduced annually in the marine environment if we are to keep this situation in balance. We have just started to study denitrification, not in the detail that you have. I certainly congratulate you on your study and I want to talk to you later about the methods.

PRATT: I was approached recently by someone who wished to make a most probable number estimation of denitrifiers in the soil samples. Do you have a medium or some suggestion as to how this can be done?

PAYNE: When we studied the conversion of aerobic to denitrifying populations, we used Tryptone-yeast extract, sea salts medium, dispensed it in small vials, and did essentially a plate count. We mixed the populations and let the medium solidify in closed vials. We incubated them in desiccators with the BBL Gas Pack anaerobic systems and found that 82% formed colonies compared with aerobic counts. Around 55% produced a gas bubble. We even went to the extent of sampling the gas in those bubbles and found it all to be nitrogen. That technique could be used if you realize that maybe only $5/8$ of the bacteria that do form colonies are going to show a visible gas bubble. This must be done with fairly rapidly growing bacteria because the agar quickly begins to split and bubbles run together. So you might have to establish a specific time in which you make your estimate and stick rigidly to it.

ZOBELL: Stan Watson mentioned that 30 million tons of nitrates are fixed each year. As a further commentary on the importance of nitrate reduction, I would mention that this represents 90 million tons of oxygen.

PAYNE: I am a believer now. I will say it a little more strongly next time.

Determination of Total Microbial Biomass by Measurement of Adenosine Triphosphate

Osmund Holm-Hansen

One of the basic objectives of biological oceanographers is to describe flux of organic carbon from its reduction by phytoplankton through all levels of the marine food chain. Such information is desired not only for specific oceanic areas in regard to exploitation of marine resources, but it is also essential for understanding the organic carbon cycle on a global scale. Considering the increased use of the oceans as a repository for most of the waste products of our civilization and the importance of marine plankton in the O_2 and CO_2 budgets of the world's atmosphere, it is essential that we understand the distribution and activity of organisms throughout the entire water column and not merely in the upper euphotic zone. These studies require knowledge of the standing stock (biomass) of various types of organisms at any depth in the water column, and the biological activity of these organisms. In this paper, I will discuss methods and instrumentation for the estimation of total microbial biomass in aquatic environments, with brief comments on the correlation between biomass and biological activity.

Previous Methods for Biomass Estimation

The determination of microbial biomass in seawater samples is very difficult by conventional methods because of (a) the large amount of detrital material (i.e., nonliving particulate matter) relative to cellular material, and (b) the difficulty of distinguishing between live and dead cells. The methods which have been commonly employed to give biomass estimates include the following.

(1) The total number of cells or organisms in any sample may be determined by direct microscopic counting of filtered samples (5) or of settled samples and use of the inverted microscope (27). If appropriate cell dimensions are also recorded, then total biomass may be calculated for all species. These methods are extremely laborious and time-consuming, and also are subject to losses associated with delicate organisms (10). When very small cells (1 to 5 μm) are being counted, the errors will be magnified because of the difficulty of distinguishing between cells and detrital material and also because much smaller volumes of sample are generally examined.

(2) A standard method for determining the concentrations of viable bacterial cells (without any information as to cell volumes) has been to streak

73

samples onto agar surfaces and count the number of colonies which grow. The errors involved in this method can be very large *(17)* and it must be concluded that this method will not yield reliable estimates of the total number of viable bacteria in any natural sample.

(3) The measurement of total organic carbon or nitrogen, or such cell constituents as RNA or DNA, cannot be used for biomass estimates, as these constituents are also found in the detrital fraction *(12, 16)*.

(4) Phytoplankton biomass is often estimated by calculation from chlorophyll measurements. The reliability of such estimates depends upon a fairly uniform ratio of cellular organic carbon to chlorophyll. This ratio, however, varies from about 20 to 200, with an average of about 75 to 100 *(12, 22)*.

(5) Various investigators have attempted to estimate biomass from kinetic studies involving increase in cell numbers as recorded by electronic particle counters *(26)*, or the assimilation of radio-labeled organic substrates *(7, 24)*. Both these methods are so dependent upon questionable assumptions that such biomass estimates must be viewed with caution. Phytoplankton biomass has also been estimated by measurement of photosynthetic uptake of ^{14}C-bicarbonate *(2)*, the results of which seem to yield reasonable values. None of these methods is applicable, however, to total microbial biomass analyses.

(6) The measurement of electron transport system (ETS) activity, combined with assumptions regarding temperature coefficients for respiration, also permits estimation of total microbial biomass *(20)*. The application of this method to total microbial respiratory activity in deep-ocean water yields reasonable results *(11, 21)*, though laboratory studies have shown that some ETS activity persists in cells which have been killed by freezing.

Adenosine Triphosphate as a Biomass Indicator

In 1963–64 John Strickland and I became interested in developing an easier and more reliable technique than the aforementioned methods to measure microbial biomass in deep-ocean samples. It seemed to us that the most promising approach to this problem was to find some fairly labile metabolic intermediate common to all cells and which would not be found in detrital material. Preliminary work indicated that RNA and DNA were not satisfactory for this application, but that adenosine triphosphate (ATP) did satisfy these requirements. All our studies since that time have demonstrated that ATP is an excellent criterion of microbial biomass and that ATP determination is a practical and easy method which is applicable both in laboratory and field studies. Some of these data, together with information on the instrumentation developed for this assay, are discussed in the following sections.

In order to be a reliable biomass indicator, ATP must meet the following criteria:

(1) It must be a constituent of all plant and animal cells which are likely to be found in the samples. As there is no evidence to the contrary, it is assumed that ATP is ubiquitous in all living organisms on earth *(18)*.

(2) The ratio of dry weight (or any other significant cellular measure such as wet weight, cell volume, and total cellular organic carbon) to ATP must be fairly

uniform in all cells. Data on freshwater and marine bacteria, algae, and zoo-plankton indicate a strikingly uniform concentration of ATP relative to cell carbon in all these diverse organisms (Fig. 1). On the basis of such data, the ratio of cellular organic carbon to ATP has been found to be close to 286, which is the value used in ecological studies to estimate biomass from ATP concentrations. The cellular dry weight or fresh weight may be estimated from the carbon value by multiplying by the factors of 2 and 10, respectively.

(3) ATP must not be found in dead cells, nor adsorbed onto detrital material. These conditions apparently are met by ATP as shown by the following data: (*a*) if suspensions of marine phytoplankton are heated at 60 C for a few minutes

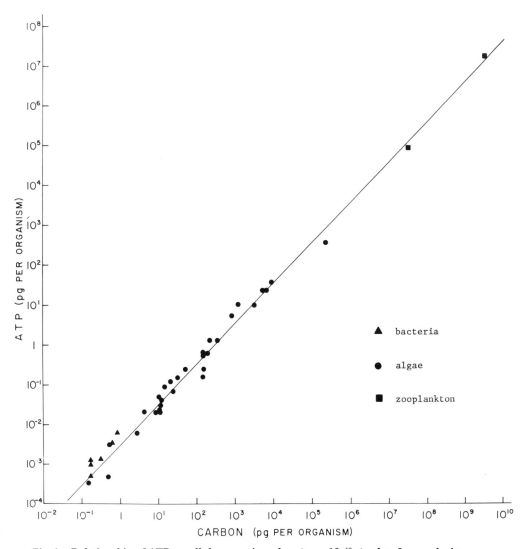

Fig. 1. **Relationship of ATP to cellular organic carbon (pg = 10^{-12} g) — data for zooplankton were obtained on the copepods *Calanus helgolandicus* and *Calanus hyperborius;* also see references (8) and (13), respectively.**

or frozen at -20 C for a few hours, there is no ATP detectable within the cellular material. *(b)* The flexibacterium, *Saprospira* sp., undergoes a natural autolysis after a certain density is reached in batch cultures. The optical density of the suspension decreases sharply shortly after the onset of autolysis and then remains at a constant value. The ATP recovered in the particulate fraction also decreases sharply with the onset of autolysis, but its concentration decreases to undetectable amounts concomitant with no viable cells in the suspension as determined by plating techniques. *(c)* When the bacterium, *Serratia* sp., is grown in batch culture, there is a sharp decrease in viable cell numbers (as determined by plating techniques) after 4 to 5 days (Fig. 2), but a significant number of cells remain alive in the suspension. The amount of ATP recovered in such a suspension indicates a fairly uniform amount of ATP per viable cell, with no significant contribution by the detrital material. *(d)* If ATP is added to a suspension of killed cells, or added to ocean samples which have been frozen, there is no significant amount of ATP recovered with the particulate matter on the filter, indicating that there is little or no adsorption of ATP onto detrital material.

(4) It is imperative that cellular ATP levels do not change drastically upon a change in environmental conditions. The most obvious stresses upon cells during sampling and filtration in field studies are changes in *(a)* light intensity, *(b)* temperature, and *(c)* pressure. Although the turnover rate of ATP is about one second or less *(25)*, phytoplankton cells are capable of maintaining fairly uniform concentrations of ATP during light/dark periods. Figure 3 shows ATP levels in a diatom when subjected to alternating periods of light and dark. Although there are fluctuations in ATP levels upon any change in light intensity, the rate of oxidative and photophosphorylation reactions apparently are sufficiently controlled so that the equilibrium values of ATP are the same regardless of light conditions. Temperature effects on ATP levels will be insignificant if samples are filtered promptly and if cells are killed immediately upon completion of filtration. The effects of decreasing pressure on samples from great depths in the oceans have not as yet been investigated.

(5) If the growth rate of an organism decreases due to a nutrient deficiency, it is important that the cellular ATP levels do not deviate markedly from that found in exponentially growing cells. Studies with silicon, nitrogen, and phosphorus, three of the mineral elements most likely to be in growth-limiting concentrations in aquatic environments, have been shown to have some effect on ATP levels *(1, 13)* but not enough to seriously affect biomass estimations based on ATP assays. Such laboratory data on nutrient-deficient algae would represent extreme conditions which probably are never obtained in nature, as the laboratory cultures were depleted of the various elements for 3 to 4 weeks. In nature, there is a continual input of mineral nutrients through regeneration processes, cellular excretion, and waste products originating in zooplankton and fish which undergo diurnal migration.

(6) The ratio of 286 between cell carbon and ATP has been obtained as described above on laboratory cultures of microorganisms. This ratio also is valid for organisms in natural water samples as indicated by the following data. *(a)* The total biomass as calculated by use of this ratio gives reasonable values when compared to the total particulate organic carbon content of the sample.

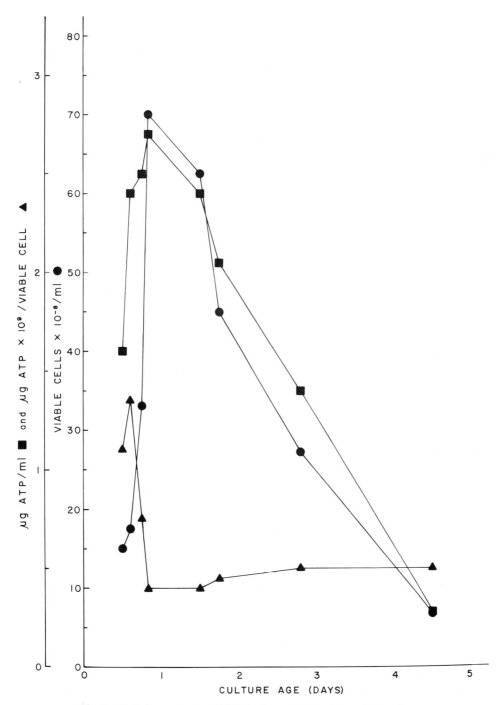

Fig. 2. **Cellular contents of ATP during a batch culture of *Serratia* sp.**

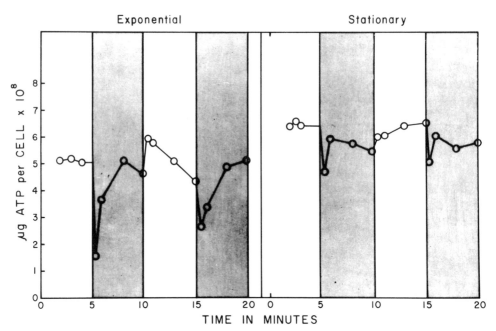

Fig. 3. Concentration of ATP in the diatom *Navicula pelluculosa* during alternate light/dark periods.

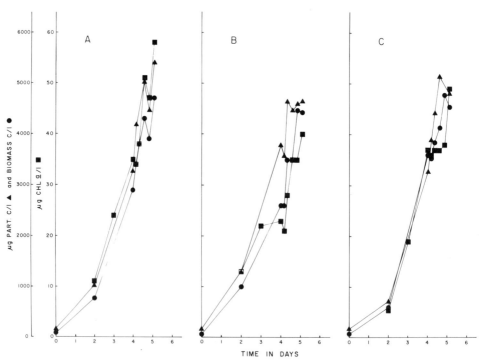

Fig. 4. Increase in chlorophyll-*a* and in cellular organic matter as determined by infrared gas analysis or estimated by ATP determination in batch cultures supplied with nitrate (A), ammonia (B), or urea (C) as the nitrogen source.

The proportion of carbon in living cells to the total carbon content generally is about 1% in deep-ocean samples and from 20 to 90% in samples from the euphotic zone. *(b)* Biomass as calculated from ATP data agrees very well with the biomass as obtained by direct microscopic counting and measurement of all recognizable cells *(6, 12)*. *(c)* A close approximation to natural conditions was achieved in some recent work *(3)* off the coast of southern California when we filled 200 l polyethylene tubs with unfiltered surface water, added nutrients, and measured various growth parameters under natural illumination for one week. The results (Fig. 4) show an excellent correlation between organic carbon as determined by infrared analyses and biomass-carbon as determined by ATP analyses, with the average ratio of carbon: ATP being 265. This ratio is very close to the value of 286 for laboratory cultures as shown by the data in Figure 1.

Procedure for ATP Determinations

(a) Sampling. The size and type of sampler to be used depend largely upon the fertility of the water being sampled. In freshwater environments and in the upper 200 m of ocean water, almost any type of clean and nontoxic sampler may be used without serious problems of contamination. For water containing relatively few organisms (e.g. ocean water below 1,000 m), it is advisable to use any bacteriological type sampler (such as plastic disposable bags) which may be sterilized or the open-cylinder type of sampler (e.g. PVC Van Dorn bottles or stainless steel Gerard-Ewing sampler) which may be scrubbed clean periodically. It is advisable to have at least one liter of sample for any deep-ocean studies, and thus it is not feasible to use many of the bacterial type samplers which yield less than one liter of sample. It is important in any deep-profiling program that one verifies that the sampler filled at the desired depth and did not misfire or leak during retrieval. This is easily done in the marine environment by comparing the salinity of the sample with the salinity from water obtained from a Nansen bottle equipped with reversing thermometers.

(b) Filtration. In nearly all field studies it is necessary to concentrate the cells in any water sample before killing and extraction of ATP. A convenient way to do this is to pass the sample through a filter that will retain essentially all organisms, including small bacteria. Membrane filters (0.4 μm pore size) are satisfactory for this purpose, although micro-fine glass fiber filters also will retain bacteria and have the advantage of much faster filtration rates. When membrane filters are used it is generally necessary to use 47 mm diameter filters, whereas with glass fiber filters the 25 mm-diameter size is more appropriate. If one uses a 47 mm-diameter glass filter, the amount of salt retained by the filter can cause interference with the subsequent luciferase reaction. One cannot eliminate this effect by first rinsing the filter with low-salt, isotonic solutions as such a treatment would result in unpredictable fluctuations in cellular ATP concentrations. Regardless of the type of filter used, it is advisable to minimize possible injury or lysis of fragile cells on the filter by not exceeding a vacuum of 25 cm Hg. Water samples, which are often prefiltered through nylon net (about 200 μm mesh size) to remove macroscopic zooplankton, should be filtered as soon as

possible to minimize any changes in ATP levels caused by changing environmental conditions.

(c) Enzyme Inactivation. As cellular levels of ATP can change very rapidly due to transphosphorylase and ATPase reactions, it is necessary to inactivate all enzymes as rapidly as possible when filtration is complete. This may be done in a variety of ways, but a rapid and convenient method is to remove the filter as soon as the last bit of water disappears into the filter and to submerge it in 5.0 ml of boiling Tris (0.02 M, pH 7.75) buffer. The sample is then heated at 100 C for 3 to 4 min.

(d) Extraction and Storage. The above heating period kills all cells and destroys the semipermeable characteristics of cell membranes. All soluble cell constituents such as ATP will thus rapidly diffuse out of the cells and ultimately will result in a uniform distribution of ATP throughout the entire suspension. After the 3 to 4 min period in the boiling water bath, the sample tube is placed in room-temperature water for 1 to 2 min and then capped and frozen at −20 C until time of analysis. Laboratory studies have shown that ATP is not hydrolyzed during the heating period and that it can be quantitatively recovered after many months of storage at −20 C. Shortly before the sample is to be analyzed, it is removed from the freezer and placed in a room-temperature water bath for 10 to 15 min.

(e) Analytical Determination of ATP. The basic reaction which is used in the bioluminescent measurement of ATP is:

$$\text{luciferin (red.)} + \text{ATP} + O_2 \xrightarrow[\text{Mg}^{++}]{\text{luciferase}} \text{luciferin (ox.)} + \text{AMP} + \text{P–P} + H_2O + hv$$

The reaction is not quite as simple as shown above, as there apparently is a decarboxylation step also involved *(19)*, but the essential aspect of the reaction, for our purposes, is that *(a)* for every molecule of ATP hydrolyzed there is one photon of light emitted (peak emission is about 560 nm) and *(b)* if everything else is in excess, the light intensity will be directly proportional to the ATP concentration. When ATP is added to the enzyme preparation, there is a rapid emission of photons, the rate of which declines in a semiexponential fashion (Fig. 5). The kinetics of this light emission decay curve is quite complex and is dependent upon the purity of the enzyme preparation.

In order to measure the amount of ATP in any sample, all one has to do is to measure the amount of light emitted per unit time and compare this value with values obtained when ATP standards are injected into the enzyme preparation. Some investigators record the peak height of emission only, but this value is very dependent upon the purity of the enzyme preparation and the efficiency of the mixing procedure. More reliable results are obtained if one integrates the amount of light emitted over a set time period (see Fig. 5). In our studies we have found a 1-min integration period to be economical in regard to time consumption per sample and reproducibility of results. A fairly simple apparatus was described by Holm-Hansen and Booth *(14)* in which the light emission

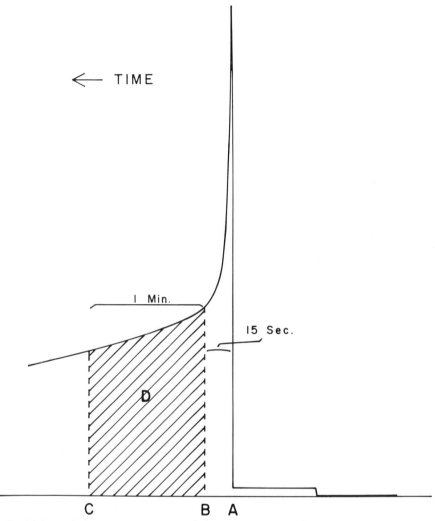

Fig. 5. **Light emission curve when a sample containing ATP is injected into the luciferin-luciferase enzyme preparation: A, time of injection of sample; B, end of 15-sec mixing period; C, end of 1.0-min light-integration period; D, area electronically integrated.**

curve was recorded on a paper strip and the area under the curve was integrated with a planimeter. We used this apparatus for some years, but it is now obsolete as we have completely automated the timing sequences and light integration period, and have a direct digital readout of the integrated value. The sample holder with sliding diaphragm is similar to that described by us, but the anode signal from the photomultiplier tube now goes to an electronic integrator with a digital readout.

We have two different methods for adding the sample to the reaction mixture. The usual way is to measure the light emission by the enzyme preparation (usually 0.2 ml) in a scintillation vial, then remove the vial from the sample holder and pipette into it 0.5 ml of the sample solution. If glass fiber filters were used in the filtration procedure, it is not necessary to remove the glass fibers from the suspension, as such particulate matter does not interfere with the quantitative determination of ATP. The sample is added to the enzyme preparation and gently swirled for 3 to 4 sec to afford complete mixing, after which the vial is replaced in the sample holder, the top is replaced, and the sliding diaphragm is pulled out. At the time when the sample is pipetted into the enzyme preparation, a foot switch is simultaneously activated which starts a 15-sec period to allow for mixing of the sample and getting it back in the sample holder; at the end of this 15-sec period, the amount of light emitted is electronically integrated for a 1.00-min period. If a permanent record of the emission curve is wanted, a recorder jack permits a paper-strip record of the light emission at the same time it is being electronically integrated. If it is desired to measure peak height, or to study the kinetics of the light emission curve, we can also do this by eliminating the 15-sec mixing period and injecting our sample through a special top on the sample holder directly into the enzyme preparation. For routine work, however, it is more convenient to use large volumes and to mix the samples by hand. Our most recent model has incorporated all components of our instrument into one unit which is about 1.2 cu ft in volume and has been designed so that it may be used on board ship.

At the present we have achieved detection of 10^{-14} g ATP per 10 μl injection or 10^{-12} g ATP/ml with partially purified enzyme. With further purification of enzyme and substrate it should be possible to increase the sensitivity by a factor of at least 10. For most biomass determinations it is feasible to use the crude, lyophilized water extract of firefly lanterns which permits detection down to 10^{-4} μg ATP/ml at a considerably reduced cost. With reduction of the voltage on the photomultiplier tube, our instrument can be used with solutions containing up to 1 mg ATP/ml.

Applications of ATP Measurements in Field Studies

There are a great many possible applications of ATP measurements in aquatic environments, but I would like to cite just a few examples to illustrate how it has been used in some of our studies:

(a) In relatively shallow profiles (from the surface to about 200 m), ATP measurements are a good indicator of phytoplankton biomass and, when used in conjunction with other data, yield information on changes in the chlorophyll: cellular organic carbon ratio or in increased biomass of heterotrophic organisms. Figure 6 shows a typical profile where the chlorophyll maximum is at 50 m, while the total particulate organic carbon and total microbial biomass are in maximal concentrations in surface waters. The estimation of the heterotrophic biomass in the water column below the main phytoplankton peak is of particular interest in studies of processes such as nutrient regeneration and oxygen consumption. Such biomass data, coupled with information on rates of metabolic activity, are essential for an overall understanding of biological activity and recycling of nutrients throughout the entire water mass.

(b) A special aspect of the above concerns the carbon cycle throughout the world's oceans and in food-chain dynamics in deep-ocean water. Figure 7 shows

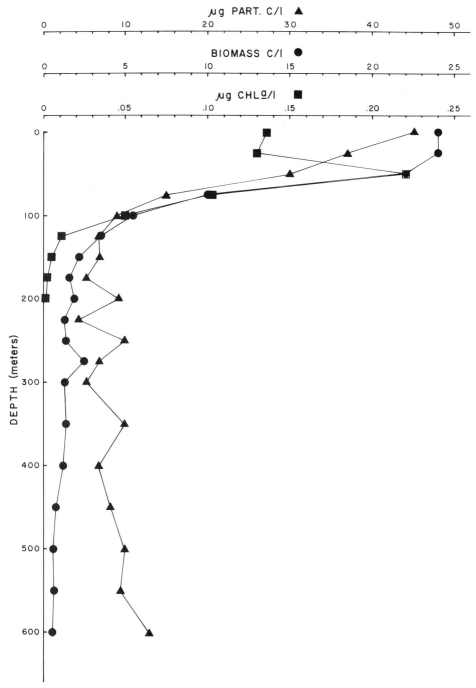

Fig. 6. **Distribution of chlorophyll-*a*, total particulate organic carbon, and biomass-carbon as calculated from ATP measurements in coastal water off the coast of southern California (32° 41′ N, 117° 35′ W).**

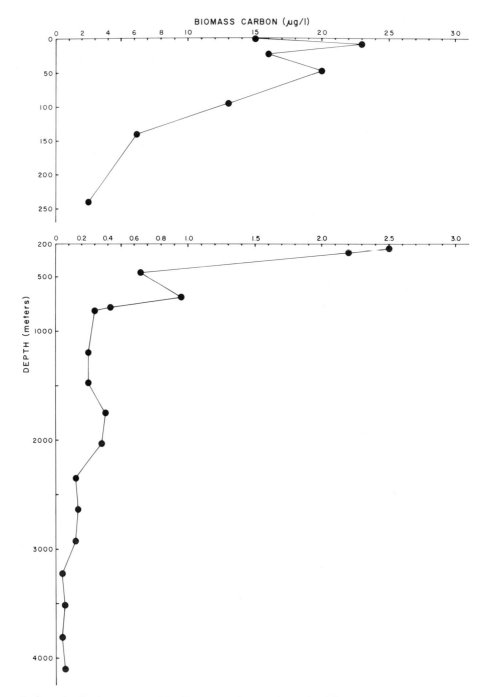

Fig. 7. **Distribution of microbial biomass-carbon as determined from ATP measurements in a deep profile north of Oahu (station position: 22° 10′ N, 158° 00′ W).**

the distribution of biomass-carbon down to 4,100 m at a station north of Oahu in the Hawaiian Islands. Below 3,000 m the biomass carbon averages about 0.06 μg C/l, which represents about 1% of the total particulate organic carbon per liter at these depths. From the surface to 100 m, the biomass carbon values account for between 50 to 90% of the total particulate organic carbon values. Such biomass estimates based on ATP determinations correlate very well with estimates made by direct microscopic analyses both in euphotic zone samples *(12)* and in deep-water studies *(6, 9)*. These estimates of microbial biomass in deep water are also quite reasonable when compared to zooplankton biomass, which is about 10% of the microbial biomass *(15)*.

(c) An estimate of total microbial biomass can be very useful in studies concerned with eutrophication of natural waters. An example of this would be our study on the standing stock and biological activity of microorganisms in relation to sewer outfalls off the coast of Southern California. Figure 8 shows the total microbial biomass in depth profiles both in and out of the effluent plume at different sewer outfalls. Although there were no significant differences between these six stations in regard to dissolved organic carbon, nitrogen, or phosphorus or in concentration of dissolved inorganic nutrients, there were marked differences in total microbial biomass as determined by ATP analyses. This increased biomass in the sewer plume was further substantiated with data on chlorophyll concentration and primary production *(4)*.

(d) In pollution studies it is often desired to know the total microbial biomass relative to the total load of suspended organic carbon in the water. In a recent paper on this subject, Ryther and Dunstan *(23)* gave data on total carbon and mentioned the desirability of knowing the standing stock of microorganisms; they had no data on this latter parameter, however, due to lack of a practical method to estimate biomass. The use of ATP for this purpose is a quick, easy, and reliable method for obtaining this kind of information.

(e) A few weeks ago I was working in the Straits of Georgia (British Columbia) on problems associated with airborne measurement of chlorophyll in surface waters. The Frazer River plume, which is very turbid with suspended sediment and which flows as a lens-shaped wedge of low-salinity water over the denser, saline straits water, showed a very high chlorophyll content on the airborne spectrometer, which was confirmed with water samples taken in and out of the plume and analyzed for chlorophyll with conventional methods. As it is possible that most of the organisms in the plume might be dead because of increasing salinity, I made ATP measurements in and out of the plume. The biomass estimates as calculated from these ATP data correlated very nicely with the chlorophyll data, indicating that the plume and associated boundaries are not only high in total particulate material but also high in microbial biomass.

(f) During the past year we have also been investigating the potential use of ATP to estimate respiration rates associated with the microbial populations sampled for ATP. Such activity estimates assume that (1) the respiration rate of most microbial forms in our samples will be fairly uniform when expressed per unit cellular carbon, (2) respiration rates will have approximately the same temperature coefficients as determined from laboratory studies, and (3) none of the organisms are nutrient-limited to such an extent that the respiration rate is

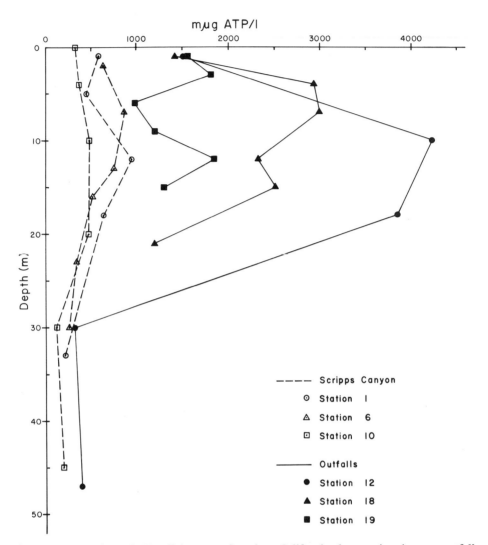

Fig. 8. Concentrations of ATP off the coast of southern California: three stations in sewer outfall plumes (continuous lines) and three stations slightly removed from the plumes (dashed lines).

markedly affected. Preliminary studies *(11)* in conjunction with Pomeroy, who measured respiration directly with oxygen electrodes, and with Packard, who estimated respiration by measurement of electron transport system activity, are promising for the potential application of ATP measurements to respiration estimates.

Acknowledgments

This research was supported in full by the U. S. Atomic Energy Commission Contract No. AT(11–1)GEN 10, P.A. 20.

Literature Cited

1. Combs, J., P. J. Halicki, O. Holm-Hansen, and B. E. Volcani. 1967. Studies on the biochemistry and fine structure of silica shell formation in diatoms. *Exp. Cell Res.* **47:**315–28.
2. Eppley, R. W. 1968. An incubation method for estimating the carbon content of phytoplankton in natural samples. *Limnol. Oceanogr.* **13:**574–82.
3. Eppley, R. W., A. F. Carlucci, O. Holm-Hansen, D. Kiefer, J. J. McCarthy, E. Venrick, and P. M. Williams. 1971. Phytoplankton growth and composition in shipboard cultures supplied with nitrate, ammonium or urea as the nitrogen source. *Limnol. Oceanogr.* **16:**741–51.
4. Eppley, R. W., A. F. Carlucci, O. Holm-Hansen, D. Kiefer, J. J. McCarthy, and P. M. Williams. 1971. Evidence for eutrophication in the sea near southern California coastal sewage outfalls. *Calif. Coop. Ocean. Fish. Invest. Reports.* In Press.
5. Fournier, R. O. 1970. Studies on pigmented microorganisms from aphotic marine environments. *Limnol. Oceanogr.* **15:**675–82.
6. Fournier, R. O. 1971. In *Data report from Gollum Station.* Honolulu: Univ. of Hawaii.
7. Hamilton, R. D., and K. E. Austin. 1967. Assay of relative heterotrophic potential in the sea: The use of specifically labelled glucose. *Can. J. Microbiol.* **13:**1165–73.
8. Hamilton, R. D., and O. Holm-Hansen. 1967. Adenosine triphosphate content of marine bacteria. *Limnol. Oceanogr.* **12:**319–24.
9. Hamilton, R. D., O. Holm-Hansen, and J. D. H. Strickland. 1968. Notes on the occurrence of living microscopic organisms in deep water. *Deep-Sea Res.* **15:**651–56.
10. Hasle, G. R. 1969. An analysis of the phytoplankton of the Pacific southern ocean: abundance, composition, and distribution during the Brategg Expedition, 1947–1948. *Norske Vidensk. Akad. I Oslo, Hvalradets Skrifter* **52:**1–168.
11. Hobbie, J. E., O. Holm-Hansen, T. T. Packard, L. R. Pomeroy, R. W. Sheldon, J. P. Thomas, and W. J. Wiebe. 1971. Distribution and activity of microorganisms in ocean water. Submitted to *Limnol. Oceanogr.*
12. Holm-Hansen, O. 1969. Determination of microbial biomass in ocean profiles. *Limnol. Oceanogr.* **14:**740–47.
13. Holm-Hansen, O. 1970. ATP levels in algal cells as influenced by environmental conditions. *Plant Cell Physiol.* **11:**689–700.
14. Holm-Hansen, O., and C. R. Booth. 1966. The measurement of adenosine triphosphate in the ocean and its ecological significance. *Limnol. Oceanogr.* **11:**510–19.
15. Holm-Hansen, O. 1970. Microbial distribution in ocean water relative to nutrients and food sources. In *Biological sound scattering in the ocean*, ed G. B. Farquhar, pp. 149–57. Washington, D.C.: Maury Center for Ocean Science.
16. Holm-Hansen, O. W., H. Sutcliffe, Jr., and J. Sharp. 1968. Measurement of deoxyribonucleic acid in the ocean and its ecological significance. *Limnol. Oceanogr.* **13:**507–14.
17. Jannasch, H. W., and G. E. Jones. 1959. Bacterial populations in sea water as determined by different methods of enumeration. *Limnol. Oceanogr.* **4:**128–39.
18. Lehninger, A. L. 1965. *Bioenergetics.* New York: Benjamin, Inc.
19. McElroy, W. D., H. H. Seliger, and E. H. White. 1969. Mechanism of bioluminescence, chemiluminescence and enzyme function in the oxidation of firefly luciferin. *Photochem. Photobiol.* **10:**153–70.
20. Packard, T. T. 1969. The estimation of the oxygen utilization rate in seawater from the activity of the respiratory electron transport system in plankton. Ph.D. dissertation, Univ. of Washington.
21. Packard, T. T., M. L. Healy, and F. A. Richards. 1971. Vertical distribution of the activity of the respiratory electron transport system in marine plankton. *Limnol. Oceanogr.* **16:**60–70.
22. Parsons, T. R., K. Stephens, and J. D. H. Strickland. 1961. On the chemical composition of eleven species of marine photoplankters. *J. Fish. Res. Bd. Can.* **18:**1001–16.

23. Ryther, J. H., and W. M. Dunstan. 1971. Nitrogen, phosphorus, and eutrophication in the coastal marine environment. *Science* **171**:1008–13.
24. Seki, H., and C. E. ZoBell. 1967. Microbial assimilation of carbon dioxide in the Japan Trench. *J. Oceanogr. Soc. Japan* **23**:8–14.
25. Shropshire, W., and K. Bergman. 1968. Light-induced concentration changes of ATP from Phycomyces sporangiophores: A reexamination. *Plant Physiol.* **43**:1317–18.
26. Strickland, J. D. H., and T. R. Parsons. 1968. *A practical handbook of seawater analysis.* Bull. 167. Fish. Res. Bd. Can. Ottawa: Queen's Printer.
27. Utermöhl, H. 1931. Neue Wege in der quantitativen Erfassung des Planktons. *Verh. Int. Verein. Theor. Angew. Limnol.* **5**:567–97.

Comments

COLWELL: How did you extract macroscopic zooplankton organisms?

HOLM HANSEN: We drop them into a large volume of boiling Tris buffer. I pick them up with 8 or 9 mm-diameter pyrex tubing and just pipette them very quickly into a large volume (5 ml) of boiling Tris. It is very important that the temperature of the extracting solution be close to 100 C. If it goes below 90 C, you will not get reliable ATP data.

COLWELL: If you had an estimation of the amount of ATP in different types of planktonic organisms and then tried to correlate this with particle size obtained with a Coulter counter, could you derive a profile of the kinds of microorganisms present based on ATP measurements?

HOLM-HANSEN: No, I do not think so, because data from the Coulter counter, even with the best model in the hands of a capable person like Sheldon, are extremely hard to interpret. All it gives you is the effective diameter of all particles, whether they be live or dead—organic or inorganic. About all it does tell you is the approximate particle volume of any dominant size class.

COLWELL: You cannot conclude anything except the total volume of cells that is present as estimated by the ATP measurement. You cannot say anything else about the populations?

HOLM-HANSEN: That is right. ATP data give you the total volume of living cells. You could get some data on size distribution of successive filtrations through filters of different porosity, but I have not bothered with that.

WATSON: Before any of us use these techniques, I think it would be valuable to know how sensitive they are. For example, in acetate oxidation, we estimated that we have to have about 200 bacterial cells per ml if we are going to use concentrations of about one micromolar and incubate for about 2 hr. Can you see one bacterium per ml or how many can you see using this technique?

HOLM-HANSEN: We filter off the particulate matter. If you have one bacterium per ml and you filter 2 l onto a glass fiber filter, you would see it. It simply depends on how much you concentrate. The absolute number of bacteria that we need is about 1,000 bacteria, using our fairly crude enzyme preparation. With our new machine we could probably push that down by a factor of 3 to 4. I could see getting down to about 100 bacteria, but I think that would be about the limit.

WATSON: So essentially, you could take a 100 ml sample in deep water and get some results.

HOLM-HANSEN: Yes, I have been doing it with 600 ml of water down to 7,000 m and getting nice results.

WATSON: What do you recommend for somebody who wants to do this type of measurement? What about using the scintillation counter? Does this present too many problems?

HOLM-HANSEN: That is how I wasted my first year. I got data, but the method is laborious and not as sensitive as this one. It takes a lot of time per sample. Depending on the geometry and the scaler, you may or may not get a linear response. My tests gave a good response with ATP and counts, but it was tough. Some people are still using scintillation counters. As a stopgap measure, if you do not have anything else, use it. But in the long term, I think one would be better off to use a better instrument.

G. JONES: How much would it cost to set up shop with this technique?

HOLM-HANSEN: That depends on whether you are going to build it yourself. This instrument was built by Booth in our laboratories. We salvaged parts here and there and did nearly all the work ourselves. We spent between $2,000 and $3,000. There is a unit on the market for about $5,000. There are many drawbacks to it, and I do not recommend it. I think a good instrument of the sensitivity I have described would probably cost about $4,000 when commercially produced. Each determination is cheap, as the enzyme costs about 4 cents a sample.

QUESTION: How would you compare your setup with DuPont's?

HOLM-HANSEN: Let me tell you two things. DuPont's biometer does not integrate. It merely gives you the maximum peak height within the first three seconds after mixing. This dictates that you have a complete and rapid mixing and also that you have uniform chemical composition of the medium week after week. That you do not have. Our instrument electronically integrates the light emission curve for a 60-sec period, which is far more reliable than peak height. Secondly, with the DuPont instrument you are limited to a 10-μl injection sample. I find this practically impossible for any kind of field work. With our instrument you can inject anything from a few microliters, and if you want to work in microamounts, up to 10 ml. I extract most of my samples from the field with 5 ml of Tris and inject 0.5 ml. We have the flexibility which DuPont does not.

The Use of Numerical Taxonomy in Estuarine Microbial Ecology

R. R. Colwell

Productivity calculations, energy budgets, and maps of food chains available in the literature define the role of bacteria and other microorganisms very incompletely. The microorganisms, mainly from lack of information, are usually treated as a vaguely defined factor included in equations to represent the "microbial parameter." It is clear that simply reporting numbers of bacteria per ml of water sample or per g of sediment is not sufficient or precise. The kinds of bacteria present in a given environment and their biochemical activities must be described for a complete understanding of the ecology of an environment. Classification and description of an environment by enumeration and correlation of distribution patterns of the species found therein is an accepted practice in "macro-ecology" and should be done in microbial ecology studies.

A research program in the microbial ecology of Chesapeake Bay has been under way since 1963. The field work has included routine sampling, at 6-week intervals, for the bacteriological examination of water, sediment, and animals (such as, oysters, crabs, etc.), for total viable, aerobic, heterotrophic bacterial flora. The methods of numerical taxonomy have been used to identify and classify bacteria in the natural populations. The methods used in the studies will be described briefly. Results of some of the numerical taxonomy analyses will be presented, as well as molecular taxonomy data verifying results obtained by computer analysis of the taxonomic data. The results to date indicate that bacteria serve well as indicators of environmental changes and this idea is offered as a working hypothesis for further studies.

Materials and Methods

The numerical taxonomy procedure followed is, first, to isolate, by random selection, cultures of bacteria from water, sediment, or animal tissue samples. The bacterial cultures are purified and tested, applying a variety of standard bacteriological tests, such as gram stain, enzyme reactions, and growth studies. By using many of the procedures published for bacterial taxonomy, as well as tests developed in our laboratory, a total of *ca* 300 characteristics is recorded for each bacterial strain. The bacteriological and computer analysis methods have been published (6, 7, 11, 12, 13, 19, 22).

The data are scored in a simplified format: 0 is recorded if a character is absent or negative, 1 if present or positive, and 3 if a test was not done or the test was not applicable. The data are key-punched for the computer using 80-column Hollerith cards. Twelve years ago, the programs were originally written for the IBM 650 computer. They are now written in PL/1 Language for the IBM Systems 360/40. A series of programs are available, comprising the Georgetown University taxonomy program library series, GTP-1 through GTP-5, that provide calculation of S-value triangles, clustering, hypothetical median organism, and serial comparison of groupings *(18, 21)*. Methods used in the DNA/DNA hybridization and isozyme experiments reported in this paper have also been published elsewhere *(5, 8, 15)*.

Results and Discussion

Numerical Taxonomy. Results of a taxonomic study of marine bacteria are shown in Figure 1. Strains sharing similarities greater than 80% form a cluster. Another clustering of less homogeneous strains, with *ca* ≥ 65% intragroup similarity, can

```
                                                              S-VALUE GRAPH

                                             1         2         3         4
                                    1234567890123456789012345678901234567890^12
PSEUDOMONAS, SP. (OX-SAWYER)
E. COLI ATCC 11303                  .
CDC 4280                            !!
CDC 4281                            !!#
VIBRIO ICHTHYODERMIS ATCC 23313     !.!!
VIBRIO ANGUILLARUM ATCC 14181       !.aa-
CDC 6670                            !!----
CDC 3637                            !!----£.
CDC 3454                            !!----=££
CDC 5002                            !!---+££*
CDC 4871                            !!--=-$££££
CDC 7606                            !!---a£££££*
CDC 8694                            !!=--$$$$££
VIBRIO PARAHAEMOLYTICUS ATCC 17802-S !!----$+$$$$£
SAK 3                               !!----$$$$$$*£
VIBRIO PARAHAEMOLYTICUS ATCC 17802-L !!-==-$$$$$£***
VIBRIO PARAHAEMOLYTICUS ATCC 17803-OP !!--=-++$$$$££*
VIBRIO PARAHAEMOLYTICUS ATCC 17803-TR !!--=-+$$$$$***##
CDC 8633                            !!--=-+$$$£$££***#
CDC 1889                            !!-===$$$£$££££*££
CDC 6202                            !!----+++$$$££$£*£££
CDC 5704                            !!--=-+$£$$£$££££***££
CDC 1334                            !!----$$££££££***££*
CDC 8198                            !!aa=-$$$£££££££££**££**
CDC 6614                            !!----$$$££$$££££*£££**
CDC 6540-L                          !!---aa--a=-===*+++=====-
CDC 6540-S                          !!---aaaaa--=-========--*
VIBRIO ADAPTATUS ATCC 19263         .............==..==.=====..
VIBRIO MARINOGILIS ATCC 14398-S     :................!:!:!.:!!....a
VIBRIO MARINOGILIS ATCC 14398-L     :................!:!:!.:!....a#
VIBRIO ALCALIGENES ATCC 14736       :.....:!:!:.!:!!!:!:!:!!:!-aa
VIBRIO MARIONPRAESENS ATCC 19648-S  :.:!!!:!:!::!:!!!:!:!!!:!:--=
VIBRIO MARIONPRAESENS ATCC 19648-L  :.;!!;aaaaaaaa------aaa;;a:-aa=
VIBRIO HALOPLANKTIS ATCC 14393-S    ..:!:.!!!;:!::a;!:!:!::!.:!aaa==
VIBRIO HALOPLANKTIS ATCC 14393-L    ..:!:.!!;:!;;a;!:!::!:!.:!aaa==#
VIBRIO MARINOVULGARIS ATCC 14394-S  ..:!:...!!:.!.:!!!:!:!!..::!aa;-=++
VIBRIO MARINOVULGARIS ATCC 14394-L  ........=!!:.!:!!!.........:!a!;a-+=#
VIBRIO MARINOVULGARIS ATCC 14395-L  ..:!:...:!!...:!:!:!!....:.!aaa;aa==$$
VIBRIO MARINOVULGARIS ATCC 14395-S  ..:!:...!!:...!:....!.....aaa;aa==$$*
VIBRIO ALBENSIS ATCC 14547-S        ..a:.;!!::!!:!!:!!!:!.!.:!aa!:!!a!;aa
VIBRIO ALBENSIS ATCC 14547-L        :.aa:aa:;a:;:!:!!;!:!a!:!!:!aa;a;aaa;aa£
VIBRIO PONTICUS ATCC 14391          .......=!!:!.:!;!!;!:!!!....:!!!:!!aa..
                                    1234567890123456789012345678901234567890^12
                                             1         2         3         4

                                                                         KEY
                                                           50 OR LESS   .
                                                               51-55    :
                                                               56-60    ;
                                                               61-65    a
                                                               66-70    -
                                                               71-75    =
                                                               76-80    +
                                                               81-85    $
                                                               86-90    £
                                                               91-95    *
                                                               95-100   #
```

Fig. 1. **Similarity values calculated for a set of marine bacteria. Strains are clustered by overall similarity (Colwell and Loveless, unpublished).**

also be noted. Several observations can be made from the example of a numerical taxonomy analysis. The set of organisms in the analysis provided in Figure 1 included bacterial strains described a number of years ago as new species. From the numerical taxonomy results, only one or two *Vibrio* species other than *Vibrio parahaemolyticus* can be detected, with significant species synonomy noted. Thus, relationships among bacterial strains can be detected and measured by the methods of numerical taxonomy.

V. parahaemolyticus, originally isolated in 1950 from victims of food poisoning in Japan, was analyzed extensively in our laboratory using numerical taxonomy and a large set of isolates. The data for *V. parahaemolyticus* were compared with those for bacterial isolates from Chesapeake Bay and, more recently, for isolates from American victims of food poisoning.* Cultures from tourists to the Far East, who had suffered *V. parahaemolyticus* food poisoning, have also been tested. Good clustering of Japanese and U.S. strains was obtained, thereby providing identification of the Chesapeake Bay isolates as strains of *V. parahaemolyticus* (Fig. 2).

The *V. parahaemolyticus* analyses proved very useful and informative, since one of the initial motivations of the Chesapeake Bay microbial ecology study was to build a data bank, putting together strain data for reference cultures, to compare with data for fresh isolates. A first success, therefore, was detecting *V. parahaemolyticus* in Chesapeake Bay and identifying this organism as the dominant bacterial species in at least one area of the bay. To confirm the identification of the Chesapeake Bay isolates, further studies were done employing additional Japanese isolates received through the courtesy of Dr. H. Zen-Yoji, Bureau of Public Health, Tokyo, Japan. Data for the additional set of Japanese isolates confirmed the diagnosis of the Chesapeake Bay isolates as a species identical to the food poisoning organisms isolated from victims in Japan (Fig. 3).

The Japanese and Chesapeake Bay strains of *V. parahaemolyticus* were compared with other species of the genus. It was found that both the Chesapeake Bay strains and the Japanese strains were *ca* 60 to 75% similar to *Vibrio cholerae* and lower in similarity to selected marine vibrios (Fig. 4). Baumann et al. (2) suggested *V. parahaemolyticus* and other marine vibrios should be placed in a new genus, *Beneckea*. Unfortunately, the genus *Beneckea* is defined as comprising chitin-digesting organisms of marine origin. All marine vibrios do not digest chitin. Hence, the separation of the marine vibrios, represented by *V. parahaemolyticus*, from *Vibrio cholerae* on this basis is not practical.

Molecular Taxonomy. The organisms grouped by computer analysis were included in a set of strains subjected to DNA base composition analyses. Using extracted, highly polymerized DNA, the percent G+C composition data revealed that organisms grouped by computer shared identical overall DNA G+C composition. Organisms not included in the *Vibrio* species clusters possessed very different percent G+C compositions (Table 1). *V. parahaemolyticus* strains revealed a G+C of 45 ±1%.

---

*Morbidity and Mortality Weekly Reports 20 **(39)**: week ending Oct. 2, 1971. Communicable Disease Center, p. 356.

S-VALUE GRAPH

```
                                               1         2         3         4         5         6
                                      1234567890123456789012345678901234567890123456789012345678901234
VIBRIO PONTICUS ATCC 14391            
VIBRIO ADAPTATUS ATCC 19263           .
VIBRIO MARINOFULVUS ATCC 14395-S       әә
VIBRIO MARINOFULVUS ATCC 14395-L      әә*
VIBRIO MARINOFULVUS ATCC 14394-L      ::$$
VIBRIO MARINOFULVUS ATCC 14394-S      ::$$#
VIBRIO HALOPLANKTIS ATCC 14393-L      ;:==+
VIBRIO HALOPLANKTIS ATCC 14393-S      ::==++#
VIBRIO MARINOPRAESENS ATCC 19648-L    ;:әә-===
VIBRIO MARINOPRAESENS ATCC 19648-S    :;әәә-===
VIBRIO ALCALIGENES ATCC 14736         !-;;;;әәә=
VIBRIO MARINOGILIS ATCC 14398-L       .әәә;әәәә-ә
VIBRIO MARINOGILIS ATCC 14398-S       .әәәәәәә--ә#
VIBRIO ICHTHYODERMIS ATCC 23313       ..........::;...
VIBRIO ALBENSIS ATCC 14547-S          .;әә;;ә;:;;әә.
VIBRIO ALBENSIS ATCC 14547-L          .;әә;әәә;ә;әә:Ɛ
VIBRIO ANGUILLARUM ATCC 14181         ......:;:;:...=:ә
CDC 6540-S                            ...::::;ә;:..-:;ә
CDC 6540-L                            .....:..;::..-.:ә*
CB CRAB 7C                            ..!...::;:....ә;.әә
CB CRAB 7Bw                           :..::::әә:әә::.;ә-ә-$
CB CRAB 11A                           :..::::::.;ә...;ә:;;-+
CB CRAB 7D                            ......:;:::.ә:;.ә-+++
CB CRAB 5D-Bs                         ......:;;::::-;ә:-=+=+Ɛ
CDC 1889                              ...:.:::-;;::;ә===--==$+
CDC 8633                              :.:::::;;-;;.:=:;-=+===++$Ɛ
VIBRIO PARAHAEMOLYTICUS ATCC 17803-TR :..:.:::-;;.:=::-=+===++ƐƐ#
VIBRIO PARAHAEMOLYTICUS ATCC 17803-OP :..:.:::-;;.:=::-===-=+++**#
VIBRIO PARAHAEMOLYTICUS ATCC 17802-L  :..:::::::-;::-=+++++ƐƐ**#
SAK 3                                 :..::::әә-;;..-:;-===$$$$Ɛ**Ɛ*
VIBRIO PARAHAEMOLYTICUS ATCC 17802-S  :.:::::;;ә::..-!;-===+$+$ƐƐ*Ɛ*Ɛ
CB CRAB 5D-Bw spr                     ::...әә;әә..;ә----ә+$$$+$ƐƐƐ$
CB CRAB 5D-Bw                         ;:::::; әәә-ә::ә-=ә=ә+-++$$ƐƐ**ƐƐ*
CDC 6202                              :.....::;::..-!;-===-==$$ƐƐƐ*ƐƐƐ*
CB CRAB 8C                            :.....әә:;;..;ә;:ә=$ƐƐ$++$+Ɛ$*Ɛ$
CDC 6614                              !.=:..::;;::..-!;-----+=-+$Ɛ*ƐƐƐ$$ƐƐ+
CDC 8198                              :.....:;ә;:..=.:-==-=+++Ɛ**ƐƐƐ++Ɛ+*
CDC 1334                              :.....:;ә;;.:-:::-==-=+$ƐƐ***ƐƐ$$Ɛ+**
CDC 5704                              :....:;ә;;.:=;;-=-----+Ɛ**ƐƐƐ=+Ɛ=Ɛ**
CB CRAB 5D-A                          ;:..::!әәәәә;;ә-;:+++$$*ƐƐ*ƐƐ*Ɛ*ƐƐ*ƐƐ*$
CDC 8694                              :.....::ә::..=::-===++$ƐƐƐƐ*Ɛ**ƐƐƐ$$ƐƐ$*
CB CRAB 11B                           :.....әәәә;..=;ә;-====+Ɛ$ƐƐƐƐƐƐƐ*ƐƐ$$$++Ɛ*
CB CRAB 12B                           ..:.:!әәәә;::=;::=++=-$+$$$$$$+$$$$=+++*Ɛ*
CB CRAB 12B spr                       .....;;әә;:::;;.,----$+$++$++++$+$-++=$Ɛ*#
CB CRAB 4A-Bw                         ......::!;;;::::;ә.=+-=++$++Ɛ+++$Ɛ+=+$=ƐɁƐ$$
CB CRAB 5A                            ......::!;;;..=:ә:====$$$$ƐƐƐ$$ƐƐ$$+$+Ɛ*ƐƐ*
CB CRAB 4A-A                          :.....::!;;;..:;!.ә-====++$++Ɛ+++$+Ɛ-=+$=ƐɁƐ$$*
CB CRAB 4A-B                          :.....::!;;;..:!;.ә-====++$++Ɛ+++$+Ɛ-=+$=ƐɁƐ$$**
CB CRAB 12A                           ......;;;;;..-!:!-===$$+$$$$$$$$Ɛ$++++=ɁƐ**ƐƐ*ƐƐ
CB CRAB 8E                            :..:::::;ә;..-:ә;ә-=+$++$$$+$$Ɛ+=$++$$+ɁƆ*$$$$$$$$
CB CRAB 12C                           .:::::::;;..-!;әә--=+++$$+$$+$Ɛ$=+++ɁƐ$ƐƐƐƐƐƐƐƐ
CB CRAB 13A                           ......::!.::...!.;:;==$+=+==+=+=+$ә==-++$ƐƐ+$++Ɛ$Ɛ
CDC 7606                              ....::!;ә:...-;;ә-+===$$$$$$=+$+ƐƐƐƐƐ$++=++=++==
CDC 4871                              :.....::!;:..=;;--====++$ƐƐ$$$$$+$$=ƐƐƐ$ƐƐ$++=$==$++-*
CDC 5002                              :.:::::;ә;:..-;ә+әәә+==ƐƐ$$$$$$$$+ƐƐƐ*$$======+=-ƐƐ
CDC 3454                              :.::::::;:..-;.-==-==$$ƐƐ*$$$+++$$ƐƐ*$$$++=======+-ƐƐ*
CDC 3637                              :.:::::!ә;:..-;.ә====---$$$+$$+-=+-$$$$=$-+=-==-+=--ƐƐƐ
CDC 6670                              ......::!;ә-әә====--$++$$$$==+=$$$++$:==----Ɂ.$.ƐƐ
CB CRAB 5B                            ::!.:..::::..:!.-ә.ә-==-$$+=+$+=+++$$$==+=ɁƐ$$+=ƐƐƐƐ*=+*==---ә-
CB CRAB 2B                            ..;!:::..........::.;ә===-=------ә-ә+;--әәәәәә+=++++++-аәәә;Ɛ
E. COLI ATCC 11303                    ...........;;әә;әә;;;;;;;+;-;:;;:-:--әә---ә;ә;:::::::-;
CDC 4281                              .:.!:.::!:;...;:әә---әә::!:----==--;ә-!-ә-:=!:;!;!;;!!..-----;!!
CDC 4280                              ..!:.::::::...:әәә--әә::!:-------;ә-:-ә-:=!:;!;!;!!........-;;!
PSEUDOMONAS, SP. (OX-SAWYER)          ........;;;::!..:.;::::::!::;;;;;;;:ә;:;!;;ₐ;ә-әә;;;ә.:;;::!!;..;;
                                      1234567890123456789012345678901234567890123456789012345678901234
                                               1         2         3         4         5         6
```

KEY
50 OR LESS	.
51-55	:
56-60	;
61-65	ә
66-70	-
71-75	=
76-80	+
81-85	$
86-90	Ɛ
91-95	*
95-100	#

Fig. 2. Numerical taxonomy analysis of Japanese and Chesapeake Bay strains of *Vibrio parahaemolyticus* (Colwell, Lovelace, and Tubiash, unpublished).

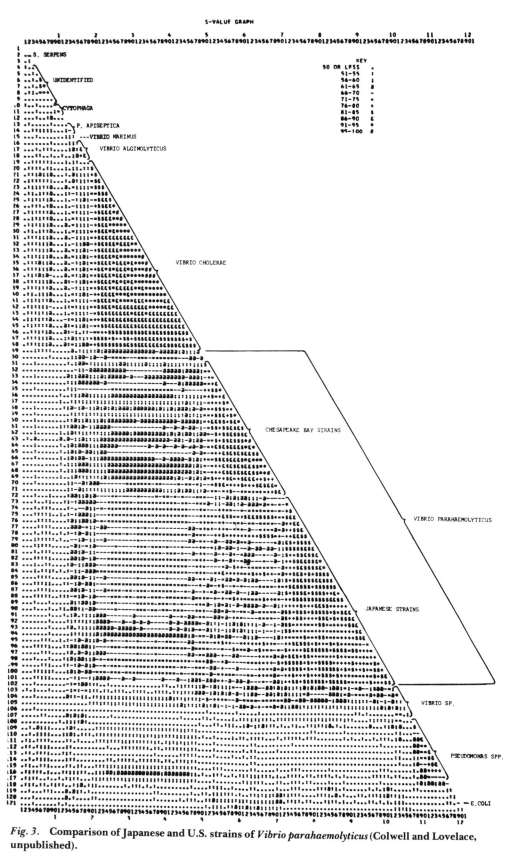

Fig. 3. Comparison of Japanese and U.S. strains of *Vibrio parahaemolyticus* (Colwell and Lovelace, unpublished).

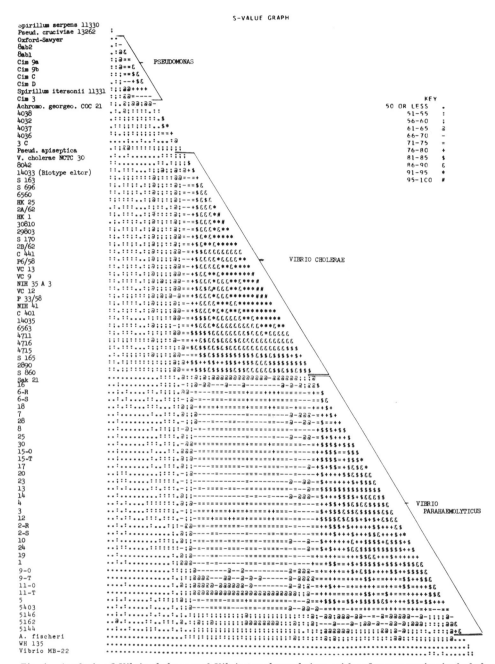

Fig. 4. Analysis of *Vibrio cholerae* and *Vibrio parahaemolyticus*, with reference strains included (Reprinted from Colwell *(7)* by permission of the American Society for Microbiology).

Table 1. DNA base composition of *Vibrio parahaemolyticus* strains isolated from blood of dead and dying blue crabs *(Callinectes sapidus)* in Chesapeake Bay

Organism	Strain Number	Tm C	Mole % G + C
Cytophaga	2C	83.3	34.2
	8F	83.5 (83.1)*	34.6 (33.8)*
Marine *Vibrio*	15	86.1 (86.5)*	41.0 (42.0)*
	5D-Bw	87.4	44.2
	7B	87.4	44.2
	5B	87.5	44.4
	5D-Bw-Spr	87.6 (87.6)*	44.6 (44.6)*
	8E	87.6	44.6
	7Bw	87.7 (87.8)**	44.9 (45.2)**
	2B	87.8	45.1
	4A-A	87.8	45.1
	5D-Bs	87.8 (87.7)*	45.1 (44.8)*
	7D	87.8	45.1
Vibrio	5D-A	88.0	45.6
parahaemolyticus	5d-A	88.0	45.6
	8C	88.0	45.6
	4A-B	88.1 (88.4)**	45.9 (46.6)**
	11B	88.1	45.9
	12B	88.1	45.9
	12C	88.1	45.9
	12A	88.2	46.1
	4A-Bw	88.3	46.4
	12B-Spr	88.3	46.4
	5A	88.5 (88.6)*	46.9 (47.0)*
	11A	88.6 (88.3)**	47.0 (46.3)**
Unidentified	7C-Grey	89.9	50.0
	7C-Wh	90.0 (89.6)*	50.5 (49.6)*
	6D	92.1 (91.8)*	55.5 (54.9)*

* One preparation, two determinations.
** Two preparations, two determinations.

Another technique for measuring more precisely the taxonomic relationships within groups is DNA/DNA reassociations. Using this procedure, a number of studies have been done by several investigators *(1, 3, 4)*. From our experiments, bacterial strains identified by computer as *V. parahaemolyticus* showed relative binding or reassociation of ⩾ 80% (Tables 2 and 3). At more stringent conditions of temperature (75C), which eliminate nonspecific binding, relative levels of intraspecific DNA duplex formation were only minimally reduced (Table 3). Marine vibrios and *V. parahaemolyticus*, at 60 C, revealed 4 to 24% reassociation. At an elevated temperature, the binding was 1 to 11% (Figs. 5 and 6). Other marine vibrios (for example, *Vibrio* strain MB-22) showed a somewhat higher

Table 2. Nucleic acid reassociation measurements. Reference strain used was *Vibrio cholerae* NIH 35 A3[a]

Strain Pairs	Relative Binding 60 C %	Relative Binding 75 C %	Thermal Binding Index (TBI)[b]
Vibrio cholerae			
NIH 35A3/35A3[c]	100	100	1.00
NIH 35A3/NIH 41	99	100	1.01
NIH 35A3/P 33/58	99	105	1.06
NIH 35A3/ATCC 14035	98	100	1.02
NIH 35A3/VC-9	99	99	1.00
NIH 35A3/C-401	100	101	1.01
NIH 35A3/ATCC 14033	87	83	.95
NIH 35A3/NCTC 2890	91	88	.97
NIH 35A3/NCTC 6563	88	87	.99
NIH 35A3/HI-1	97	95	.98
NIH 35A3/SLH 29803	96	95	.99
NIH 35A3/2A/62	97	94	.97
NIH 35A3/NCTC 4711	85	81	.95
NIH 35A3/NCTC 4715	86	85	.99
NIH 35A3/NCTC 4716	85	83	.98
NIH 35A3/NCTC 8042	86	83	.97
NIH 35A3/NCTC 30	82	74	.90
NIH 35A3/S 860	86	84	.98

[a] Reprinted from Citarella and Colwell (5) by permission of the American Society for Microbiology.
[b] TBI = relative binding at 75 C/relative binding 60 C.
[c] Source of radioactive DNA fragments.

(24%) reassociation with *V. parahaemolyticus*. Thus, it is evident that the identifications achieved by numerical taxonomy were confirmed, since *V. cholerae*, *V. parahaemolyticus*, and *V. marinus* were observed to be separate species.

Direct comparison of percent similarity (i.e., phenotypic similarity) with percent reassociation (i.e., molecular genetic similarity) demonstrated a significant correlation of 0.92 (7).

Isozyme patterns of *Vibrio* sp. were examined to determine whether this approach might offer a rapid and concise method for identifying species of bacteria occurring in estuaries (Figs. 7 and 8). *V. parahaemolyticus* isolated from blue crabs in Chesapeake Bay and Japanese strains obtained from victims of *V. parahaemolyticus* food poisoning were found to have similar esterase patterns. Organisms, known from computer analyses not to be *V. parahaemolyticus*, demonstrated quite different isozyme patterns. Thus, a rapid method for distinguishing groups of estuarine microorganisms is provided by this approach. The esterase pattern for strains implicated in *V. parahaemolyticus* food poisoning suf-

Table 3.   Nucleic acid reassociation measurements. Reference strain used was *Vibrio parahaemolyticus* SAK-4 [a]

Strain Pairs	Relative Binding 60 C %	Relative Binding 75 C %	Thermal Binding Index (TBI)[b]
Vibrio parahaemolyticus			
V. p.-4/V.p.-4[c]	100	100	1.00
V. p.-4/V.p.-1	86	84	.98
V. p.-4/V.p.-5	88	86	.98
V. p.-4/V.p.-13	84	84	1.00
V. p.-4/V.p.-15	88	84	.95
V. p.-4/V.p.-17	81	82	1.01
V. p.-4/V.p.-23	87	84	.97
V. p.-4/NIH 35A3	15	1	.07
V. p.-4/ATCC 14033	16	4	.25
V. p.-4/NCTC 4715	14	1	.07
V. p.-4/-5144	3	0	.00
V. p.-4/-5146	4	0	.00
V. p.-4/-5162	7	2	.28
V. p.-4/MB-22	24	11	.46
V. p.-4/PS-207	9	1	.11
V. p.-4/MP-1	4	0	.00
V. p.-4/MB-1	5	1	.20

[a] Reprinted from Citarella and Colwell *(5)* by permission of the American Society for Microbiology.
[b] TBI = relative binding 75 C/relative binding 60 C.
[c] Source of radioactive DNA fragments.

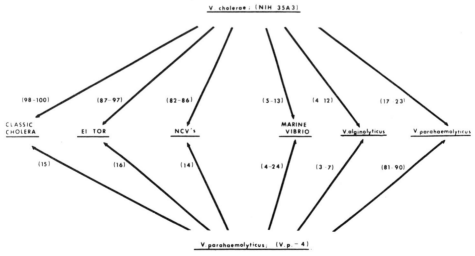

Fig. 5.   **Summary of relative binding values obtained for reassociation reactions at 60 C. See Figs. 6 and 7 for individual values. (From R. V. Citarella 1970. Polynucleotide sequence relationships among selected *Vibrio* species. Ph.D. dissertation, Georgetown University Washington, D.C.)**

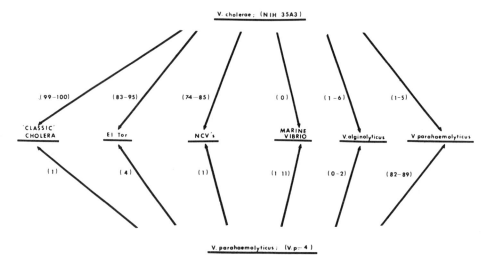

Fig. 6. Summary of relative binding values obtained for reassociation reactions at 75 C. See Figs. 6 and 7 for individual values. (From R. V. Citarella, Ph.D. Dissertation, Georgetown University).

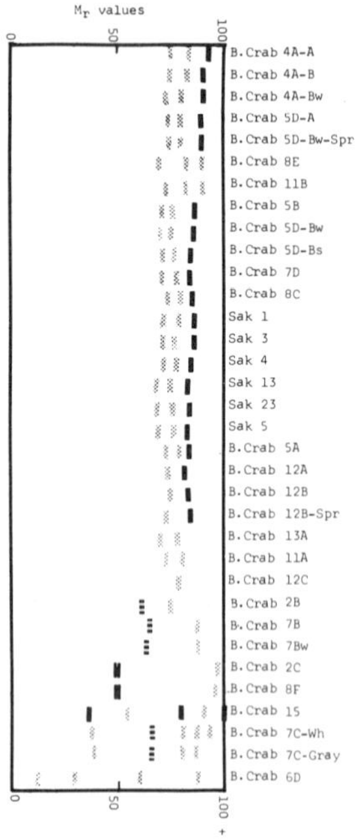

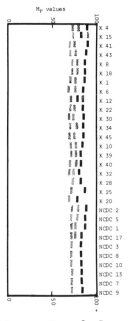

Fig. 7. Esterase patterns obtained for *V. para-haemolyticus* cultures isolated from blue crabs in Chesapeake Bay. Strains 4A-A through 12 C have been identified as *V. parahaemolyticus.* (Colwell and Wan, unpublished).

Fig. 8. Esterase patterns for Japanese strains (K-4 through K-20) and U.S. strains (NCDC-2 through NCDC-9) of *V. parahaemolyticus* (Colwell and Wan, unpublished).

fered by American tourists traveling to the Far East (NCDC strains) is shown in Figure 8.

Environmental Studies. The reliability and reproducibility of the methods used to identify and classify estuarine and marine bacteria have been amply demonstrated. It remained, then, to apply these methods to practical problems in estuarine and marine microbiology.

One such problem involved the survival of *V. parahaemolyticus* in Chesapeake Bay through the winter. *V. parahaemolyticus* does not survive refrigerated storage; yet, the organism persists in Chesapeake Bay despite freezing temperatures during the winter. Another problem concerned the distribution of *V. parahaemolyticus* in Chesapeake Bay.

Three areas of Chesapeake Bay were sampled routinely from 1964 to the present. The first area, Eastern Bay, was commercially productive for shellfish. The second location, Marumsco Bar on the eastern shore of Maryland, was not, because of serious oyster mortalities. A third area, the Rhode River, is not a commercial oyster harvesting location. Seasonal changes in the microbial flora were evident during the eight-year study. More importantly, distinct differences in the composition of the flora in each of the sampling areas were noted. *Vibrio* species were implicated in the oyster mortalities occurring in Marumsco Bar by analysis of the microbial ecology data, since significant differences were noted in the abundance of the major bacterial groups in these areas. *Vibrio* was the most abundant species in the Marumsco Bar region (Table 4).

Table 4. Generic distribution of bacterial types in samples tested from Marumsco Bar and Eastern Bay in Chesapeake Bay[1,2]

Taxonomic Group	Marumsco Bar	Eastern Bay
Vibrio sp.	46%	20%
Pseudomonas sp.	22%	9%
Achromobacter sp.	13%	17%
Corynebacterium sp.	5%	7%
Cytophaga/Flavobacterium sp.	3%	30%
Micrococcus/Bacillus sp.	0	6%
Enterics[3]	7%	3%
Other[4]	3%	8%
Total in sample	125	104

[1] Reprinted from Lovelace, Tubiash, and Colwell *(20)* by permission of the National Shellfisheries Association.
[2] Total number of strains studied was 229.
[3] Enterics: *Enterobacter* sp.; *Proteus* sp.
[4] Other: *Caulobacter* sp.; *Saprospira* sp.; *Spirillum* sp.

Vibrio species were isolated during the winter in all areas, but were and how vibrios "over-winter" was not clear. Data accumulated during sampling trips to Rhode River showed that the total viable count (TVC) of bacteria in water and sediment was high in summer months, without a severe drop in the winter (Figs. 9 and 10). However, *Vibrio* sp. and *V. parahaemolyticus* counts declined in the

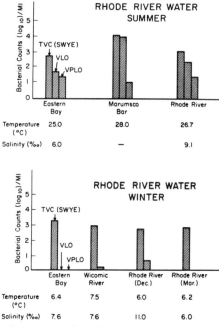

Fig. 9.   Bacterial counts of water samples: TVC(SWYE), total viable counts on a simulated sea-water, yeast extract, peptone medium (see *13*); VLO, *Vibrio* sp.; VPLO: *Vibrio parahaemolyticus* (Kaneko and Colwell, unpublished).

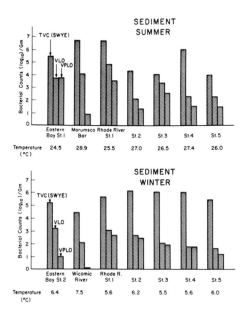

Fig. 10.   Bacterial counts of sediment samples (Kaneko and Colwell, unpublished).

Table 5. Bacterial counts per gram wet weight of plankton samples obtained from the Rhode River area of Chesapeake Bay (Kaneko and Colwell, unpublished data)

Sample Number	Location	Date	Temp. °C	TVC*		VLO	VPLO	Coliform	E. coli
				YE	SWYE				
1	Rhode River	Dec. 1970	5–6	5.5×10^5	5.5×10^5	1.2×10^2	4.7×10^1	2.3×10^1	0.8
2	Rhode River	Dec. 1970	5–6	1.1×10^6	1.1×10^6	9.3×10	9.3×10^1	2.1×10^1	0.0
3	Eastern Bay	Mar. 1971	6–7	6.8×10^6	2.4×10^6	9.9×10^3	0.0	6.8×10^2	2.2×10^1
4	Rhode River	Mar. 1971	5–6	–	2.4×10^5	1.2×10^3	0.0	–	–
5	Rhode River	Apr. 1971	8–9	5.8×10^4	1.5×10^5	6.3×10^2	0.5	1.0×10^2	1.3×10^1
6	Rhode River	Apr. 1971	8–9	6.3×10^5	3.3×10^5	6.3×10^3	0.0	6.4×10^2	2.0×10^1
7	Rhode River	Apr. 1971	8–9	3.9×10^5	4.4×10^5	4.4×10^4	0.0	4.4×10^2	4.4×10^1

* TVC: Total viable counts.
YE: Yeast extract, peptone medium prepared with fresh water.
SWYE: Yeast extract, peptone medium prepared with simulated seawater.
VLO: *Vibrio* sp.
VPLO: *Vibrio parahaemolyticus.*

water column in winter. Sediment counts of *V. parahaemolyticus* dropped much less dramatically, staying relatively high during winter months (Fig. 10). Thus, it was concluded that *V. parahaemolyticus* survives the winter in sediment, which serves as a reservoir for the organism. In addition, an association of *V. parahaemolyticus* with the zooplankton in Chesapeake Bay was discovered (Table 5). The population cycles of *V. parahaemolyticus* thus appeared to be both temperature- and "host"-dependent *(16)*.

Total viable counts of bacteria, whether from animals, water, or sediment, provide only one facet of the information needed to evolve a theory of microbe-microbe and microbe-environment interactions. It is clearly important to determine the kinds, as well as the total numbers, of bacteria. Ten years ago, samples of oysters collected from several areas in Puget Sound for bacteriological studies were examined for total viable counts *(9)*. The total viable counts were found to be in the range from 10^3 to 10^5 per ml mantle fluid or per g of homogenized tissue (Fig. 11). Little variation in total viable counts of bacteria was observed for oysters collected from several areas. *Pseudomonas, Vibrio, Achromobacter,* and *Flavobacterium* were the most common species. *Micrococcus* and *Bacillus* species were a minor component. *Pseudomonas* and *Vibrio* were the dominant organisms *(10)* (Table 6).

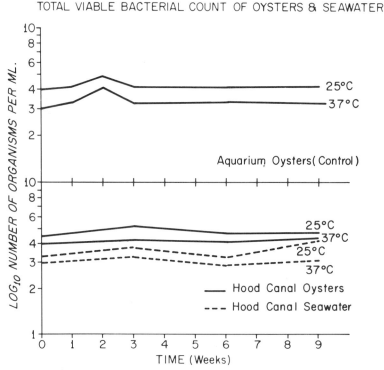

Fig. 11. **Total viable counts of oysters and seawater in Hood Canal, Washington, and in the University of Washington aquarium (Reprinted from Colwell and Liston *(9)* by permission of the Williams and Wilkins Company).**

Table 6. Generic distribution of organisms from oysters in controlled and natural environments[a]

| | Source of Oyster | | | | | |
Organism	Saltwater Aquarium	Hood Canal	Willapa Bay	Oyster Bay	Total	Percent
Pseudomonas-Vibrio	20	27	18	14	79	52.0
Achromobacter	2	5	1	0	8	5.3
Flavobacterium	6	7	5	8	26	17.1
Corynbacterium	1	3	0	1	5	3.3
Alcaligenes	2	1	0	0	3	2.0
Micrococcus	6	5	2	3	16	10.5
Bacillus	4	1	2	0	7	4.6
Enterococci	1	0	1	0	2	1.3
Miscellaneous	1	1	1	3	6	3.9
Sample Total	43	50	30	29	152	100

[a] Reprinted from Colwell and Liston *(9)* by permission of the Williams and Wilkins Company.

Bacteria associated with invertebrates of Chesapeake Bay have been similarly enumerated and identified *(20)* (Table 7). The total numbers of bacteria associated with oysters of Chesapeake Bay were found to be the same as in Puget Sound, *ca* 10^3 to 10^5 per ml fluid or per g (wet weight) gill tissue, with counts in sediment of 10^4 to 10^5 per gram and *ca* 10^2/ml in the water column. Thus, it can be concluded that the oyster, under normal conditions in the natural habitat, harbors a commensal bacterial flora and maintains a relatively constant total viable, aerobic, heterotrophic bacterial population. Although the viable populations varied very little in total numbers, the kinds of bacteria isolated from animals harvested from two of the areas in Chesapeake Bay differed significantly. *Vibrio* species were dominant in the Marumsco Bar samples (Table 4). In Eastern Bay, a variety of bacteria were found, with *Pseudomonas, Vibrio,* and *Achromobacter* species comprising a major part of the flora. *Vibrio* species isolated from Marumsco Bar samples, in the location where severe mortalities have occurred over

Table 7. Total viable aerobic heterotrophic bacterial populations from water, mud and animals of Marumsco Bar and Eastern Bay (Colwell, Lovelace, and Tubiash, unpublished data)

Location	Date	Water	Mud	Gills	Mantle Fluids
Marumsco Bar	8–15–66	4×10^2	8.2×10^4	5.2×10^4	5.0×10^4
Marumsco Bar	8–29–66	4.5×10^2	1.0×10^5	3.0×10^4	5.0×10^4
Marumsco Bar	9–19–66	7.3×10^2	6.1×10^5	7.0×10^3	3.5×10^4
Marumsco Bar	10–13–66	2.5×10^2	7.0×10^4	6.1×10^4	4.8×10^4
Marumsco Bar	3–21–67	6×10^2	9×10^4	2.5×10^3	3.5×10^4
Eastern Bay	11–7–66	4.0×10^2	9.0×10^5	8.0×10^4	2.6×10^5
Eastern Bay	1–17–67	1.1×10^2	8.8×10^5	1.5×10^3	9.2×10^4
Eastern Bay	3–2–67	4.6×10^2	4.2×10^5	4.3×10^4	9.2×10^4

the past several years, were capable of destroying oyster larvae.* The larvae, when infected with *Vibrio* isolated from Marumsco Bar, were killed and subsequently liquefied by the proteolytic enzymes of the bacteria. With a capability of infecting and killing oyster larvae, these bacteria must affect significantly the life cycles of the animals by contributing to the mortality and loss of commercial harvests of oysters.

Electron Microscopy. The major groups of bacteria isolated from Chesapeake Bay samples of animals, water, and sediment have also been studied by electron microscopy. An ultrathin section of a strain of *V. parahaemolyticus* is shown in Figure 12. The organism is gram-negative and rod-shaped. Like many vibrios, it produces round bodies in later growth stages *(17)*. During the course of examination of marine and estuarine vibrios, *V. marinus,* a psychrophile, was found to have an unusual mode of cell division. The organism does not exhibit concomitant invagination of the wall and the membrane *(14)* (Fig. 13).

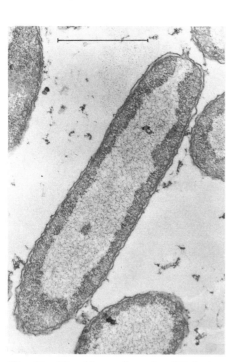

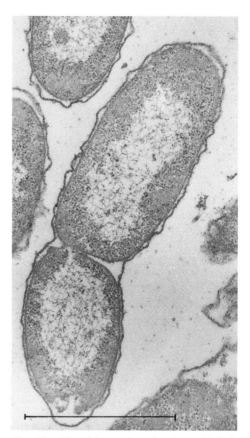

Fig. 12. **Ultrathin section of a *Vibrio parahaemolyticus* strain. (Fig. 12–24, Colwell and Chapman, unpublished).**

Fig. 13. **Ultrathin section showing cell division in *Vibrio marinus* (17).**

* Tubiash and Colwell, unpublished data.

The organism shown in Figure 14 is typical of the *Pseudomonas* species found in Chesapeake Bay. It is a short, gram-negative rod with an irregular outer cell envelope. Another *Pseudomonas* species, shown in Figure 15, demonstrates less irregularity of the cell envelope.

The *Achromobacter* species shown in Figure 16 is much larger than *Pseudomonas*. The *Achromobacter* and *Acinetobacter* species commonly encountered in the estuary are gram-negative, rod-shaped organisms. Cell division of an unusual *Achromobacter* isolated from Chesapeake Bay, shown in Figure 17, is more like gram-positive bacteria in its fine structure, but the organism is a gram-negative bacterium in its taxonomic features and cell chemistry.

Cytophaga species are among the commonly isolated bacteria in Chesapeake Bay, notably in those areas relatively free from pollution. The *Cytophaga* are morphologically distinct. An ultrathin section of a *Cytophaga* isolate (Fig. 18) reveals many mesosomes, or peripheral bodies. The *Cytophaga* are motile by means of a gliding motility that cannot be explained by presence of external appendages. However, an external capsulelike polysaccharide material has been found to be associated with cell walls of *Cytophaga*.

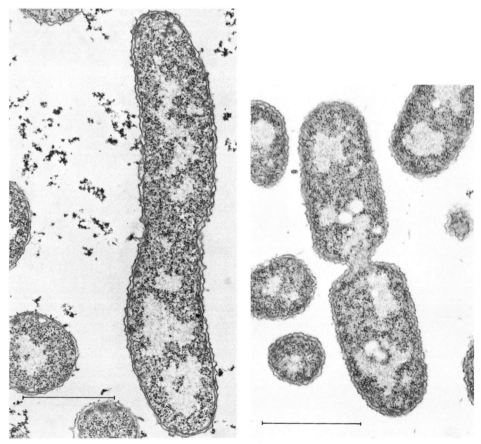

Fig. 14. **Pseudomonas** species.   Fig. 15. **Pseudomonas** species.

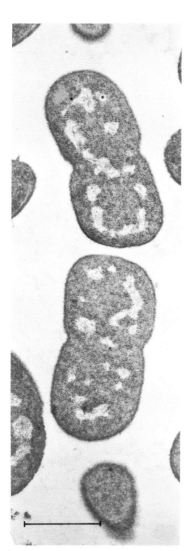

Fig. 16. *Achromobacter* species.

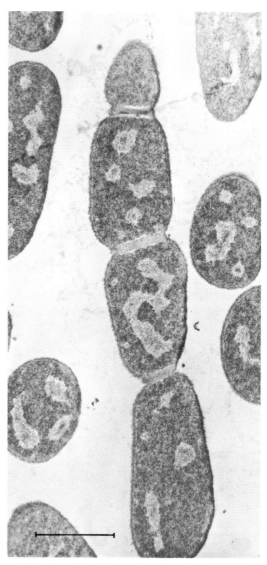

Fig. 17. Ultrathin section showing cell division by an *Achromobacter* species.

Members of the genus *Spirillum,* the spiral-shaped bacteria, are common inhabitants of the estuary. An ultrathin section of a *Spirillum* isolate is shown in Figure 19. The micrograph demonstrates the outer cell envelope and the plasma membrane.

A less significant fraction of the bacterial populations of Chesapeake Bay, in terms of total numbers, are the micrococci. *Micrococcus* species are not dominant in the water column in the deep sea or in the estuary. However, *Micrococcus* can be isolated from sediment or in parts of an estuary where eel grass and other aquatic flora are present. Species of *Micrococcus* isolated from Chesapeake Bay samples have not been distinctive, in comparison with terrestrial forms, and the micrococci usually grow well on media prepared with fresh water (Fig. 20).

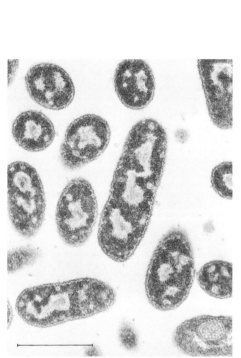

Fig. 18. *Cytophaga* species.

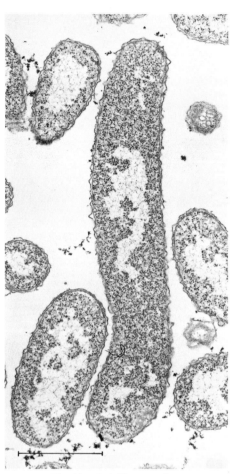

Fig. 19. *Spirillum* species.

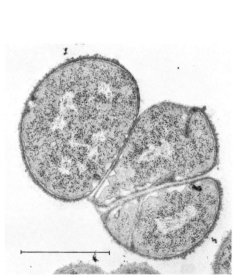

Fig. 20. *Micrococcus* species.

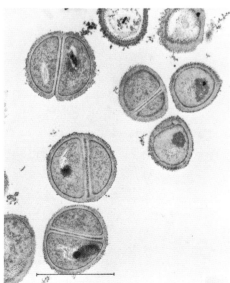

Fig. 21. Ultrathin section showing cell division by a *Micrococcus* species.

Cell division of the estuarine micrococci was found, in the few cases examined, to be much like that of terrestrial species, i.e., in two planes, with one plane of division perpendicular to the other (Fig. 21).

There are, as would be expected, a variety of unidentified organisms that can be isolated from estuarine regions. One of these, most probably a *Rhizobium* species, is shown in Figures 22 and 23. *Rhizobium* and *Azotobacter* may well prove

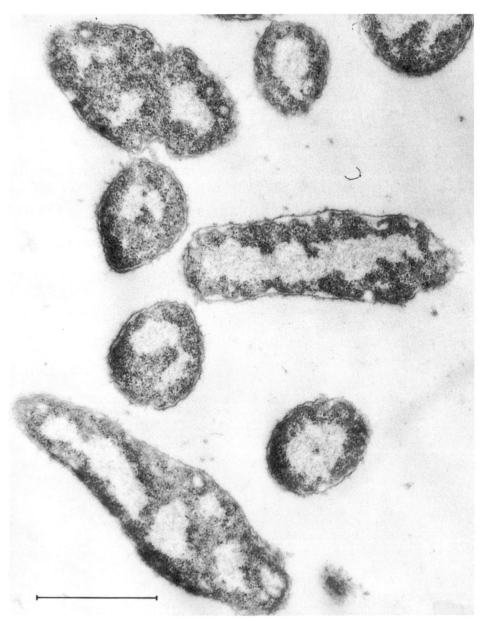

*Fig. 22. **Rhizobium** species.*

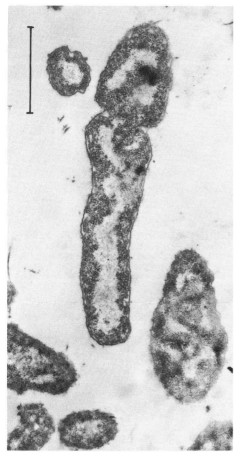

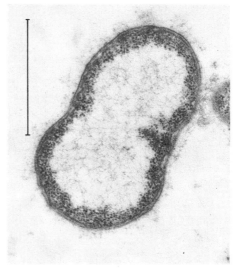

Fig. 23. Rhizobium **species.**   *Fig. 24.* **Unidentified bacterium.**

to be significant in the microbial ecology of the estuary. However, very little data exist on these genera in relation to estuarine microbial ecology.

Bacteria exhibiting odd morphological forms were also encountered. A "pear-shaped" bacterium isolated from Chesapeake Bay is shown in Figure 24. It is an unidentified gram-negative bacterium.

Conclusion

The application of numerical taxonomy to microbial ecology will prove to have enormous value in elucidating interactions of microbial populations. Clearly, the practical use of this technique in analyzing the Chesapeake Bay microflora has been demonstrated. From our studies, it is logical to predict that by determining the kinds of bacteria present and their enzymatic capabilities and correlating this information with physicochemical data, subtle changes occurring in the estuary will be detected quickly, before major changes ensue. Too frequently, the latter are easily recognized by the unaided senses of sight and smell, and are often not easily remedied or, in some instances, are irrevocable. Subtle compositional shifts in the bacterial flora may well prove to be a very sensitive biological indicator of environmental change. By incorporation of microbial ecology data in computer simulations of estuarine ecology, the predictive capacity of the models can be expected to be significantly increased.

Acknowledgments

The author gratefully acknowledges the excellent technical assistance of Miss T. E. Lovelace and Mrs. L. Wan. Dr. R. V. Citarella and Mr. T. Kaneko also contributed to the research. Dr. G. B. Chapman and Mrs. S. Zane were most helpful with the electron microscopy.

The work was, in part, supported by grants from the Bureau of Commercial Fisheries, Biological Laboratory, Oxford, Md. (National Marine Fisheries Service, National Oceanic and Atmospheric Administration), National Science Foundation Grant No. GB 18274, Sea Grant Project GH-91, and Contract N00014-69-A0220-0006 between Georgetown University and the Office of Naval Research.

Literature Cited

1. Anderson, R. S., and E. J. Ordal. 1972. Deoxyribonucleic acid relationships among marine vibrios. *J. Bacteriol.* **109**:696–706.
2. Baumann, P., L. Baumann, and M. Mandel. 1971. Taxonomy of marine bacteria: The genus *Beneckea. J. Bacteriol.* **107**:268–94.
3. Brenner, D. J., G. R. Fanning, K. E. Johnson, R. V. Citarella, and S. Falkow. 1969. Polynucleotide sequence relationships among members of Enterobacteriaceae. *J. Bacteriol.* **98**:637–50.
4. Britten, R. J., and D. E. Kohne. 1968. Repeated sequences in DNA. *Science* **161**:529–40.
5. Citarella, R. V., and R. R. Colwell. 1970. Polyphasic taxonomy of the genus *Vibrio:* Polynucleotide sequence relationships among selected *Vibrio* species. *J. Bacteriol.* **104**:434–42.
6. Colwell, R. R. 1964. A study of features used in the diagnosis of *Pseudomonas aeruginosa. J. Gen. Microbiol.* **37**:181.
7. Colwell, R. R. 1970. Polyphasic taxonomy of the genus *Vibrio.* Numerial taxonomy of *Vibrio cholerae* and *Vibrio parahaemolyticus. J. Bacteriol.* **104**:410–33.
8. Colwell, R. R., V. I. Adeyemo, and H. H. Kirtland. 1968. Esterases and DNA base composition analysis of *Vibrio cholerae* and related vibrios. *J. Appl. Bacteriol.* **31**:323–35.
9. Colwell, R. R., and J. Liston. 1960. Microbiology of shellfish. Bacteriological study of the natural flora of Pacific oysters *(Crassostrea gigas). Appl. Microbiol.* **8**:104–109.
10. Colwell, R. R., and J. Liston. 1959. A bacteriological study of the natural flora of Pacific oysters *(Crassostrea gigas)* when transplanted to various areas in Washington. *Proc. Nat. Shellfish. Assoc.* **50**:181–188.
11. Colwell, R. R., and J. Liston. 1961. Taxonomic relationships among the pseudomonads. *J. Bacteriol.* **82**:1–14.
12. Colwell, R. R., M. L. Moffett, and M. D. Sutton. 1968. Computer analysis of relationships among phytopathogenic bacteria. *Phytopathology* **58**:1207.
13. Colwell, R. R., and W. J. Wiebe. 1970. "Core" characteristics for use in classifying aerobic, heterotrophic bacteria by numerical taxonomy. *Bulletin Georgia Acad. Sci.* **28**:165–85.

14. Felter, R. A., R. R. Colwell, and G. B. Chapman. 1969. Morphology and round body formation in *Vibrio marinus*. *J. Bacteriol*. **99:**326–35.
15. Hogan, M. A., and R. R. Colwell. 1968. DNA base composition and esterase patterns of bacteria isolated from deep-sea sediments. *J. Appl. Bacteriol*. **32:**103–11.
16. Kaneko, T., and R. R. Colwell. 1972. Ecology of *Vibrio parahaemolyticus* and related organisms in Chesapeake Bay. *J. Bacteriol*. (in press).
17. Kennedy, S. F., R. R. Colwell, and G. B. Chapman. 1970. Ultrastructure of a psychrophilic marine vibrio. *Can. J. Microbiol*. **16:**1027–31.
18. Liston, J., W. Wiebe, and R. R. Colwell. 1963. Quantitative approach to the study of bacterial species. *J. Bacteriol*. **85:**1061.
19. Lockhart, W. R., and J. Liston (eds.). 1970. *Methods for numerical taxonomy*. Bethesda, Md.: Amer. Soc. Microbiol. Publ.
20. Lovelace, T. E., H. Tubiash, and R. R. Colwell. 1968. Quantitative and qualitative commensal bacterial flora of *Crassostrea virginica* in Chesapeake Bay. *Proc. Nat. Shellfish. Assoc*. **58:**82.
21. Oliver, J. D., and R. R. Colwell. 1972. A computer program designed to follow seasonal fluctuations in bacterial populations. *Bacteriol. Proc*.
22. Sokal, R. R., and P. H. A. Sneath. 1963. *Principles of numerical taxonomy*. San Francisco. W. H. Freeman and Co.

Comments

ZOBELL: How many characteristics of these bacteria must you tell to the computer before it can tell you its taxonomical position?

COWELL: One can use as few as 60 characteristics, but it is preferable to have 100 to 150 coded characteristics, including morphological, physiological, and biochemical features.

ZOBELL: Are most of these cultural and biochemical characteristics?

COWELL: We found out very early in our work that not just morphology or physiology but a balance of morphology, physiology, biochemistry, serology, etc., must be provided.

ZOBELL: When are we going to start using biochemical characteristics rather than morphology, physiology, and biochemistry? The guanine-cytosine (GC) determination is one step in that direction.

COLWELL: "Base line work" has had to be done in numerical taxonomy in order, first, to describe microbial groups and, then, to determine correlative characteristics for efficient and rapid identification of bacteria. In other words, what are sought are those linked genetic characteristics present with rather high frequency in a given species. These are identifying criteria. To use only a dehydrogenase reaction or any other single characteristic simply will not work. For example, *Serratia marcescens* is a well-known bacterial species which produces prodigiosin, a characteristic, brick-red pigment with an identifiable absorption spectrum. It is known that *Serratia marcescens* will grow at 37 C and possesses the Embden-Meyerhoff pathway for carbohydrate metabolism. It is neither seawater-requiring nor psychrophilic. Yet a microorganism has been isolated which produces prodigiosin; the characteristic brick-red pigment was identified chemically. This microorganism is gram-negative and possesses a GC ratio of *ca* 31%. *Serratia marcescens* possesses a GC ratio of *ca* 58%. The new isolate is obligately psychrophilic, does not grow above 18 C and requires seawater for growth. So far as can be determined, the Embden-Meyerhoff pathway simply does not function in the marine strain. Hence, if one relies on a single characteristic, the identification will be erroneous. A plexus of several characteristics, at the minimum, is required to achieve a reliable and reproducible identification. Also, one need not be impeded by the requirement for 60 to 100 characteristics. A lot of information can be gained simply by a few manipulations, such as streaking on a variety of media, examination under the microscope, and test reactions observed in a half a dozen special media.

QUESTION: Of the 60 characteristics, how dependent are they in terms of culturing conditions, pH, temperature, and redox potential? Are these tests fairly independent of the microenvironment in the actual culturing conditions?

COLWELL: A rigorously standardized procedure is followed. In general, however, results at 25 C compare well with results at 10 C for most species. In the main, 90 to 95% of the enzymes function at lower temperatures (for most marine and estuarine bacteria); hence, the same test results are obtained at both temperatures.

Furthermore, experiments have been done whereby ten bacterial species were subcultured into ten different media, as diverse as brain-heart infusion broth and seawater without added nutrient. The ten species were subcultured several times in ten different media, so that ten "substrains" for each of ten original strains were obtained. Numerical taxonomy of the entire set of 100 strains was then done to see how much variation was induced. There was better than 90% similarity within species. Fluctuation was very low.

PAYNE: Dr. Colwell might like to mention that she published all these procedures in the Bulletin for the Georgia Academy of Science.

COLWELL: Yes, Dr. Wiebe and I put together a standardized format for characterizing and identifying the aerobic, heterotrophic, marine bacteria. This is available, along with details for isolating DNA for base composition determinations.*

OLIVER: In defense of the graduate students, I would like to say that we do not run 60 tests; we run close to 300 tests per organism.

COLWELL: That is correct. Actually, 60 is a minimum number of characteristics to be determined for each strain. The maximum to date is 350. On the average, 200 characteristics are scored for computer analysis.

QUESTION: How independent are they though? Can you not run correlations? Can you not relate some of these tests to show that they are related?

COLWELL: We have studied that point and have eliminated some tests. However, we are referring here to tests such as lysine decarboxylase, lecithinase, lipase, etc. Nevertheless, some tests may be correlated in that the same enzyme is being examined. The testing schedule is not rigid. As new information is obtained, the tests and test methods will, naturally, be changed.

In the early phases of the work, we did include nearly every test published for bacteria. Some rather poor tests, such as growth in litmus milk and on potato, were used, but these are not at all satisfactory. The nitrate reduction test, for instance, as presently done, will reveal that half of the bacteria tested are positive, i.e., half of the strains will reduce nitrate and half will not. Hence, in this case it is nondiscriminatory. Eventually, we hope to eliminate those tests that are useless in terms of separating groups or are too difficult to interpret correctly.

*Colwell and Wiebe, 13.

The Application of Scanning Electron Microscopy to Estuarine Microbial Research*

R. L. Todd,[1] W. J. Humphreys,[2] and E. P. Odum[3]

The scanning electron microscope, although extensively employed in metallurgy, structural composition, and other diverse industrial investigations for the past fifteen years, has only recently found application in biological research. In microbiology its application has been limited to studies of the surface structures of actinomycetes *(13)*, bacterial spores *(1)*, protozoa *(12)*, and streptomyces *(2)*. The scanning electron microscope has been used to investigate the phenomena of antibiotic-induced alterations in the morphology of bacterial cells *(9, 10)*. Axenic cultures of several bacterial species have also been examined by this procedure *(3, 6, 8)*. Reports of *in situ* observation of microbes utilizing the scanning electron microscope has been limited to two publications *(4, 5)*. These investigators were able to demonstrate the presence of fungal filaments and bacterial cells in a natural soil matrix.

This report is a discussion of the sensitivity of the scanning electron microscope in the detection of microorganisms from the estuarine environment. The future application of this technique for research in this area would allow for *in situ* observation with the minimum distortion of organism and environment. Such an application of this procedure may prove superior to existing methods, such as bright-field, dark-field, or phase-contrast microscopy, for such observations.

Characteristics of the Scanning Electron Microscope

Scanning electron microscopy requires a beam of electrons with an extremely narrow diameter which is focused onto the specimen and scanned across it in a small rectangular raster. Individual points on the surface of the specimen, when struck by the electron beam, emit secondary electrons in quantities characteristic for the individual points. The secondary electrons from each individual point are then captured, amplified, and displayed as a spot on the face of a cathode-ray tube that is being scanned in exact synchrony with the specimen. The brightness of each spot so formed on the tube corresponds to the quantity of secondary electrons emitted from a corresponding spot on the specimen.

* Contribution No. 36 from the Eastern Deciduous Forest Biome, US-IBP.
[1] Research Associate, Institute of Ecology.
[2] Director, Electron Microscopy Laboratory.
[3] Director, Institute of Ecology.

For each point on the scanned specimen there is a corresponding point on the face of the scanned cathode-ray tube. Thus, the image is built up point by point on the cathode-ray tube from small spots of varying brightness. Variations in brightness cause the image to have contrast of a continuous-tone type and the image appears to have a three-dimensional aspect much the same as images seen on television screens. Magnification is simply a ratio of the width of the image displayed on the cathode-ray tube to the width of the area scanned on the specimen. The image is exhibited on a second cathode-ray display unit as well. The first unit has a long persistence phosphor for visual examination while the second cathode-ray tube has an extremely short persistence phosphor which is more suited for photography. The image can be videotaped or directly linked to a suitable computer process unit (see Figs. 1 and 2).

Scanning electron microscopy does not require any special sectioning or replication of the specimen as does transmission electron microscopy. A specimen to be examined can be up to 12 mm in diameter and 4 to 5 mm in height. The specimen is affixed to a "stub" or specimen holder (see Fig. 3), which is then coated with a layer of a heavy metal alloy (about 20 nm thick) by means of a vacuum evaporation technique to minimize "charging" effects. The stage

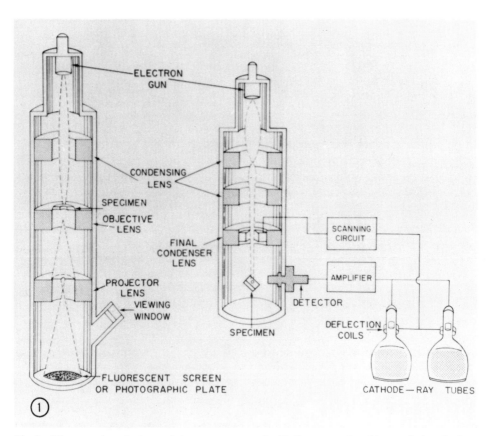

Fig. 1. **The scanning electron microscope compared with the conventional transmission electron microscope.**

Fig. 2. Cambridge stereoscan electron microscope, model Mark IIA.

Fig. 3. Specimen holders or "stubs" used in the Cambridge Stereoscan electron microscope: a and b are general views of the stub; c is a sediment sample affixed to the stub by double-adhesive cellulose tape; d is a similar sample metal coated with a gold-palladium alloy for viewing in the scanning electron microscope; e is a section of membrane filter mounted by double-adhesive cellulose tape; and f is a similar sample that has been metal-coated.

can be tilted so that the electron beam will strike it at an angle of 0 to 90°. An additional advantage is in the fact that the stage can be rotated up to 360° as well as manipulated in three planes.

The scanning electron microscope has a range of magnification from 20 to 50,000 diameters. The maximum resolution is theoretically 10 nm and a resolution of 25 nm can be routinely obtained. The depth of focus obtained is at least 300 times that of conventional light microscopy. A more detailed discussion of the characteristics and applications of the scanning electron microscope can be found in the article by Kimoto and Russ *(7)*.

Operation of the Scanning Electron Microscope

A Cambridge Stereoscan Electron Microscope (Mark IIA, Cambridge Instruments Company, Ltd., London, England) was utilized for the studies summarized in this report. Specimens were examined and photographed at an accelerating voltage of 10 KV. The specimen platform was adjusted so as to allow the electron beam to strike the stage at an angle of 30 to 45°. Photographs were obtained with Type 55 P/N Polaroid sheet film. (Polaroid Corporation, Cambridge, Mass.).

Observations with the Scanning Electron Microscope

Odum and de la Cruz *(11)* published a study on the composition and origin of detritus in an estuary dominated by *Spartina alterniflora* at the University of Georgia Marine Institute at Sapelo Island. This investigation emphasized the importance of microbial colonization of detritus in the estuarine food chain. Organic detritus from this salt-marsh estuarine ecosystem was further examined with the aid of the scanning electron microscope.

Suspended particles contained in a mid-ebb tidewater sample were divided into the following size classes. A "fine" fraction referred to material which passed a 74-mesh but was retained by a 200-mesh screen, and a "nanno" fraction passed through the 200-mesh screen but was retained by a 0.45 μm membrane filter *(11)*. These samples were allowed to air-dry on the membrane filter or wire screen. Specimen stubs (Fig. 3), to which was affixed double adhesive cellulose tape, were gently pressed against the surface on the dried samples. The specimen holder was then coated with a gold-palladium alloy, to an approximate thickness of 20 to 40 nm, by means of vacuum evaporation. The specimen

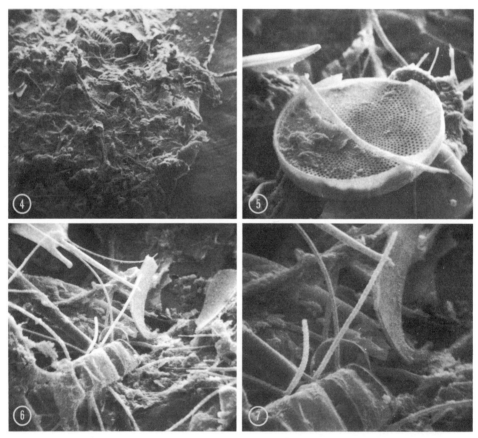

Fig. 4. "Fine" (74–200 mesh size) estuarine detritus fraction.

Fig. 6. Scanning electron micrograph of a "fine" estuarine detritus fraction (220 x). Note the phytoplankton residue overlaid with plant residue.

Fig. 5. Scanning electron micrograph of a diatom species *Coscinodiscus* observed in a "fine" estuarine detritus fraction (220 x).

Fig. 7. Scanning electron micrograph of a "fine" estuarine detritus fraction (440 x).

was placed in the scanning chamber of the microscope and Figures 4–11 illustrates the results observed.

A micrograph of a particle from the "fine" detritus fraction is shown in Figure 4. A marine diatom *Coscinodiscus* was observed on the surface of this fraction (Fig. 5). The scanning electron microscope allows an excellent observation of the topography of a specimen. Figure 6 and Figure 7 illustrate this feature of a "fine" detritus particle.

The frequency at which diatoms were observed in the scanning electron fields increased as the size of the fraction decreased. Figure 8 is a typical field of a "nanno" detritus fraction. Numerous intact diatoms, such as several *Nitzschia* species, were noted in these fractions (Fig. 9). In addition, numerous unidentified diatom particles were found (Figs. 10 and 11).

Estuarine water samples were collected at the Sapelo Island Marine Institute

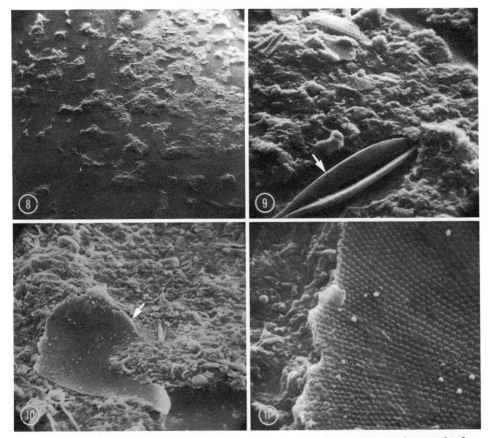

Fig. 8. Scanning electron micrograph of a "nanno" (< 200 mesh size) estuarine detritus fraction (22 x).

Fig. 9. Scanning electron micrograph of a diatom species tentatively observed in a "nanno" estuarine detritus fraction (220 x).

Fig. 10. Scanning electron micrograph of diatom fractions observed in a "nanno" estuarine detritus fraction (440 x).

Fig. 11. Scanning electron micrograph of an unidentified phytoplankton species (880 x).

and filtered by 0.45 μm membrane filter. The filter was allowed to air-dry. Upon returning to the laboratory, sections of these filters were mounted on specimen stubs, were metal-coated with a gold-palladium alloy, and examined in the scanning electron microscope. Figure 12 illustrates a general view of a section of membrane filter treated in this manner. At higher magnification, many diatoms are distinguishable. Figure 13 is a electron micrograph of several diatom particles; Figure 14, *Nitzschia closterium* and Figure 15, a *Chaetoceros* species. Figure 16 is an unidentified particle formation.

Figures 17, 18, and 19 are representative electron micrographs of filtrates of water samples from a marine model system. After passage of the desired sample through a membrane filter (0.45 μm), it was "fixed" by a 30 min soak in 0.1% glutaraldehydate in 0.1 M cacodylate buffer pH 7.2. This was immediately followed by a 90 min fixation in 1% glutaraldehydate in 0.1 M cacodylate. The filter was then washed for 60 min in fresh cacodylate buffer. After affixing the air-dried specimen to a specimen holder by means of cellulose tape, it was metal-

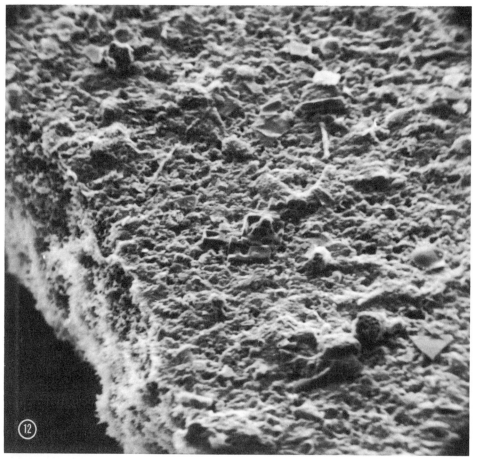

Fig. 12. **Scanning electron micrograph of a section of membrane (0.45 μm) filter with a "load" of estuarine filtrate (220 x).**

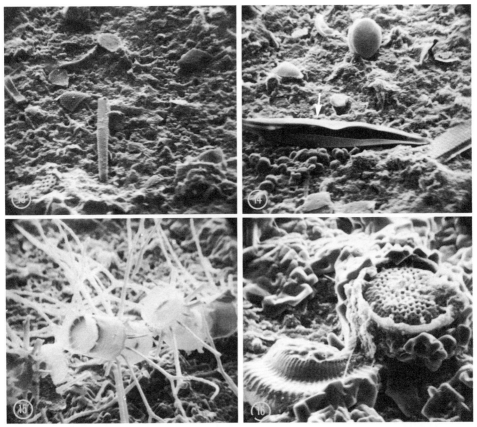

Fig. 13. Scanning electron micrograph of an estuarine filtrate (440 x). Note various unidentified diatom forms.

Fig. 14. Scanning electron micrograph of a diatom species tentatively identified as a *Nitzschia* observed in an estuarine filtrate (440 x).

Fig. 15. Scanning electron micrograph of a diatom tentatively identified as a *Chaetoceros* species observed in an estuarine filtrate (440 x).

Fig. 16. Scanning electron micrograph of an estuarine filtrate (880 x).

coated with a gold alloy and examined in the scanning electron microscope. Figure 17 is a control membrane filter; Figure 18 and Figure 19 are scanning electron micrographs of the surface of membrane filters through which the samples were passed. The arrows in these two latter figures indicate bacterial cells were present in the original sample.

Photomicrographs of an axenic culture of a typical marine bacterium, *Arthrobacter*, comprise Figures 20 and 21. Broth cultures of this bacterium were concentrated by membrane filtation. The filter was then "fixed" by the previously described glutaraldehydate procedure and metal (gold)-coated. Figure 20 is the same culture after 5 hr of additional incubation prior to filtration. The arrows indicate numerous long, twisted, rod-shaped cells which are characteristic of this genus under these growth conditions.

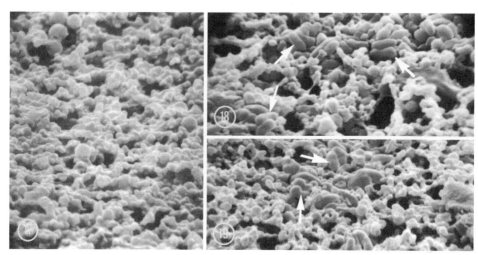

Fig. 17. Scanning electron micrograph *Fig. 18 and 19.* Scanning electron micrograph of of a membrane filter (0.45 μm "pore" aquatic bacteria contained in a 0.45 μm membrane size) (2,570 x). filtrate (Fig. 18, 2,520 x; Fig. 19, 2,320 x).

Through the cooperation of Drs. Baldensperger, Guarraia, Gal, and Humphreys (unpublished observations), Figures 22 and 23 are photomicrographs of a marine bacterium (a sulfate reducer) cultivated on solid media on a membrane filter. Sections of this filter-culture system were mounted, fixed, metal-coated, and examined in the scanning electron microscope.

Summary

Formulation of this report was not only for the purpose of emphasizing a research need in a distinct branch of microbial ecology but also to speculate on the potential applications of scanning electron microscopy as both a comprehensive and a descriptive research tool in ecological investigations in biological science. An application of this procedure to future research should provide very informative results.

In this report, we have summarized some very preliminary observations using the scanning electron microscope to examine organic detritus, phytoplankton, and other marine microflora in axenic and natural cultures. In all incidents, this method of examination has rendered observations of these micro-ecosystems superior to those by any previously employed method. The scanning electron microscope should prove invaluable in future ecological orientation research projects, such as the microbial colonization of natural substrates and their response to "stressed" environmental changes by observing resulting fluxes in the endogeneous microflora.

There are, as with all known methods, disadvantages to using scanning electron microscopy for observing microbes. Adverse effects to be aware of are the potential artifacts resulting from observing a specimen which had to be metal-

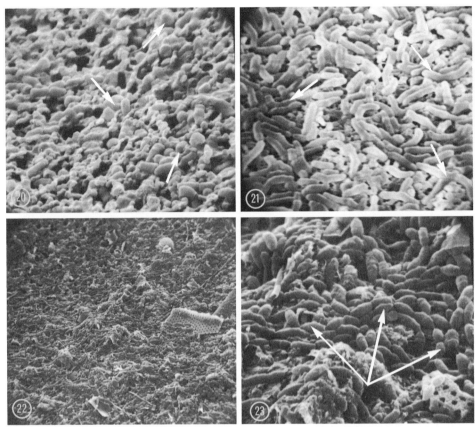

Fig. 20. Scanning electron micrograph of a 0.45 μm membrane filtrate of an axenic culture of a lag-phase culture of an *Arthrobacter* species. Note cellular morphology of short rods. (2,230 x).

Fig. 22 and 23. Scanning electron micrographs of a sulfate-reducing bacterium which has been cultivated on a membrane filter. Note the appearance of long chains of cells.

Fig. 21. Scanning electron micrograph of a log phase *Arthrobacter* culture filtrate. The culture (5 hr older than Fig. 20) displays a characteristic morphology of long, bent rods. (2,200 x).

(Fig. 22, 470 x; Fig. 23, 2,650 x). (Baldensperger, Guarraia, Gal, and Humphreys, unpublished figures).

coated as a prerequisite to examination. The specimen must be subjected to a high vacuum (10^{-4} to 10^{-5} torr) if an electron beam is to be used to illuminate the specimen for observation.

The use of scanning electron microscopy is rapidly becoming a very useful addition to the battery of procedures and techniques at the disposal of the microbial ecologist. Its expanded use will render those *in situ* observations that are so vitally needed to expand our limited knowledge on the "mode of action" of one of our more complex ecosystems—the microbial environment.

Acknowledgments

We thank Robert J. Reimold, Marine Institute, Sapelo Island, Georgia, for his cooperation in the collection of the detritus samples; Ben Spurlock and Janet Johnson, Electron Microscopy Laboratory, for their excellent technical assistance in preparing and examining specimens; R. Eugene Turner, Department of Zoology, for collection of specimens used in Figures 12–16; and L. J. Guarraia for allowing us to include Figures 22–23 in this report.

This investigation was supported by NIEH Training Grant 5 T01 ES00074-05. Additional support for this study was supplied by the Deciduous Forest Biome project, International Biological Program, funded by NSF (subcontract 3477) under Interagency Agreement AG-199, 40-193-69 with Oak Ridge National Laboratory.

Literature Cited

1. Bulla, L. A., G. St. Julian, R. A. Rhodes, and C. W. Hesseltine. 1969. Scanning electron and phase-contrast microscopy of bacterial spores. *Appl. Microbiol.* **18:**490–95.
2. Dietz, A., and J. Mathews. 1969. Scanning electron microscopy of selected members of the *Streptomycetes hygroscopicus* group. *Appl. Microbiol.* **18:**694.
3. Fass, R. J., J. Carleton, C. WatanakunaKorn, A. S. Klainer, and M. Hamburger. 1970. Scanning-beam electron microscopy of cell wall-defective staphylococci. *Infec. Immun.* **2:**504–15.
4. Gray, T. R. G. 1967. Stereoscan electron microscopy of soil microorganisms. *Science* **155:**1668–70.
5. Hagen, C. A., E. J. Hawrylewicz, B. T. Anderson, V. K. Tolkacz, and M. L. Cephus. 1968. Use of the scanning electron microscope for viewing bacteria in soil. *Appl. Microbiol.* **16:**932–34.
6. Kammer, G. M., J. D. Pollack, and A. S. Klainer. 1970. Scanning-beam electron microscopy of *Mycoplasma pneumoniae. J. Bacteriol.* **104:**499–502.
7. Kimoto, S., and J. C. Russ. 1969. The characteristics and applications of the scanning electron microscope. *Amer. Sci.* **57:**112–33.
8. Klainer, A. S., and C. J. Betsch. 1970. Scanning-beam electron microscopy of selected microorganisms. *J. Infec. Dis.* **121:**339–43.
9. Klainer, A. S., and R. L. Perkins. 1970. Antibiotic induced alterations in surface morphology of bacterial cells: A scanning-beam electron microscope study. *J. Infec. Dis.* **122:**323–28.
10. Klainer, A. S., and R. L. Perkins. 1971. Normal and abnormal morphology of microorganisms: A scanning-beam electron microscope study. *J. Amer. Med. Ass.* **215:**1655–57.
11. Odum, E. P., and A. A. De la Cruz. 1967. Particulate organic detritus in a Georgia salt marsh estuarine ecosystem. In *Estuaries*, ed. G. Lauff, pp. 383–88. Washington, D.C.: AAAS.
12. Small, E. B., and D. S. Marszalek. 1969. Scanning electron microscopy of fixed, frozen, and dried protozoa. *Science* **163:**1064–65.
13. Williams, S. T., and F. L. Davis. 1967. Use of a scanning electron microscope for the examination of Actinomycetes. *J. Gen. Microbiol.* **48:**171–77.

Comments

QUESTION: You mentioned that we are looking at dead material with the scope. Is there any way you can distinguish between that which was living before the sample was taken and that which is detritus?

TODD: Since all the specimens are metal-coated prior to examination, it is difficult to distinguish living from dead material. We have, however, examined several uncoated specimens but in all cases the specimen, under the electron beam, will reflect or bounce the electrons off the surface, resulting in a very bright image. Such a phenomenon is referred to as "charging." The purpose of the heavy metal coating is to prevent this situation by draining the primary electrons striking the surface off to ground. It might be possible to observe specimens which have not been coated by the use of a selective filter system, but with present methods it is not possible to make such an observation.

HOLM-HANSEN: Would there be any advantage to using a nuclear pore filter versus a Millipore filter? Millipore looks like it is more porous.

TODD: Since this is the only type of membrane filter I have examined, I would not have any idea. At high magnification the surface structure of the filter does make observation of small particles (such as bacterial cells) difficult.

Section Three
HETEROTROPHIC ACTIVITY

The *raison d'être* of *in situ* Microbial Activity Studies

Lawrence J. Albright and James W. Wentworth

Significant advances in the study of the physiology, biochemistry, and genetics of microbes have been made in the past few years and we are now more aware than ever of the great physiological versatility of marine microbes, especially bacteria. Pure culture studies have made us aware of the *potential* of microbes to grow and modify their physicochemical environment and in turn to be modified by it. However, what a microbe does or is potentially capable of doing within the pure laboratory culture is not necessarily what this cell is doing in nature.

Environmental conditions differ markedly from laboratory conditions because, although we attempt to duplicate the environment within the laboratory, in by far the majority of cases this is not possible. For example, within the laboratory it is very difficult to study the microenvironment although this is an extremely important part of the overall microbial ecosystem. Perhaps a suitable approach or tool could be developed for studying this microniche *in situ*.

Another problem which has concerned microbiologists for a long time is that we suspect we are simply not able to grow and isolate many of the microbes present within the marine environment. We thus study mainly organisms which grow and reproduce readily under defined laboratory conditions. However, are we as microbiologists studying an organism or a process which is important in the marine environment? Should we not go to the environment and attempt *in situ* measurements of various processes within it and try to correlate these with laboratory experiments?

There are a variety of techniques and numerous types of equipment available for attempting to duplicate microbial ecosystems within the laboratory. Several types of equipment are light incubators, polythermostats, and fermentors. However, it is much more difficult at the present time to use laboratory techniques and equipment under *in situ* field conditions. Methodology in most cases has not been developed. Older techniques such as the direct count, plate count, and the use of artificial baits have been used for decades. However, can we not apply many of the techniques developed by the physiologist, the biochemist, and the geneticist to microbial ecology?

Situations which might conceivably be applied to studying the effects of the environment upon microbes under *in situ* conditions are the following:

(a) Radioactive labeling of a relatively stable component of a cell, such as the DNA genome, the cell wall, or the cell membrane, and "tracing" this labeled cell once it has been placed in the estuarine microenvironment. By suitably designing an experiment one could perhaps obtain an estimate of the growth rate of this cell within the microniche knowing that genomes replicate in a semiconservative fashion.

(b) Lysogenic marine bacteria could be "labeled" with a temperate bacteriophage and placed within the microecosystem. The prophage would be passed on to the resulting progeny and this "handle" could perhaps then be used to trace the cells within the environment.

(c) One could, conceivably, abortively transduce a marine bacterium with a bacteriophage and use this "handle" subsequently to trace this one cell, but not its progeny, within the environment.

Hopefully, the use of techniques such as these to label particular cells would allow one to determine what is happening to the cells within the microecosystem. We are, in particular, thinking of factors such as predation and cell division rates as well as immigration and emigration rates within the microniche.

There is a large body of literature derived from laboratory studies on the response of microbial mutants to physical and chemical environments. Could one not remove indigenous microbes from the estuarine milieu, mutate them, and replace them in the environment? Hopefully, by screening sufficient numbers and varieties of mutants, along with their response to the microbial environment, one could obtain an appreciation of the genetic potential and its expression required by microbes to survive and grow in nature.

As opposed to the study of the effects of the environment upon microbes, the study of the effects of microbes on the environment may require somewhat different approaches. That is to say, microbiologists will have to develop more sophisticated means of studying the environment surrounding the microbes. Can one treat an aqueous or sediment system as a biochemist treats an enzyme preparation? That is, can classical enzyme kinetics such as the Michaelis-Menten approach be applied to uptake and respiration of radioactive labeled metabolites by the natural microflora? Data obtained from the Lineweaver-Burk modification of a Michaelis-Menten plot allow determination of factors such as turnover time of a substrate and the maximum velocity of an enzymic reaction. When one applies such a situation to uptake of a common nutrient, such as glucose, by the natural microecosystem, could not one calculate turnover time of the nutrient in question, the maximum velocity of uptake, and even an estimate of the concentration of the nutrient present in the environment under study? Application of this technique briefly mentioned here will be discussed later in this session by Dr. Wright.

Hopefully, we as marine microbiologists will in the future take many of the techniques, ideas, and tools of the physiologist, biochemist, and geneticist and suitably apply them to *in situ* microbial ecosystems. We believe that the data and results will be worth the effort.

Comments

COLWELL: I would like to add to the suggestions that you offered. I believe it would be useful to isolate *Pseudomonas* or *Vibrio* species, for example, from a number of estuaries around the country or the world, and measure the amount of DNA-DNA reassociation among the isolates. We could thereby estimate the amount of evolution or molecular divergence or convergence taking place within the genome of estuarine microorganisms as they function or survive in environments each of the estuaries provide. Each estuary would be very similar, but, of course, each would have its own subtle differences.

ALBRIGHT: I quite agree. The point I am trying to stress is that we have a large body of data and ideas which have built up in other disciplines throughout the last two decades. We should be using these ideas and techniques. It is going to take time to adapt or modify the techniques but they are very powerful, and I believe that they can be quite useful.

Distribution and Activity of Proteolytic Bacteria in Estuarine Sediments[1]

Ronald K. Sizemore, L. Harold Stevenson, and Bruce H. Hebeler

Protein macromolecules are often too large and complex to be utilized directly as nutrients by bacteria. Consequently, the production of extracellular protease by a bacterium would facilitate the utilization of these compounds by that organism; likewise, this activity could also function in general nutrient regeneration.

The protein degrading ability of marine bacteria has been reported and is well known *(10, 13)*. Despite the recorded observations, the ecological significance of the extracellular hydrolysis of proteins in marine or estuarine environments remains largely unexplored. The most extensive work with proteolytic marine bacteria has been concentrated in two areas; namely, the involvement of bacteria in spoilage in sea food *(12)* and the biochemical properties of extracellular proteases produced by selected organisms *(1, 3, 6)*.

In an attempt to understand the ecological implications of the extracellular protein degradation promoted by estuarine bacteria, our laboratory is studying the distribution and activity of proteolytic bacteria in the estuarine environment as well as the effects of various parameters on protein hydrolysis by bacteria. The objectives of the research reported in this communication are to determine (1) the population levels of proteolytic bacteria in estuarine samples, (2) the location of such bacteria in the environment, (3) whether facultative anaerobes are important protease producers, and (4) if cell-free proteolytic activity can be detected in estuarine samples.

General Methods

Study Area. The area of study was located in the North Inlet estuary near Georgetown, South Carolina. The estuary is a semitropical salt marsh with little freshwater inflow. The salinity is normally greater than $30\%_{00}$. Sediment samples were obtained from an intertidal mud flat on the inland side of Debidue Island at low tide. Water samples were collected in Debidue Creek. For a complete description of the area, refer to Vernberg *(11)* in this volume.

Sampling, Isolation, and Cultivation. ZoBell's marine agar 2216 (Difco Laboratories) was the substrate employed for the isolation, cultivation, and storage of

[1] Contribution No. 61 of the Belle W. Baruch Coastal Research Institute.

133

the bacteria. All cultures were incubated at 25 C. Water samples were collected with a modified Johnson-ZoBell sampler,* and sediment cores were obtained with 1.5 cm sterile glass cylinders. After collection, both water and sediment samples were stored in an ice chest and processed within one hour. Viable counts were determined by first diluting the specimens with four-salts solution (8) and then plating, using the spread-plate technique. Cultures were screened for protein degrading ability by spot inoculating pure cultures of the isolates onto casein agar double-layer plates (9). Cultures were subjected to anaerobic conditions using the BBL Gaspak anaerobic system (Baltimore Biological Laboratories). The BBL methylene blue anaerobic indicator strip was used to estimate the reduction potential in the anaerobic jars.

Results

Relative Abundance of Proteolytic Bacteria. Laboratory stocks of 204 estuarine bacteria and random isolates picked from spread-plates inoculated with estuarine samples were screened for caseinolytic activity. A comparison of the percentage of proteolytic organisms in each group is shown in Table 1. Of the stock cultures tested, 44% of the sediment organisms and 49% of those obtained from water proved to be proteolytic. The random isolates demonstrated a higher percentage of proteolytic organisms; 56% of those obtained from sediment and 62% of the isolates from water cleared the casein agar plates.

Table 1. Percentage of proteolytic organisms isolated from estuarine samples

Source		Number Tested	Number Proteolytic	Percentage Proteolytic
Laboratory stocks	sediment	93	41	44%
	water	111	54	49%
Random isolates	sediment	117	66	56%
	water	89	55	62%

Variations in the relative abundance of proteolytic bacteria on a seasonal basis were also examined. The same area was sampled over a 9-month period to assay for proteolytic organisms (Fig. 1). A slight decrease in the relative abundance of protein-degrading bacteria was observed in December and June. Similar decreases have been reported in samples obtained from Long Island Sound (7).

Distribution of Proteolytic Bacteria. Sediment cores were examined to determine if a vertical variance in the distribution of proteolytic bacteria could be detected. Three adjacent cores were collected and sectioned into 2 cm increments. One set of subsamples was placed in tared weighing bottles for subsequent determination of dry weight. The second series was used to measure

* Millwood, unpublished.

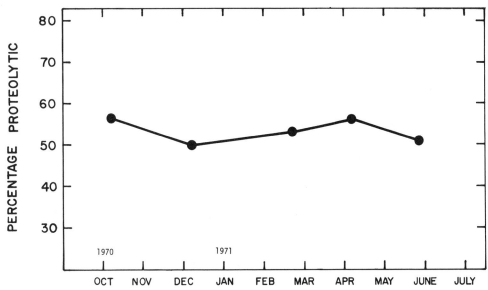

Fig. 1. **Seasonal variation in the percentage of proteolytic organisms taken from a 10 cm sediment core. Each point represents 200 to 1,200 colonies.**

proteolytic activity by an assay technique to be described later in this paper, and the third set was used to determine viable counts. In an attempt to insure a random sample, every colony appearing on the spread-plates was examined for caseinolytic activity. Figure 2 illustrates the number of heterotrophic and caseinolytic bacteria cultured per gram dry weight from each of the core sections. The largest populations were observed in the surface slime layer with a decrease at the deeper depths. The relative percentage of casein-degrading organisms among the total heterotrophic population cultures is shown in Figure 3. Each bar illustrates the mean value obtained from the analysis of six samples obtained on four occasions. The vertical line in each bar represents the standard deviation. A consistently higher percentage of caseinolytic bacteria was found in the intermediate depths. Although the surface slime contained the highest number of organisms, a lower percentage of the strains was proteolytic. The persistent occurrence of a larger proportion of proteolytic organisms in the intermediate depths may suggest a selective advantage in this area.

Facultative Anaerobes. The low oxygen content of salt marsh sediments has been documented by many workers *(12, 13)* and the North Inlet estuary is no exception. Anaerobic conditions were evident in the sampling areas and black layers of sediment were observed 4 to 8 cm below the surface. Consequently, the sediment was tested for organisms capable of surviving and functioning in an area of fluctuating oxygen content. Sediment samples were diluted and plated in duplicate onto marine agar. After 8 hr exposure to air, one set was placed in anaerobic jars and incubated for 7 days. The other set was kept under aerobic conditions for a like period. The colonies appearing on both sets of plates were

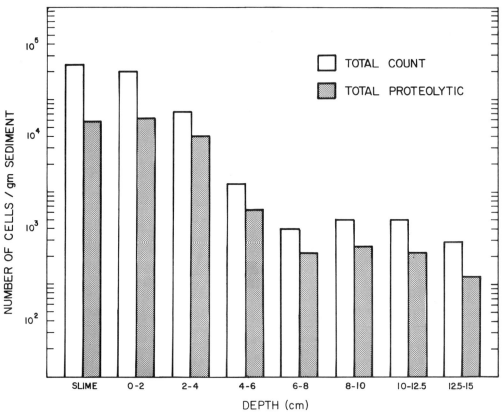

Fig. 2. Comparison between the total number of aerobic heterotrophic organisms and the number of proteolytic microorganisms in the vertical stratum of a sediment core taken on May 20, 1971.

counted and the number of aerobic and facultative organisms per gram of sediment was calculated (Fig. 4). The greatest population of both types was obtained from the surface slime; however, the highest proportions of facultative anaerobes were recovered from the deeper sections. The facultative population was observed to be less than 10% of that detected on the aerobic plates prepared using sediments from the upper 4 cm and greater than 40% in the lower 10 to 15 cm.

Each colony developing on the plates incubated in the BBL jars was transferred to replicate double-layer casein agar plates. One set of assay plates was incubated aerobically and the other set anaerobically. Following eight days of incubation, the plates were examined for zones of clearing around each subculture. Results of two such experiments employing samples obtained from various sediment locations are summarized in Figure 5. In every case, a consistently higher percentage of the facultative organisms hydrolyzed casein when cultured aerobically. It was also noted that the zones of clearing in the protein indicator layer were generally much larger on the aerobic plates than were observed on plates incubated in a reduced oxygen atmosphere.

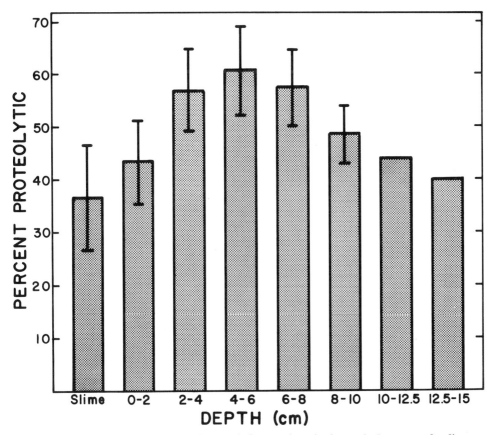

Fig. 3. Variation in the percentage of proteolytic organisms in the vertical stratum of sediment. At least 500 colonies from each sample were tested; thus, each bar represents data obtained from a minimum of 3,000 colonies.

Measurement of **in situ** *Proteolytic Activity.* Attempts were made to measure cell-free proteolytic enzymes in estuarine materials. Unfortunately, the results obtained were inconclusive. Crude enzyme solutions were prepared by diluting one part fresh sediment with two parts 0.1 M phosphate buffer (pH 8). The slurry was thoroughly, but gently, mixed and filtered using a 0.45 μm membrane filter (Nalgene filter unit). Controls were prepared using phosphate buffer in place of the enzyme solution. Initially, assays performed using hemoglobin (1), casein (4), and azocasein (2) as substrates did not detect appreciable proteolytic activity. Occasionally, indications of enzymatic activity were observed, but consistent quantitative measurements could not be obtained. In an attempt to employ a more sensitive assay of proteolysis, the technique of Lin et al. (5), employing dimethylated casein as substrate, was selected and modified. The maximum incubation time was extended from 15 to 16 min and the reaction temperature was increased from 22 to 37 C. Either ninhydrin solution or trinitrobenzenesulfonic acid was used for the colorimetric determination of free amino groups following incubation. High background readings were fre-

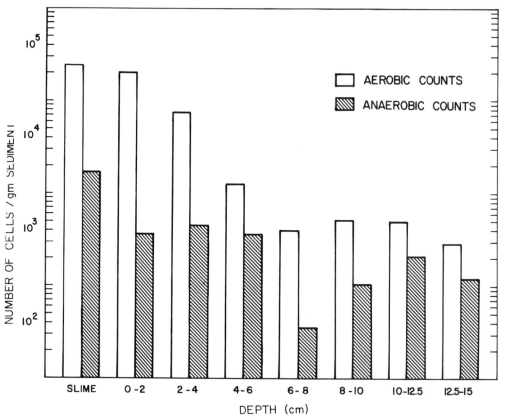

Fig. 4. Comparison between the number of aerobes and facultative anaerobes in vertical sections of a sediment core.

quently obtained in the assay system because of the presence of soluble protein-like material in the sediment.

Figure 6 illustrates an example of data obtained from sediment filtrates when dimethyl casein was employed as the substrate. Although this sample had a high background reading, certain trends were evident. The intermediate samples (2–6 cm) showed consistently higher enzymatic activity than that obtained from surface or deeper samples. This trend was repeated in a majority of the samples tested. Another type of information gained with this technique is illustrated in Figure 7, which shows the activity measured after 30 min incubation. The highest activity detected in this set of samples was obtained from the 2–4 and 10–12.5 cm sections. The low activity in the 6–8 cm section was correlated with an unusually low population of proteolytic bacteria.

Summary and Discussion

Forty-four to sixty-two percent of the bacteria isolated from the North Inlet estuary were capable of extracellular protein degradation. The number of proteolytic bacteria recovered per gram of sediment ranged from 1.2×10^3 to

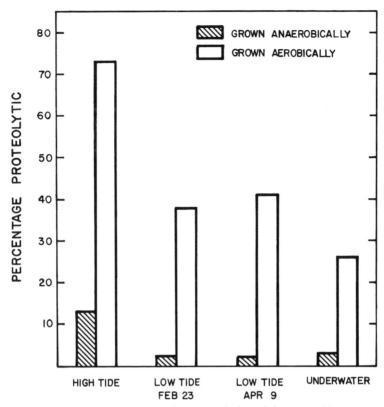

Fig. 5. Percent facultative bacteria showing proteolytic activity anaerobically and aerobically.

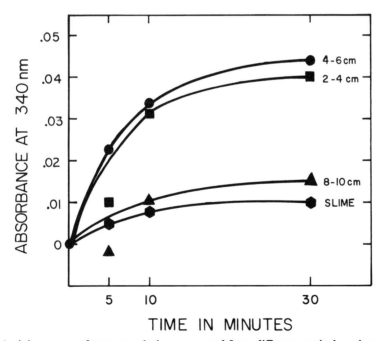

Fig. 6. Activity curves of enzyme solutions prepared from different vertical sections of a sediment core; colorimetric reactions in this experiment were done with trinitrobenzenesulfonic acid.

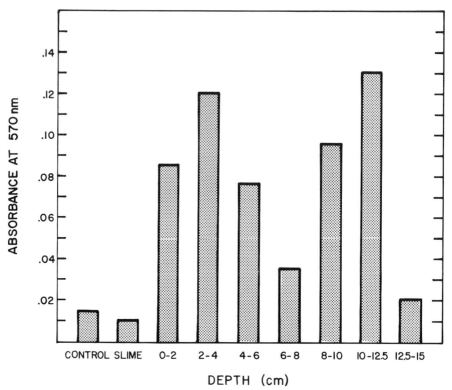

Fig. 7. **Hydrolysis products of proteolysis measured by absorbance after 30 min of enzymatic reaction — enzyme solutions were prepared from different vertical sections of sediment cores; colorimetric reactions in this experiment were made with ninhydrin solution.**

6.5×10^4. Approximately the same number of organisms was found in various localities within the estuary and during different seasons; however, differences in the vertical stratification within the upper 15 cm of sediment were observed. The greatest population of proteolytic bacteria was recovered from the surface slime; however, a consistently higher percentage of proteolytic organisms was obtained from a region 2 to 8 cm below the surface. A large proportion of the facultative anaerobes were capable of protein degradation when cultured aerobically. Lack of oxygen, however, greatly reduced proteolytic activity.

A higher percentage of proteolytic bacteria was observed among organisms isolated from water samples than among those obtained from sediment samples. The production of an exoenzyme by an organism suspended in water would not seem to be a favorable characteristic because of the rapid dilution of the exzyme in the water column. Protease production by these organisms is probably important only in the colonization of suspended protein material. Indeed, it may be that extracellular hydrolytic enzymes produced by bacteria in the sediment are involved only in the microenvironment surrounding the producing cell.

Low levels of cell-free proteolytic activity were detected in the sediment. While we have focused our attention on those bacteria capable of extracellular protein hydrolysis, there is no assurance that the enzymatic activity measured in filtrates prepared from mud slurries represents true extracellular enzymes or even enzymes of bacterial origin. Generally, the greatest activity was correlated with that region of the sediment from which the highest percentage of proteolytic bacteria was obtained. It is hypothesized that the slight proteolytic activity in the lower 10 to 15 cm results from a relatively low number of protease-producing bacteria and, perhaps more importantly, the prevalence of an anaerobic environment. Even though the population of proteolytic bacteria in the upper surface area was high, the low measurable cell-free activity in that region could be due to leaching of the enzymes by tidal action. The intermediate depths appeared to present the optimal combination of abundance of protease-producing organisms, oxygen availability, and lack of tidal leaching, allowing maximum efficiency of extracellular protein hydrolysis.

Acknowledgment

This work was supported in part by a grant from the Belle W. Baruch Foundation.

Literature Cited

1. Burgum, A. A., J. M. Prescott, and R. J. Harvey. 1964. Some characteristics of a proteolytic system from *Phymatotrichum omnivorum. Proc. Soc. Exp. Biol. Med.* **115**:39–43.
2. Charney, J., and R. M. Tomarelli. 1947. A colorimetric method for the determination of the proteolytic activity of duodenal juice. *J. Biol. Chem.* **171**:501–505.
3. Griffin, T. B., and J. M. Prescott. 1970. Some physical characteristics of a proteinase from *Aeromonas proteolytica. J. Biol. Chem.* **245**:1348–56.
4. Kunitz, M. 1947. Crystalline soybean trypsin inhibitor. 11. General properties. *J. Gen. Physiol.* **30**:291–310.
5. Lin, Y., G. E. Means, and R. E. Feeney. 1969. The action of proteolytic enzymes on N, N-dimethyl proteins. Basis for a microassay for proteolytic enzymes. *J. Biol. Chem.* **244**:789–93.
6. Litchfield, C. P., and J. M. Prescott. 1970. Regulation of proteolytic enzyme production of *Aeromonas proteolytica.* 1. Extracellular endopeptidase. *Can. J. Microbiol.* **16**:17–22.
7. Murchelano, R. A., and C. Brown. 1970. Heterotrophic bacteria in Long Island Sound. *Mar. Biol.* **7**:1–6.
8. Quigley, M. M., and R. R. Colwell. 1968. Properties of bacteria isolated from deep-sea sediments. *J. Bacteriol.* **95**:211–20.
9. Sizemore, R. K., and L. H. Stevenson. 1970. Method for the isolation of proteolytic marine bacteria. *App. Microbiol.* **20**:991–92.
10. Skerman, T. M. 1963. Nutritional patterns in marine bacterial populations. In *Symposium on marine microbiology,* ed. C. H. Oppenheimer, pp. 685–98. Springfield: C. C. Thomas, Publisher.
11. Vernberg, F. J. 1972. Introduction. The estuarine ecosystem. In *Belle W. Baruch library in marine science. Vol. 1. Estuarine microbial ecology,* ed., L. H. Stevenson and R. R. Colwell, pp. 3–8. Columbia: Univ. of South Carolina Press.
12. Wood, E. F. J. 1967. *Microbiology of oceans and estuaries. Elsevier oceanography series, Vol. 3.* New York: Elsevier Publishing Co.
13. ZoBell, C. E. 1946. *Marine microbiology.* Waltham: Chronica Botanica Co.

Comments

BUCK: Can you tell us anything about the taxonomy of the proteolytic isolates?

SIZEMORE: We know very little about the taxonomy of these organisms. The information we do have is confined to the laboratory stock cultures of the proteolytic organisms. Of the laboratory stocks, approximately 45% were gram-positive rods with some cocci represented. The others were primarily gram-negative rods.

BUCK: Were they pigmented?

SIZEMORE: A few of the more highly proteolytic bacteria were pigmented, but in general they were not. About 50% were white or off-white. The next largest group, 12%, were yellow pigmented.

ZOBELL: It escaped me what criteria were used for anaerobic growth. I raised this question because you undoubtedly know that the ability of many microorganisms to grow in the absence of oxygen is influenced by the organic substrate, especially the carbohydrates. They are also affected by the presence or absence of nitrate, sometimes sulfate, and by the redox potential.

SIZEMORE: The medium used for the initial isolation of anaerobes was your formulation of Difco Marine Agar 2216. I do not think it has any carbohydrate included. Double-layer plates were used to test for proteolytic activity. Skim-milk agar was overlayed with marine agar in these experiments. Some carbohydrates could be present through diffusion. The only parameter used to indicate anaerobic conditions was the methylene blue indicator supplied with the BBL anaerobic system.

WATSON: What is the pH of these sediments?

SIZEMORE: This was a difficult problem in the assay technique. Actually, I measured pH of the buffered solution before I added it to the substrate. I found that the upper sediment was on the neutral side, about 7, but deeper, the pH increased to about 8.

WATSON: What about as you went down?

SIZEMORE: I did not work any lower than 10 to 15 cm.

WATSON: I would have thought that it would have gone the other way.

SIZEMORE: Yes, I would, too. I had to check the pH before I ran the assay because all of my enzymatic assays were pH-dependent. I found that the upper sediment, at least the slurry I filtered, was usually about 7.5 after filtration. But, this was after mixing a pH buffer with the sediment. With the deeper sediment slurries, the filtrate was about pH 8.

WATSON: Have you ever tried to measure the total *in situ* oxygen consumption by these organisms to determine what role they are playing in the overall decomposition of the material in the sediments?

SIZEMORE: No, sir, but that is an interesting problem.

ANTHONY: I would like to make two points, one a question, and the other a general observation-suggestion. I am curious to know whether you have sufficient information to establish the actual change in distribution with depth with any statistical confidence. I ask, because we were very interested in this with respect to lake sediments for a time. We were very impressed with the early work which seemed to show, like a disinfection curve, organisms declining with depth.

We came to a conclusion that this was not really so. Exposed sediments offer some opportunities for studying the relationship of organisms to depth that are not available to people studying sediments that are completely submerged. I would suggest here that there is a great opportunity for inversion of the sediments. I think it would be useful in any kind of study like this to invert sections of the sediment and sample them after various periods of time. There has always been a question about the migration, in a sense, through the sediment, not only of the living but also of the nonliving components. In fact, at times, I have been inclined to the notion that sediments may be acting a bit like a chromatogram. When you think of things like crusts of manganese deposits, you are almost forced to consider that they are acting this way. I think if you are working with exposed sediments, your sampling procedure should include ones that have been inverted before you started studying. I just make this suggestion as a general observation.

Interactions of Amino Acids and Marine Bacteria[1]

Carol D. Litchfield

In order to understand better the activities and function of microorganisms in the marine environment, it is important that we have some understanding of the cycling of the organic matter which affects the microbial population. Several years ago, while in the Department of Oceanography at Texas A & M University, we initiated studies of heterotrophic activity in the marine environment. We were not so much interested in primary productivity or general utilization, but instead addressed ourselves to the question of the fate of specific organic molecules in seawater and sediments.

Amino acids are of great importance as a source of organic nitrogen for obligate heterotrophic bacteria. Until recently, however, the qualitative and quantitative aspects of amino acid concentration in the sea have not been known. We were especially interested in these compounds in view of reports on the isolation of amino-acid-requiring bacteria from the sea. Later, as we continued our investigations, we became involved in studies on the extracellular enzymes, especially proteases, which many of the marine bacteria produce and which could provide a means for the continual replenishment of amino acids to the marine environment. It is the development, over the last several years, of these ideas and the techniques we have used for the analysis of the amino acid distribution in seawater and sediments, the presence of extracellular proteases in these sediments, and how these factors might interrelate with the obligate heterotrophic microbial population that is the subject of this paper.

Analysis for Amino Acids

Microbiological Assay. One of the first orders of priority was to establish the concentration of the compounds selected for study in the marine environment. Earlier work by Tatsumoto et al. *(42)* and Park et al. *(34)* had demonstrated the presence of amino acids in waters of the Gulf of Mexico. Their procedure involved the use of ferric chloride coprecipitation *(3)* to remove the amino acids from solution. After desalting, the amino acids were quantitated on an amino acid analyzer. Subsequent work *(10,* and Litchfield and Hood, unpublished data) demonstrated that coprecipitation was not a quantitative means for recovering the amino acids. We were already using microbiological assay to test for vitamins, purines, and pyrimidines *(27, 28),* so we developed further auxo-

145

trophic mutants of *Serratia marinorubra* to amino acids, as had been done previously by Belser *(4, 5)*. Indeed, we obtained one mutant for threonine which was very specific in that it could not grow on homocystine, cysteine, serine, methionine, aspartic acid, or glutamic acid. We established the dose-response curve and proceeded to test various samples from the Gulf of Mexico and the bays and estuary at Port Aransas, Texas. Unfortunately, in over 30 samples we failed to find any indication of threonine *(28)*. This was partly due, no doubt, to the rather high levels of threonine required by the auxotrophic mutant (0.5–10 mg/l seawater), and later work has indeed shown this to be one of the less concentrated amino acids when it is present *(10, 12, 23, 1)*. In fact, when aliquots of several of the samples were tested by the dansylation method, they did indeed fail to show any indication of a DNS*-threonine derivative *(30)*.

Obviously, the bioassay system is laborious, slow, and not particularly adapted for shipboard use. The inherent advantage of bioassay — its extreme specificity — is related to its main disadvantage — requirement of rather high concentrations of the amino acid for growth. Should anyone wish to utilize the bioassay procedure, it would seem to me that a valuable improvement in technique would be to adapt the concept, first applied by Gold *(14)* and currently in use by Carlucci and Silbernagel *(9)* for B$_{12}$ determinations, of incorporation of a radioactively labeled substrate. With these bacterial strains, glycerol serves as the preferred carbon source *(4)* and would be best adapted for such uses. In this manner, increases in growth not detectable by spectrophotometric methods could be detected and thus increase the sensitivity by perhaps an order of magnitude.

Amino Acid Derivitization Methods. We still needed information on what amino acids or organic nitrogen sources the obligate heterotrophs were using. We felt that there must be some microbiological significance to the amino acids being reported in the literature *(12, 13, 23, 33, 34, 37, 39, 44)*. Part of our rationale for this attitude was based on two projects we currently had under way in Dr. J. M. Prescott's laboratory in the Department of Biochemistry at Texas A & M University. The first project was a study of the extracellular products of a continuous culture of *Chlorella sorokiniana* and the effects of excreted amino acids on the bacterial flora. The second project concerned an investigation of the metabolic control, in the marine bacterium *Aeromonas proteolytica*, of extracellular endopeptidase synthesis which we had observed to be affected by very low levels of asparagine.

We therefore attempted the development of a rapid screening procedure for amino acids by which we naively reasoned we could determine which amino acids were present. From those data, we hoped to predict whether bacteria were present with requirements for the missing amino acids.

The amino acid content of marine samples is extremely low. It seemed reasonable, therefore, to apply the concepts of lipid and hydrocarbon analysis, i.e., nonpolar substances can be recovered from aqueous systems with nonpolar solvents. Thus, if one could convert the amino acids in any sample to less water-soluble, more nonpolar, compounds, one should then be able to concentrate and

---

* 1-dimethylaminonaphthalene-5-sulfonyl (dansyl).

extract the altered amino acids all at once. We sought to develop a method which would fit the following criteria: (1) the amino acids should be quickly concentrated without interference by the overwhelming amount of salt; (2) the concentration-extraction procedure should not leave any one amino acid in the aqueous phase; (3) the procedure should be rapid; (4) it should require no special equipment so that it could be used or at least initiated in the field; and (5) it should be quantitative. Faced with these requirements, we decided to investigate the newly available amino acid derivitization reagents shown in Figure 1. All three

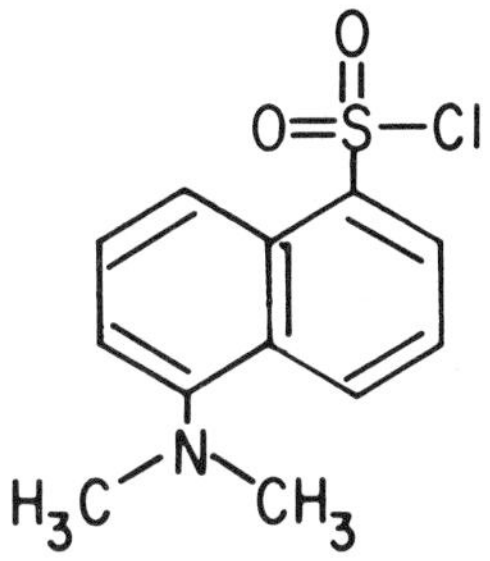

1, DIMETHYLAMINONAPTHALENE-5-SULFONYL CHLORIDE

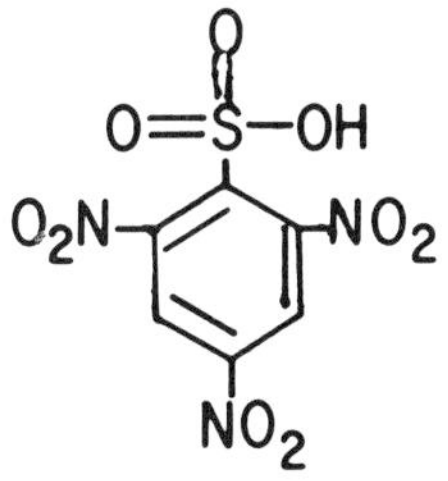

2,4,6, TRINITROBENZENESULFONIC ACID

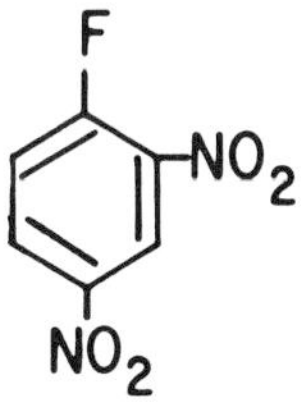

2,4, DINITROFLUOROBENZENE

Fig. 1. Structures of some amino acid derivitization reagents.

reagents are reactive in aqueous systems, produce generally nonpolar derivatives, and are based on the reactivity of the F^-, Cl^-, or OH^- with a H^+ on a nonprotonated amino group.

1. DINITROPHENYLATION   Following Sanger's use of DNFB* *(40)* for N-terminal amino acid analyses, Palmork *(33)* published his method for dinitrophenylation of amino acids in seawater with subsequent analysis by circular paper chromatography. However, there are several disadvantages to this procedure so far as marine samples are concerned: (1) 2–5 mg of the amino acids must

---

*Dinitrofluorobenzene.

be present; (2) the reaction conditions must be carefully controlled at pH 8.9 and 40 C *(35)*; and, (3) most seriously, the reaction results in several water soluble DNP†-amino acids: DNP-arginine, DNP-cysteic acid, DNP-histidine (mono-, di-, and imidazole derivatives), mono-DNP-cysteine, DNP-ornithine, mono-DNP-lysine, and O-DNP-tyrosine *(35)*.

2. TRINITROBENZENE SULFONATE DERIVITIZATION   Another more recently available reagent is TNBS,‡ also used for N-terminal amino acid analyses of peptides *(6, 19, 24, 32, 41)*. Unfortunately, it too requires heating, at 50 C *(19, 24)* for reaction, is very sensitive to hydrolysis by light *(19, 24)*, and the solvent of choice, to date, is *n*-butanol which is rather difficult and slow to evaporate. The more serious problem, however, is the extraction by butanol of a significant amount of salt. Thin-layer chromatography (TLC) is generally unaffected by salts contaminating the sample, but the amounts resulting from *n*-butanol extraction overload even the capacity of TLC. The result on TLC is extensive streaking of the derivatives. This may be avoided by electrolytic or ion-retardation chromatography to free the sample from the salts, but if one elects to use these methods, you do not quantitatively recover your samples and you may encounter extensive delays in the analyses. However, the TNBS technique is more quantitative. Recoveries of around 77% have been achieved by us with no apparent differences in reaction/extraction efficiency for the various amino acids *(29)*. A potential advantage of this method is that reportedly one can recover the amino acids and peptides following dilute acid hydrolysis *(19)*. If this should prove true, it would be an extremely valuable tool for obtaining and analyzing low molecular-weight peptides which have so far eluded our analytical procedures.

3. DANSYLATION   Dansylation, or DNS, also has advantages and disadvantages. The major disadvantage is that it is not quantitative; salts other than phosphate apparently interfere with the reaction *(48)*. However, the procedure is simple *(16)*, requires little amino acid (1×10^{-9} M), *(15)*, requires no special equipment (hence, is adaptable for field studies), and up to 12 samples of 1–5 l each of seawater can be extracted per day. The derivatives are stable to mild alkalai *(15)*, and TLC plates when stored in the dark of a lab drawer have been stable for up to 1 year. Because of these advantages, most of our studies have been completed using the dansylation procedure.

The details have been published previously *(30)*, but the method can be summarized as: adjustment of the sample pH to 9.8, addition of DNS-Cl (20 mg/50 ml acetone/seawater or 10 g sediment), incubation at room temperature in the dark for 18–24 hr, acidification to pH 2.5–3.0, extraction (5 times for total of $2 \times$ the volume of the sample), with anhydrous diethyl ether, pooling of the extracts and evaporation to dryness, TLC of samples redissolved in 95% ethanol.

Several points, however, need more detailed clarification. The initial method of Gray and Hartley *(16)* used a $NaHCO_3$:NaOH buffer system to maintain the pH at 8.5–9.0. We *(29)* and others *(48)* have found more recently that better yields of DNS-amino acids are obtained when the pH of the reaction mixture is

---

†Dinitrophenyl.
‡Trinitrobenzenesulfonic acid.

maintained at pH 9.5–9.8. In field studies, we have found it sufficient to use pHydrion paper and NaOH to adjust the sample pH. The other critical step concerns the reaction itself. For optimal derivitization, the reaction must proceed in the dark. This is most conveniently accomplished by wrapping the flask, bottle, tube, etc., in aluminum foil and storing away from direct light. To insure mixing, it is also advisable, particularly with sediment samples, to agitate the sample after addition of the DNS-Cl. It should be done during the first 20–30 min after addition of the reagent, but it is not essential and a ship will often provide sufficient movement to insure adequate mixing. One further word of caution on the technique as applied to sediments. It is advisable to conduct the acidification/extraction steps in a vessel with at least 20 times the volume of the sample to be acidified. This is especially true with carbonate sediments as the CO_2 resulting from the acidification to pH 2.5 will cause excessive bubbling and much of the sample will be lost to the floor or even the ceiling.

The resulting DNS-amino acids are separated by two-dimensional TLC on Adsorbosil (Applied Science Laboratories, State College, Penn.) or on the commercially available pre-coated Keiseilgel G plates, with or without fluorescent indicator. The solvent system of choice depends on the silica gel preparation in use, but I have found the following to be the most successful:

Solvent A   Benzene:pyridine:acetic acid
            (75:25:2 v:v:v) used with Adsorbosil
            (80:40:3 v:v:v) used with precoated plates
Solvent B   Chloroform:benzyl alcohol:acetic acid
            (75:30:2 v:v:v)
Solvent C   Methyl acetate:isopropanol:35% ammonium hydroxide
            (45:35:20 v:v:v)
Solvent D   Chloroform:ethanol (95%)
            (90:10 v:v) used with Adsorbosil
Solvent E   Chloroform:methanol:acetic acid
            (105:30:5 v:v:v) used with precoated plates

When the first dimension is developed in Solvent A, the second dimension is chromatographed in Solvent B; when Solvent C is used first, then either Solvent D or E is used for the second dimension. The total running time for two dimensions depends on temperature, humidity, and silica gel preparation, but a minimum of 2 hr and a maximum of 4 hr have been consistently observed. Visualization is achieved with a 260 nm wavelength lamp which causes the DNS-amino acids to fluoresce a characteristic yellow color.

DNS-Amino Acids from Aqueous Systems. Using the DNS procedure, we have analyzed algal culture filtrates, pond and seawater samples *(30)*, and sediments *(29)*. The results of some of these analyses are shown in Table 1. Analyses for amino acids in natural water samples show that the predominant amino acids are: aspartic acid, arginine, glycine, serine, alanine, and isoleucine. The other amino acids, which were identified as DNS derivatives, are found on a scattered basis with no discernible pattern. Inasmuch as small molecular weight peptides would also react with DNS-Cl, and could conceivably be the basis for several of

Table 1. Dansyl amino acids from natural waters

DNS-Amino Acids	Pond	Shamrock Cove	Station 8	Station 9	Jetty Channel
Aspartic Acid	+	+	+	+	+
Arginine	+	+	+	+	+
Glycine	+	+	+	+	−
Serine	+	+	−	+	+
Alanine	+	+	−	−	+
Isoleucine	+	−	−	+	+
Asparagine	−	+	+	−	D
Glutamic Acid	+	−	−	−	+
Leucine	+	−	+	D	−
ε-DNS-Lysine	+	+	−	−	+
Ornithine	+	D	+	−	D
Valine	D	+	−	−	+
Citrulline	D	+	−	−	D
Phenylalanine	−	+	−	D	−
Threonine	+	D	−	−	D
Tryptophan	−	−	D	+	−
Histidine	D	D	−	−	−
Di-DNS-Lysine	D	−	D	−	D
Methionine	D	D	D	−	D
Proline	D	D	D	−	D
Tyrosine	−	D	−	D	D
O-DNS-Tyrosine	−	D	−	−	−
Glutamine	−	−	−	−	−
Ninhydrin +	+	−	−	+	+
Number Unknown Spots	3(3)	3(3)	1(1)	8(3)	5(2)

+ = DNS-amino acid identified in both solvent systems.
D = DNS-amino acid identified in only one solvent system.
− = DNS-amino acid not found.

the unidentified spots listed in Table 1, all samples were subjected to acid hydrolysis in melting point tubes according to the method described by Blackburn (7). The resulting hydrolysates were then separated by TLC along with unhydrolyzed samples and comparisons were made of the number of fluorescent spots. Each plate was then sprayed with ninhydrin-cadmium reagent (21) and examined for evidence of newly formed or unreacted α-amino groups. Only those samples treated with HCl were found to contain ninhydrin-positive spots, and these all remained at the origin. In Table 1, the values in parentheses, listed with the number of unidentified spots, refer to the number of these unidentified spots remaining after acid hydrolysis. Generally, in those samples which were ninhydrin-positive, there was a reduction in the number of unknowns. It is not too surprising that there were fluorescent spots whose identity could not be determined, because only 25 DNS-amino acids were available for identification purposes. The reaction, extraction, and replicate runs on two-dimensional TLC in two sets of solvent systems for these five samples should take from 5 to 7 days, assuming one has standard TLC equipment available. In this respect, as well as the simplicity of the procedure, the DNS method has obvious advantages over other techniques currently in use.

DNS-Amino Acids from Sediments. In order to obtain some concept of the changes in the amino acid composition of a sedimentary core, typical examples of the types of distribution patterns are shown in Figures 2, 3, and, 4. In Figure

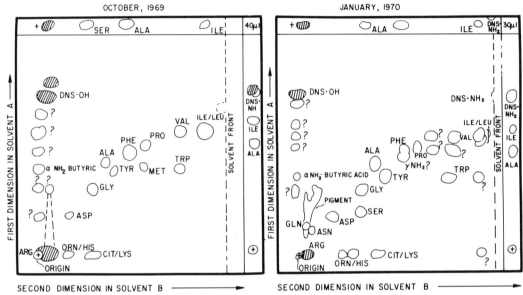

Fig. 2. TLC of the DNS-amino acids extracted from the surface area of cores at station 1.

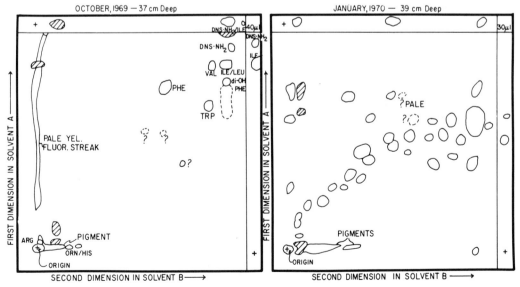

Fig. 3. TLC of the DNS-amino acids extracted from the middle area of cores from station 1.

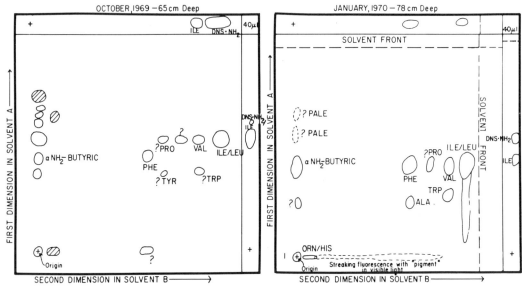

Fig. 4. **TLC of the DNA-amino acids extracted from the bottom area of cores from station 1.**

2, the DNS-amino acid pattern from surface sediment samples taken at the same station but three months apart is shown. It is immediately evident that the pattern is more complex in the January sample, consisting of all of the standard DNS-amino acids and several unidentified spots as well. Next, in Figure 3, from a sample slightly deeper in the same cores, one notices that fewer amino acids were found in the October sample, but the pattern was almost unchanged in the January sample. Finally, the DNS-amino acids in the bottom samples of these two cores showed greater similarity than at either of the other two positions. The amino acids here appeared to be predominantly the branched chain and aromatic amino acids; perhaps these are more stable in the environment or are less often required by microorganisms for growth. Throughout all three positions, however, α-amino butyric acid was identified.

As I mentioned earlier, one criterion we had included in our list was that the amino acid derivitization procedure should be quantitative. The results shown in Table 2, however, indicate that, at least in the marine environment, neither the reaction nor the extraction is consistent and therefore quantitative. When labeled amino acids were added to one liter of seawater and then dansylated, the recovery varied from 31% for valine to 10% for glutamic acid. Other data *(29)* indicate that various extraction solvents make little overall difference. Reports in the literature *(11, 17, 18, 36, 48)* on the 95% or greater yield from dansylation have been based on the attainment of a maximum absorbance at 265 nm and not on the recovery of a ^{14}C-labeled compound as the dansyl derivative. Our data have been confirmed by Hobbie and Crawford (personal communication). Chau and Riley *(10)* found similar results when labeled amino acids were added to seawater and treated with DNFB. Apparently, the high concentration of Na^+ does interfere with the reaction; some evidence for this has recently been reported in a statement that it is better to substitute phosphate

Table 2. Percent recovery ^{14}C-labeled amino acids from seawater

Amino Acid	Percent Recovery Previously Dansylated Amino Acid†		Percent Recovery of Amino Acid as Dansyl Derivative**	
	Diethyl Ether*	Ethyl Acetate*	Diethyl Ether	Ethyl Acetate
Alanine	35	40	22	—
Glycine	60	40	16	21
Glutamic Acid	—	35	10	30
Serine	5	35	—	—
Valine	35	40	31	—

† 10 μg labeled DNS-amino acid/liter seawater.
* Comparison of solvent extraction efficiency.
** 4 μg labeled amino acid/liter seawater.

buffer for Gray's original NaHCO$_3$:NaOH buffer *(16)*, but no further data were given *(48)*.

Bacterial Utilization of Excreted Algal Amino Acids. Although still lacking the desired quantitative data, part of our goal was achieved and some concept of the amino acid distribution patterns in sediments was attained. Meanwhile, we had been isolating and testing for bacteria with specific amino acid requirements from each of the sample systems. From the studies on the extracellular products of continuous culture *Chlorella sorokiniana*, we had identified not only most of the known amino acids *(30)*, but also several bacteria with amino acid requirements (Table 3). Four isolates were classified by using numerical taxonomic

Table 3. Amino acid utilization by bacteria isolated from *Chlorella sorokiniana*

Bacterial Culture	Growth in Knop's Medium Plus Glycerol	Growth in Spent-Algal Culture Filtrate Plus Glycerol	Amino Acids	
			Bacterial Utilization	Present in Spent-Algal Medium
Flavobacterium sp. AF-7	No	Yes	Leu, Pooled Amino Acids	All
Flavobacterium sp. AF-14	No	Yes	Thr, Pro, Val, Ile, Phe, Lys	All
Pseudomonas sp.* AF-18	No	Yes	Glu, Met, Gly, Ala, Leu, Phe, Lys, His, Trp	All
Flavobacterium sp.* AF-19	No	Yes	Ser, Met, Gly, Ala	All

* Amino acids served as both carbon and nitrogen sources.

programs *(26)* and *Flavobacterium* sp. AF-7 and AF-14 were found to require an additional carbon source for growth on amino acid media. By contrast, *Pseudomonas* sp., strain AF-18, and *Flavobacterium* sp., AF-19, were able to grow on spent-algal culture filtrate alone. The cultures were washed and inoculated into filter-sterilized spent-algal medium that was originally Knop's medium as modified by Vela and Guerra to contain additional EDTA* *(43)*. An organic carbon source, glycerol, was added to one-half of the tubes but not to the other half. Even when supplemented with glycerol, Knop's medium alone would not support the growth of any of the four isolates, but proved an excellent culture medium after algal growth took place in it. The amino acids that could serve as nitrogen sources for the growth of the four strains are listed in Table 3. In all cases, all of the amino acids were found to be present in the spent-algal culture filtrate when analyzed by the DNS procedure *(30)* or on an amino acid analyzer *(38)*.

Microbiology of Marine Sediments

A thorough microbiological and biochemical investigation of one set of samples was undertaken at the Marine Science Laboratories in Menai Bridge, North Wales. Sediments from the Irish Sea were selected for a study of the distributions with depth of the following: (1) the amino acids; (2) the bacterial populations which might utilize amino acids; (3) the free proteolytic enzymes present in the sediments and the possibility that they were of microbial origin. Seasonal changes in these three factors were also investigated using cores taken along a transect from the Isle of Man to Ireland in October 1969 and January 1970 *(29)*. Treatment of each core is diagramed in Figure 5.

Sediment samples were suspended in sterile aged seawater for the initial 1:10 dilution. The suspensions were then plated onto the surface of 0.1% peptone in a mixture of 20% mud extract and 80% aged filtered seawater plus agar or onto 0.1% peptone in aged filtered seawater agar. Major differences in the total numbers of isolates were noted on the different media *(25,* and Litchfield and Floodgate, unpublished data). The numbers of bacteria in the sediments were rather low and appeared to vary with the season but not too significantly with the depth (Table 4). The colony counts for October 1969 were complicated by the rather extensive growth of fungal colonies. Initially, we felt that the logical explanation for this was contamination, despite the appearance of a dilution effect. Since then, in June 1970 and in October 1970, similar bacterial and fungal counts have been recorded by Floodgate (personal communication) from surface samples at these three stations.

It would appear that there is some type of microbial succession or balance in these sediments that is reflected in the seasonal counts and is probably the result of some unidentified biochemical factors. Therefore, the fact that a rather distinctive amino acid pattern was also observed from surface samples at these three stations should not be ignored.

Approximately 260 random isolates were selected from the October and January plates and tested for their requirements for organic nitrogen sources.

*Ethylenediamine tetraacetic acid.

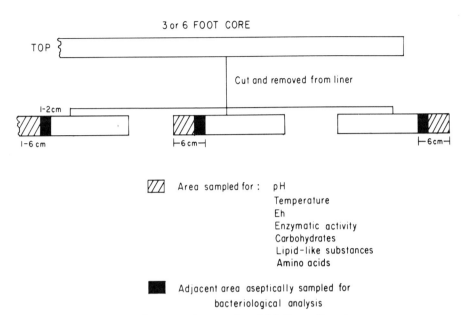

Fig. 5. Sampling procedure used on Irish Sea sedimentary cores.

According to the results shown in Figure 6, 10 to 14% of the selected isolates were found to require some organic nitrogen compound.

Autochthonous Bacteria and Amino Acid Sediment Composition. In an attempt to correlate these requirements with the previously noted distribution patterns of amino acids, the obligate heterotrophs were studied further to determine their nitrogen requirement. Eighteen amino acids were grouped roughly according to metabolic pathways and tested for a growth response by the obligate

Table 4. Distribution of aerobic heterotrophic bacteria in Irish Sea sediments

Station	October			January		
	Water Depth (meters)	Depth in Core	Total Bacteria per g	Water Depth (meters)	Depth in Core	Total Bacteria per g
1	227	Surface	3.2×10^5	249	Surface	6.3×10^5
		37 cm	2.6×10^4		39 cm	2.8×10^5
		65 cm	7.3×10^2		78 cm	3.5×10^5
2	201	Surface	<10	198	Surface	1.8×10^5
		36 cm	NG*		42 cm	7.5×10^4
		63 cm	NG		109 cm	2.8×10^4
3	183	Surface	<10	174	Surface	1.4×10^4
		35 cm	NG		33 cm	9.3×10^3
		65 cm	NG		108 cm	6.5×10^4

* NG = no bacterial colonies.

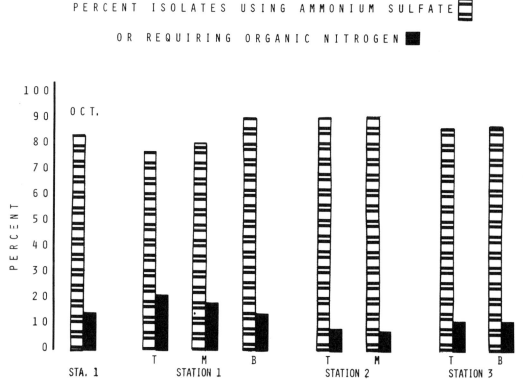

Fig. 6. **Relative percentages of obligate and facultative heterotrophic bacterial isolates from sediments.**

heterotrophs. In Table 5 are shown the results obtained for the first group of amino acids—alanine, aspartic acid, glutamic acid, glycine, serine, and threonine. From the columns labeled "general requirements," it is evident that there were many strains with a nonspecific requirement for organic nitrogen. Only two isolates were found to have a requirement for one, or all, of the amino acids in this group, and they were both from the surface of the core taken from station 1 in October 1969.

The branched chain amino acids were grouped together for testing as amino acid group 2. The results found when these amino acids were the sole nitrogen source for the isolates are shown in Table 6. Again, the number of strains with a specific requirement was very small, but one isolate was cultured from each core. We are currently investigating more thoroughly the minimal amino acid requirements for J-217 and find that valine, of these three amino acids, is the better nitrogen source, but the best growth appears when all three amino acids are present at a 1 mg/ml level or higher.

A rather heterogeneous group of amino acids was combined for testing as group 3, and the results are shown in Table 7. We have not yet had time to pursue this set further, but it would certainly appear that perhaps some rather specific requirements for these strains are reflected in the absence of most of

Table 5. Distribution of amino acids and bacterial requirements

Station	Location in Core	Group 1* Amino Acids Present† (October)	Bacteria with General Requirements (October)	Bacteria with Specific Requirements (October)	Group 1* Amino Acids Present† (January)	Bacteria with General Requirements (January)	Bacteria with Specific Requirements (January)
1	Top	Ser, Ala, Gly	2	O-18	Thr, Ser, Ala, Gly	3	0
	Mid	None	0	O-17	Glu, Ser, Ala, Asp	4	0
	Bottom	None	0	0	Ala	1	0
2	Top	Gly, Asp	0	0	None	0	0
	Mid	None	0	0	All Doubtful	1	0
	Bottom	Ser, Ala, Gly, Asp	0	0	None	0	0
3	Top	None	0	0	All	3	0
	Mid	Ala	0	0	All	0	0
	Bottom	Ser, Gly, Ala	0	0	Ala, Thr	2	0

* Glutamic acid, alanine, aspartic acid, glycine, serine, threonine.
† Only those amino acids identified in both solvent systems considered positive.

Table 6. Distribution of isoleucine/leucine and valine and bacterial requirements

Station	Location in Core	October Group 2* Amino Acids Present†	October Bacteria with General Requirements	October Specific Requirements	January Group 2* Amino Acids Present†	January Bacteria with General Requirements	January Specific Requirements
1	Top	+	2	0	+	3	0
	Mid	+	0	0	+	4	J-133
	Bottom	+	0	0	+	1	0
2	Top	+	0	0	Ile/Leu	0	J-182
	Mid	Ile/Leu	0	0	+	1	0
	Bottom	+	0	0	+	0	0
3	Top	Ile/Leu	0	0	Ile/Leu	3	0
	Mid	+	0	0	+	0	0
	Bottom	+	0	0	+	2	J-217

* Isoleucine/leucine and valine.
† Only those amino acids identified in both solvent systems considered positive.

Table 7. Distribution of amino acids and bacterial requirements

Station	Location in Core	October Group 3* Amino Acids Present†	October Bacteria with General Requirements	October Bacteria with Specific Requirements	January Group 3* Amino Acids Present†	January Bacteria with General Requirements	January Bacteria with Specific Requirements
1	Top	His, Met	2	0	Met, Pro	3	0
	Mid	Arg, His	0	0	Met, Pro	4	J-51 J-52
	Bottom	None	0	0	None	1	0
2	Top	Arg, His	0	0	Arg	0	J-73
	Mid	None	0	0	Arg, His	1	0
	Bottom	Arg, His, Met	0	0	None	0	0
3	Top	Arg	0	0	Arg, His, Met, Pro	3	0
	Mid	Arg	0	0	Arg, His, Pro, Met	0	0
	Bottom	Met	0	0	Arg	2	0

* Arginine, cysteine, histidine, methionine, proline.
† Only those amino acids identified in both solvent systems considered positive; no DNS-Cys tested for.

the group 3 amino acids from the sediments. The final pool of amino acids, group 4, resulted in growth of four strains with requirements for some or all of these compounds (Table 8). Under further investigation is strain J-218, which apparently will grow on tyrosine when it is present at 1 mg/ml medium concentration. Low amino acid concentrations, again as with J-217, seem to favor a mixture of amino acids. In related studies, the addition of yeast extract to the media does not appear to increase the amounts of growth of either J-217 or J-218.*

Extracellular Proteases. Inasmuch as most marine bacteria appear to produce proteases in laboratory culture, the next question logically was whether these enzymes might provide a source of the amino acids in the sediments, and, related to this, whether we could account for "free" proteases in sediments based on a bacterial origin. Of our random isolates, 33% did produce clearing on a 1% casein agar plate when amido black was used to detect hydrolysis (Litchfield and Floodgate, unpublished data).

A total of 85 isolates were tested for the production of extracellular proteases when grown in a 0.2% casein-seawater medium at 15 C under either constant agitation or under stationary conditions. A total of 17 strains produced clearing of the gelatin on film strips (Tri-X-Pan) or showed hydrolytic activity when tested by a modified Anson's hemoglobin procedure (8) (Table 9), but there was no direct correlation between obligate organic nitrogen requirements and protease(s) production. In fact, the major enzyme producer, J-193, had no requirement for amino acids and grew just as well on ammonium salts. During the growth cycle of each strain, protease production was monitored via the qualitative filmstrip method which can only detect gelatinase activity. From Table 9, we can see that there are obvious substrate differences. J-217 (which has the requirement for branched chain amino acids) produced substances which hydrolyzed gelatin but showed no activity against hemoglobin. However, J-231, J-75, J-83, and J-11 were all negative according to the gelatin test (note no time listed for initiation of protease excretion), but showed definite hydrolysis with the hemoglobin substrate.

For comparison purposes, the proteolytic activities of the sediments are also listed in Table 9. Within 20–30 min after a core was returned to the deck of the ship, samples were slurried with 1×10^{-3} M phosphate buffer, pH 8.0. These slurries were then spotted onto filmstrips or cut into a 1% casein agar plate. All samples were run in duplicate and determinations of hydrolysis were made after 3–4 hr and after 20–24 hr incubation at 37 C. By this procedure, gelatin hydrolysis was visible on the filmstrips after washing off the sample and allowing the film to dry. Casein hydrolysis was detected by flooding the plates with amido black. A buffer control was added to all tests. Even at the end of 24 hr incubation at 37 C, there were no visible signs of bacterial growth. Plates held for up to a week, however, were completely overgrown by fungal and bacterial colonies. Except for the four samples appropriately marked in Table 9, all other samples were positive for both gelatin and casein hydrolysis after 20–24 hr.

*Litchfield, unpublished data.

Table 8. Distribution of amino acids and bacterial requirements

Station	Location in Core	October Group 4* Amino Acids Present†	October Bacteria with General Requirements	October Bacteria with Specific Requirements	January Group 4* Amino Acids Present†	January Bacteria with General Requirements	January Bacteria with Specific Requirements
1	Top	Phe, Trp, Tyr	2	0	Phe, Trp, Tyr	3	0
	Mid	Phe, Trp, Tyr	0	0	Lys, Phe	4	0
	Bottom	Phe	0	0	Phe, Trp, Tyr	1	J-91
2	Top	Phe	0	0	Lys	0	0
	Mid	Phe	0	0	Phe, Lys	1	0
	Bottom	Phe, Trp, Tyr	0	0	Phe, Trp	0	0
3	Top	Phe	0	0	Phe, Trp, Tyr	3	J-4 J-33
	Mid	Phe	0	0	Phe, Tyr	0	0
	Bottom	Phe, Trp	0	0	Trp	2	J-218

* Lysine, phenylalanine, tryptophan, tyrosine.
† Only those amino acids identified in both solvent systems considered positive.

Table 9. Production of extracellular proteases

Organism	Qualitative Gelatinase* Positive	Quantitative Hemoglobin Units†	Requirement for Organic Nitrogen	Sediment Sample Gelatin	Casein	Location in Core
O-17	Late Log	4.5	Yes	0	+	1 − top
O-29	Mid Log	5.0	Yes	0	+	1 − top
J-56	−	0.0	Yes	0	0	1 − mid
J-231	−	3.3	No	0	0	1 − mid
J-139	Late Log	5.5	No	0**	±	1 − bot
J-193	Late Log	11.0	No	0	±	1 − bot
J-74	Mid Log	3.5	No	0	±	2 − top
J-75	−	0.8	No	0	±	2 − top
J-83	−	1.0	No	0	±	2 − top
J-183	Late Log	2.1	Yes	0	±	2 − top
J-171	Late Log	2.3	No	0**	±	2 − bot
J-3	Late Log	0.5	No	0**	±	3 − top
J-11	−	1.4	Yes	0	±	3 − top
J-26	Late Log	1.0	No	0	±	3 − top
J-30	−	0.0	Yes	0	±	3 − top
J-84	Mid Log	1.0	No	0	±	3 − top
J-217	Late Log	0.0	Yes	0**	0	3 − bot
J-218	Late Log	1.2	Yes	0	0	3 − bot
J-220	Mid Log	2.3	No	0	0	3 − bot

†Quantitative hemoglobin assay consisted of 5 ml modified Anson's hemoglobin *(8)* mixed with 1 ml culture filtrate after equilibration to 37 C. The reaction was allowed to proceed for 10 min at 37 C, was stopped by the addition of 10 ml of 5% TCA; the precipitate was filtered through Whatman #3 paper after incubation for a further 30 min at 37 C. Absorbancy of the filtrate was determined by reading at 280 nm on an Hitachi-Perkin Elmer Spectrophotometer, model 139. Hemoglobin units are defined as the change in optical density/ml culture filtrate/10 min assay × 10.

* Qualitative gelatin assays were conducted by placing one drop of the culture medium onto the gelatin-coated side of Tri-X Pan film (Kodak) and incubating for 30 min in a closed, water-saturated atmosphere, petri plate at 37 C. The film was washed under running tap water and allowed to air dry before determination of hydrolysis, which is the visible disintegration of the gelatin coating.

** Test negative after 24 hrs.

Possible Ecological Significance of Exoenzymes. Given these rather interesting facts, the next question is to try to determine if they have any meaning in the ecological survival and growth of these bacteria. With a view toward answering this question, we are undertaking a study of *(a)* minimum nutrient levels for these bacteria; *(b)* the uptake and turnover of amino acids in sediments; and *(c)* the influence of extracellular proteases on *(a)* and *(b)*.

In preliminary work, which is still very much in the experimental stage, the problem of the potential importance of proteases to natural microbial populations is being examined. One approach is to take a complex protein, currently casein, and suspend low concentrations of it in a dialysis bag. To this bag is then added a semipurified protease preparation. The dialysis bag is submerged in a seawater-glycerol medium into which the obligate heterotroph is inoculated. We have selected J-217 (which requires branched chain amino acids) and J-218 (aromatic or basic amino acid requirement). Rather good growth has occurred as a result of the dialyzable peptides and amino acids resulting from the proteolytic digestion of the casein (Fig. 7). The control flasks consisted of one with autoclaved protease plus casein, and another with just seawater-glycerol. No growth occurred in either of the control flasks inoculated with either J-217 or J-218. The next step, of course, is to determine if these same strains can grow as a result of their own proteases and what effect different protein substrates within the dialysis bag might have on this parameter.

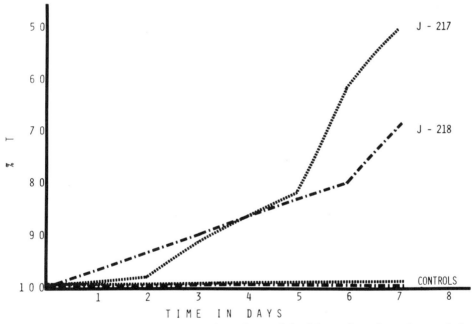

Fig. 7. **Growth of amino acid requiring bacteria on dialyzable products from *A. proteolytica* protease hydrolysis of casein.**

Conclusions

In conclusion, it is becoming increasingly evident that the amino acids which are present in seawater or sediments are not there as biochemical accidents. Work by Hamilton and Greenfield *(20)*, Williams and co-workers *(1, 45, 46, 47)*, Hobbie, et al. *(21, 22)*, and others on the uptake and turnover times of amino acids, as well as preliminary work in my laboratory, indicate that the amino acids are probably there as a result of a steady state between introduction from excretion and enzymatic hydrolysis on the one hand and utilization by the microbial flora on the other hand. The residual levels which we can measure may also be acting to control metabolic activities via such classical mechanisms as feedback inhibition or catabolite repression, and also thereby control growth. Some evidence has already been presented concerning the effect of various concentrations of L-asparagine on endopeptidase synthesis by *Aeromonas proteolytica (31)*. In addition, the presence of extracellular enzymes probably does have an effect on the microenvironment in enabling either the producer organism to compete or enabling the population as a whole to maintain its balance following predation or other environmental stresses.

Acknowledgments

I would like to thank Dr. D. W. Hood, Dr. J. M. Prescott, Dr. G. D. Floodgate, and Dr. A. L. S. Munro for their help and suggestions during my research on amino acids and bacteria throughout the last several years. I would also like to acknowledge financial assistance from the Rutgers Research Council for support of portions of the work reported here.

Literature Cited

1. Andrews, P., and P. J. LeB. Williams. 1971. Heterotrophic utilization of dissolved organic compounds in the sea. III. Measurement of the oxidation rates and concentrations of glucose and amino acids in sea water. *J. Mar. Biol. Ass. U.K.* **51:**111–25.
2. Atfield, G. N., and C. J. O. R. Morris. 1961. Analytical separations by high-voltage paper electrophoresis. Amino acids in protein hydrolysates. *Biochem. J.* **81:**606–14.
3. Bader, R. G., D. W. Hood, and J. B. Smith. 1960. Recovery of dissolved organic matter in sea-water and organic sorption by particulate material. *Geochim. Cosmochim. Acta* **19:**236–43.
4. Belser, W. L. 1958. Possible application of a bacterial bioassay in productivity studies. *Spec. Sci. Rept. Fish.* **279:**55–58.
5. Belser, W. L. 1959. Bioassay of organic micronutrients in the sea. *Proc. Natl. Acad. Sci. U.S.* **45:**1535–42.
6. Binkley, F., F. Leibach, and N. King. 1968. A new method of peptidase assay and the separation of three leucylglycinases of renal tissues. *Arch. Biochem. Biophys.* **128:**397–405.
7. Blackburn, S. 1966. Micro techniques for amino acid analysis and peptide separation based on high-voltage electrophoresis. In *A laboratory manual of analytical methods of protein chemistry*, vol. 4, ed. P. Alexander and H. P. Lundgren, pp. 78–102. London: Pergamon Press Ltd.
8. Burgum, A. A., J. M. Prescott, and R. J. Hervey. 1964. Some characteristics of a proteolytic system from *Phymatotrichum omnivorum. Proc. Soc. Exp. Biol. Med.* **115:**39–43.
9. Carlucci, A. F., and S. B. Silbernagel. 1966. Bioassay of sea water. I. A ^{14}C-uptake method for the determination of concentrations of vitamin B_{12} in sea water. *Can. J. Microbiol.* **12:**175–83.
10. Chau, Y. K., and J. P. Riley. 1966. The determination of amino acids in sea water. *Deep-Sea Res.* **13:**1115–25.
11. Crowshaw, K., S. J. Jessup, and P. W. Ramwell. 1967. Thin-layer chromatography of 1-dimethylamino-naphthalene-5-sulphonyl derivatives of amino acids present in superfusates of cat cerebral cortex. *Biochem. J.* **103:**79–85.
12. Degen, E. T., J. H. Reuter, and K. N. F. Shaw. 1964. Biochemical compounds in offshore California sediments and sea waters. *Geochim. Cosmochim.* Acta **28:**45–66.

13. Erdman, J. G., E. M. Marlett, and W. E. Hanson. 1956. Survival of amino acids in marine sediments. *Science* **124**:1026.
14. Gold, K. 1967. Microbiological assays of sea water using radioisotopes—an assay for vitamin B$_{12}$ measured by ^{14}C assimilation. In *Estuaries,* ed. G. H. Lauff, pp. 341–45. Washington, D.C.: AAAS.
15. Gray, W. R. 1967. Dansyl-chloride procedure. In *Methods in Enzymology,* vol. 11, ed. S. P. Colowick and N. O. Kaplan, pp. 139–50. New York: Academic Press Inc.
16. Gray, W. R., and B. S. Hartley. 1963. A fluorescent end-group reagent for proteins and peptides. *Biochem. J.* **89**:59p.
17. Gros, C. 1967. Microanalyse des acides aminés et peptides. Réactivité des acides aminés et peptides avec le chlorure de dansyle et séparation des dérivés dansylés formés. *Bull. Soc. Chim. Fr.* **10**:3952–54.
18. Gros, C., and B. Labouesse. 1969. Study of the dansylation reaction of amino acids, peptides and proteins. *European J. Biochem.* **7**:463–70.
19. Habeeb, A. F. S. A. 1966. Determination of free amino groups in proteins by trinitrobenzenesulfonic acid. *Anal. Biochem.* **14**:328–36.
20. Hamilton, R. D., and L. J. Greenfield. 1967. The utilization of free amino acids by marine sediment microbiota. *Zeit. für Allg. Mikrobiol.* **7**:335–42.
21. Hobbie, J. E., and C. C. Crawford. 1969. Bacterial uptake of organic substrate: New methods of study and application to eutrophication. *Verh. Internat. Verein. Limnol.* **17**:725–30.
22. Hobbie, J. E., and C. C. Crawford. 1969. Respiration corrections for bacterial uptake of dissolved organic compounds in natural waters. *Limnol. Oceanogr.* **14**:528–32.
23. Hobbie, J. E., C. C. Crawford, and K. L. Webb. 1968. Amino acid flux in an estuary. *Science* **159**:1463–64.
24. Lin, Y., G. E. Means, and R. E. Feeney. 1969. The action of proteolytic enzymes on N,N-dimethyl proteins. *J. Biol. Chem.* **244**:789–93.
25. Litchfield, C. D. 1971. Distribution of heterotrophic bacteria in marine sediments. *Bacteriol. Proc.,* p. 56.
26. Litchfield, C. D., R. R. Colwell, and J. M. Prescott. 1969. Numerical taxonomy of heterotrophic bacteria growing in association with continuous culture *Chlorella sorokiniana. Appl. Microbiol.* **18**:1044–49.
27. Litchfield, C. D., and D. W. Hood. 1965. Microbiological assay for organic compounds in sea water. I. Quantitative assay procedures and biotin distribution. *Appl. Microbiol.* **13**:886–94.
28. Litchfield, C. D., and D. W. Hood. 1966. Microbiological assay for organic compounds in sea water. II. Distribution of adenine, uracil and threonine. *Appl. Microbiol.* **14**:145–51.
29. Litchfield, C. D., A. L. S. Munro, L. C. Massie, and G. D. Floodgate. 1971. Analysis by dansylation of amino acids in marine sediments. Submitted to *Limnol. Oceanogr.*
30. Litchfield, C. D., and J. M. Prescott. 1970. Analysis by dansylation of amino acids dissolved in marine and fresh waters. *Limnol. Oceanogr.* **15**:250–56.
31. Litchfield, C. D., and J. M. Prescott. 1970. Regulation of proteolytic enzyme production by *Aeromonas proteolytica.* I. Extracellular endopeptidase. *Can. J. Microbiol.* **16**:17–22.
32. Okuyama, T., and K. Satake. 1960. On the preparation and properties of 2,4,6-trinitrophenyl-amino acids and -peptides. *J. Biochem.* **47**:454–66.
33. Palmork, K. H. 1963. The use of 2,4-dinitro-1-fluorobenzene in the separation and identification of amino acids from sea water. *Acta Chem. Scand.* **17**:1456–57.
34. Park, K., W. T. Williams, J. M. Prescott, and D. W. Hood. 1963. Amino acids in Redfish Bay. *Texas Inst. Mar. Sci.* **9**:59–63.
35. Pataki, G. 1966. *Techniques of thin-layer chromatography in amino acid and peptide chemistry,* rev. ed. Ann Arbor, Michigan: Ann Arbor Sci. Publ.
36. Pataki, G., and E. Strasky. 1966. Quantitative thin-layer chromatography. V. Direct fluorometry of amino acids as their dinitrophenyl, dimethyl aminonaphthalenesulfonyl-, and phenylthiohydantoin-derivatives. *Chimica* **20**:361–63.

37. Plunkett, M. A. 1957. The qualitative determination of some organic compounds in marine sediments. *Deep-Sea Res.* **5**:259–62.
38. Prescott, J. M., S. H. Wilkes, and C. D. Litchfield. 1968. Amino acid interrelationships in mixed cultures of algae and bacteria. Final Report. Air Force Contract F4 1609-67-C-0101.
39. Rittenberg, S. C., K. O. Emery, J. Hülsemann, E. T. Degens, R. C. Fay, J. H. Reuter, J. R. Grady, S. H. Richardson, and E. E. Bray. 1963. Biogeochemistry of sediments in experimental Mohole. *J. Sed. Petrol.* **33**:140–72.
40. Sanger, F. 1945. The free amino groups of insulin. *Biochem. J.* **39**:507–15.
41. Satake, K., T. Okuyama, M. Ohashi, and T. Shinoda. 1960. The spectrophotometric determination of amine, amino acid, and peptide with 2,4,6-trinitrobenzene 1-sulfonic acid. *J. Biochem.* **47**:654–60.
42. Tatsumoto, M., W. T. Williams, J. M. Prescott, and D. W. Hood. 1961. Amino acids in samples of surface sea water. *J. Mar. Res.* **19**:89–96.
43. Vela, G. R., and C. N. Guerra. 1966. On the nature of mixed cultures of *Chlorella pyrenoidosa* TX71105 and various bacteria. *J. Gen. Microbiol.* **42**:123–31.
44. Webb, K. L., and L. Wood. 1966. Improved techniques for analysis of free amino acids in sea water. In *Automation in analytical chemistry, Technicon symposium*, pp. 440–44. Medical White Plains, New York: Technicon Corp.
45. Williams, P. J. LeB. 1970. Heterotrophic utilization of dissolved organic compounds in the sea. I. Size distribution of population and relationship between respiration and incorporation of growth substrates. *J. Mar. Biol. Ass. U.K.* **50**:859–70.
46. Williams, P. J. LeB., and C. Askew. 1966. A method of measuring the mineralization by microorganisms of organic compounds in sea water. *Deep-Sea Res.* **15**:365–75.
47. Williams, P. J. LeB., and R. W. Gray. 1970. Heterotrophic utilization of dissolved organic compounds in the sea. II. Observations on the response of heterotrophic marine populations to abrupt increases in amino acid concentration. *J. Mar. Biol. Ass. U.K.* **50**:871–81.
48. Zanetta, J. P., G. Vincendon, P. Mandel, and G. Gombos. 1970. The utilization of 1-dimethylamino-napthalene-5-sulfonyl chloride for quantitative determination of free amino acids and partial analysis of primary structure. *J. Chromatog.* **51**:441–58.

Comments

WATSON: I think the ecological significance of extracellular enzymes is very thought-provoking. I have often wondered how one would go about studying the effects of such enzymes. Has anybody ever tried this with the chemostat? It would seem to me that extracellular enzymes would be diluted out in the water column quite rapidly. But, in a chemostat, you could actually determine how many organisms per ml or per l you might need in order to decide if this could be of possible ecological significance.

LITCHFIELD: As I mentioned, I hope to visit Jannasch at Woods Hole and learn some of the techniques of chemostat culture. I think that this might be one method. Another is to use some of the techniques we have been using in protein chemistry for years, for example, ultrafiltration. One could take large samples of seawater, concentrate it, and run disc gel electrophoresis to find out how many enzymes are present. As far as we know, this could be 1 enzyme or it could be 100. We have no concept as to whether there is 1 or 200 or 2,000 enzymes. It is completely untouched in terms of ecological significance.

WATSON: The evolutionary significance is also very interesting. It does not seem logical to me for an organism to produce an extracellular enzyme in seawater.

You would not think that this characteristic would survive. I would almost think that the organisms that do this are those which come from the sediment and those which have come into the marine environment from the soil.

LITCHFIELD: I would suspect that the sediments would be much more interesting in this respect because you see where the enzymes can be trapped either by the clay or other particles and still be available to the cell. In an aqueous environment, however, they go right into the ocean.

WATSON: Going back to particles, like styrene balls, is there any evidence that these exocellular enzymes are trapped or absorbed onto a surface?

LITCHFIELD: As far as I know, no one has looked at this. In some cases, we have found one organism, a *Bacillus,* that produced higher enzyme yields when we had a suspended protein substrate than when we had a dissolved protein substrate. Whether this has to do with growth, excretion, or absorption we really were not capable of determining at that time. It would be interesting to find out; but at this point, I do not think anyone knows.

HOLM-HANSEN: You mentioned something to the effect that bioassays are not practical for any type of field work. I assume you wish to apply your statement only to your amino acid work, because certainly bioassays are very valuable to other types of work. In regard to amino acids and the quantitative determination of dissolved amino acids in seawater, would you comment on the use of Chelex-100, the copper resin?

LITCHFIELD: I think Claude Crawford is perhaps more experienced in this. We tested it with the *Chlorella* culture filtrates. We found recoveries, particularly of glycine, for example, up to and exceeding 130%. There seems to be a great deal of contamination, the source of which we are not certain. Also, the procedure is extremely time-consuming. The technician who was running these tests for the culture filtrates was lucky to get five samples done in a week. Granted, we were not set up to have 20 columns going, but desalting via this procedure seems to be a great handicap.

CRAWFORD: Dr. Ken Webb at the Virginia Institute of Marine Science actually processed samples that I took. I know he had a well-trained technician and a lab set up for this. It took about 24 hr per sample.

HOLM-HANSEN: I have not done this personally, but I have watched Peter J. Williams in our laboratory for the last six months. It seems to me like a very simple procedure. Essentially, all you do is pass the filtrate through a column of Chelex-100 which hangs up all amino acids and eliminates all the salts. Then, you merely lift from the column, concentrate the affluent, and then add it to an amino acid analyzer to get a quantitative recovery. In talking with Williams, I got the impression that he obtained close to 100% recovery. With the exception of one or two amino acids, the only one I remember he had trouble with was, I think, alanine. In this instance he apparently had some contamination of the column with salt. This contamination is something that could be eliminated by exercising a little more care.

LITCHFIELD: We found that regeneration of the column is the most time-consuming step. It takes about 24 hr to regenerate the column after each run.

HOLM-HANSEN: Why not throw it away?

LITCHFIELD: Expense. I think the method of choice in the future is going to be through gas-liquid chromatography. Pocklington has worked up a very good procedure for gas-liquid chromatography of amino acids in seawater. It looks much more quantitative, reproducible, and easier to use than anything that we have to date. I expect to see a lot of people using it.

HOLM-HANSEN: What kind of concentration does this method use in the beginning?

LITCHFIELD: He is directly forming the trimethylsilyl ethers in seawater.

WATSON: We worked some with decontamination of the copper column. I think this is a minor part of the job, and we do not worry about it. It was the amino acid analysis that gave us problems. We could only run three a day. This was the time-consuming part. You could set up dozens of the columns. I wonder if anybody in this group has tried gas chromatography for amino acid analysis. If they have, would they comment on how sensitive the GLC is compared to the amino acid analyzer. It certainly would be a lot more rapid.

LITCHFIELD: This is Pocklington's work. Basically, you are limited on the amino acid analyzer by the sensitivity of the ninhydrin method. You can only go so far with ninhydrin. Whereas with forming the trimethylsilyl ethers or other derivatives, you obtain 10 to 100 times greater sensitivity.

ALEXANDER: I thought that water interfered in a dansylation process.

LITCHFIELD: I am not really that familiar with his procedure; his data on recoveries look very good. Also, he does some concentration with flash evaporation, but I do not know too much more. I assume he is still continuing some of his studies.

OLIVER: Was your TLC with silica gel?

LITCHFIELD: The Adsorbosil is a type of silica gel and the other was Keiseilgel which is silica gel GF254. We buy the precoated plates which save a lot of time.

OLIVER: If your optimum protease activity appears in late log phase in general, how do you conclude that this is an extracellular protease and not just the result of lysis of the cells?

LITCHFIELD: Because the cells are still in the log phase of growth. You find that proteases resulting from lysis appear during stationary growth stage. You can correlate this with the number of cells. In our studies, the initiation of protease activity usually comes when the cultures are just beginning to get into the log phase. Maximum levels occur in the late log phase so that there is little lysis in comparison to the protease level. Also, the concentration of protease does *not* increase with the age of the stationary cultures.

Utilization of Dissolved Organic Compounds by Microorganisms in an Estuary[1]

C. C. Crawford, J. E. Hobbie, and K. L. Webb

Ecology has progressed from a descriptive to a quantitative phase in which not only numbers of organisms but also the rates of growth, of transfer of material, and of chemical transformation are considered. Studies of the ecology of aquatic microorganisms have followed the same track, but we cannot yet claim that all the types and capabilities of these organisms are known. While it is important to determine population size, it is even more important, from an ecological point of view, to measure rates of microbial activity.

A number of techniques are available for studying both the numbers and rates of activity of the heterotrophic bacteria. Numbers of bacteria can be determined by using plate counts or direct microscopic counts. One way to study the activity of the heterotrophic community is by oxygen changes. This can be large-scale, as in an entire summer's changes in the hypolimnion of a lake or changes in the bottom water of a stratified estuary. There can also be small-scale studies, such as changes in a bottle. However, there are problems of population change when using bottles so one has to perform experiments of extremely short duration. In many environments, such as oceans and oligotrophic lakes, microbial activity is not high enough to be detected by present methods; thus, the plankton must be concentrated. Pomeroy and Johannes *(9)* and others have developed gentle methods of concentrating the microplankton from hundreds of liters into only a few tens of milliters in order to measure activity via oxygen consumption.

Another useful method of study employs radioisotopes. At first glance it might be enough merely to add an organic substrate such as glucose- or acetate-^{14}C. However, if the rates of uptake in nature are really what is sought, and if the indigenous bacteria, for example, react the same way as in culture, then the uptake kinetics follow Michaelis-Menten equations and would therefore increase (up to a point) with increases in added substrates.

When uptake measurements were first made on natural plankton, it was found that even though a mixed population was involved, the uptake resembled that of a pure culture and followed Michaelis-Menten equations *(8)*. A maximum velocity of uptake could be measured and was considered as a heterotrophic potential. Wright and Hobbie *(15)* extended these techniques to fresh water

[1]Contribution No. 21 from the Pamlico Marine Laboratory.

and found that the turnover time (T_t) and a sum representing the Michaelis constant (K_m) plus the substrate concentration in nature (S_n) could also be determined. This rather circuitous route was necessary to get around the lack of a measure of the natural substrate concentrations (S_n), which made it impossible to calculate the actual uptake in nature.

The next step was to measure $^{14}CO_2$ produced during the experimental incubations (4). While this can be done in a number of different ways, a wick absorption technique can be used for large numbers of samples. The amount respired reached as high as 60% of the amount taken up and varied with substrate (amino acids) or position of the labeled carbon (glucose-6- vs. glucose-1- vs. glucose-U-^{14}C).

Substrate concentrations in nature are extremely difficult to measure as they are present in such small quantities (parts per billion). When the concentrations are known, they may be combined with uptake studies to give actual rates of uptake or flux.

The amount of the isotopic carbon remaining in the microorganisms is a measure of the particulate carbon that has been produced. It is becoming increasingly clear that bacteria are important as a source of particulate organic material in natural waters. For instance, Kuznetsov (7) reported that bacteria produced as much or more biomass than the phytoplankton in the Rybinsk reservoir. Also, we have found that bacteria in the Pamlico River estuary in North Carolina produce significant quantities of particulate material through uptake of amino acids (2).

The method used to measure heterotrophic activity, described by Hobbie and Crawford (4, 5), consists of adding different amounts of a ^{14}C-labeled substrate to each of a series of subsamples. The substrates were diluted so that at least 0.01 μc could be added in order to obtain sufficient uptake for counting. After incubation (with shaking, in the dark, at the temperature of the sample) the plankton and respired CO_2 were collected and their ^{14}C counted by liquid scintillation. Usually only 5% or less of the added substrate is taken up.

The procedure, suggested by Smith (10), uses sealed vessels for recovery of the respired CO_2. A 5 ml sample of estuarine water was placed in a 25 ml Erlenmeyer flask and the substrate added with a micropipette. The flask was immediately sealed with a rubber serum stopper fitted with a suspended plastic cup (Kontes Co., Vineland, N.J., No. K 882320; Fig. 1) containing a 25 mm × 51 mm piece of accordion-folded chromatography paper (Whatman No. 1). While uptake was found to be linear over time (4), incubation periods (2–6 hr depending upon temperature) were kept to the minimum necessary to give adequate uptake of ^{14}C to count efficiently. After incubation, 0.2 ml of a 2 N solution of H_2SO_4 were injected into the water through the serum stopper. This both stops biological activity and lowers the pH sufficiently to drive off the dissolved CO_2. Next, still working through the serum stopper, 0.2 ml of phenethylamine was slowly added to the folded paper and the flask returned to the shaker for about 50 min at room temperature. The paper was then immediately placed in 15 ml of a toluene-based scintillation counting cocktail. The plankton was filtered onto Millipore filters (0.45 μm), rinsed with 10 ml of distilled water, air dried,

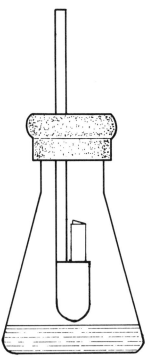

Fig. 1. **Apparatus used to recover respired $^{14}CO_2$ onto filter paper moistened with phenethyl-amine.**

and counted in the same counting cocktail. Wolfe and Shelske *(14)* found that dried Millipore filters became clear in a toluene-based cocktail and could be efficiently counted. For each test of each amino acid, four substrate concentrations in duplicate were used as well as one control flask. The control consisted of natural water in which the plankton had been killed by the addition of 0.2 ml of 2 N H_2SO_4 before addition of the labeled substrate at the highest concentration. In all, nine flasks were used for each test. For amino acid analysis, water samples were desalted in a column of Chelex-100 resin in the Cu^{++} form. The eluate from this column was lyophilized and applied to a Technicon Autoanalyzer *(12)*.

Calculations of uptake kinetics were similar to those described by Wright and Hobbie *(15)* with a slight alteration to correct for respiration. The equation for velocity was given by Parsons and Strickland *(8):*

$$v = \frac{cf(S_n + A)}{C\mu t}$$

where v is the velocity of uptake, c is the counts from the sample, f is a factor to correct the isotopic discrimination (which will be ignored here), S_n is the natural substrate concentration, C is the counts from one μc of C^{14} in the counting assembly used, μ is the fraction of a μc used, t is the incubation period in hours, and A is the concentration of added substrate. This equation is combined with a modification of the Lineweaver-Burk form of the Michaelis-Menten equation:

$$\frac{(S_n + A)}{v_n} = \frac{K_t}{V_{max}} + \frac{(S_n + A)}{V_{max}}$$

where K_t is the transport constant and V_{max} is the maximum velocity of uptake to give the equation:

$$\frac{C\mu t}{c} \text{ or } \frac{(S_n + A)}{v_n} = \frac{K + S}{V_{max}} + \left(\frac{1}{V_{max}}\right)A$$

The value $C\mu t/c$ is graphed on the ordinate against A on the abscissa and the resulting straight line is computed by a regression analysis (Fig. 2). The data were acceptable if the correlation coefficient (r) was above that required for the 95% probability level or, for an eight-point line, $r = 0.71$ or greater. The inverse of the slope is V_{max} or the maximum velocity of uptake. The intercept on the abscissa is $(K_t + S_n)$ or a sum representing the transport constant (K_t) plus the natural substrate concentration (S_n).

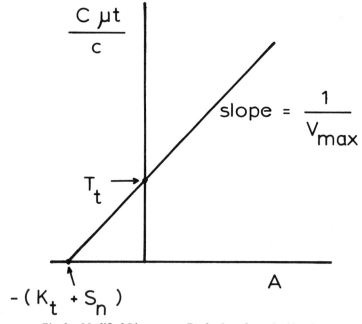

Fig. 2. **Modified Lineweaver-Burk plot of uptake kinetics.**

Early forms of this technique, which had no provision for either respiration correction or substrate analysis, yielded only maximum velocity data. Such techniques were used to study the uptake of glucose in the Pamlico River estuary in 1966–67 and in 1969 (Fig. 3). These studies indicated a peak of V_{max} in August which coincided with a large peak of primary productivity due to a high dinoflagellate population. The increased bacterial activity was probably due to the

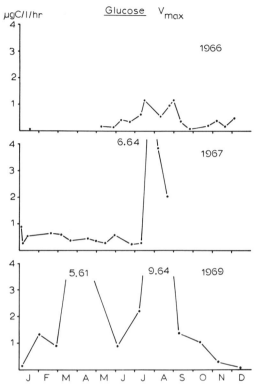

Fig. 3.   **Uptake of glucose in Pamlico River estuary, 1966, 1967, 1969.**

effects of high water temperature and increased dissolved organic material resulting from the phytoplankton bloom.

Using these same techniques, South Creek,. a slow-moving tributary of the Pamlico River, was tested at stations having salinities ranging from 9 to 13 ppt (Fig. 4). At about 3 km on this figure, the effluent from a sewage lagoon serving about 600 people enters the river. At about 18 km, wastes from a phosphate mine are discharged. Neither discharge creates any physical or chemical change in the river and bacterial counts showed no significant change. Yet, there are two distinct peaks in the V_{max} values during July and lesser peaks during November.

Based on the uptake of glucose, the Pamlico River estuary was found to be one of the richest or most microbially active bodies of water ever measured; the V_{max} for glucose was 600 to 96,000 $\times$ 10^{-4} μg C/l/hr. Hobbie and Wright (6) measured V_{max} values in Sweden of 3.2 $\times$ 10^{-4} μg C/l/hr in an oligotrophic lake in Lappland and 24 to 400 $\times$ 10^{-4} μg C/l/hr in Lake Erken at different seasons. Also in Sweden, Allen (1) found a V_{max} of 60,000 $\times$ 10^{-4} μg C/l/hr in the highly polluted Lötsjön. In addition, the oceans are areas of very low microbial activity. For example, the western north Atlantic had V_{max} values of 2.0 to 12.0 $\times$ 10^{-4} μg C/l/hr* and the tropical Pacific had values of 5.6 to 540 $\times$

*J. E. Hobbie, personal communication.

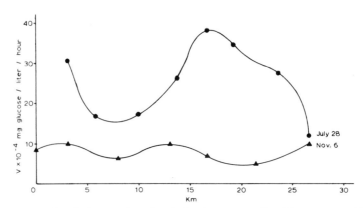

Fig. 4. **Uptake of glucose along South Creek near Pamlico River estuary, 1966.**

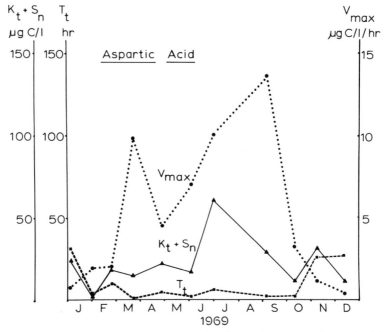

Fig. 5. **The kinetic parameters of the uptake of aspartic acid, South Creek, 1969.**

10^{-4} μg C/l/hr *(3)*. The Antarctic Ocean, the western Mediterranean, and the eastern Atlantic had no detectable uptake when tested by methods used in this study.†

In 1969, the uptake of amino acids and amino acid concentrations were also measured in the estuary, thus allowing the calculation of actual velocities of uptake or flux in addition to the kinetic parameters V_{max}, T_t, and $(K_t + S_n)$. The highest levels of activity occurred in the warm summer months. A representative annual cycle is that for aspartic acid (Fig. 5). Here the turnover time,

†C. C. Crawford, unpublished data.

T_t, was longer during the midwinter months than the rest of the year. In addition, V_{max} was low during these winter months and increased during warmer weather. Peaks of activity, indicated by a large increase in V_{max}, coincided with the peaks of production in the early spring (March) and late summer (August and September). Uptake of other amino acids followed similar cycles.

The amino acids could be divided into two quite distinct groups according to their high or low V_{max} values. The group with a higher V_{max} included arginine, aspartic acid, tyrosine, alanine, and glutamic acid. On those dates when the uptake of serine and glycine was measured, these amino acids were also found to be in this group of high V_{max}. Those amino acids with low V_{max} were proline, valine, lysine, threonine, methionine, leucine, and isoleucine. Phenylalanine was in the lower group in all cases, except in August when it had a very large peak of 17.69 μg C/l/hr. The June value of 0.70 μg C/l/hr was about average throughout the year. These groupings showed no patterns since both the high and low groups have all types of amino acids represented (aliphatic, acidic, basic, etc.) and all sizes of molecules.

Actual velocity of uptake v_n and transport constant (K_t) can be obtained from the parameters of uptake kinetics if one has the actual concentration of the substrate in nature (S_n) (Fig. 6). Since the S_n values are such a small fraction of the $(K_t + S_n)$ obtained from the original kinetic graphs, a graphic representation of K_t alone is virtually identical to the $(K_t + S_n)$ graph.

Those amino acids with the highest V_{max} (tyrosine and arginine) have the lowest actual velocities of uptake and usually show a relatively small S_n and a large K_t. Four amino acids, aspartic acid, glutamic acid, alanine, and threonine, generally showed the highest actual velocities of uptake and were usually re-

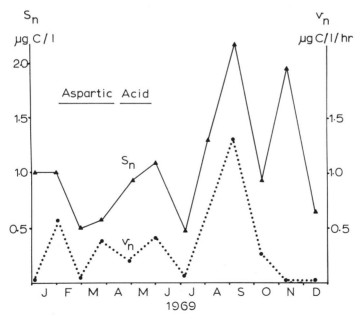

Fig. 6. Concentration (S_n) and natural uptake velocity (v_n) of aspartic acid, South Creek, 1969.

sponsible for over half of the total uptake. The actual velocities of uptake (v_n) represented small fractions of the maximum velocities, 1% to 10% usually, and varied widely both from substrate to substrate and from month to month for the same substrate. For this reason V_{max} should not be used as an estimate of actual velocities of uptake by bacteria.

The net uptake, or the rate of production of particulate organic material attributable to the action of the microbes upon a given substrate, depends not only on velocity of uptake but also on the rate of loss of carbon through respiration, which may be as high as 50% for some substrates. Previous workers (11, 15) failed to measure this loss of $^{14}CO_2$ due to respiration and therefore their estimates of total uptake in both fresh water and seawater are conservative.

The percentage of amino acid carbon taken up and subsequently respired was relatively constant for a given substrate regardless of temperature, incubation period, or substrate concentration. For example, for aspartic acid tested on 29 April 1969 the respiration, measured in duplicate at four concentrations, ranged from 52% to 55% with a mean of 53.75% (SD = 1.16). Percentages respired of the total carbon taken up ranged from 13% (leucine) to 50% (glutamic acid and aspartic acid); the majority of the values lay in the range from 20 to 40%. These respiration rates are in general agreement with the 22% average reported by Williams (13) for oceanic bacteria acting upon an amino acid mixture. While glutamic acid and aspartic acid are both taken up very rapidly, their respiration causes a large loss of carbon and, as a result, the incorporation of these amino acids into particulate organic material is not as high as others. Leucine and valine are in the low ranges of rate of uptake, but due to their low respiratory loss they show the highest rate of incorporation into particulate organic matter. About 60% to 70% of the particulate organic material produced from dissolved amino acids could be attributed to alanine, leucine, valine, serine, glycine, aspartic acid, and glutamic acid. Particulate production averaged 0.79 μg C/l/hr for the year and ranged from 0.06 (December) to 2.37 μg C/1/hr (September) (Fig. 7).

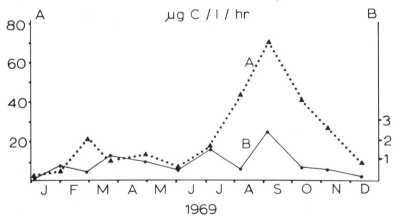

Fig. 7. **Primary production (A) and bacterial production (B), South Creek, 1969.**

By comparing the rates of algal and bacterial production one can easily see that the bacteria are an important source of particulate organic material (Fig. 7). It is remarkable that the production of particulate organic material by bacteria is equal to about 10% of the production by algae during the summer months, especially since this bacterial activity was measured for only twelve amino acids.

In summary, we have presented some results of a method that gives measurements of activity of heterotrophic estuarine microorganisms. Ultimately, the concentrations of all substrates being taken up as well as the uptake rates should be measured. However, it is clear that this is an enormous task. Instead, we must be content with getting information from one or several substrates and making some assumptions about the remainder.

It is obvious that flux data, such as those presented here, must be combined with information on populations and their changes before we can truly define the role of heterotrophic bacteria in estuaries. Even this will be only a first step toward an understanding of the environmental interactions and importance of aquatic microorganisms in these interactions.

Literature Cited

1. Allen, H. L. 1967. Chemo-organotrophic utilization of dissolved organic compounds by planktic algae and bacteria in a pond. M.S. thesis, Michigan State University.
2. Crawford, C. C. 1971. The utilization of dissolved free amino acids by estuarine microorganisms. Ph.D. thesis, North Carolina State University.
3. Hamilton, R. D., and J. E. Preslan. 1970. Observations on heterotrophic activity in the eastern tropical Pacific. *Limnol. Oceanogr.* **15**:395–401.
4. Hobbie, J. E., and C. C. Crawford. 1969. Respiration corrections for bacterial uptake of dissolved organic compounds in natural waters. *Limnol. Oceanogr.* **14**:528–32.
5. Hobbie, J. E., and C. C. Crawford. 1969. Bacterial uptake of organic substrate: New methods of study and application to eutrophication. *Verh. Internat. Verein. Limnol.* **17**:725–30.
6. Hobbie, J. E., and R. T. Wright. 1968. A new method for the study of bacteria in lakes: Description and results. *Mitt. Internat. Verein. Limnol.* **14**:64–71.
7. Kuznetsov, S. I. 1968. Recent studies on the role of microorganisms in the cycling of substances in lakes. *Limnol. Oceanogr.* **13**:211–24.
8. Parsons, T. R., and J. D. H. Strickland. 1962. On the production of particulate organic carbon by heterotrophic processes in seawater. *Deep-Sea Res.* **8**:211–22.
9. Pomeroy, L. E., and R. E. Johannes. 1968. Occurrence and respiration of ultraplankton in the upper 500 meters of the ocean. *Deep-Sea Res.* **15**:381–91.
10. Smith, D. E. 1967. Location of the estrogen effect on uterine glucose metabolism. *Proc. Soc. Exp. Biol. Med.* **124**:747–49.
11. Vaccaro, R. F., and H. W. Jannasch. 1967. Variations in uptake kinetics for glucose by natural populations in seawater. *Limnol. Oceanogr.* **12**:540–42.
12. Webb, K. L., and L. Wood. 1967. Improved techniques for analysis of free amino acids in seawater. In *Automation in analytical chemistry, Technicon symposium 1966*, Vol. 1, White Plains, New York: Mediad.
13. Williams, P. J. LeB. 1970. Heterotrophic utilization of dissolved organic compounds in the sea. *J. Mar. Biol. Ass. U.K.* **50**:859–70.
14. Wolfe, D. A., and C. L. Shelske. 1967. Liquid scintillation and Geiger counting efficiences for carbon-14 incorporated by marine phytoplankton in productivity measurements. *J. Consiel, Consiel Perm. Intrenat. Exploration Mer.* **31**:31–37.
15. Wright, R. T., and J. E. Hobbie. 1966. Use of glucose and acetate by bacteria and algae in aquatic ecosystems. Ecology **47**:447–64.

Comments

ZOBELL: I believe that you must be assuming CO_2 fixation by autotrophic bacteria or else there is an input of organic matter from some other sources in the immediate environment. How could you possibly have more bacterial biomass than that produced by primary production?

CRAWFORD: In the beginning remark that I made, I pointed out the general importance of bacteria and indicated that there were indeed autotrophic bacteria fixing CO_2. There is a tremendous amount of organic material washed into this environment from the shallow areas that are very rich in rooted aquatics. Additionally, there is some agricultural activity in the area. We feel sure that the largest single input of particulate organic material in this estuary is from the rooted aquatics in the shallow areas.

WATSON: The question of how important bacteria are in the ocean is a very intriguing one. I hope more people comment on it. I have always assumed that the bacteria were not responsible for recycling more than 1–10% of the total organic carbon produced by plants. This is not based on any experimental evidence; rather, it is the result of some "after-hours reasoning." It is very difficult to believe bacteria could be responsible for recycling 50% of the organic matter, at least in the oceans as a whole. Dr. ZoBell, you must have thought about this for a long time. How important are bacteria in the ocean?

ZOBELL: I have spent a few minutes thinking about this and working on it. The results of years of work were reviewed independently by Strickland and Seki. They came to the conclusion that my results were conservatively low. I am not going to take the time to tell you the methods I used or the assumptions that had to be put into the thinking and calculations, but the arithmetic is as follows. Annual production of bacterial biomass is of the order of 10^6–10^7 tons per year in the oceans at large. This is a lot of bacterial biomass. Ten million tons is quite a bit of animal food. However, when compared with a primary production of the order of 10^{10} tons per year, it is only a small fraction. These data are for the oceans as a whole. The situation turns around with regard to the difference between bacterial biomass production and primary production in deep waters. Plant production is virtually nil below 1,000 meters. Photosynthetic plants and their bodies are progressively less important and bacterial synthesis of protoplasm is increasingly more important below 1,000 meters. So at greater depths, we have preponderantly more bacterial biomass. More particulate matter is being produced by bacteria than by any other source, because most of the animals producing biomass are dependent upon the detritus that settles down to them for the photosynthetic zone. The bacteria are using dissolved organic matter, colloidal organic matter, and waste organic matter, and they are converting it into particulate material that can be used as a source of food. My figures were, of course, in the order of magnitude calculations. Strickland said we may be low by one order of magnitude. I do not think we could possibly be off by more than this.

WATSON: What percent of the energy captured by the plants is being utilized by the bacteria? Is it more than 1%?

ZOBELL: I do not think so because there are so many predators, so many kinds of animals that are eating plant material. I do not think it is anything like 1%.

WATSON: The bacteria and the fungi use 50% of the energy that is captured by the plants in the terrestrial environment. A different situation is true in the marine environment because there are so many intermediate animals in the ecosystem. Therefore, bacteria really have a minor role in the ocean compared to terrestrial environment.

HAMILTON: Since you used scintillation counting, why do you use counts per minute and not disintegrations per minute?

CRAWFORD: The machine I am working with is not dependable in both channels.

HAMILTON: I gather you are summing your counts per minute from CO_2 collections. Are the counts observed on the filters?

CRAWFORD: That is correct, but that is after correcting the counts from the machine for the differences in efficiencies for counting the two types of filters.

HAMILTON: You call your calculated values a transport constant. The constant you calculate from that kind of summation is a transport plus, or set of enzymes constants plus, the release of CO_2. I question that you can call that a transport constant any longer.

CRAWFORD: I agree. It was originally called the transport constant because the transport system is the first system that the bacterium uses when it comes in contact with a substrate by the bacterium, the bottleneck inside the bacterium could be the constant that is read.

HAMILTON: It is a constant, but you do not know which constant it is. Perhaps utilization constant would be a better term.

ALBRIGHT: You cannot even use that term. It is not utilization either. It could be transport or utilization. We do not know.

CRAWFORD: What would you call it? We cannot call it anything at this point.

HAMILTON: I would just refer to it as "the constant." Then, you can state in your papers that you do not know exactly what it is. If you call it K_t or K_m, you will get papers back with reviewers' red marks all over it.

WRIGHT: I have one comment on the amount of organic matter in the sea fixed by algae and used by bacteria. Taking into consideration the production of excretory products by the algae, a higher portion of their production can be utilized by bacteria, that is, it is 1 to 10% of the total algal production. This represents soluble organic compounds that are usually of fairly low molecular weight. Bacteria can readily use these. If you add that to the total amount available, it would get above 1%.

ZOBELL: I agree with you, and I should have emphasized my calculations. They do not take into account the amount of carbon fixed by algae that may be lost. We know that this is appreciable. Much of that, I believe, can be used by bacteria to increase the bacterial biomass. I omitted a very important item that I should have mentioned to correct my figures. My calculations and assumptions do not take into account the bacterial biomass in the intestinal tract of animals that ingest the organic matter. Also, I did not take into account the microflora associated with integument of animals. This may be very appreciable, but I could not put it into the calculations because the estimates were not good enough to record how much assimilation takes place in association with animals, either on the integument or in the gut. We do know that the animals, like the copepods which may have a thick covering of chitin, have a very dense population of chitinoclastic bacteria associated with them. The chitin alone that is produced by Crustaceans represents a figure higher than I gave you first for the annual production of bacterial biomass. I do not know how much of this would be used in the production of bacterial biomass. This is a factor that would increase the conservative figures—perhaps by an order of magnitude. Neither the excretion of organic matter by photosynthetic plants nor the metabolism that would take place in association of animals was considered in my experiments.

ALBRIGHT: Your associate, Dr. Hobbie, has made some rather interesting speculations concerning the use of this technique for comparing various waters, polluted and unpolluted. Would you comment on this?

CRAWFORD: The figure illustrating the sampling down the creek demonstrated the ability of this technique to detect very subtle changes in the organic and microbial content of the water. It showed little peaks at the two points of pollution. For that reason, it should be very useful in comparing areas of the same body of water—sampling upstream and downstream from an affluent. We are looking for the effects of an effluent on the water. It could also be standardized to be used as commonly as a BOD measurement.

WATSON: When looking for pollution, would you expect inhibition or an increase? In one figure, you show a tremendous inhibition.

CRAWFORD: You find the same thing with BOD. Sometimes a higher BOD is obtained, and sometimes lower. It would depend on whether the pollution was mercury salts or human sewage, for example.

WATSON: I do not think that it is a good index of pollution, because you can get inhibitions as well as stimulation. I do not know how you would interpret it. Last year when we were working on the Savannah River, we used this technique and I do not think it made any sense. It went all over the map.

CRAWFORD: Would not the same thing hold true for something as commonly used as the BOD? That also goes all over the map at times.

WATSON: Not as much.

COLWELL: To support Dr. ZoBell's comment about the association of bacteria and plankton, I would estimate that roughly 10 to 16% of the net weight of plankton consists of bacteria. This is a substantial source of carbon.

Monosaccharide and Disaccharide Interactions on Uptake and Catabolism of Carbohydrates by Mixed Microbial Communities

Lindsay W. Wood

Uptake of simple monomers (amino acids, monosaccharides, and miscellaneous other compounds) have been examined for mixed microbial communities by Allen *(2, 3)*, Wetzel *(35)*, Crawford *(8)*, and others. In most estuarine and freshwater studies, the use of the kinetics methods of Wright and Hobbie *(36)*, as later modified to correct for $^{14}CO_2$ loss by Hobbie and Crawford *(17)*, yielded a figure for velocity at saturation of the uptake sites, a maximum concentration of substrate $(K + S)$, and a turnover time represented by $K + S/V$. To obtain true velocity from potential values, S (the natural substrate concentration) must be determined by an independent method. Then K, (constant) and flux (v) can be calculated *(8, 18)*. However, basic assumptions of the methodology enabling comparison of potentials have been tested only on isolates from the community *(3, 36)*. Methodological assumptions of concern in this paper include (1) similar responses by the organisms in the community to the substrate in question and (2) the applicability of first-order saturation kinetics to mixed enzyme and substrate systems. Only Crawford *(7, 8)* has attempted to examine interactions of substrates in the community as a whole with amino acid additions.

To determine whether other potential carbohydrate substrates in aquatic systems affect uptake and the extent of this interaction, the uptake of four ^{14}C-labeled sugars alone and in the presence of seven unlabeled saccharides were tested. If interactions between carbohydrates were observable to the extent that the velocity of uptake of the labeled material was influenced, V, $K + S$, and $K + S/V$ resulting from measuring this velocity would be altered. This alteration would be a direct result of another compound on the velocity of uptake of the substrate in question at the addition levels used in the experiment. If interactions are found, comparisons of the potentials of the heterotrophic parameters of Wright and Hobbie *(36)*, in the same community at different times or between communities, should be approached with caution.

Methods

The work reported here, for convenience, relied quite heavily on an aquarium microcosm with a sandy sediment. To stimulate a natural system, oligochaetes and other invertebrates, water inocula, and plant material both from Hanlan's Point, Toronto Harbour, and a phycologically endowed goldfish bowl were added. Charcoal-filtered, dechlorinated water from Lake Ontario was initially used to fill the aquarium (15 gal). Subsequently, distilled water was used to replace water lost by evaporation or sampling. This system was allowed to stabilize for three months prior to the first sampling. A 1 l flask dipped into the microcosm served as a sampling apparatus, and the sample was processed within 15 min of collection.

Because it could be argued that a mixed community of a stagant aquarium held little relationship to those found in nature, some experiments were done with water samples taken from Lake Ontario. These samples were collected one-half mile west of the mouth of the Humber River with a clean plastic bucket. The contents were then poured into an insulated glass bottle for transport to the laboratory. Processing of the samples was completed within 1 hr of collection.

The procedures of Hobbie and Crawford (17), which allow for correction of the net uptake (36), formed the basis for all our experimental work. Briefly, 25 ml flasks, to which uniformly labeled ^{14}C-organic substrate (1–4 μc/l) had been added at various concentrations, were inoculated with 5 ml of water from one of the study systems. Prior to the addition of the microbial community, the control flask was fixed with 0.2 ml of 2 N H_2SO_4 or the same amount of a solution containing 60g/l trichloroacetic acid (TCA) plus 1 mg/l $HgCl_2$ in 10% formalin. Because the H_2SO_4-fixed preparations showed a loss of activity over time as well as changes in blank counts depending upon the amount of labeled material used and the filtration pressure, H_2SO_4 was replaced after 3 May 1971 by the mercuric TCA in formalin which did not exhibit these drawbacks.

The flasks were sealed with a serum stopper through which a plastic cup containing a 2 × 5 cm piece of folded Whatman No. 1 chromatography paper had been inserted. The closed systems were incubated at *in situ* temperatures on a shaking table for 1 to 1.5 hr. Following incubation, the reaction was stopped by injection of 0.2 ml fixative. The fixative served both to stop the reaction and to lower the pH. A second injection was employed to place 0.2 ml of phenethylamine on the chromatography paper wick for absorption of $^{14}CO_2$.

A 45 min shaking followed the injections to allow for entrapment of $^{14}CO_2$. The wicks were then removed and placed in liquid scintillation vials, while the fixed communities were filtered onto 0.45 μm membrane filters (Millipore Corp.). The membranes were washed with 10 ml water, placed in a second set of scintillation vials, and air dried at room temperature. After drying both sets of vials received 15 ml of Omnifluor (New England Nuclear Corp.) and were counted in the Packard Tricarb Scintillation Counter (model 3320). Quench was corrected by the use of an internal standard wick. This factor resulted from experiments comparing entrapment of $^{14}CO_2$ and $^{14}CO_2$ in solution. Studies of the interactions of carbohydrates were performed using unlabeled stock solutions adjusted so that 5 to 20 μl would give the desired concentration. A value (v^*) which was calculated for some of these experiments represents the velocity of uptake of the labeled material. The other experiments utilized a linearization of the Michaelis-Menten equation.

The Hanes double reciprocal plot (9) as modified by Wright and Hobbie (36), was used for the calculation. This equation,

$$\frac{C\mu t}{c} = \frac{t}{f} = \frac{K+S}{V} + A\left(\frac{1}{V}\right),$$

utilizes the incubation time *(t)* over the fraction of added isotope taken up *(f)* as the *Y* axis and the amount of added substrate *(A)* as the *X* axis to allow calculation of *V* (the inverse of the slope), turnover time *(Y* intercept) and maximum substrate concentration *(−K + S* at the negative *X* intercept). Because this equation emphasizes some values over others *(10)*, the data were calculated using linear regression techniques and experiments not significant at the 3% level were rejected. Where questions of difference of slope or elevation were questionable as to significance, analysis of covariance was used *(28)*.

Results

Experiments with radioactive maltose, sucrose, glucose, and fructose showed that uptake over the 1 to 1.5 maximum incubation time was constant at all concentrations used (Table 1). Only fructose showed any complications of change in uptake over time (Fig. 1), and all levels of addition showed the break at about 1.3 hr incubation time. Consequently, every effort was made to keep fructose incubation shorter than 1.3 hr. The respiration pattern of fructose indicated that a minimum of 0.5 hr was required for the respiration to become steady. Thus, a 30-min incubation period was required for the labeled and unlabeled fructose to come to a steady state.

Table 1. Final substrate concentrations in flasks after adding various amounts of labeled stock solution to 5 ml of suspended microbial community

Substrate	μg per Liter	μl Added
Maltose	0.66	5
	1.32	10
	1.98	15
	2.64	20
Sucrose	6.43	5
	12.86	10
	19.30	15
	25.73	20
Fructose	2.30	5
	4.61	10
	6.91	15
	9.22	20
Glucose	0.63	5
	1.27	10
	1.90	15
	2.54	20

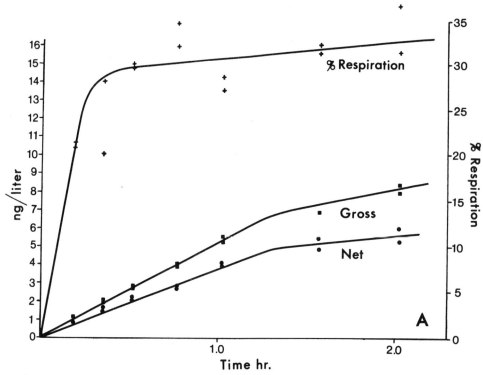

Fig. 1. Gross (corrected for respired substrate) and net (within cell retention of the substrate) uptake (ng/l) of fructose by the community in the aquarium microcosm. Only the curves for the addition of 9.22 µg/l are shown.

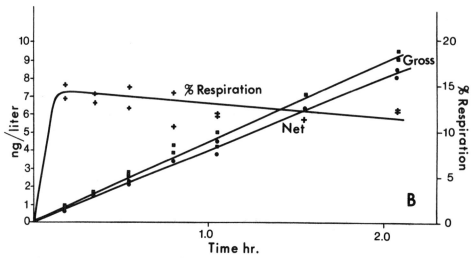

Fig. 2. Uptake of glucose (ng/l) by the community in the aquarium microcosm. Only the curves for the addition of 2.54 µg/l are shown.

Glucose reached the steady state much quicker, in less than 0.18 hr (Fig. 2), but, contrary to fructose, the respiration declined with time. Uptake of glucose was constant over the 2-hr period for all concentrations used.

Maltose uptake showed a similar linearity of uptake over a 2-hr incubation period (Fig. 3) and, like fructose, seemed to require 0.5 hr for a steady state to develop. Because the data points beyond 1 hr were variable, incubation times for this substrate were limited to 1 hr.

Sucrose showed good linearity of uptake over a 2-hr period and reached a steady state in substrate respiration after 0.35 hr of incubation (Fig. 4). By using the concentrations shown and performing all four of the preceding experiments at the same time on the same community, the respiration of sucrose was seen to reach a steady state intermediate in time between its constituents, glucose and fructose.

Comparison of the rates of uptake of the labeled material *(v*)* at 1 hr showed no relationship between sucrose and the monosaccharide components. Both the net *v** (rate of uptake of the labeled material retained within the cells of the microbial community) and the gross *v** (net *v** corrected for respiratory loss) were greater for sucrose than for either glucose or fructose, or the two combined. The respiration value for sucrose at 1 hr was 18%, which was intermediate between those for glucose (13%) or fructose (31%). Maltose, however, showed a net *v** of 3.8 ng/l/hr, which was slightly below that for glucose (4.0), and a gross *v** of 4.4, which was equal to that of glucose. The respiration was higher (15% at 1 hr) than that for glucose. Thus, no relationship appeared to exist between uptake of sucrose and its monosaccharide components, but maltose utilization appeared to be related to glucose uptake.

Further support for a lack of correspondence of sucrose uptake with the uptake of its constituent monosaccharides came from a comparison of the heterotrophic parameters, turnover time *(t)*, potential velocity of uptake *(V)*, and percent respiration (Table 2). The turnover time for sucrose was less than for its constituents, the gross *V* was greater than for the constituents, singly or added together, and the *K + S* value was much greater than for the constituents. Contrary to the preceding data, the percent respiration for sucrose in this experiment (16%) was less than either fructose (which agreed with the 19 July data) or glucose (which was much higher than the 19 July data).

Since there was not a correspondence between sucrose uptake and the uptake of the hexoses that form the disaccharide, the possibility of disaccharide uptake in addition to that of monosaccharides became a central question. Two alternative explanations existed for oligosaccharide uptake. Either the disaccharides were broken at the surface of the cell with a loss of a Kcal/mole and the resulting monosaccharides taken up, or the disaccharide was consumed as a whole molecule which was then split within the components of the microbial community, conserving that Kcal/mole as well as possible energy savings related to the activation of the transport system.

Data in support of the latter hypothesis came from studies with maltose showing that labeled maltose uptake *(v*)* was not inhibited by the presence of cellobiose, lactose, glucose, galactose, or fructose (Table 3). The addition of 40 μg/l of unlabeled maltose increased net labeled maltose uptake by 25.6% while 40 μg/l of sucrose increased net maltose uptake by 56.3%. As shown by the lack of inhibition of maltose uptake and its stimulation by the presence of sucrose, there existed (in mixed communities) systems whereby disaccharides were directly

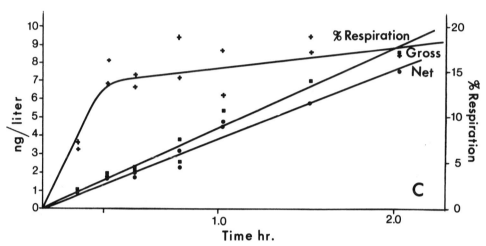

Fig. 3. Uptake of maltose (ng/l) by the community in the aquarium microcosm. Only the curves for the addition of 2.64 μg/l are shown.

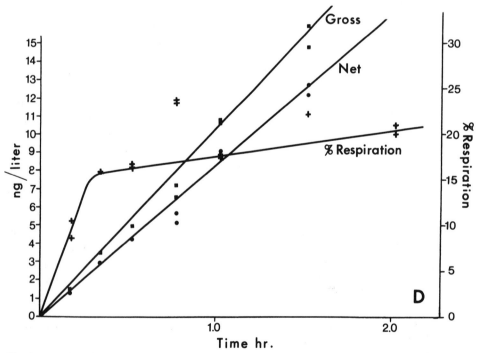

Fig. 4. Uptake of sucrose (ng/l) by the community in the aquarium microcosm. Only the curves for the addition of 6.43 μg/l are shown. All determinations done on 19 July 1971.

Table 2. Uptake of sucrose-U-^{14}C, and its component monosaccharides, fructose-U-^{14}C and glucose-U-^{14}C, using aquarium water

Substrate*	t (hr)	V (μg/l/hr)	$-K + S$ (μg/l)	Percentage Respiration	r
Sucrose-U	7.12	9.70	69.08	16	0.959
Glucose-U	8.14	1.02	8.27	24	0.971
Fructose-U	16.98	0.47	8.00	27	0.999

*2 March 1971; incubation temperature 22 C.

Table 3. Uptake (v^*) of labeled maltose-U-^{14}C (1.98 μg/l additions) by the microbial community of Lake Ontario water in the presence of 40 μg/l additions of other unlabeled carbohydrates

Substrate*	v^* (ng/l/hr)		Percentage Respiration
	Gross	Net	
Maltose	276	173	37
+Maltose	369	187	34
+Sucrose	673	467	32
+Cellobiose	265	173	35
+Lactose	271	166	37
+Fructose	282	174	39
+Glucose	244	195	36
+Galactose	266	163	36

*Incubation and water temperature 14.5 C.

taken up as entire molecules at low substrate concentrations. Also, like sucrose, maltose uptake was not associated with monosaccharide uptake.

Preliminary investigations into the nature of the system by which maltose was taken up suggested an allosteric affinity for the system. Calculation of the actual velocities and the substrate concentrations indicated a sigmoid rather than a rectangular hyperbola function was associated with the uptake of maltose and alleviated the possibilities of random and compulsory two-substrate systems (4, 5, 6). These two systems can be solved by additions of a single concentration of substrate B while varying substrate A (9).

The maltose system could not be solved by single additions of sucrose. Such additions resulted in straight lines indicative of diffusion (3, 36) similar to the line M and line M + S (Fig. 5). However, when the addition of unlabeled sucrose was varied along with the addition of labeled maltose, solvable kinetic expressions resulted. Thus, there is a direct dependence of the uptake of maltose on the presence of sucrose. The relationships showed an insignificant difference in slope ($F = 5.967$; 1,6 df) for the 1–4 μg/l additions of sucrose, but a significant increase in elevation when compared with the 10–40 μg/l additions ($F = 14.704$,* 1,7 df). Quite obviously, there was a significant difference in slope for the 100–400 μg/l additions of sucrose, (Table 4). These data further supported the sigmoid function hypothesis, as tangents to various sections of a sigmoid curve

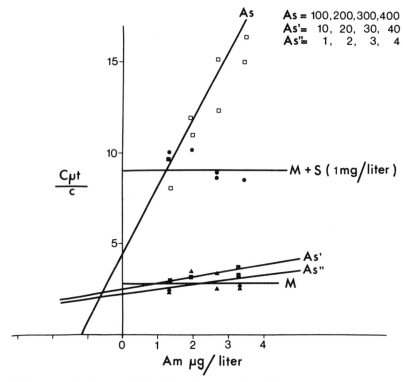

Fig. 5. Maltose uptake by a probable allosteric system involving sucrose as one of the possible effector substrates.

should show the changes in the $K + S$ figures observed. Such relationships could only be obtained if the additions (in this case, maltose) were kept sufficiently close together to allow tangent slopes leading to V to be calculated. This was achieved with a difference of 1.98 μg/l between the lowest and highest levels of addition.

Table 4. Maltose uptake as influenced by various sets of additions of sucrose

Sucrose Addition (μg/l)	t (hr)	V (μg/l/hr)	$-K + S$ (μg/l)	Percentage Respiration	r
100–400	4.45	0.29	1.28	33	.939
10–40	2.52	3.37	8.51	36	.969
1–4	2.27	3.02	6.84	37	.985

Interrelationships of the two substrates, as might be expected, would also be complicated by the amount of substrate catabolized. Maltose alone (36% respiration) and maltose with a single addition of 1 mg/l sucrose (34%) were both close to the values of 33% for the 100–400 μg/l additions of sucrose and 36% for the 10–40 μg/l additions of sucrose. The expected relationship of greater maltose catabolism without sucrose, and less with 1 mg/l sucrose was not found.

As these experiments with Lake Ontario water and the aquarium microcosm indicated, the presence of other carbohydrates (both monosaccharides and disaccharides) may not repeatably yield the simplified pattern for a single sub-strate uptake hypothesis; thus, inhibition experiments at various addition levels were performed. The first series were designed to determine the effects of different levels of unlabeled carbohydrates on uptake of a labeled carbohydrate (v*).

Labeled fructose uptake was decreased by all of the carbohydrates tested, with the possible exception of sucrose (Table 5). All of the substrates, except for galactose, showed a definite effect at 40 μg/l. Galactose decreased net uptake at 40 μg/l and gross uptake at 200 μg/l. Both net and gross uptake of labeled material showed similar inhibition patterns, differing in extent, depending upon the amount of labeled substrate catabolized. These data showed a stimulation of labeled fructose oxidation by more unlabeled fructose, glucose, cellobiose, and maltose at the 40 μg/l addition level. Lactose was stimulatory to ^{14}C-fructose oxidation at the 200–400 μg/l level, and none of the carbohydrates was inhibitory, except for possibly galactose at 200 μg/l concentrations.

Glucose also showed similar patterns of decreasing net and gross ^{14}C-uptake figures. Unlike fructose, lactose was stimulatory to glucose uptake at 400 μg/l, and fructose may have shown enhancement at 40 μg/1 (Table 6). Cellobiose,

Table 5. Inhibition of the uptake of labeled fructose 14 C (net $(N)v$* = 925; gross $(G)v$* = 1026 ng/l/hr and respiration %R = 10%) in the presence of other carbohydrates (8 August 1971)

| Concentration (μg/l) | 40 | | | 200 | | | 400 | | | 1,000 | | |
Substrate	Nv*	Gv*	%R	Nv*	Gv*	%R	Nv*	Gv*	%R	Nv*	Gv*	%R
Fructose	273	330	17	202	256	21	194	226	21	176	220	20
Glucose	585	717	18	546	661	17	617	747	17	615	727	15
Galactose	883	997	12	787	849	9	681	782	13	680	766	11
Sucrose	740	848	13	782	900	13	780	869	10	790	899	12
Maltose	679	787	14	643	734	12	575	672	12	578	685	16
Cellobiose	624	747	17	594	700	15	477	568	16	606	665	9
Lactose	774	859	10	732	848	14	593	689	14	658	807	10

Note: Values represent means of replicate determinations minus the control for the addition level.

Table 6. Uptake of labeled glucose (v*) in the presence of other carbohydrates by the community in the aquarium microcosm (9 August 1971)

| Concentration (μg/l) | 40 | | | 200 | | | 400 | | | 1,000 | | |
Substrate	Nv*	Gv*	%R	Nv*	Gv*	%R	Nv*	Gv*	%R	Nv*	Gv*	%R
Fructose	652	732	11	533	604	12	571	650	12	506	569	11
Glucose	161	196	17	88	116	24	96	111	14	76	97	21
Galactose	514	545	6	483	512	6	384	404	5	318	343	7
Sucrose	424	452	6	608	690	12	527	589	11	493	530	7
Maltose	453	512	11	373	426	13	327	379	14	277	322	14
Cellobiose	399	442	10	270	300	10	223	255	12	341	386	12
Lactose	618	684	10	555	615	10	703	825	15	465	513	9

Note: The control values are net $(N)v$* = 575; gross $(G)v$* = 658 ng/1/hr, 13% respiration (%R) and natural substrate concentration = 7 μg/1.

sucrose, and maltose definitely slowed net uptake at 40 $\mu g/l$, and this group was joined by galactose when the gross data were considered.

Contrary to fructose, all additions of carbohydrates decreased catabolism of labeled glucose, at 40 $\mu g/l$ additions. Unlabeled glucose greatly enhanced oxidation at the same level. At 400 $\mu g/l$, however, labeled glucose enhancement by unlabeled glucose was joined by maltose and lactose; the latter again became inhibitory at 1 mg/l.

Uptake of labeled sucrose showed an intermediate picture between those of its constituent hexoses in being inhibited at 40 $\mu g/l$ of maltose and glucose while enhancement at this level was observed for galactose, fructose, and lactose (Table 7). Cellobiose definitely decreased gross labeled sucrose uptake at 40 $\mu g/l$. In keeping with the interaction between sucrose and maltose, the least inhibition of sucrose by this substrate was at 200 $\mu g/l$.

Table 7. Uptake of labeled sucrose (v^*) in the presence of other carbohydrates for the community of the aquarium microcosm (6 August 1971)

| Concentration ($\mu g/l$) | 40 | | | 200 | | | 400 | | | 1,000 | | |
Substrate	Nv^*	Gv^*	$\%R$	Nv^*	Gv^*	$\%R$	Nv^*	Gv^*	$\%R$	Nv^*	Gv^*	$\%R$
Fructose	779	898	13	849	969	12	747	855	13	754	877	14
Glucose	484	70	15	462	495	6	349	390	10	423	454	7
Galactose	820	933	12	689	783	12	691	759	9	531	558	5
Sucrose	100	127	20	54	72	25	54	68	20	20	44	31
Maltose	356	389	8	455	488	7	382	415	8	320	342	7
Cellobiose	579	626	8	397	431	8	380	405	6	346	369	6
Lactose	797	916	13	759	880	14	713	823	13	598	690	13

Note: The values for sucrose alone are net $(N)v^* = 638$; gross $(G)v^* = 729$ ng/l/hr and 12% respiration *(%R)*.

Respiration of labeled sucrose was stimulated by more sucrose, fructose, and glucose at 40 $\mu g/l$, and decreased by galactose, lactose, and maltose at the same level. Glucose became inhibitory of sucrose catabolism at 200 $\mu g/l$.

Maltose utilization was similar to fructose in that all added carbohydrates decreased uptake at 40 $\mu g/l$, while (unlike fructose) cellobiose, glucose, and galactose slowed uptake more than the similar addition of unlabeled maltose. Fructose, sucrose, and lactose, the other carbohydrates, showed less inhibition than the addition of a similar amount of the unlabeled substrate. Of direct interest was the pattern at a concentration of 200 $\mu g/l$ where all of the carbohydrates were more inhibitory than maltose, and at 1 mg/l, where maltose was the greatest inhibitor of labeled maltose uptake and oxidation (Table 8).

Again, the respiration values were similar to the fructose pattern at 40 $\mu g/l$, where all of the carbohydrates tested, except sucrose, enhanced maltose catabolism. However, at 1 mg/l lactose, cellobiose, and fructose, along with sucrose, decreased the oxidation of maltose.

From the data, an interaction of substrates potentially present in the milieu was seen to affect the velocity of uptake of added labeled material. The concentration of only one of the substrates in the system was available. Glucose determination by the method of Hicks and Carey *(15)*, using biochemical and fluorescent techniques, showed 7.0 $\mu g/l$ of this substrate to be present at the time the

Table 8. Uptake of labeled maltose *(v*)* in the presence of other unlabeled carbohydrates by the microbial community of the aquarium microcosm (7 August 1971)

Concentration (µg/l)	40			200			400			1,000		
Substrate	Nv*	Gv*	%R	Nv*	Gv*	%R	Nv*	Gv*	%R	Nv*	Gv*	%R
Fructose	75	95	22	81	101	19	78	98	20	91	116	17
Glucose	56	72	23	46	60	23	32	44	27	28	39	28
Galactose	66	83	21	65	81	20	62	76	19	54	68	26
Sucrose	122	148	17	101	120	15	83	98	15	95	110	13
Maltose	73	90	19	113	136	17	70	87	20	16	25	34
Cellobiose	51	64	20	42	53	21	46	55	18	28	33	15
Lactose	93	118	21	89	108	18	71	86	17	138	164	15

Note: The values for maltose alone are net *(N)v** = 182; gross *(G)v** = 221 ng/l/hr, and 18% respiration *(%R)*.

experiments were performed. Taking this into consideration would not alter the patterns observed for *v**; for ease of comparison with the other experiments where the substrate concentrations were unknown, the data were not corrected. In either case, the inescapable conclusion could only be that *v**, net or gross, from which the parameters V, $K + S$, and $K + S/V$ (as turnover time) were calculated, was directly related to interactions with other carbohydrates potentially available to the community. Because of this interaction, the calculations of V should show typical noncompetitive or mixed inhibition patterns.

Experiments on the effects of unlabeled additions to communities of microorganisms confirmed the hypothesis of mixed or noncompetitive inhibitions. Fructose (Table 9) showed noncompetitive or mixed inhibition in the presence

Table 9. Influence of other carbohydrates on gross uptake of fructose by mixed microbial community of aquarium microcosm and that of Lake Ontario inshore water

Level of Addition (µg/l)	Substrate	System + Date + Temp.	t (hr)	V (µg/l/hr)	−K + S (µg/l)	Percentage Respiration	r
0	Fructose	Aquarium	7.00	0.471	7.998	27	0.999
1,000	Sucrose	2 Mar. 1971 21 C	133.28	0.048	6.437	35	0.994
0	Fructose	Aquarium	4.76	0.748	3.568	16	0.985
400	Sucrose	20 Mar. 1971	6.20	0.811	5.028	17	0.990
400	Lactose	21 C	6.81	0.618	4.207	17	0.994
400	Maltose		7.09	0.819	5.811	16	0.976
0	Fructose	Aquarium	6.15	2.214	13.615	28	0.973
200	Galactose	7 May 1971	9.65	1.082	16.445	29	0.855
200	Cellobiose	21 C	4.83	1.250	6.038	29	0.965
200	Sucrose		5.35	1.246	6.672	30	0.986
0	Fructose	Lake Ont.	30.53	0.297	9.076	23	0.940
40	Lactose	27 May 1971	27.39	0.296	8.109	20	0.981
40	Maltose	7.5 C	36.96	0.435	16.070	25	0.964
40	Sucrose		35.24	0.408	14.408	26	0.941
40	Cellobiose		42.00	0.335	14.066	28	0.880

of 1 mg/l sucrose and 400 μg/l lactose. However, for the 400, 200, and 40 μg/l additions of sucrose, the regression lines met in the first rather than the second or third quadrant (31). Sucrose additions of 1 mg/l inhibited glucose, but showed an interaction in the second quadrant at 400 μg/l additions (Table 10). Lactose at 400 μg/l additions similarly intersected glucose alone in the second quadrant.

Table 10. Influence of other carbohydrates on the gross uptake of glucose by the mixed microbial community of an aquarium microcosm

Level of Addition (μg/l)	Substrate	System + Date + Temp.	t (hr)	V (μg/l/hr)	−K + S (μg/l)	Percentage Respiration	r
0	Glucose	Aquarium	8.49	0.881	7.471	24	0.979
1,000	Sucrose	2 Mar. 1971					
		21 C	106.77	0.073	7.762	22	0.974
0	Glucose	Aquarium	4.81	1.269	6.099	18	0.938
400	Sucrose	20 Mar. 1971	4.96	0.685	3.401	18	0.938
400	Lactose	21 C	9.01	0.745	6.714	19	0.958
400	Maltose		5.06	1.668	8.447	18	0.997

Sucrose, which by itself can be shown to produce the rectangular hyperbola for uptake by mixed microbial communities, showed a response similar to those of the monosaccharides (Table 11). Plotting the data for experiments with 1 mg/l additions of monosaccharides according to the method of Hanes (9), three distinct $K + S$ values were obtained: 16.43 for galactose, 35.50 for glucose, and 50.63 for fructose. These results basically agreed with those for monosaccharides and at other levels of addition of unlabeled material, in that the most common result was change in slope producing either mixed or noncompetitive inhibition. Even sucrose was noncompetitively inhibited by sucrose in natural communities when 40 μg/l of unlabeled sucrose was added to the experimental flasks. These results confirm the indications of the previous set of experiments that, in natural systems as well as laboratory microcosms, the uptake of any particular carbohydrate substrate can be affected by the presence of other carbohydrates also present in the system.

Discussion

From the data presented here and by Hamilton and Austin (14), the parameters of V and $K + S$ should be interpreted carefully for natural systems, whether the net (36) or the gross (respiration corrections of Hobbie and Crawford, 17) values are determined. Of greater utility is the method for determination of flux of substrates in nature (8), necessitating the determination of natural substrate concentrations by different methods. Amino acid analysis techniques (34) and biochemical assay techniques (15) are available. The dilution bioassay, first proposed by Hamilton and Austin (14), and used by Allen (1), for bioassay

Table 11. Influence of other carbohydrates on gross uptake of sucrose by the mixed microbial community of an aquarium microcosm and that of Lake Ontario inshore water

Level of Addition (μg/l)	Substrate	System + Date + Temp.	t (hr)	V (μg/l/hr)	$-K+S$ (μg/l)	Percentage Respiration	r
0	Sucrose	Aquarium	7.12	9.70	69.08	16	0.959
1,000	Lactose	2 Mar. 1971	5.82	7.38	42.95	12	0.986
1,000	Fructose	21 C	15.89	3.19	50.63	14	0.994
1,000	Glucose		136.48	0.26	35.50	19	0.982
1,000	Galactose		100.19	0.16	16.43	28	0.970
0	Sucrose	Aq. H$_2$O	36.80	1.29	47.58	19	0.945
400	Lactose	2 Mar. 1971	172.82	0.17	28.81	23	0.863
400	Maltose	21 C	42.93	1.06	45.45	20	0.991
0	Sucrose	Lake Ont.	5.36	2.86	15.345	33	0.920
40	Sucrose	19 Mar. 1971	3.71	4.22	15.665	31	0.937
40	Maltose	15 C	5.53	13.58	75.165	33	0.855

with pure cultures of bacteria *(19)* should only be used with extreme caution. Indeed, that the problems encountered in this study are present in other systems was confirmed by the work of Allen *(1)*. He found changes in $K+S$ intersection of the diluted sample at points less than, greater than, and equal to the undiluted control for acetate. The work of Vaccaro, et al. *(32)* generally agreed with the Hobbie-Wright technique, but as a rule the bioassay technique showed less glucose than the enzymatic technique, indicating the possibility of similar interaction problems.

The concentrations of added substrates were undoubtedly much higher than natural inputs to aquatic ecosystems. Placing flux (rate of uptake) equal to rate of generation with natural substrate being a minimum, Vaccaro et al. *(32)* reported generation rates of, from less than 0.001 to 0.006 μg/l/hr for glucose in the Atlantic Ocean. Similarly, in Toronto Harbour water, with natural glucose concentrations in the range of those for the Atlantic Ocean, generation rates of 0.199 to 0.811 μg/l/hr were obtained *(38)*. The work of Hobbie, Crawford, and Webb *(18)* and Crawford *(8)* showed similar low figures for amino acid generation in estuarine waters. However, in Toronto Harbour sediments, the concentrations used were well within generation values of 0.626 to 4.702 mg/l/hr *(38)*. Thus, if similar mechanisms for sediment heterotrophic activity (bacteria, wherever they occur, representing multiple substrate and multiple enzyme systems), these experiments could well apply to at least this region of aquatic ecosystems.

The uptake of maltose, by what appears to be a two-substrate system having affinities to allosteric systems *(25)* and requiring sucrose, suggests that multiple substrate uptake mechanisms might be present in microbial communities. That multiple substrate uptake systems are present in some bacteria for carbohydrates was established by Egan and Morse *(11)*. They found that a single genetic locus in *Staphylococcus aureus* controlled the uptake of at least eight carbohydrates. Dixon and Webb (9) stated that multiple substrate systems are quite common and using the Lineweaver-Burk plot (1/v vs 1/S), results for some of the data similar to those of Frieden (12, 13) were obtained. If so, the equation:

$$\frac{VS}{K+S} = v = \frac{V}{1 + \frac{K}{S}} \tag{1}$$

presented by Wright and Hobbie (36) might better be expressed:

$$v_A = \frac{V_A}{1 + \frac{K_A}{[A]} + \frac{K_B}{[B]} + \frac{K_{AB}}{[A][B]}} \tag{2}$$

for a two-substrate reaction, or, more generally:

$$v_A = \frac{V_A}{1 + X_A + X_B + L_{AB}} \tag{3}$$

where $X_A = K_A/(A)$, or an activity coefficient (K_A) for a substrate concentration A, and $L_{AB} = K_{AB}/(A)(B)$ or an interaction L_{nm} between X's with subscripts n and m. Expansion to cover a reaction with i substrates and m interactions produces the general model:

$$V_A = \frac{V_A}{1 + \Sigma X_i + \Sigma L_m} \tag{4}$$

where m is exponentially related to i and has the constraint of:

$$L_{kj} = L_{jk} \tag{5}$$

where j and k are two specific substrates in the total of i substrates. In all cases, the holding of all but one substrate constant will produce equation (1) for that substrate, but unless all inhibitory factors are minimized and stimulatory factors maximized, V will not be maximal.

For isolated enzymes, substrate concentrations, activity coefficients, and V_{max} have been studied for multiple substrate systems (4, 5, 6, 13) so that the system holds for simple substrate-enzyme relations. For multiple substrate and multiple enzyme systems, Wright and Hobbie (36) have shown the multiple substrate system can be reduced to a simple equation by varying only one of the substrates.

Because the amounts of the other substrates present in the milieu of microbial communities are unknown for the most part, and the interaction possibilities are enormous, multiple substrate systems are probably operating, as the data presented here tend to suggest. The heterotrophic parameters used for inter- or intracommunity comparisons by Wood (37), Hobbie (16), Wetzel (35), and Allen (2, 3) would then be of little use.

Extension of the two-substrate system hypothesis to other groups of compounds was indicated by Crawford (7, 8), who showed that amino acid uptake was also possibly affected by multiple substrate systems in some cases. He found, for example, that amino acid uptake in estuarine water was competitively inhibited by others of similar structure. However, stimulation of aspartic acid uptake by 100 μg/l additions of alanine, and serine by 100 μg/l of tyrosine was also found. Apparently much of this work at 100 μg/l additions was similar to the (v^*) work reported here for carbohydrates.* Therefore, only a rough comparison can be made of actual interactions between the substrates.

Organisms living in a dilute medium of energy sources, such as the microbial community in nature, would have a distinct advantage if multiple substrate uptake systems were operative. If the requirement of 1 ATP/mole was required for activation of multiple substrate systems as well as for the transport of thiogalactoside (20), activation required 5.5 Kcal (= ΔF). If 1 mole of glucose was transported, this provided a net gain of 668.8 Kcal (= ΔH), or with a single disaccharide, 1,343.1 Kcal. However, if the multiple substrate hypothesis is correct, then a possible mechanism for control of individual species in the microbial community exists through biochemical feedback.

Excretion of various amino acids has been reported by Johannes and Webb (21, 22), Webb and Johannes (33), and others (27) for various marine animals. Work with tubificid oligochaetes has indicated that carbohydrate is released by these freshwater forms.** Thus, should two substrate systems be operating, partial or complete control could be exerted by relative concentrations of the excretions that might best be considered as a possible energy subsidy to the microbial community. The community composition and the relative abundance of the species would then be related to both the requirements of the transport system, and to relative concentrations of the substrates.

Finally, the diauxic, or catabolite repression, hypothesis of Monod (24), Magasanik (23), or Strumm-Zollinger (29, 30) is of interest. It asserts that the carbohydrates, glucose, mannose, sucrose, or manitol repress catabolism of another group consisting of maltose, arabinose, sorbitol, and inositol. The data from the communities studied here would indicate that, at low concentrations such as those found in nature, the repression of maltose uptake or oxidation was only partial and depended upon relative concentrations of those carbohydrates of the group containing glucose, sucrose, and fructose in relation to maltose. Of distinct interest was the stimulation of maltose catabolism in the community by glucose and fructose at all addition levels except 1 mg/l, where fructose was inhibitory. Similarly, maltose was found to stimulate fructose oxidation at 40 and 1,000 μg/l additions, inhibit glucose oxidation at 40 μg/l, and stimulate the catabolism of this substrate at 400 and 100 μg/l. Contrary to theory, respiration of sucrose was inhibited at all addition levels of maltose. Even in the Lake Ontario community, maltose stimulated fructose oxidation and, if anything, inhibited sucrose oxidation. The reaction of the other substrates on maltose oxidation in Lake Ontario water showed similar mixed results—inhibition by sucrose and glucose, but stimulation by fructose. Consequently, the diauxic metabolism phenomenon was possibly a consequence of laboratory procedures using pure, or "reconstituted natural communities," and does not operate in nature, or at low (under 1 mg/l) concentrations of the other carbohydrates.

* Crawford, personal communication.
** Wood, Chua, and Brinkhurst, unpublished data.

Acknowledgments

Research was supported by grants from N. R. C. (Canada), and the Department of Energy, Mines, and Resources. The technical assistance of Miss Susan Ketchell and the comments of Dr. R. D. Hamilton on the manuscript are gratefully acknowledged.

Literature Cited

1. Allen, H. L. 1968. Acetate in fresh water: Natural substrate concentrations determined by dilution bioassay. *Ecology* **49**:346–49.
2. Allen, H. L. 1969. Chemo-organotrophic utilization of dissolved organic compounds by planktonic algae and bacteria in a pond. *Int. Rev. Ges. Hydrobiol.* **54**:1–33.
3. Allen, H. L. 1971. Primary productivity, chemo-organotrophy, and nutritional interactions of epiphytic algae and bacteria on macrophytes in the littoral of a lake. *Ecol. Monogr.* **41**:97–127.
4. Cleland, W. W. 1963. The kinetics of enzyme-catalyzed reactions with two or more substrates or products. I. Nomenclature and rate equations. *Biochim. Biophys. Acta* **67**:104–37.
5. Cleland, W. W. 1963. The kinetics of enzyme-catalyzed reactions with two or more substrates or products. II. Inhibition: Nomenclature and Theory. *Biochim. Biophys. Acta* **67**:173–87.
6. Cleland, W. W. 1963. The kinetics of enzyme-catalyzed reactions with two or more substrates or products. III. Prediction of initial velocity and inhibition patterns by inspection. *Biochim. Biophys. Acta* **67**:188–96.
7. Crawford, C. C. 1967. The heterotrophic uptake of dissolved amino acids by bacteria in natural waters. M.S. Thesis, North Carolina State University.
8. Crawford, C. C. 1971. Utilization of dissolved free amino acids by estuarine microorganisms. Ph.D. dissertation, North Carolina State University.
9. Dixon, M., and E. C. Webb. 1964. *Enzymes.* London: Longmans Green and Co. Ltd.
10. Dowd, J. E., and S. Riggs. 1965. A comparison of estimates of Michaelis-Menten kinetic constants from various linear transformations. *J. Biol. Chem.* **240**:863–69.
11. Egan, J. B., and M. L. Morse. 1965. Carbohydrate transport in *Staphylococcus aureus.* I. Genetic and biochemical analysis of a pleiotropic transport mutant. *Biochim. Biophys. Acta* **97**:310–19.
12. Frieden, C. 1957. The calculation of an enzyme-substrate dissociation constant from the overall initial velocity from reactions involving two substrates. *J. Amer. Chem. Soc.* **79**:1894–96.
13. Frieden, C. 1959. Glutamic dehydrogenase. III. The order of substrate addition on the enzymatic reaction. *J. Biol. Chem.* **234**:2891–96.
14. Hamilton, R. D., and K. E. Austin. 1967. Assay of relative heterotrophic potential in the sea: The use of specifically labeled glucose. *Can. J. Microbiol.* **13**:1165–73.
15. Hicks, S. E., and F. G. Carey. 1968. Glucose determination in natural waters. *Limnol. Oceanogr.* **13**:361–63.
16. Hobbie, J. E. 1966. Glucose and acetate in freshwater: Concentrations and turnover rates. *Chemical environment in the aquatic habitat,* ed. H. L. Gotterman and R. S. Clymo, pp. 245–51. Proc. I.B.P. Symposium. Koninklijke Nederlandse Akademic van Wetenschappem, Amsterdam and Nieuwersluis.
17. Hobbie, J. E., and C. C. Crawford. 1969. Respiration corrections for bacterial uptake of dissolved organic compounds in natural waters. *Limnol. Oceanogr.* **14**:528–32.
18. Hobbie, J. E., C. C. Crawford, and K. L. Webb. 1968. Amino acid flux in an estuary. *Science* **159**:1463–64.
19. Hobbie, J. E., and R. T. Wright. 1965. Bioassay with bacterial uptake kinetics: glucose in fresh water. *Limnol. Oceanogr.* **10**:471–74.
20. Kepes, A., and G. N. Cohen. 1962. Permeation. In *The bacteria,* Vol. 4, ed. I. C. Gunsalus and R. Y. Stanier, pp. 179–221. New York: Academic Press Inc.

21. Johannes, R. E., and K. L. Webb. 1965. Release of dissolved amino acids by marine zooplankton. *Science* **150**:76–77.
22. Johannes, R. E., and K. L. Webb. 1970. Release of dissolved organic compounds by marine and fresh water invertebrates. In *Organic matter in natural waters*, ed. D. W. Hood, pp. 257–73. Institute of Marine Science, Occasional Publication No. 1. College, Alaska: Umiversity of Alaska.
23. Magasanik, B. 1961. Catabolite repression. Cold Spring Harb. *Symp.* **26**:249–54.
24. Monod, J. 1947. The phenomenon of enzymatic adaption and its bearings on problems of genetics and cellular differentiation. *Growth* **11**:223–89.
25. Monod, J., J. Wyman, and J. Changeux. 1965. On the nature of allosteric transition: A plausible model. *J. Mol. Biol.* **12**:88–118.
26. Morowitz, H. J. 1968. *Energy flow in biology.* New York: Academic Press Inc.
27. Prosser, C. L., and F. Brown. 1961. *Comparative animal physiology.* Philadelphia: W. B. Sanders Co.
28. Snedecor, G. W., and W. C. Cochran. 1967. *Statistical methods.* Ames, Iowa: Iowa State Univ. Press.
29. Strumm-Zollinger, E. 1966. Effects of inhibition and repression on the utilization of substrates by heterogeneous bacterial communities. *Appl. Microbiol.* **14**:654–64.
30. Strumm-Zollinger, E. 1968. Substrate utilization in heterogeneous bacterial communities. *J. Water Poll. Cont. Fed.* **40**:R213–R229.
31. Thomas, G. B. 1960. *Calculus and analytical geometry.* Reading, Mass.: Addison Wesley Publishing Co., Inc.
32. Vaccaro, R. F., S. E. Hicks, H. W. Jannasch, and F. G. Carey. 1968. The occurrence and role of glucose in sea water. *Limnol. Oceanogr.* **13**:356–60.
33. Webb, K. L., and R. E. Johannes. 1967. Studies of the release of dissolved free amino acids by marine zooplankton. *Limnol. Oceanogr.* **12**:376–82.
34. Webb, K. L., and L. Wood. 1966. Improved techniques for analysis of free amino acids in sea water. In *Automation in analytical chemistry, Technicon symposium,* pp. 440–44. Medical White Plains, New York: Technicon Corp.
35. Wetzel, R. G. 1968. Dissolved organic matter and phytoplankton productivity in marl lakes. *Mitt. Internal. Verein. Limnol.* **14**:261–70.
36. Wright, R. T., and J. E. Hobbie. 1966. Use of glucose and acetate in aquatic ecosystems. *Ecology* **47**:447–64.
37. Wood, L. W. 1970. The role of estuarian sediment microorganisms in the uptake of organic solutes under aerobic conditions. Ph.D. dissertation, North Carolina State University.
38. Wood, L. W., and K. E. Chua. In preparation. Glucose flux at the sediment-water interface of Toronto Harbour, Lake Ontario, with reference to pollution stress.
39. Wood, W. A. 1955. Pathway of carbohydrate degradation in *Pseudomonas fluorescens.* Bacteriol. Rev. **19**:222–33.

Some Difficulties in Using ^{14}C-Organic Solutes to Measure Heterotrophic Bacterial Activity

Richard T. Wright

One of the most significant developments in the aquatic sciences during the 1950s was the successful application of ^{14}C-labeled bicarbonate to measure the primary productivity of planktonic algae. Although a number of technical problems exist, the approach has become firmly established as a powerful tool in acquiring knowledge about aquatic ecosystems. Stimulated by primary productivity studies, information on dissolved organic compounds in natural waters began to accumulate. It soon became obvious that ^{14}C might again provide the means for measuring the activity of planktonic microorganisms—this time, the rate at which they were utilizing dissolved organic matter for energy and cell growth. Some early attempts at measuring heterotrophic production were unsuccessful, *(22, 23)* and a major breakthrough was needed. This came in 1962, when Parsons and Strickland published data on the heterotrophic uptake of ^{14}C-labeled glucose and acetate by planktonic microorganisms of inshore Pacific waters. Analyzing their data, they showed that uptake related to substrate concentration followed the first-order kinetics for the enzyme-substrate relationship originally described by Michaelis and Menten. Subsequent research *(30, 31)* built on this basic observation, using the kinetic relationships and derived values of the enzyme-substrate models.

It is now ten years since the Parsons and Strickland paper appeared. Many workers have adopted the approach of using ^{14}C-organic solutes to measure heterotrophic bacterial activity. I suggest that we have gone beyond the stage of working out basic methods and have entered a more mature stage of applying them to some difficult problems. We are beginning to get answers to some longstanding questions. This being the case, I thought it worthwhile to bring together in this presentation a number of problems and difficulties of which many of us using the approach have become aware. In doing so, I hope to stimulate an increased interest in working with natural heterotrophic activity, together with a healthy caution about exact interpretation of results.

Problems Related to the Basic Use of the Enzyme Kinetic Model

The most fundamental point to be made here is that a model is employed to give a rational scheme for interpreting data from the activities of a community of planktonic microorganisms. Any time a model is adopted or constructed,

199

there is the danger of assuming that the model is reality; thereafter, the data are coaxed to fit it. A number of basic assumptions derive from the nature of the model.

1. The uptake of a solute by organisms using it shows first order, or saturation kinetics. That is, as the organisms are presented with a range of increasing concentrations of substrate, their uptake reaches a rate which cannot be increased by further increases in substrate concentration. This assumption makes it possible to calculate a number of kinetic parameters, as illustrated in the Lineweaver-Burk variation of the Michaelis-Menten relationship shown in Figure 1. The equation shown in Figure 1, first applied to the uptake of organic solutes by Wright and Hobbie (30), is the form used by most current workers.

The problem with this assumption is the same problem other investigators have faced in their work with membrane transport. That is, we do not fully understand the uptake process at the level of cell structure and function. No single structure-function model seems to fit the known phenomena of transport across cell membranes. It seems clear, however, that transport across the membrane involves specific proteins and is often highly selective in its action on a group of similar organic molecules (see 19 for a review of the concepts).

2. Use of the enzyme kinetic model assumes that all of the important users of the substrate in the plankton are responding in the same way to variations in solute concentration being presented to them, much as an axenic culture of microorganisms might respond. This assumption is essential, yet there is no clear justification for it except that the method usually works and therefore the assumption is not proved false. This problem will be discussed later.

3. Use of the model requires that the "catalytic agents," the transport systems

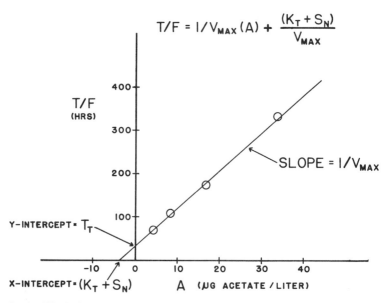

$$T/F = 1/V_{MAX}(A) + \frac{(K_T + S_N)}{V_{MAX}}$$

Fig. 1. The modified Lineweaver-Burk plot of Michaelis-Menten kinetics, as derived in Wright and Hobbie (30).

of the planktonic microbes, do not change in concentration during the measurement period. Thus, bacterial growth and/or induction of transport systems are undesirable. The measurement period should be brief relative to the time required to change these parameters. The most straightforward way to test this problem is to hold substrate concentration constant and measure uptake over a time sequence. If the process is linear over the usual measurement time, the assumption is at least not falsified. Again, the answer is a very pragmatic one, but the data support, rather than refute, the assumption *(24, 30)*.

4. Bacterial transport systems, when investigated under carefully controlled conditions, have revealed such phenomena as a common transport system for a group of similar substrates, as well as various kinds of inhibition of uptake by other molecules (e.g., *16*). Natural waters contain a very great variety of organic solutes, yet these must be worked with one at a time when using labeled solutes for obtaining transport kinetic data. The degree of interference of the measured uptake by, for example, similar molecules competing for transport sites, is very difficult to assess, especially in the case of groups of amino acids (see later comments on the work in *4*). The basic problem, then, is that use of the kinetic model assumes an experimental medium with known concentrations of all solutes. This is not possible for natural waters.

5. In any given test concentration of substrate, the substrate concentration itself should not change significantly. Natural substrate concentrations are generally so low that the added, labeled substrate usually raises the sample substrate concentration severalfold, and the amount of this taken up or respired during a measurement can be controlled by varying the incubation time or the substrate amounts. The more eutrophic or polluted the body of water, the greater is the potential for this problem, for the larger bacterial populations have a greater potential for significantly grazing down the substrate.

6. Use of the enzyme kinetic model is justified only for the basic transport phenomenon—the initial entry of the solute into the cell—and then only if no significant transport in the opposite direction is occurring. This latter possibility is tied to the existence of "pools" of substrate unincorporated into new cell material or metabolic pathways. Work by Kay and Gronlund *(17)* indicates that pool concentrations are extremely low unless substrate is provided at luxury levels.

A much more serious problem relates to the method as applied by most workers until quite recently. The basic approach—clearly showing its origin within primary productivity methodology—was to add substrate, incubate, then filter through a membrane filter. The radioactivity remaining on the filter was called uptake; respiration of the substrate was ignored or minimized. Careful work by Hamilton and co-workers *(8)* showed that respiration of labeled substrate during the period of measurement was likely to be quite important, and they called the method into question because it did not deal with this problem. Fortunately, Hobbie and Crawford *(14)* adapted a method from cell physiology which measures the $^{14}CO_2$ respired during the incubation period. Therefore, the present revised method accounts for both assimilated and respired substrate, giving total substrate transported into the cell. The only fraction unaccounted for would be substrate taken in and subsequently lost to the cell in the form of another organic compound, perhaps of lower energy value.

Figure 2 shows the results of a kinetic measurement of glucose uptake, indicating that the separate phenomena of respiration and assimilation follow Michaelis-Menten kinetics, as does the total transport as the sum of the two. The importance of these two phenomena will be touched on later.

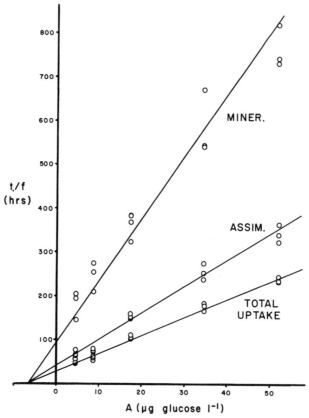

Fig. 2. Microbial activity on glucose for surface water (temp. 5 C) from upper Klamath Lake, Ore., 11 March 1970.

Problems Related to Variables in Aquatic Systems

1. Natural Substrate Concentration. The natural concentration of any low molecular weight organic compound is so low that it can only be measured with difficulty and often a great deal of imprecision. As methods have become more refined, natural concentrations reported in the literature have tended to be increasingly lower. It may soon be safe to say that values for substrates like glucose or the amino acids that exceed 10 μg/l are either suspect or temporary phenomena which ought to be explained. The value in measuring natural substrate along with uptake is that these can be combined to give the flux of a solute through the system *(13)*. Because concentrations are so low, it becomes necessary to employ ^{14}C-compounds with high specific activity in order to be working in the same order of magnitude as the compounds occur in nature. It is possible to use ^{14}C-organics as true tracer substances, if one is prepared to use very large water samples for the measurement. For example, uniformly labeled ^{14}C-glucose is available at a specific activity of 320 $\mu c/\mu m$, which is greater than 80% ^{14}C. To add 0.1 μc (220,000 dpm) to a sample of 50 ml volume, one must add .06 μg glucose, giving a concentration of 1.2 $\mu g/l$. This is too high to be a tracer. Increasing the sample size tenfold would give 0.12 $\mu g/l$ of added glucose, but processing large numbers of samples of this size would be very difficult. Williams and his co-workers *(3, 28)* use a tracer approach to measure mineralization turnover time in marine waters, but their number of samples is reduced by not having to employ a kinetic series in order to find what they are measuring. One further point is that many of the more interesting compounds are not available at such high specific activities. Glycollate, for example, comes no higher than 12% ^{14}C.

Another problem related to the naturally occurring substrates is that there are so many of them that no single worker can deal with even the more obvious ones at any one time. Furthermore, there appears to be no obvious way of knowing what the more important substrates are. For these reasons, the use of ^{14}C-organics can not possibly do for heterotrophic productivity what ^{14}C-bicarbonate has done for primary productivity. Yet, the study of how the various organic substrates are utilized in natural waters, together with work on their origin and natural concentrations, constitutes an intriguing and challenging segment of knowledge that is just beginning to be appreciated. Much of this could not be investigated without using ^{14}C-organic compounds.

2. Natural Populations. In attempting to measure heterotrophic activity in natural waters, any investigator must know that he is dealing with a heterogeneous mixture of organisms. Crude separations may be made with plankton nets or filters, and when this has been done, it has been found that most of the heterotrophic activity in the water is associated with the smallest particles *(29, 31)*. There seems at this time little reason to doubt that the heterotrophic activity detected at substrate concentrations close to natural concentrations is due to the bacteria. A number of investigators have isolated bacteria with transport abilities very similar to those found in natural populations *(8, 30)*. A few algae seem to have specific transport systems *(11)*, but the usefulness of these systems and the importance of these algae compared with the bacteria seem to be quite limited. The natural phytoplankton does seem to have a capacity for taking up ^{14}C-organics, but the amount is related more or less directly to substrate concentration and is, at low concentrations, extremely small relative to the phytoplankton biomass present *(2, 31)*. Nevertheless, a dense algae population can at times remove enough of the labeled solute to make a measurement of bacterial transport difficult or impossible.

Even if the heterotrophic activity in question is bacterial, there remains the basic objection that the bacterial flora is a heterogeneous mixture which could hardly be expected to respond to a given substrate in a uniform way *(25)*. The fact that the bacterial flora often does yield a reasonable kinetic plot opens up a number of intriguing ecological questions. For example, do all forms successfully using a given substrate have similar transport abilities? Or, does a good kinetic response indicate the presence of a predominant species? The work of Vaccaro and Jannasch (25) and Vaccaro (26) in showing a rapid response of the microbial flora to enrichment with organic solutes does not provide an answer to these questions.

Many attempts at measuring heterotrophic activity in very oligotrophic waters have failed to reveal the usual kinetic response. This is especially true for the open oceans (25, 26). There are at least two possible explanations for this state of affairs: the bacteria are present and active but the technique used is not sensitive enough to measure their activity, or perhaps the bacteria present are functionally inactive in the transport of a given solute. Vaccaro and Jannasch (25) obtained an atypical sigmoid curve relating uptake to concentration in a large number of samples from the open Pacific. The natural bacterial flora responded rapidly to enrichment with unlabeled substrate, showing a good kinetic response within 12 hr. Because the response was too rapid to be explained by bacterial multiplication, these results support the idea that at least some members of the bacterial community were functionally "turned off."

One final problem becomes apparent when doing replicates of heterotrophic uptake at a single concentration of solute. The replicates often show disturbing variability, perhaps best explained by heterogeneity in the distribution of the microbial flora. This would be unlikely unless some of the microbes were associated with larger particles, such as detritus. Along this line, Wright and Hobbie (31) presented evidence that acetate-utilizing bacteria were associated with larger particles while glucose-users were not.

3. Natural Environments. For heterotrophic uptake studies, the easiest community to work with among aquatic systems is the planktonic community. In this medium organisms can be removed easily by filtration, and labeled CO_2, derived from the applied substrate, can be recovered.

The epiphytic bacterial community is considerably more difficult to work with, due to the impossibility of separating the bacteria from their sites of attachment. One approach is to provide an artificial substratum, remove the attached flora, and measure uptake by suspending the removed bacteria in filtered natural water (2). Another approach would be to enclose the natural plant material and its epiphytic community in filtered water and uniformly labeled substrate, and measure only mineralization by recovering the labeled CO_2 released by the community.

One very important bacterial environment is the sediment. Wood (29) employed a 100/1 dilution of estuarine sediment with sterile artificial estuarine water to measure assimilation and respiration of organic solutes, treating the sample as if it were a plankton suspension. Harrison et al. (10) applied uniformly labeled organic solutes to a sediment slurry, and after incubation recovered the labeled CO_2 generated by microbial mineralization. The latter method has the advantage of minimizing the disturbance of the physical state of the sediment, while Wood's approach lends itself to a measurement of assimilation as well as mineralization. Both approaches allow a kinetic analysis of the microbial activity, and hence are superior to the method suggested by Kadota et al. *(15)*, which employs labeled substrate at only one concentration.

Problems Related to the Mechanics of Measurement

1. Substrate Purity. In applying the respiration correction method suggested by Hobbie and Crawford *(14)*, disturbingly high "blank" values are sometimes obtained from a given substrate *(10, 27)*. This could have at least two possible origins: (1) volatilization of the substrate and subsequent absorption by the phenylethylamine used to absorb the labeled CO_2, or (2) perhaps $^{14}CO_2$ itself is a contaminant in the ^{14}C-organic solute being used. The latter seems to be the more logical explanation, based on (1) the wide variety of organic solutes that have given these results and (2) a series of experiments conducted to differentiate between the two possibilities. In these experiments, the ^{14}C-compound giving rise to high blanks in samples containing labeled glucose, acetate, and glycollate at all times behaved like the $^{14}CO_2$ employed as a control in experiments testing pH effects, time-course of appearance of the activity, and percent recovery in samples of sterile water. The $^{14}CO_2$ could conceivably be present in very small quantities as a contaminant in the organic compound when originally purchased. More likely, it could result from decomposition of the labeled organic compound during heat sterilization or storage. An extremely small amount of breakdown — whether through strictly chemical processes or through radiolysis — can create a disturbingly high blank, from several hundred dpm/μc on up. Williams and Askew *(27)* suggest acidifying before autoclaving ^{14}C-glucose to prevent this problem. This is sound advice, and perhaps should be augmented with gas purging, especially if the substrate is to be filter-sterilized or frozen instead of autoclaved.

2. Fixation. Various methods have been used to end the period of incubation. Parsons and Strickland *(21)* used formaldehyde, and prepared a "blank" sample for each substrate concentration by adding the formalin at the start of the incubation period to one of several replicates. Wright and Hobbie *(31)* found that Lugol's acetic acid fixative was superior to formalin, in that the counts retained on the filter were uniformly higher. This is the fixative of choice for preservation of phytoplankton for counting. Working with glycollate, Wright *(32)* found that substantially higher counts were obtained by ending incubation with rapid filtration, omitting the fixative. However, all of these studies predate the respiration correction of Hobbie and Crawford *(14)*, which involves a new set of problems. With the respiration correction, the fixative has the double function of killing the microorganisms and also creating acid conditions favoring the conversion of all inorganic carbon to the gaseous state to be picked up by the filter paper soaked in phenylethylamine, a CO_2-absorbant. Hobbie and Crawford used 0.2 ml of a 2 N H_2SO_4 solution to a 5 ml sample. I have used 0.5 N H_2SO_4, adding on a 1:50 ratio of fixative to sample. This gives a pH of around 2.2 in fresh water, and in some preliminary tests I found no difference between samples fixed in this way and those filtered immediately without fixation. Williams and Askew *(27)* use 50% phosphoric acid to fix seawater samples for mineralization measurements, on a 1:100 ratio. Obviously, the choice of fixative and its strength are factors that will have to be determined by the investigator attempting these measurements.

3. Liquid Scintillation Counting. The addition of liquid scintillation counting to the methodology of heterotrophic studies has made it possible to count large numbers of samples under carefully controlled counting conditions. The respiration fraction of uptake, involving phenylethylamine-soaked filter paper, requires liquid scintillation counting. Williams and Askew *(27)*, however, presented a method for measuring respired $^{14}CO_2$ by trapping it in barium hydroxide, filtering the precipitate on a glass fiber filter, and measuring activity with a planchet counter. This method is quite laborious and is clearly a method of necessity rather than one of choice. The assimilation fraction, that proportion of activity caught on a membrane filter, may be measured with liquid scintillation or with a planchet-type 2-Π counter (proportional or Geiger). Much higher efficiencies are obtained with liquid scintillation counting, and this is obviously the method of choice.

Both membrane filters and the phenylethylamine-soaked pieces of filter paper cause quenching of ^{14}C-activity contained on them, requiring a correction in order to arrive at actual activity as dpm. The usual drop in efficiency is from 6 to 10% but can be higher if the samples contain any water or are colored by phytoplankton pigments. Since most of the counts in both phases of measurement are associated with the solid phase (membrane filter or filter paper), the most accurate correction can be obtained with the channels ratio method *(5, 18, 20)*. With this correction method, I found 100% recovery of known amounts of $H^{14}CO_3$ added to sterile samples and subsequently absorbed on phenylethylamine-soaked filter papers suspended in serum bottles. Hobbie and Crawford *(14)*, using internal standardization by adding a ^{14}C-toluene spike, had to apply a correction factor of 1.23 in order to relate counts on filter paper strips to the expected activity as measured by gas-phase counting of known samples.

4. Incubation Conditions. The temperature of incubation is very important. With Q_{10} measurements ranging from two on upward *(4, 30)*, a few degrees Celsius difference between samples or between incubation temperature and that of the environment can significantly change the data obtained. The time of incubation should be the minimum needed to get significant radioactivity at the lowest substrate concentration used.

In view of the temperature-dependency of activity—both seasonal and Q_{10} effects—incubation time should generally bear an inverse relationship to sample temperature. With experience, one learns to obtain the optimum measurement conditions by manipulating sample volume, activity added, and incubation time. One additional incubation condition is agitation of the samples. Some workers employ gentle agitation during incubation, others do not. Curiously, no one seems to have tested the effect of agitation as a variable.

Results and Discussion

Having considered in some detail many of the problems connected with attempts at measuring heterotrophic activity, it might be useful to consider some results from various studies in the light of these problems. One of the most critical questions facing an investigator looking at his data is whether the data really lend themselves to a Michaelis-Menten kinetic scheme. The modified Lineweaver-Burke equation shown in Figure 1 generates a straight line, which lends itself nicely to statistical analysis. A combination of least squares regression on the data and the correlation coefficient for the regression gives adequate information for judging the data. The data should be considered acceptable if the correlation coefficient is above that required for the 95% probability level. This kind of analysis, along with the quench curve corrections and calculations of T/F (see Fig. 1), is obviously best done with the help of a computer. Computer programs using the FOCAL language for a DIGITAL PDP-12 are available for this analysis and can be obtained by writing to the author. This approach to the kinetic data was applied to a total of 84 measurements of heterotrophic uptake kinetics carried out over 1969–70 at Klamath Lake, Oregon. Four substrates were used: glucose, acetate, glycine, and glycollate. A total of 54 of the measurements were statistically acceptable. Glucose was the most reliable substrate to use, 19 of 21 being acceptable. Glycollate gave the least number of acceptable measurements, 9 of 21. Hamilton and Preslan *(9)*, working with the assimilation fraction of uptake in Pacific waters, and using a wide variety of substrates, found 48 of 120 measurements to give acceptable kinetic results. Glucose, aspartic acid, glutamic acid, glycine, and arginine produced useful data on eight or more of the twelve measurements made for each substrate. Some of the results were unacceptable because of very low activity. Crawford *(6)* found acceptable kinetic results in 150 of 190 kinetic measurements in estuarine waters, using 15 amino acids and glucose.

No less important to the investigator is the question of what the results mean once they have been given statistical approval. The kinetic values obtained as shown in Figure 1 are the result of the collective activities of a heterogeneous system; hence, they can not be called "constants." Vaccaro and Jannasch *(25)* suggest the use of the word "parameters." One of these parameters is V_{max}, the theoretical maximum transport rate due to the bacterial population at the time of the measurement and at the sample temperature, the rate which would be obtained if substrate concentration were high enough to completely saturate the transport mechanisms of the natural population. Wright and Hobbie *(31)* argued that this parameter might be useful for estimating the heterotrophic potential and the biomass of the population. Although V_{max} has been shown to be proportional to biomass in pure cultures *(8, 30)*, V_{max}/cell varies greatly for various marine bacteria *(7)*, casting doubt on the usefulness of V_{max} for estimating the bacterial biomass in natural populations. Further, enrichment studies *(26)* show that when substrate levels are raised experimentally, perhaps to the level where V_{max} is indicated, the V_{max} also gradually increases as would be expected if bacteria multiply in response to added substrate. In a sense, one could probably say that the true heterotrophic potential of a natural population is a function of substrate concentrations and the species of bacteria which happen to be present, and therefore V_{max} is more a measure of what the existing population is *capable* of doing, provided substrate levels do not change. Where natural substrate concentration is known, the actual transport rate can be calculated and is always some fraction of V_{max} (e.g., 31% for glucose as reported, *30*). Therefore, perhaps the most one could say about V_{max} is that it is certainly a function of the existing population, it is sensitive to changes in temperatures (independent of cell increases), and it responds to changes in substrate concentration, presumably as the population itself changes in numbers and species composition. Its greatest value, then, will be in comparing seasonal and spatial effects in a given body of water, comparing the importance of different substrates, and comparing heterotrophic activity in different bodies of water.

Another of the calculated kinetic parameters is $(K_t + S_n)$, a value combining a transport "constant" and the natural substrate concentration. As Wright and Hobbie *(31)* pointed out, perhaps the chief value of this parameter is to set upper limits on both of these. In those studies which have measured S_n independently while also measuring $(K_t + S_n)$ kinetically, the S_n values have always been in the same order of magnitude as $(K_t + S_n)$ and usually lower, in spite of the considerable imprecision involved in both kinds of measurement *(13, 4, 6)*. The few attempts at measuring K_t for bacteria isolated from natural waters have revealed some bacteria with K_t values close to $(K_t + S_n)$ values, as well as some which are considerably higher *(12, 7, 24)*. Some very interesting bioassays have been worked out using the measurement of $(K_t + S_n)$ under a variety of conditions *(12, 24, 1)*.

The third parameter obtained from the kinetic analysis is T_t (turnover time), theoretically the time involved in a complete utilization of the existing substrate at the existing transport rate. This again is a value probably of most use for making comparisons, for it describes a hypothetical situation: uptake operating at a constant rate which requires a steady state in substrate concentration. Therefore, the substrate must be supplied precisely as rapidly as it is utilized. Our knowledge of the sources of dissolved organic substrate for the heterotrophic bacteria is much too fragmentary at this time to make any further comment on this problem.

Harrison, et al. *(10)* have pointed out that the respiration fraction of uptake, when considered independently of assimilation, is actually the process called mineralization by other workers—the process of degradation of organic compounds to CO_2 and other inorganic units. The kinetic approach to heterotrophic activity, then, also gives information on the mineralization turnover time for a given organic solute, provided uniformly labeled substrate is used. Williams and co-workers (e.g., *28*) obtain this information with labeled organics by assuming that the labeled solute is being used at such low concentrations relative to natural concentrations that it acts as a true tracer and does not influence the uptake process. This assumption contributes some uncertainty to the values obtained unless it is shown empirically that the process is indeed not influenced by the substrate concentrations used to make the measurement.

Under some conditions, the applied substrate concentration range influences the nature of the kinetic results obtained. Figure 3, from Wright and Hobbie

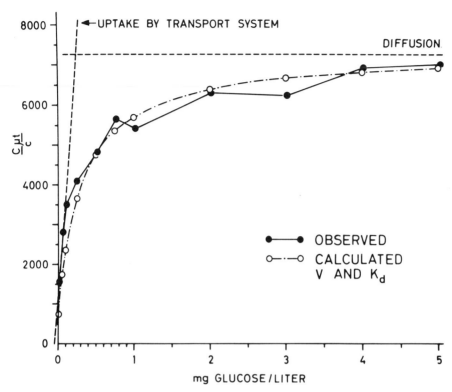

Fig. 3. **Uptake of glucose by a natural community of algae and bacteria, showing a strong influence on kinetics by the algae (Lake Erken, Sweden, 1 m depth, 14 Feb. 1964). (Reprinted from Wright and Hobbie *(31)* by permission of the Ecological Society of America.)**

(31), shows uptake data from a freshwater lake, obtained over a broad concentration range and plotted with the modified Lineweaver-Burke equation. Clearly, the results are not consistent with Michaelis-Menten kinetics. Figure 4 shows a similar plot obtained by combining axenic cultures of a bacterium and an alga. The point made by this work is that algae can interfere with uptake measurements and may be responsible for some of those occasions when the measured uptake does not meet acceptable statistical requirements. This problem can be minimized by using as low substrate concentrations as possible when algal populations are dense, so as to be in the concentration range where bacterial uptake is effective and algal uptake is not *(31)*.

Another example of substrate concentration influence is found in Harrison et al. *(10)*. Figures 5 and 6 show mineralization in lake sediment measured over seven orders of magnitude. This was done in an attempt to establish a concentration range and incubation time that would allow a kinetic analysis of mineralization. Two interesting phenomena are evident. At the lowest concentrations, the substrate was mineralized so rapidly that it was beginning to be depleted. The time course of mineralization at the higher concentrations suggests that induction of additional transport and mineralization enzymes might be occurring in response to the unnaturally high glucose levels. On the basis of these

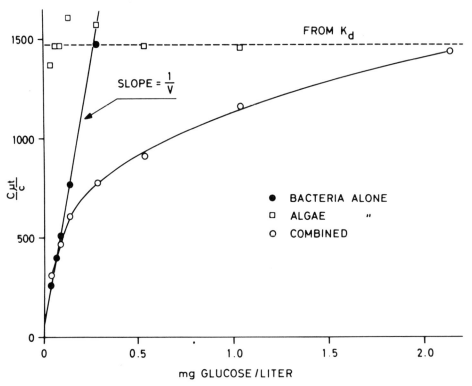

Fig. 4. **Uptake of glucose by a bacterium and an alga separately, and the two combined, in a mineral medium. (Reprinted from Wright and Hobbie (31) by permission of the Ecological Society of America.)**

findings, a concentration range of 1–5 μg/ml sediment and an incubation time of 5 min were chosen for kinetic measurements. The findings also suggest that measuring mineralization at only one substrate concentration, as suggested by Kadota et al. *(15)*, would give highly questionable results.

Earlier, the problem of competitive effects on uptake by other similar substances was mentioned. An example of this can be seen in Figure 7, which indicates that glycollate and lactate share a common transport system for natural populations of bacteria. This phenomenon occurs with pure cultures when tested in a similar manner. When measuring glycollate transport kinetics, therefore, the kinetic parameters of $(K_t + S_n)$ and T_t are probably increased by any lactate present, while V_{max} would be unaffected. This would be the situation if lactate were acting as a competitive inhibitor for glycollate uptake and metabolism.

The amino acids as a group of substrates present a very discouraging picture when all of the potential competitive effects are considered. Burnison *(4)* demonstrated clear-cut competitive effects between a number of amino acids in a study of heterotrophic utilization of amino acids in Klamath Lake, Oregon. No doubt similar competitive effects influenced the results of other studies of amino acid heterotrophy, casting some serious doubts on the precision of the transport values reported *(6, 13)*.

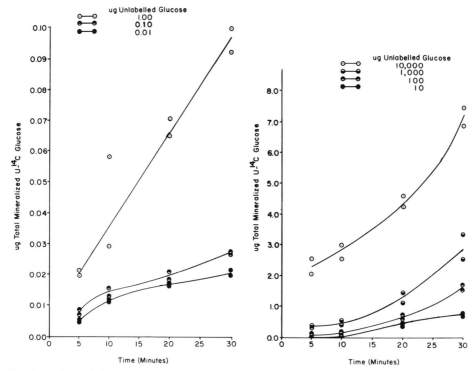

Fig. 5. **Effect of increased amounts of un-labeled glucose on the mineralization of labeled glucose; each reaction vessel contained 2 ml of substrate in distilled water and 5 ml of slurried sediment from upper Klamath Lake. (Reprinted from Harrison, Wright, and Morita *(10)* by permission of the American Society for Microbiology.)**

Fig. 6. **Effect of increased amounts of un-labeled glucose on the mineralization of labeled glucose; Each reaction vessel contained 2 ml of substrate in distilled water and 5 ml of slurried sediment from upper Klamath Lake. (Reprinted from Harrison, Wright, and Morita *(10)* by permission of the American Society for Microbiology.)**

The more recent studies of heterotrophic activity include respiration values as well as the more frequently used assimilation fraction of transport *(4,6,14,27)*. These studies and my own results show that respiration (mineralization) is usually a very significant fraction of total uptake. Figure 8 shows data from Klamath Lake for a set of amino acids. The substrates are ranked in order according to the percentage respired and the data indicate that more than 50% of uptake may be subsequently respired for some substrates. Glutamic acid, for example, seems to be a preferred energy source *(4)*, which seems logical enough in view of the easy entry of glutamate into the Krebs cycle. Of all the substrates reported, the highest mineralization percentage values come from glycollic acid. Using both singly and doubly labeled ^{14}C-glycollate, I have found an average of 72% of uptake going to complete mineralization. Clearly, the respiration correction introduced by Hobbie and Crawford is not just a refinement but should be an essential part of any future heterotrophic studies using the kinetic approach.

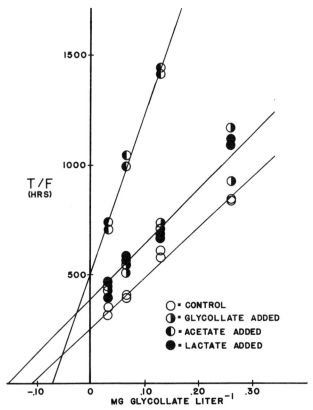

Fig. 7. Effect of unlabeled compounds on the uptake of labeled glycolate by natural plankton from 5 m, gravel pond, 16 Aug. 1968; glycollate, acetate, and lactate added at a concentration of 100 μg/1. Modified Lineweaver-Burk plot as derived in Wright and Hobbie *(30)*.

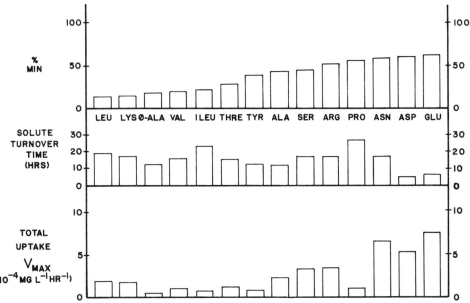

Fig. 8. Uptake, mineralization, and turnover of amino acids in surface water temp. 5 C of upper Klamath Lake, 25 Feb. 1970.

Conclusion

I doubt that I have exhausted all of the difficulties, problems, and objections connected with the use of ^{14}C-organic solutes in studying heterotrophy in aquatic ecosystems. The main thrust of the paper is to inform and help other workers maintain an awareness of the limitations and dangers of the model system being employed, and perhaps to encourage them to overcome some of the obstacles that stand in the way of gathering information in this critical area of study. Obviously, the basic approach is not the kind of technique one is eager to recommend to all aquatic ecologists. Those using it now should be careful not to push it too far. Yet, the basic approach is gradually giving us insight into what the heterotrophic bacteria are doing in nature, and there may be no other way to get that information.

Acknowledgments

The author is grateful for support from the National Science Foundation through Faculty Fellowship 69210 and Grant GB 7741. Thanks are also due to Dr. Richard Y. Morita, Oregon State University, for encouragement and additional support through F.W.Q.A. Program No. 16110.

Literature Cited

1. Allen, H. L. 1968. Acetate in fresh water: Natural substrate concentrations determined by dilution bioassay. *Ecology* **49**:346–49.
2. Allen, H. L. 1969. Primary productivity, chemo-organotrophy, and nutritional interactions of epiphytic algae and bacteria on macrophytes in the littoral of a lake. Ph.D. dissertation, Michigan State University.
3. Andrews, P., and P. J. Le B. Williams. 1971. Heterotrophic utilization of dissolved organic compounds in the sea. III. Measurement of the oxidation rates and concentrations of glucose and amino acids in sea water. *J. Mar. Biol. Ass. U.K.* **51**:111–25.
4. Burnison, B. K. 1971. Amino acid flux in a naturally eutrophic lake. Ph.D. dissertation, Oregon State University.
5. Bush, E. T. 1963. General applicability of the channels ratio method of measuring liquid scintillation counting efficiencies. *Anal. Chem.* **35**:1024–29.
6. Crawford, C. C. 1971. The utilization of dissolved free amino acids by estuarine microorganisms. Ph.D. dissertation, North Carolina State University.
7. Hamilton, R. D., K. M. Morgan, and J. D. H. Strickland. 1966. The glucose uptake kinetics of some marine bacteria. *Can. J. Microbiol.* **12**:995–1003.
8. Hamilton, R. D., and K. E. Austin. 1967. Assay of relative heterotrophic potential in the sea: The use of specifically labelled glucose. *Can. J. Microbiol.* **13**:1165–73.
9. Hamilton, R. D., and J. E. Preslan. 1970. Observations on heterotrophic activity in the eastern tropical Pacific. *Limnol. Oceanogr.* **15**:395–401.
10. Harrison, M. J., R. T. Wright, and R. Y. Morita. 1971. Method for measuring mineralization in lake sediments. *Appl. Microbiol.* **21**:698–702.
11. Hellebust, J. A. 1970. The uptake and utilization of organic substances by marine phytoplankters. In *Organic matter in natural waters*, ed. D. W. Hood, pp. 225–56. Occasional Institute of Marine Science, Publication No. 1. College, Alaska: University of Alaska.
12. Hobbie, J. E., and R. T. Wright. 1965. Bioassay with bacterial uptake kinetics: Glucose in freshwater. *Limnol. Oceanogr.* **10**:471–74.

13. Hobbie, J. E., C. C. Crawford, and K. L. Webb. 1968. Amino acid flux in an estuary. *Science* **159**:1463–64.
14. Hobbie, J. E., and C. C. Crawford. 1969. Respiration corrections for bacterial uptake of dissolved organic compounds in natural waters. *Limnol. Oceanogr.* **14**:528–32.
15. Kadota, H., Y. Hata, and H. Miyoshi. 1966. A new method for estimating the mineralization activity of lake water and sediment. *Mem. Res. Inst. Food Sci. Kyoto Univ.* **27**:28–30.
16. Kay, W. W., and A. F. Gronlund. 1969. Amino acid transport in *Pseudomonas aeruginosa. J. Bacteriol.* **97**:273–81.
17. Kay, W. W., and A. F. Gronlund. 1969. Amino acid pool formation in *Pseudomonas aeruginosa. J. Bacteriol.* **97**:282–91.
18. Lind, O. T., and R. S. Campbell. 1969. Comments on the use of liquid scintillation for routine determination of ^{14}C activity in production studies. *Limnol. Oceanogr.* **14**:787–89.
19. Pardee, A. B. 1968. Membrane transport proteins. *Science* **162**:632–37.
20. Parmentier, J. H., and F. E. L. Ten Haaf. 1969. Developments in liquid scintillation counting since 1963. *Int. J. Appl. Radiat. Isotop.* **20**:305–34.
21. Parsons, T. R., and J. D. H. Strickland. 1962. On the production of particulate organic carbon by heterotrophic processes in sea water. *Deep-Sea Res.* **8**:211–22.
22. Rodhe, W. 1962. Sulla produzione di fitoplancton in laghi transparenti di alta montagna. *Mem. Ist. Ital. Idrobiol.* **15**:21–28.
23. Saunders, G. W. 1958. The application of radioactive tracers to the study of lake metabolism. Ph.D. dissertation, University of Michigan.
24. Vaccaro, R. F. 1969. The response of natural microbial populations in seawater to organic enrichment. *Limnol. Oceanogr.* **14**:726–35.
25. Vaccaro, R. F., and H. W. Jannasch. 1966. Studies on heterotrophic activity in seawater based on glucose assimilation. *Limnol. Oceanogr.* **11**:596–607.
26. Vaccaro, R. F., and H. W. Jannasch. 1967. Variations in uptake kinetics for glucose by natural populations in seawater. *Limnol. Oceanogr.* **12**:540–42.
27. Williams, P. J. Le B. 1970. Heterotrophic utilization of dissolved organic compounds in the sea. I. Size distribution of population and relationship between respiration and incorporation of growth substrates. *J. Mar. Biol. Ass. U.K.* **50**:859–70.
28. Williams, P. J. Le B., and C. Askew. 1968. A method of measuring the mineralization by microorganisms of organic compounds in sea-water. *Deep-Sea Res.* **15**:365–75.
29. Wood, L. W. 1970. The role of estuarian sediment microorganisms in the uptake of organic solutes under aerobic conditions. Ph.D. dissertation, North Carolina State University.
30. Wright, R. T. 1970. Glycollic acid uptake by planktonic bacteria. In *Organic matter in natural waters*, ed. D. W. Hood, pp. 521–36. Institute of Marine Science, Occasional Publication No. 1. College, Alaska: Univ. of Alaska.
31. Wright, R. T., and J. E. Hobbie. 1965. The uptake of organic solutes in lake water. *Limnol. Oceanogr.* **10**:22–28.
32. Wright, R. T., and J. E. Hobbie. 1966. The use of glucose and acetate by bacteria and algae in aquatic eco-systems. *Ecology* **47**:447–64.

Comments

HAMILTON: I have a few remarks regarding procedural details. First I would like to recommend that the use of dioxane-based counting cocktails be encouraged. Such cocktails completely dissolve most membrane filters and thus eliminate a host of problems associated with scintillation counting of heterogeneous systems.

WRIGHT: That is good for the Millipore filter but you are still stuck with filter paper.

HAMILTON: Agreed. Perhaps there are some membrane filters which could be used as solid support for CO_2-trapping fluids, yet which could be dissolved in suitable cocktails.

WRIGHT: You would probably have to go away from phenylethylamine. I think that dissolves the Millipore filters.

HAMILTON: My second remark is concerned with errors which are apparently due to the filtration process. A few years ago, it was noted that the results from primary productivity determinations varied according to the amount of sample filtered. In essence, the larger the aliquot filtered the less the activity per unit volume that was recovered. The effect is so common that we now avoid filtering in our primary productivity determinations by simply acidifying the sample, sweeping off the unused $^{14}CO_2$, and counting the liquid sample in a suitable cock-tail. Results obtained in this manner agreed well with those predicted by filtration correction curves. We have observed very similar effects in the determination of heterotrophic uptake. Indeed the error can be quite serious; approaching factors of 10 between the use of a 1 ml sample as opposed to the use of a 20 ml sample. The occurrence and magnitude of the error appear to vary both with the substrate and the environment under investigation. In this case, we cannot take advantage of the procedure used for primary productivity, and I am afraid that the repetitious filtrations necessary to establish a correction curve are the only immediate solution. Dr. Holm-Hansen and I have discussed this matter and I would now ask him to present his views.

HOLM-HANSEN: I do not know all the answers to these questions. We have been doing a lot of productivity work. I do not know the cocktail we have been using, but we get no quenching by using glass fiber filters. This solves many of the problems to which you have alluded. In regard to the problem of the artifacts due to filtering during productivity, I find it hard to comprehend. I would really have to look at the experimenter and his data to judge that. For our work this past year, we have been looking at not only the stuff on the filter, but also the amount excreted into the medium. After acidification and removal of all the inorganic carbons, we combust it by UV oxidation of the organic materials and measure the products by scintillation counting. By our techniques, I think we would have seen any artifacts due to breaking of the cells during filtration. I have not seen them. You mentioned using a cocktail that could take some water; we have done this, too. This might be all right if you are dealing with highly productive water, but if you are working in oligotrophic waters, the sensitivity and the amount of incorporation are not sufficient to dispense with filtering. You must filter unless you want a few counts over the background. I think it will take a little more investigation before we really know all the answers. Certainly there are enough peculiar data from various investigators in the literature to suggest another look at what is happening during filtration. We should get that ironed out if we can. I think it is going to be explained by some artifact. I think we will be able to get around it, if it is indeed real.

HAMILTON: You may be correct. The effect, however, is important enough that I believe anyone using standard techniques for heterotrophic uptake ought to try filtering different aliquots of their samples. This will simply tell the investigator whether the effect is occurring in their environment and with their substrate.

WATSON: It is certainly very difficult to know what some of these measurements mean as far as biological productivity goes. We have been working on urea decomposition, and we have obtained some very interesting results off the coast of Africa. The concentration of urea on the shelf off the NW coast of Africa is about two microgram atoms per liter. We measured the rate of urea decomposition and we could completely account for all the nitrogen which was being used by plant production in that area. In that particular area, which may be unique, it looked like the complete growth rate of plants was being controlled by the rate at which urea was broken down.

ZOBELL: What is the source of urea in this case?

WATSON: This is quite interesting also. We did stations on the shelf where the water was less than 200 m deep. The urea concentration was relatively constant throughout the water column. There is no indication that the phytoplankton were using urea. If the phytoplankton were using the urea, then it should have been less in the euphotic zone. We went offshore and we got a very sharp decrease in urea concentration at 100 m. This would indicate that there is something in the euphotic zone that is probably producing the urea. On the shelf, there is no mixing and you cannot detect this. I do not know what is producing it. Do copepods excrete urea? I do not know. Whatever was producing it was in the euphotic zone, and it was also obvious that the phytoplankton were not using it directly. They were dependent upon the bacterial breakdown of urea. Most phytoplankton in inshore waters can use urea, but very few of them in offshore water can actually use urea. So they depend upon bacteria to break it down.

HOLM-HANSEN: I think you would look logically to the fish as a source of urea in this case.

WATSON: Yes, but I would think that they would be below 100 m. On offshore stations there is a very sharp decrease. It dropped a whole order of magnitude below 100 m. I do not know what it is. It may be fish.

WRIGHT: It would take a fantastic amount of fish biomass to produce that much urea.

WATSON: What about copepods, do they excrete urea?

WRIGHT: Not too much of it, I do not think.

HOLM-HANSEN: If you follow the fish with sonic devices, you will get a very high concentration in surface waters at night. They come up to feed every night and then they go down. The concentration factor might not bear any relation to the concentration of vertebrates or copepods in the water column during the day. I think this point could easily be overcome.

ZOBELL: In regard to the experiments concerned with the conversion of carbon into particulate biomass, it seems to me that there is a built-in error in extrapolating such very good experimental results to field conditions. I mean, what does this conversion contribute to the food cycles in the sea? As I understand these techniques, the data are gained from rather young cells. What happens if they are left for several hours or several days until autolysis starts to take place and the biomass breaks down and is returned to a soluble state before it is in-gested by animals? Under these conditions, the biomass is not getting in the animal food chain, but it may be available for bacteria. Some of the bacterial products may be used by other bacteria until metabolic products accumulate to the extent that they are no longer used over. In the sea, I do not believe the bacteria are picked off and eaten by an animal just as soon as the cells pass the log phase.

WRIGHT: I think it would be very difficult to establish any direct lines of energy flow here. I am sure that what you say is true. There are all sorts of little internal cycles that are going on, and the organic matter may be taken up and put right back as dissolved organic matter. Then, on the other hand, it might be miner-alized.

ZOBELL: I did not mean to imply that this detracts one bit from the excellence of your work concerning how much is converted into particulate matter or how much is mineralized. I am concerned with how much gets into the food chain and can be counted in this inventory.

Section Four
ENVIRONMENTAL EFFECT

Salinity and Temperature Interactions and Their Relationship to the Microbiology of the Estuarine Environment[1]

*Richard Y. Morita, Larry P. Jones,
Robert P. Griffiths, and Thomas E. Staley*

An estuary, unlike the open ocean, is a dynamic system that is continuously undergoing changes in its physical and chemical properties. Marked fluctuations in the levels and kinds of nutrients, temperature, and salinity are normally found in estuarine environments. These changes occur principally as a result of freshwater intrusions, tidal changes, evaporation, seasonal variations, and diurnal insolation. Moreover, imposition by man of thermal, fecal, and industrial wastes has also affected these properties.

In terms of the ecology of aquatic environments, bacteria serve a number of functions, including the mineralization of refractile organic matter, the assimilation of low levels of dissolved organics, and possibly the contribution of essential growth factors to organisms of other trophic levels. In the estuarine environment, it would be expected that all of these microbial activities would be significantly affected by changes in termperature and salinity.

There are numerous reports on the affects of either temperature or salinity on bacteria. However, a temperature response or a salt requirement by a marine bacterium does not validate inferences from the laboratory to the marine environment. Rather than make suppositions about *in situ* microbial activities based on laboratory data, we have been investigating the interrelationship of temperature and salinity on the growth and rates of assimilation of organics by marine psychrophilic bacteria. This report represents results of some of these investigations.

The Salinity Effects on Maximal Growth Temperature

In 1968, Stanley and Morita (6) demonstrated that the maximum growth temperature of *Vibrio marinus* MP-1 was affected by changes in salinity. A similar response was exhibited by three other bacteria isolated from marine sources (Fig. 1). With two strains of *V. marinus*, a difference of about 10 C was found between the maximal growth temperature at the highest and lowest salinities at which growth occurred. The difference was less marked with the two bacterial isolates from the Antarctic. The maximal growth temperature for MP-1 at a salinity of approximately 7‰ was 10 C. However, at a salinity of 35‰ the maximal growth temperature increased to 21.2 C. In all salt-requiring bacteria

[1] Published as special report No. 339, Oregon Agricultural Experiment Station.

221

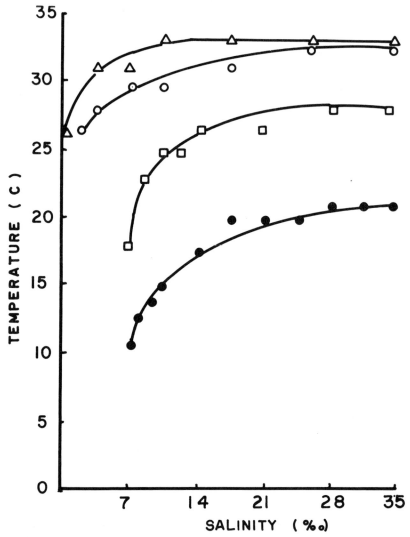

Fig. 1. **Effect of salinity of the growth medium on the maximal growth temperature of certain bacteria isolated from the marine environment. In addition to artificial seawater, the medium contained 2.5 g of polypeptone, and 10 ml of vitamin solution dissolved in 1 l of distilled water. The pH was adjusted to 7.4. Symbols: ●,** *Vibrio marinus* **MP-1; □,** *V. marinus* **PS 207; ○, 20-PFD/2; △, D30-10WBD/3. (Reprinted from Stanley and Morita** *(6)* **by permission of the American Society for Microbiology.)**

that have been tested in this laboratory (including *Halobacterium cutirubrum*), we have found that salinity influences the maximal growth temperature.

Further studies were undertaken to determine whether these results were due to an osmotic phenomenon or due to requirements for specific salts. It can be seen from Table 1 that when sodium and lithium chlorides were added to a basal medium containing known quantities of essential salts, growth occurred at 20 C. However, with magnesium chloride the maximal growth temperature was depressed to 16.3 C. No growth was obtained when potassium, rubidium, or ammonium chloride was present. However, in a completely defined medium, lithium chloride could not completely replace the sodium chloride requirement.

Table 1. Effect of different cations on the growth temperature of *Vibrio marinus* (MP-1[a])

Cation Added as the Cl Ion	Maximal Growth Temp (C)
None	No growth
Na[+]	20.0
Li[+]	20.0
K[+]	No growth
Rb[+]	No growth
NH$_4$[+]	No growth
Mg[++]	16.3

[a] The salts were added to the basal medium to give a cation concentration of 0.40 M. Basal medium contained, per liter of distilled water: polypeptone, 2.5 g; NaCl, 0.1 g; K$_2$HPO$_4$, 0.1 g; MgSO$_4$ 7H$_2$O, 0.1 g; trace element solution, 10 ml; vitamin solution, 10 ml. The pH was adjusted to 7.4. (Reprinted from Stanley and Morita (6) by permission of the American Society for Microbiology.)

In an effort to clarify further the effectiveness of NaCl in raising the maximal growth temperature, MP-1 was grown in different concentrations of sodium chloride in a chemically defined medium (Fig. 2). The lowest concentration of sodium chloride at which growth took place was 0.15 M. At this concentration, the maximum growth temperature was 11.5 C. The maximal growth temperature thereafter increased with increasing concentrations of sodium chloride to reach a maximum of 20.0 C between 0.35 and 0.60 M. Above this concentration there was a reduction of the maximum growth temperature.

Salinity Effects on Catabolism of Glucose

The effects of salinity on the maximal growth temperature were examined by Griffiths and Morita (2) in terms of nutrient uptake and catabolism. In observing the effects of cations and anions on the uptake of uniformly labeled glucose, they found that the cations that did not permit growth did not permit active uptake of glucose. It was concluded that it was quite possible that previous researchers (6) had actually been looking at a specific cation requirement for nutrient uptake.

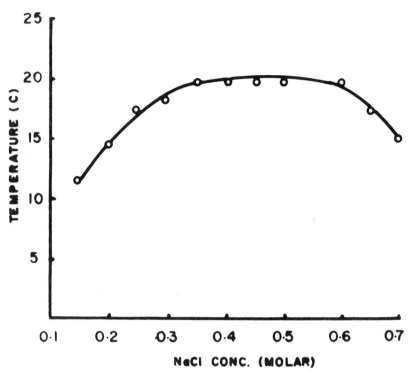

Fig. 2. **Effect of NaCl concentration on the maximal growth temperature of *Vibrio marinus* MP-1 in a defined medium. The growth medium was the same as the stock culture medium except that polypeptone was replaced by glucose (2.5 g/l) and KNO₃ (1.0 g/l). (Reprinted from Stanley and Morita (6) by permission of the American Society for Microbiology.)**

Griffiths and Morita *(2)* also looked at the effects of varying salinity on the uptake of glucose. As the salinity was increased from 0.15 M NaCl to 0.30 M NaCl, there was a rapid increase in the amount of glucose taken up. In this salinity range, the change in the observed disintegrations per minute (DPM) per 0.1 M NaCl was 4.33×10^3. As the sodium chloride concentration was increased from 0.3 to 0.9 M, there was a steady decrease in the observed DMP but this was in the order of $\Delta 0.67 \times 10^3$ DPM/$\Delta 0.1$ M NaCl. The first narrow region of rapid change includes the minimum growth salinity (at 15 C). This same rapid shift in uptake was not seen, however, in salinity range of maximum growth, i.e., 0.7 M NaCl. The rapid shift in uptake patterns seen in the salinity range for minimum growth suggests that shifts in respiratory patterns were due to changes in salinity.

Specifically labeled glucose and radiorespirometry were used to follow the effects of salinity on respiration. Of the various labels used, those labeled on the first and sixth carbons of glucose gave some of the most interesting results. As the salinity was increased from 0.15 M to 0.30 M NaCl, there was a rapid decrease in the amount of $^{14}CO_2$ evolved. A gradual change in cell respiration was seen as the salinity was increased from 0.30 M to 0.90 M NaCl. Even more important was the fact that the relative amount of $^{14}CO_2$ evolving from the sixth carbon increased at a faster rate (with increasing salinity) than that from the first carbon. As a result, the $C_6:C_1$ ratios changed from 1.0 at 0.25 M NaCl to 0.19 at 0.90 M NaCl. It may be of significance that the $C_6:C_1$ ratio is one at a point just below the minimum salinity for growth. Thus, it was postulated that salinity was altering the relative participation of the pentose pathway resulting in a shift in the respiratory pattern.

It was assumed that all of the glucose was catabolized via the Embden-Meyerhof and pentose pathways. They argued that if the amount of CO_2 evolved from the third and fourth carbons of glucose was approximately equal and much greater than that evolved from the first and sixth carbons, then the Embden-Meyerhof pathway was being utilized for glucose catabolism. However, any significant difference in the levels of CO_2 evolved from the first and sixth carbons reflects the relative participation of the pentose pathway in the catabolism of glucose. If these assumptions are true, it would appear that the salt concentration affected the amount of glucose catabolized through the pentose pathway.

Salinity-Temperature Effects on Protein Synthesis

In an effort to elucidate further the effects of salinity-temperature interaction on the physiology of named microorganism, Cooper and Morita *(1)* examined the effects of salinity-temperature interactions on the net protein synthesis, RNA synthesis, and viability of *V. marinus.*

Cultures of *V. marinus* were grown to mid-log phase, harvested, washed by filtration, and resuspended in fresh medium. Growth curves of this organism were shown to be nearly identical at salinities of 25 through 35‰. However, growth was negligible at lower salinities. Proline and uracil, radioactive precursors to protein and RNA respectively, were added to the medium at temperatures from 15 to 25 C. The cell suspensions were assayed for protein or RNA synthesis. At a salinity of 25‰, protein synthesis occurred at 15 and 20 C but was significantly depressed at 25 C. Similar results were obtained at salinities of 30 through 40‰.

Protein synthesis and precursor uptake by whole cells was determined at one-degree temperature intervals between 20 and 25 C at various salinities. These results are shown in Figure 3. At a salinity of 25‰, protein synthesis was similar at 20 and 21 C, but a marked decrease in protein synthesis occurred at 22 C. Precursor uptake into whole cells continued after protein synthesis had decreased. However, this marked decrease in protein synthesis at 22 C was not observed at a salinity of 30‰. Instead there occurred a gradual decrease in protein synthesized with increasing incubation temperature.

At a salinity of 35‰, no decrease in protein synthesis was seen until after 20 min incubation at 24 C. Again precursor uptake continued after protein synthesis had decreased. It can be concluded that the cell was taking up proline, but the proline was not being incorporated into the protein within the cell.

At a salinity of 40‰, however, precursor uptake decreased before protein synthesis decreased at the incubation temperatures of 24 and 25 C. Here a different effect was probably preventing the incorporation of labeled proline into protein.

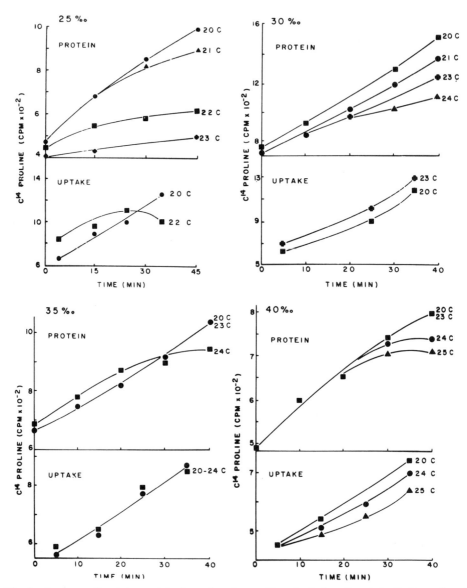

Fig. 3. Incorporation into protein (top of each graph) and whole-cell uptake (bottom of each graph) of ^{14}C-proline by *V. marinus* MP-1 incubated in SSPU medium at various salinities. Volume of cell suspension was 1.0 ml. (Reprinted from Cooper and Morita *(1)* by permission of the American Society of Limnology and Oceanography, Inc.)

Thus, the data would indicate that at salinities of 25 through 35‰, precursor entered the cells and was available for protein synthesis. However, thermal affects probably were preventing the uptake of label into the cells at a salinity of 40‰.

To determine if a thermal lesion occurred in translation or in transcription, concurrent RNA and protein synthesis studies were performed at salinities of 15 to 40‰. In Figure 4, it is clearly shown that at a salinity of 20‰, the amount

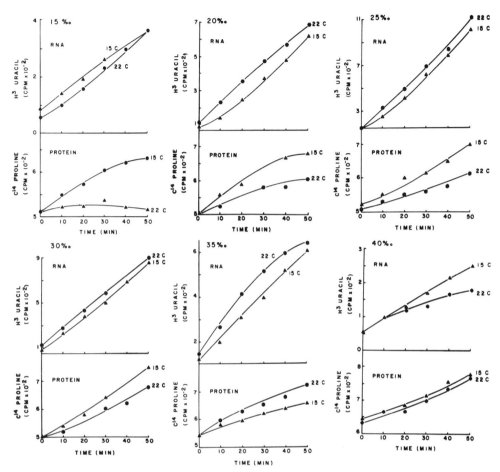

Fig. 4. Incorporation of ³H-uracil into RNA (top of each graph) and ¹⁴C-proline into protein (bottom of each graph) during incubation of *V. marinus* MP-1 in SSPU medium at various salinities. Volume of cell suspension was 0.5 ml. (Reprinted from Cooper and Morita *(1)* by permission of the American Society of Limnology and Oceanography, Inc.)

of RNA synthesized during the first 40 min was greater at 22 than at 15 C. At this salinity, protein synthesis was less at 22 than at 15 C. Identical results were observed with salinities of 25 and 30‰.

At a salinity of 35‰, both protein and RNA synthesis were greater at 22 than at 15 C, whereas at a lower salinity, protein synthesis was less at 22 than at 15 C. These studies indicate that the thermal lesion occurred at the translation level of protein synthesis.

At salinities of 15 and 40‰, however, RNA synthesis was less at 22 than at 15 C. This would indicate that extremes of temperature and salinity also decrease the total amount of label incorporated into RNA.

Temperature-Salinity on Glutamic Dehydrogenase Synthesis

Staley and Morita *(5)* have shown that temperature and salinity can influence the synthesis of glutamic dehydrogenase (GDH) in *V. marinus.* The scheme involving GDH synthesis was selected because of the strategic position of the enzymes in the TCA cycle.

Cells used for the experiments were first grown at 15 C for 48 hr in a defined glucose-ammonium medium (GAM). They were harvested, washed in a phosphate-salts buffer, and resuspended. Aliquots of the washed cell suspension were added to flasks containing glutamate medium (GM) at a variety of salinities and were equilibrated to given temperatures. After shifting the glucose-grown cells to the GM, portions were removed at appropriate time intervals. The cells were lysed by sonic treatment and the cell-free extracts assayed for GDH activity. Cells not added to GM were taken as the control samples (i.e., no exposure to glutamate). In this way Staley and Morita (4) were able to measure the effects of temperature and salinity on the synthesis of the enzyme over a period of time. Other studies by Staley (4) indicate that the synthesis of GDH is also affected by temperature so that a temperature-salinity interaction comes into play.

As indicated in Figure 5, the rate of GDH synthesis in cells grown for 48 hr at 15 C and shifted to GM at 15 C at various salinities was apparently greatest in 0.4 M NaCl. When cells were exposed to sodium chloride concentrations below the optimal for GDH synthesis (i.e., 0.26 M), the cells showed a reduced ability to synthesize GDH. Since the substrate uptake rate, (Fig. 6) as well as the rate of protein synthesis (Fig. 7), was maximal at this salinity, the organisms were quite capable of getting the substrate into the cells as well as synthesizing GDH. Net synthesis occurred at a very low rate. A correlation of Figure 5 with Figures 6 and 7 cannot be made. This probably indicates that the salinity effect on the synthesis of this protein (GDH) is not related to the salinity effect on uptake of glutamic acid or the synthesis of net protein by the cell. In other words, precisely how salinity affects the synthesis of GDH remains to be established.

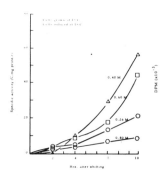

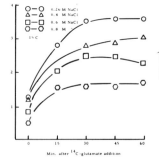

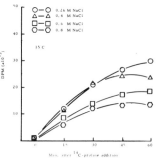

Fig. 5. **Synthesis of glutamic dehydrogenase in** *V. marinus* **when cells were shifted from GAM to GM medium containing various molarities of NaCl at 15 C.**

Fig. 6. **Time course of ^{14}C-glutamate (0.2 μc/ml cell suspension) uptake by** *V. marinus* **cells in GM medium (3.0 mM glutamate) at 15 C at various NaCl concentrations.**

Fig. 7. **Time course of ^{14}C-proline (0.05 μc/ml cell suspension) incorporated by** *V. marinus* **in GM medium (3.0 mM glutamate) at 15 C at various NaCl concentrations.**

Concluding Remarks

We have shown that salinity plays an important role in governing the temperature range of growth of certain bacteria and we have tried to elucidate the mechanisms responsible for the effect observed in temperature-salinity interactions. From a more practical point of view, it is of interest to speculate on the physiological reaction of a marine bacterium subjected to salinity and temperature variations and to the nutritionally richer waters of the estuary. Weimer and Morita *(7)*, using an inducible enzyme system, showed with *Vibrio* MP-41 that the activity of the exoenzyme, gelatinase, was adversely affected by varying salt concentrations up to 15%.

Greatest activity was observed in the absence of salts and this activity decreased with increasing salinity. Increasing the salt concentration above 3.6%, however, caused little further decrease in activity. Thus, increasing the salinity in the presence of cooler water could put a limit on the potentialities of this exoenzyme.

Conversely, what would happen to a marine bacterium subjected to an influx of relatively warm, fresh water? In Figure 8, Haight and Morita *(3)* show that there is leakage of protein, RNA, DNA, and amino acids into the surrounding medium when an organism (in this case, *V. marinus* MP-1) is subjected to temperatures above 20 C. Such events could occur when organisms present in a salt wedge are forced to the surface in an estuary. The data in Figure 9 demonstrate how organisms can be subjected to changes in salinity and temperature resulting in the lysis of certain marine microorganisms such as the psychrophile *V. marinus*

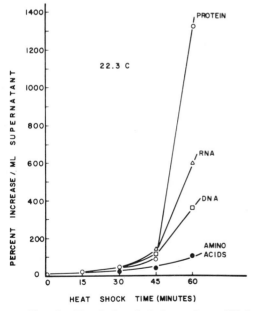

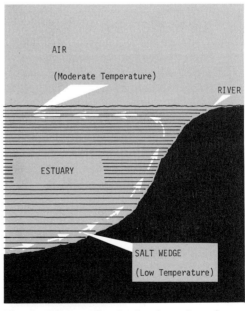

Fig. 8. Heat-induced leakage from *Vibrio marinus* MP-1 at 22.3 C. (Reprinted from Haight and Morita *(3)* by permission of the American Society for Microbiology.)

Fig. 9. Diagramatic sketch of a salt wedge surfacing in an estuary.

MP-1. Lowering salinity also enhances leakage of intracellular material from the cell. However, this phenomenon is not completely detrimental to the ecology of the estuary. In fact, it probably adds to the overall productivity by releasing vitamins, enzymes, purines, and pyrimidines—materials known to affect growth in other organisms.

For these and other reasons, tides are important in the recycling process and in the general ecology of estuaries. The movement of tides not only affects the temperature and salinity of the environment but also the amount and types of nutrient compounds found in the water. Tides contribute large amounts of nutrients to the offshore environments, thus allowing them to enter the food cycles essential to productivity in all waters.

Morita and Kalber* took samples of water at 1-hr intervals during tide changes in a mud flat near Freeport, Texas, on two occasions. The overall regeneration or utilization of phosphate and ammonia was measured by the difference between the initial figure and that present in the water after a 15-min incubation period. The data clearly demonstrate that the regeneration or utilization of ammonia and phosphate in this environment varied from hour to hour and that more ammonia and phosphate were found in bottom water than in surface water (Fig. 10). These data also indicate that within an intertidal area, the water is heterogenous in terms of ammonia and phosphate content and also in utilization and regeneration of ammonia and phosphate. The mud flat thus represents a dynamic situation where the ammonia and phosphate content of the water is constantly undergoing changes.

It is important to note, moreover, that although there have been many studies enumerating microorganisms in a given estuarine environment, numbers alone do not tell us the rate of *in situ* activity taking place. The presence of any physiological group of bacteria only indicates that the potential for microbial activity exists. This activity may not be expressed, however, until the conditions are suitable. It is important to understand that conditions of flux exist at all times in a complex environment such as an estuary.

With this in mind, we believe that the Wright-Hobbie kinetic approach, discussed by Dr. Wright in this symposium, is very useful in helping to understand the total environment and the activities taking place therein. Our work on Upper Klamath Lake in Oregon** and in the Antarctic† illustrates some applications of a kinetic approach. This approach takes into consideration the indigenous microflora in relation to the other environmental conditions such as temperature, salinity, and primary nutrients.

We thus conclude that marine microorganisms are always vulnerable to salinity and temperature changes that may significantly affect their normal functions and consequently affect the environment. A knowledge of the results of these changes on metabolic potentiality (i.e., formation of adaptive enzymes and exoenzymes, rate of metabolism, etc.) of marine organisms under near-shore environmental conditions is of paramount importance if we are to understand and more wisely participate in the dynamic economy of the oceans.

* Unpublished data.
** Burnison, Gillespie, Harrison, Wright, and Morita, unpublished data.
† Gillespie, Jones, and Morita, unpublished data.

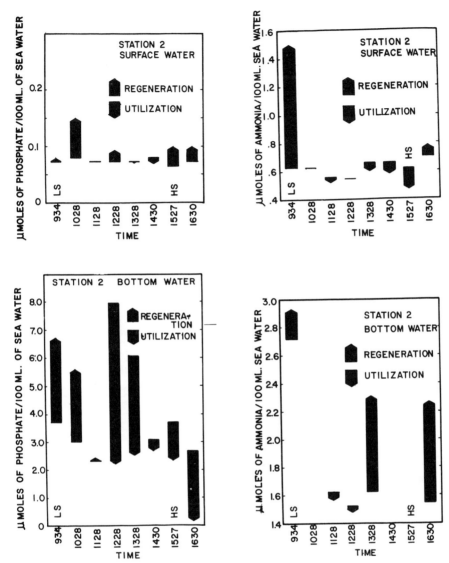

Fig. 10. Regeneration and utilization of phosphate and ammonia in various samples of seawater at Station II (mud flat near Freeport, Texas) during a tidal cycle. Water samples were taken and the analyses made at times indicated (0-time). Analyses were also made on subsamples of water after incubation for 15 min. The difference in phosphate and ammonia levels between analyses after 15-min incubation and 0-time gives the amount of regeneration (+ value) and utilization (− value). Incubation was in the dark at approximately the same temperature of the water. HS = high slack water; LS = low slack water.

Acknowledgment

The research reported in this paper was supported in part by the Oceanography Section, NSF, NSF grant GA-28521 and in part by PHS Training Grant 5 T1 GM 704–08 MIC from the National Institute of General Medical Sciences.

The senior author wishes to express his sincere thanks to his former and current students and associates for some of the data presented in this paper.

Literature Cited

1. Cooper, M. F., and R. Y. Morita. In press. Interaction of salinity and temperature on net protein synthesis and viability of *Vibrio marinus. Limnol. Oceanogr.*
2. Griffiths, R. P., and R. Y. Morita. 1971. The effects of salinity on glucose uptake and catabolism in the marine psychrophile, *Vibrio marinus. Bacteriol. Proc.*, p. 133.
3. Haight, R. D., and R. Y. Morita. 1966. Thermally induced leakage from *Vibrio marinus*, an obligately psychrophilic marine bacterium. *J. Bacteriol.* **92**:1388–92.
4. Staley, T. E. 1970. Effects of salinity and temperature changes on the induction of glutamic dehydrogenase in the marine psychrophilic bacterium, *Vibrio marinus* MP-1. Ph.D. dissertation, Oregon State University.
5. Staley, T. E., and R. Y. Morita. 1971. Effect of sodium chloride on the induction of glutamic dehydrogenase in a marine psychrophile. *Bacteriol. Proc.*, p. 19.
6. Stanley, S. O., and R. Y. Morita. 1968. Salinity effect on the maximal temperature of some bacteria isolated from the marine environment. *J. Bacteriol.* **95**:169–73.
7. Weimer, M. S., and R. Y. Morita. 1968. Effect of hydrostatic pressure and temperature on gelatinase produced by an obligately psychrophilic marine vibrio. *Bacteriol. Proc.*, p. 34.

An Ecological Survey of Open Ocean and Estuarine Microbial Populations. I. The Importance of Trace Metal Ions to Microorganisms in the Sea

Galen E. Jones

Geochemical Considerations

The theory as to the origin of the earth is still an unsettled question. Some investigators suggest a condensation from hot gas as the origin and others, an accretion from cold gas. Dating from radioactive breakdown indicates that the earth originated about 4.7 billion years ago. Whatever the process of formation was, geochemical fractionation of elements has continued on a slow time scale ever since, and the earth has not reached a chemical equilibrium (1).

Gibson (8) has indicated three major partings of the elements. First, the primitive earth, warmed either by its heat of formation or by radioactive decay processes, separated into three concentric layers known as the core, the mantle, and the crust. Knowledge of the core and the mantle is still indirect and fragmentary even though they comprise most of the mass of the earth. The crust, however, has been extensively analyzed.

Most of the original crust of the earth was molten magma. Even though the crust of the earth is solid, the inner layers are still partly molten and molten rock or magma can be observed in active volcanoes. When liquid magma solidified, a second parting of the elements occurred. In the cooling process, a series of minerals forming igneous rocks crystallized in a definite order, and the final residue consisted mainly of liquid water with some gaseous nitrogen. The oceans and the atmosphere are products of the second parting.

During the third parting of the elements, the igneous rocks were slowly destroyed by physical, chemical, and biological processes (1). The combined action of water, various other natural solutes, living organisms, the action of blown sand in some regions, and frost in others have denuded exposed rock surfaces. The products of weathering were carried by rivers to the ocean and formed sediments on the ocean floor. By the process of diagenesis, sediments were compacted to form sedimentary rocks. Some sediments became buried so deeply that they were profoundly altered by the high temperature and pressure, in which case they are called metamorphic rocks. Ancient sediments that have been changed in this way are often difficult to tell from igneous rocks.

Seventy-one percent of the surface of the earth is covered by the ocean, which has been described as the dilute solution of practically every known element. The ocean has a mass of 1.42×10^{21} kg and a density of 1.03 at 0 C (1). In the beginning, the ocean probably had a different composition from what we know

233

today because of the many cycles of erosion and sedimentation. These two processes have been in dynamic equilibrium for at least 500 million years so that the composition of the ocean has remained approximately the same for that length of time.

Measurements of the water cycle show that 0.024% of oceanic water or 3.34×10^{17} kg falls on land as rain, and 63% of this evaporates again: thus, rivers return 0.37×10^{17} kg of water per year to the sea. There is, however, a significant difference between the solids contained in river water and in seawater. The average river water contains 120 ppm total solids, but river waters have a wide range of total solids and their total composition is not consistent. Seawater contains 35,000 ppm total solids and has a constant composition.

The main ions present in river water vary with climate and local rock formations. In most temperate regions, the most common solutes are Ca and HCO_3 (1). In tropical regions, rivers contain less salts but more organic matter with measurable amounts of Al, Fe, and Si. In arid regions, rivers are very salty and the most common solutes are Na, Cl, and SO_4.

Concentration of Trace Elements by Living Organisms

Although most of the elements in the periodic table are present in the electrolyte solution seawater, 99.9% of the dissolved solids in seawater consist of ten ionic species, namely Cl, Na, SO_4, Mg, Ca, K, HCO_3, Br, Sr, and H_3BO_3 (7). The major ions in seawater are the relatively conservative or unreactive elements, which is why they have tended to be retained by the ocean (9). Table 1 shows the concentrations of these major ions in seawater. The trace or minor elements in the ocean, which comprise less than 0.1% of the dissolved solids, are involved in some of the more important inorganic and biochemical reactions of the marine environment.

Table 1. Major ionic constituents of seawater

Constituent	Concentration, g/kg of 35 ppt Seawater	Percent Composition	Mole/Liter
Na^+	10.76	30.61	0.47015
Mg^{++}	1.294	3.69	0.05357
Ca^{++}	0.413	1.16	0.01024
K^+	0.387	1.10	0.00996
Sr^{++}	0.008	0.03	0.00015
Cl^-	19.353	55.05	0.54830
$SO_4^=$	2.712	7.70	0.02824
HCO_3^-	0.142	0.36	0.00234
Br^-	0.067	0.18	0.00083
H_3BO_3	0.026	0.07	0.00043
Totals	35.162	99.95	1.12421

Large numbers of marine plants and animals concentrate trace elements against the gradient by many hundreds to thousands of times (1, 19, 23, 24, 28,

32, 33). The large concentration factors of trace elements are most obvious perhaps for marine organisms, especially for algae which obtain all of their nutrients directly from seawater. Table 2 indicates the biogeochemistry of principal trace elements in the marine environment by comparing their concentrations in igneous rocks, seawater, marine algae, and marine animals. Two sets of analyses are available for marine algae and marine animals: one set was derived

Table 2. Biogeochemistry of principal trace elements in the marine environment

Element	Igneous Rock[a] ppm[g]	Seawater[b] ppb[h]	Marine Algae[a] ppm	Marine Algae[c] ppm	Marine Animals[a] ppm	Marine Animals[d,e] ppm	Toxicity[f]
Ag	0.07	0.11	0.25	3.3– 14.1	3 – 11	0.03– 1.3	H
Al	82,000	10	60	118 –5,000	10 – 50	53 – 2,700	M
As	1.8	3	30		0.005 – 0.3		M–H
Au	0.004	0.004	0.012		0.0003– 0.008		Sl
B	10	4,600	120		20 – 50		M
Ba	425	30	30	34.5– 262	0.2 – 3		M
Be	2.8	0.0006	0.001	1.1– 8.4	0.0003– 0.002		VH
Bi	0.17	0.02		2.0– 7.7	0.04 – 0.3		M
Cd	0.2	0.11	0.4		0.15 – 3	0.03– 73	M
Ce	60	0.0052					M
Co	25	0.1	0.7		0.5 – 5	0.16– 17.0	H
Cr	100	0.05	1.0	2.8– 30	0.2 – 1	ND 2.0	H
Cs	1	0.5	0.07				–
Cu	55	3	11	25 – 210	4 – 50	0.09– 270	VH
Fe	56,300	10	700		400	1.6–22,000	Sl
Ga	15	0.03	0.5		0.5		M
Gd	5.4	0.0024			0.06		Sl
Hg	0.08	0.03	0.03		0.04 – 0.3		VH
Li	20	170	5		1		Sl
Mn	950	2	53	3.8– 118	1 – 60	0.52– 2,100	M
Mo	1.5	10	0.45		0.6 – 2.5		M
Ni	75	2	3	2.7– 48	0.4 – 25	0.12– 850	H
P	1,050	70	3,500		4,000 –18,000	1,400 –11,000	–
Pb	12.5	0.03	8.4	8– 900	0.5	0.4 42	H
Rb	90	120	7.4		20		Sl
Re	0.005	–	0.014		0.0005– 0.006		–
Sb	0.2	0.5	–		0.2		M
Se	0.05	0.4	0.8				M–H
Sn	2	0.8	1.0	8– 101	0.2 – 20		H
Ti	5,700	1.0	12–80	7.8– 940	0.2 – 20		–
V	135	2.0	2.0	1.2– 5.7	0.14 – 2		M
W	1.5	0.1	0.035		0.0005– 0.05		M
Zn	70	10	150	75 – 480	6 – 1,500	0.04– 940	M
Zr	165	0.022	20		0.1 – 1.0		M

[a] Data summarized from Bowen *(1)*.

[b] Data summarized from Goldberg *(9)*.

[c] Data summarized from Riley and Roth *(24)*.

[d] Data summarized from Riley and Segar *(23)*.

[e] Data summarized from Segar, Collins, and Riley *(28)*.

[f] Toxicity

   VH = very high

   H = high

   M = moderate

   Sl = slight

   – = no toxicity.

[g] ppm = parts per million.

[h] ppb = parts per billion.

from Bowen (1) and the other from Riley's laboratory (23,24,27). The following general conclusions can be derived from Table 2 and the cited literature.

1. All of the elements studied are more or less concentrated with the exception of Cl, which is definitely rejected; Na is weakly rejected. The concentration factors for Br, F, Mg, Na, and S are of the order of 1 and are higher for all other elements.

2. The order of affinity for living material among the cations is generally: tetravalent and trivalent elements > divalent transition metals > divalent Group II A metals > univalent Group I metals. The tetravalent and trivalent subgroups have somewhat different affinities for marine algae and marine animals:

Marine Algae: Fe > Al > Ti > Cr > Ga

Marine Animals: Fe > Al > Ga > Cr > Ti

Similar discrepancies exist among organisms in their affinities for the divalent transition metals. The exact concentration factors are impossible to calculate due to the range of values for each type of organism. The order of concentration factors is not completely correlated with the order of stability of complexes formed with organic ligands.

3. The order of affinity of living matter for anions is: nitrate > trivalent anions > divalent anions > univalent anions. The unique position of nitrate is complicated by other forms of nitrogen available in the ocean, particularly ammonium, organic nitrogen, and dissolved dinitrogen which may be available to different members of the plankton community.

Among trivalent anions, phosphate is concentrated by very large factors, rivaling nitrate in this respect. Arsenate and vanadate are less strongly concentrated (1).

Some of the trace elements are essential for organisms and others do not have a definitely established physiological function. For example, Harvey (10, 11) has reported evidence indicating suboptimal amounts of Fe and Mn in seawater may adversely affect the growth of marine phytoplankton. Other trace elements in seawater do not appear to be limiting to the growth of phytoplankton in the sea, although Provasoli (21) has reported laboratory data suggesting that they are nutritionally important. Yet, significant aspects of the chemistry of the sea are regulated by the life therein (22).

The abundance of trace elements in marine bacteria is unknown. The trace element analyses for bacteria are generally limited (1).

Active Uptake of Trace Elements

It is obvious that the uptake of ions by cells can be accounted for by assuming that another process operates apart from simple diffusion. This process, called active uptake, is closely linked with metabolic processes inside the cell. The energy necessary for uptake is provided by the metabolic processes against a concentration gradient, but the exact mechanism is still a subject of some conjecture. The maintenance of the concentration of immobile anions in the cell is clearly one function. A larger temperature coefficient is necessary for active uptake than for uptake by diffusion. A rise in temperature of 10 C increases the rates of absorption by 100% for active uptake and by 20% for diffusion processes. Complications arise in determining the exact effect of temperature due to increased rates of growth, cell division, and other metabolic processes. Substances that inhibit respiration, such as cyanide, also inhibit active uptake of ions. Active uptake does not involve the appearance of exchanged ions in the medium. The anions that are rapidly consumed by metabolic reactions in the cell, such as NO_3, SO_4, HCO_3, and PO_4, are of particular importance (1).

Small amounts of heavy metal ions are necessary for the vital physiology of living cells. Almost any element for the nutrition of a living cell in seawater has a concentration at which it is deficient, and a slightly higher concentration at which it is in the optimum range, leading to luxury consumption and finally, to a higher concentration for toxicity and lethality. This type of an effect of an element on a cell is known as an oligodynamic effect.

Mechanisms of Toxicity of Trace Elements

The most important mechanism of toxic action is considered the poisoning of enzymes. The more electronegative metals, such as Cu, Hg, and Ag, have a great affinity for amino, imino, and sulfhydryl groups which are reactive sites on many enzymes and are readily chelated by organic molecules. There has been some success in attempts to correlate the toxicities of metals with such factors as their electronegativities (6, 31), the insolubility of their sulfides (29), or the order of stability of their chelated derivatives (12, 13, 30). The order of electronegativities for the divalent metals is as follows: Hg > Cu > Sn > Pb > Ni > Co > Cd > Fe > Zn > Mn > Mg > Ca > Sr > Ba. The order of solubility products of the sulfides is as follows: Hg > Cu > Pb > Cd > Co > Ni > Zn > Fe > Mn > Sn > Mg > Ca. The order of stability of chelates is as follows: Hg > Cu > Ni > Pb > Co, Zn > Cd > Fe > Mn > Mg > Ca (14, 18). Attempts to correlate these factors with biological toxicity have never proved complete for any given system; yet, there is a general correlation. Considering large numbers of enzymes in living cells, variations in toxicity are hardly surprising. Bowen (1) has delineated other possible modes of toxic action. Sadler and Trudinger (25) have related these directly to reactivity with microbial cells.

The effect of metal ions on enzyme activity may be arrested by substances called inhibitors, many of which are well-known poisons. Inhibitors such as cyanides, sulfides, azides, carbon monoxide, and ethylenediaminetetraacetic acid (EDTA) specifically inhibit enzymes containing metals by forming coordination complexes with the metals. Other inhibitors, such as Pb, Hg, and "heavy metal" ions probably act by competing with essential metals for active sites on enzymes.

It is highly possible that the more active trace metals on the avidity series act by substituting for Ca and Mg which are lower down on the series and displace these structural ions in the cell wall and other metabolic construction of the cell.

Trace Elements in Marine Ecology

Saunders *(26)* noticed the importance of the interaction of dissolved organic matter in seawater on the influence and availability of trace metals. He pointed out that the chelation of trace metals may affect biological systems in four ways. First, the organic chelate may affect the availability of a trace metal ion by lowering its effective concentration in the water system below the level or requirement by a particular organism. Thus, a trace metal required for a particular biological system may become completely unavailable or only partially available but limiting to the normal growth and reproduction of the organism. Second, if there is an excess of a trace metal that is toxic to a particular organism, chelation may reduce the metal ion concentration below the level of toxicity of that organism. Third, chelation may remove the metal ion that is antagonistic to a metal poison, effectively increasing the relative concentration of the poison to a level at which it becomes toxic. Fourth, if a trace metal tends to precipitate and become essentially unavailable to an organism, chelation may maintain that particular element complexed in a soluble state and thus keep the effective level of concentration up above that which would prevail if chelation did not occur.

Terrestrial bacteria do not survive well in seawater and a major mechanism in the lethality of seawater for coliform bacteria is heavy metal toxicity *(16, 27)*. Bacteria indigenous to the marine environment have a number of salient characteristics but two stand out: their tolerance to heavy metal ions *(17)* and their ability to utilize low levels of energy sources *(15)*.

The mechanisms by which trace elements are concentrated and the manner in which they are held in bacteria are unknown or incompletely understood *(20)*. The cells of the marine bacterium, *Arthrobacter marinus*, become considerably enlarged (15–20 μm) or megalomorphic in laboratory media where Ni ions exceed the chelating ability of the media *(4)*. At appropriate concentrations, Ni ions are believed to interfere with the cell division processes of the cell, but not with growth *(3)*. Vacuolation is intense in the megalomorphic cells, but the total protoplasm is equivalent in Ni-treated and untreated flasks *(5)*. Other metal ions cause aberrant growth in *A. marinus*, but none cause the giant, elliptical megalomorphs produced by Ni ions *(4)*. Ni does not appear to penetrate the membrane of *A. marinus* but rather affects the peptidoglycan layer of the cell wall.

Bioliths or residues from organic matter accumulate many trace elements. Crude oil may be rich in Mo, Ni, or V, and asphalt is often rich in Ni *(1)*. Breger *(2)* has suggested that Ni and V form strong complexes with porphyrins in organic debris that accumulate in petroleum. A large number of elements are accumulated by coal but not by every seam *(1)*.

It has been assumed that microorganisms play a very minor role in geochemical processes because they are highly sensitive to the presence of heavy metals. The concept of heavy metal sensitivity has developed from laboratory studies involving pure cultures of microorganisms growing exponentially with short generation times, in defined or complex laboratory media, and with high energy sources. Under these conditions, the concept has specific validity. However, extrapolating laboratory observations to microorganisms in the sea or other natural habitats should be done with caution. Pure cultures do not occur under natural conditions and growth rates are very slow compared to the highly favorable laboratory conditions with irregular generation times and a wide array of factors that must be considered and undoubtedly modify the organism's response. Due to the extended time of natural chemical processes, there is ample time for the selection of strains optimally adapted for growth under the particular conditions in any given ecological situation. Many factors may modify the response of a particular organism to the presence of a heavy metal. Considerations of this type suggest that a reappraisal of the sensitivity of microorganisms to heavy metals under natural conditions may be necessary. It is likely that microorganisms in the sea and elsewhere may have performed a number of vital functions in geochemical processes.

The sea is chemically complex. The most salient aspects of this chemical milieu for life in the oceans are the micro amounts of trace elements and organic compounds which are a reflection of the great drains on these substances by microbial life. Marine microorganisms are well adapted to the marine environment.

Literature Cited

1. Bowen, H. J. M. 1966. *Trace elements in biochemistry.* New York: Academic Press Inc.
2. Breger, I. A. 1963. *Organic geochemistry.* New York: Pergamon Press.
3. Cobet, A. B. 1968. The effect of nickel ions on *Arthrobacter marinus*, a new species. Ph.D. dissertation, University of New Hampshire.
4. Cobet, A. B., C. Wirsen, and G. E. Jones. 1970. The effect of nickel on a marine bacterium, *Arthrobacter marinus* sp. nov. *J. Gen. Microbiol.* **62:**159–69.
5. Cobet, A. B., G. E. Jones, J. Albright, H. Simon, and C. Wirsen. 1971. The effect of nickel on a marine bacterium: Fine structure of *Arthrobacter marinus. J. Gen. Microbiol.* **66:**185–96.
6. Danielli, J. F., and J. T. Davies. 1951. Reactions at interfaces in relation to biological problems. *Adv. Enzymol.* **11:**35–89.
7. Garrels, R. M., and M. E. Thompson. 1962. A chemical model for sea water at 25 C and one atmosphere total pressure. *Amer. J. Sci.* **260:**57–66.
8. Gibson, D. T. 1949. The terrestrial distribution of the elements. *Quart. Rev. Chem. Soc.* **3:**263–91.
9. Goldberg, E. D. 1965. Minor elements in sea water. In *Chemical oceanography*, vol. 1, ed. J. P. Riley and G. Skirrow, pp. 163–96. New York: Academic Press Inc.
10. Harvey, H. W. 1939. Substances controlling the growth of a diatom. *J. Mar. Biol. Ass. U. K.* **23:**499–520.
11. Harvey, H. W. 1947. Manganese and growth of phytoplankton. *J. Mar. Biol. Ass. U. K.* **26:**562–79.
12. Horsfall, J. G. 1956. *Principles of fungicidal action.* Waltham, Mass.: Chronica Botanica Co.
13. Hunter, J. G., and O. Vergnano. 1953. Trace element toxicities in oat plants. *Ann. Appl. Biol.* **40:**761–77.
14. Irving, H., and R. J. P. Williams. 1948. Order of stability of metal complexes. *Nature* **162:**746–47.
15. Jannasch, H. W. 1967. Growth of marine bacteria at limiting concentrations of organic carbon in sea water. *Limnol. Oceanogr.* **12:**264–71.
16. Jones, G. E. 1964. Effect of chelating agents on the growth of *Escherichia coli* in seawater. *J. Bacteriol.* **87:**483–99.
17. Jones, G. E. 1971. The fate of freshwater bacteria in the sea. *Develop. Ind. Microbiol.* **12:**141–51.

18. Mellor, D. P., and L. Maley. 1948. Order of stability of metal complexes. *Nature* **161**:436–37.
19. Noddack, I., and W. Noddack. 1940. Die Haufigkeiten der Schwermetalle in Meerestieren. *Ark. Zool.* **32A**:1–52.
20. Perlman, D. 1965. Microbial production of metal-organic compounds and complexes. In *Advances in applied microbiology*, vol. 7, ed. W. W. Umbreit, pp. 103–38. New York: Academic Press Inc.
21. Provasoli, L. 1963. Organic regulation of phytoplankton fertility. In *The sea*, vol. 2, ed. M. N. Hill, pp. 165–219. New York: Interscience.
22. Redfield, A. C., B. H. Ketchum, and F. W. Richards. 1963. The influence of organisms on the composition of seawater. In *The sea*, vol. 2, ed. M. N. Hill, pp. 26–77. New York: Interscience.
23. Riley, J. P., and D. A. Segar. 1970. The distribution of the major and some minor elements in marine animals. I. Echinoderms and Coelenterates. *J. Mar. Biol. Ass. U.K.* **50**:721–30.
24. Riley, J. P., and I. Roth. 1971. The distribution of trace elements in some species of phytoplankton grown in culture. *J. Mar. Biol. Ass. U.K.* **51**:63–72.
25. Sadler, W. R., and P. A. Trudinger. 1967. The inhibition of microorganisms by heavy metals. *Mineralium Deposita* **2**:158–68.
26. Saunders, G. W. 1957. Interrelationships of dissolved organic matter and phytoplankton. *Bot. Rev.* **23**:389–410.
27. Scarpino, P. V., and D. Pramer. 1962. Evaluation of factors affecting the survival of *Escherichia coli* in sea water. VI. Cysteine. *Appl. Microbiol.* **10**:436–40.
28. Segar, D. A., J. D. Collins, and J. P. Riley. 1971. The distribution of the major and some minor elements in marine animals. II. Molluscs. *J. Mar. Biol. Ass. U.K.* **51**:131–36.
29. Shaw, W. H. R. 1954. Toxicity of cations toward living systems. *Science* **120**:361–63.
30. Shaw, W. H. R. 1961. Cation toxicity and the stability of transition-metal complexes. *Nature* **192**:174–75.
31. Sommers, E. 1960. Fungitoxicity of metal ions. Nature **187**:427–28.
32. Vinogradov. A. P. 1953. The elementary composition of marine organisms. *Sears Found. Mar. Res. Mem. II.* New Haven: Yale Univ.
33. Webb, D. A. 1937. Studies on the ultimate composition of biological material. Part II. Spectrographic analysis of marine invertebrates with special reference to the chemical composition of their environment. *Sci. Proc. R. Dubl. Soc.* **21**:505–39.

Comments

HOLM-HANSEN: Have you interpreted the metal ion effects only in a toxic meaning? Do you rule out the possibility that things like 8-hydroxyquinoline have a beneficial effect through making available some microelement that was not normally available?

G. JONES: I mentioned that concept earlier. That was one of the possible mechanisms listed by Saunders *(26)*. Chelators will make metals available in some instances.

HOLM-HANSEN: But you have not ruled out the possibility?

G. JONES: Oh no, not at all. That is part of the whole emphasis in the importance of heavy metals in the sea.

HOLM-HANSEN: You could separate these two major effects in laboratory experiments.

G. JONES: We have in many instances. We get both types of effects as indicated in the following paper. Also, one of the most important aspects is the concentration of both metal ions and organic matter.

HOLM-HANSEN: Not so much concentration, but the ionic activity of the species.

G. JONES: Right, we are working on that, too, whether the metal ions are complexed or free—label or nonlabel.

PRATT: We attempted to pursue the idea that heavy metals were causing die-off of organisms such as *Escherichia* and *Salmonella*. In areas where there is an enrichment of organic matter, there is an increase of sulfide due to hydrogen sulfide production by sulfate-reducing bacteria. I felt that the sulfide might protect organisms against the heavy metal effects. So we set up quite a series of experiments using several strains of *Salmonella* and *Escherichia*. We started with about 1,000 cells per ml. The result was that it took about four days for the organisms to die in the natural seawater. When we added copper and mercury at about one-half part per million to the cells, they died off promptly. We could protect them by the addition of sulfide. In our natural seawater, the die-off did not seem to be related to the metal ions any longer. Perhaps we should test seawater for its toxicity due to metal ions by using sulfide, or perhaps something like cysteine, to see if a longer survival of organisms could be obtained. Actually, we were not even convinced that the organisms were dying in our seawater at an accelerated rate. In fact, they died off in distilled water at about the same rate as they were dying in seawater.

G. JONES: We have done this with distilled water. They did not die off quite as fast as in seawater, but they did die off rapidly in distilled water. We attributed that mortality to a lack of organic matter, as much as anything. They did not have the energy source and they were not able to metabolize. But, I would like to comment on the sulfide. Transition metals in natural seawater are in the range of 1 to 10 ppb for most of the ones that we are considering here. Mercury is even less concentrated than that. They may not be concentrated enough to complex with sulfide in some instances.

PRATT: At alkaline pH, the amount of metal that you need to get the sulfide complex is very, very low.

TODD: I am curious to know if you have noted the monster cell formation in species of *Arthrobacter* other than the one you have investigated.

G. JONES: Not really; megalomorph formation is restricted to *A. marinus* among the *Arthrobacter*. We have noticed some tendency for other genera of marine bacteria to enlarge in high concentrations of nickel ions in marine media but nothing as dramatic as in *A. marinus*.

TODD: Not as dramatic as the one species you have here? Just a quick question on your specimen preparation and fixation. What did you fix it in?

G. JONES: We used 0.2% gluteraldehyde and air-dried.

TODD: Did you concentrate the broth culture in any way?

G. JONES: We centrifuged the cells.

An Ecological Survey of Open Ocean and Estuarine Microbial Populations. II. The Oligodynamic Effect of Nickel on Marine Bacteria

Edward R. Gonye, Jr., and Galen E. Jones

Marine bacteria are indigenous to an environment characterized by small amounts of organic matter and large amounts of inorganic salts. The effect of major ions on marine bacteria has received considerable attention *(4–6, 18–21, 23–28, 31, 32)*. The effect of minor ions on marine bacteria has received less attention even though the effect may be more profound *(10–15)*. Heavy metal ions in the sea have been implicated as principal inhibitory factors for the survival of terrestrial bacteria entering the sea *(10, 29)*.

As trace metals enter the ocean from rivers, estuaries, marshes, and other coastal environments, the response of microorganisms in these environments to heavy metals may not be similar to those in the open ocean. Work in this laboratory has been proceeding to determine the effect of specific metals such as copper *(17)* and nickel on marine bacteria *(2, 3, 8)*. The purpose of the work reported in this paper is to determine the response of the heterogeneous, heterotrophic marine bacterial population from different ecological areas, such as the open ocean, coastal and estuarine environments, to various concentrations of nickel ions in an organically enriched medium.

Materials and Methods

Sampling. Seawater samples were obtained using Cobet bacteriological water samplers *(13)*. From depths up to 200 m, the seawater samples were transferred immediately upon return to the surface into sterile milk dilution bottles and appropriate aliquots passed through 0.45 μm membrane filters (Millipore Filter Corporation, Bedford, Massachusetts) aboard ship.

The membrane filters were placed on marine nutrient agar plates (Fisher Scientific Company) that were supplemented with increasing molarities of reagent grade $NiCl_2$ (Fisher Scientific Company). The marine nutrient agar medium was prepared with synthetic seawater adjusted to a salinity of 26 ± 1 ppt *(16)*. In the open ocean, samples were obtained from surface, 25, 50, 75, 100, and 200 m with six treatments for each depth. The sampling procedure was completed in less than one hour. The membrane filters on the marine nutrient agar medium were incubated at temperatures approximating the *in situ* temperature of the sampling area. Colonies developing on the membrane filters were enumerated every other day for periods up to six weeks or until maximum populations were attained.

243

Ocean water samples were collected on 20 January through 26 January 1971, from the *R. V. Knorr* of the Woods Hole Oceanographic Institution. Sampling areas included seven stations on the northern edge of the Gulf Stream, approximately 350 mi east of the New Jersey coast and 400 mi south of Woods Hole, Massachusetts (Fig. 1). The *R. V. Jere A. Chase* was employed to collect coastal and estuarine water samples. Coastal water samples were collected on 31 March 1971 from four stations approximately 4 to 8 mi from the New Hampshire coast near the Isles of Shoals (Fig. 1). Estuarine water samples were collected from four stations in the Great Bay–Little Bay estuarine complex on the southeastern coast of New Hampshire on 12 April 1971 (Fig. 1).

Temperature profiles for 200 m of seawater were provided by direct recorded reading of XBT_s at each open ocean station. Salinity determinations for open ocean seawater were obtained from Dr. Gordon H. Volkmann of the Woods Hole Oceanographic Institution. Coastal and estuarine salinity determinations were determined by titration at the Jackson Estuarine Laboratory *(30)*.

Results

Open Ocean Area. The temperature and salinity data for stations in the area of the Gulf Stream are presented in Table 1. Temperature profiles indicating water colder than 15 C at 200 m were considered external to the northern edge of the Gulf Stream, whereas temperature profiles warmer than 15 C at 200 m were internal to the northern edge of the Gulf Stream. The hydrographic cast was considered down to a depth of 400 m prior to making a firm decision regarding its position relative to the Gulf Stream. Most stations indicated relatively stable salinities in the water column down to 200 m. Salinity extremes were observed at station no. 3 where the surface salinity was 36.296 and the 100 m salinity was 34.949.

The heterotrophic microbial population was sparse but rather consistent with

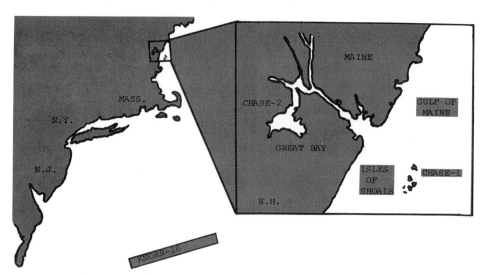

Fig. 1. Sampling areas where microbiological casts were obtained.

Table 1. Hydrographic data for Gulf Stream stations on *Knorr* 18 cruise

Station No.	Date	Location	Depth in Meters	Temp. (C)	Salinity (ppt)	Gulf Stream Position
1	1/20/71	37°08′N	surface	14.1	—	Outside
		71°42.5′W	25	13.9	—	"
			50	13.9	—	"
			75	13.9	—	"
			100	13.8	—	"
			200	15.5	—	"
2	1/23/71	37°12′N	surface	17.7	35.940	Outside
		71°51.3′W	25	17.7	35.853	"
			50	17.7	35.912	"
			75	17.2	35.929	"
			100	16.2	36.115	"
			200	10.0	35.298	"
3	1/23/71	36°44.8′N	surface	22.0	36.296	Inside
		71°42′W	25	21.5	36.199	"
			50	19.4	35.989	"
			75	17.5	35.980	"
			100	16.6	34.949	"
			200	15.6	35.691	"
4	1/25/71	37°08.2′N	surface	12.4	35.013	Outside
		71°55.5′W	25	12.2	34.167	"
			50	12.4	34.412	"
			75	12.5	34.706	"
			100	13.7	35.026	"
			200	14.0	35.419	"
5	1/25/71	36°48′N	surface	22.2	36.222	Inside
		71°51.6′W	25	22.2	36.264	"
			50	22.2	36.212	"
			75	21.7	—	"
			100	21.3	36.265	"
			200	16.9	36.171	"
6	1/26/71	38°19′N	surface	21.6	36.240	Inside
		68°02′W	25	21.6	—	"
			50	21.6	36.215	"
			75	18.7	35.815	"
			100	18.5	35.819	"
			200	15.5	—	"
7	1/26/71	38°29′N	surface		35.861	Outside
		68°11.2′W	25		35.833	"
			50		36.019	"
			75		35.684	"
			100		35.839	"
			200		35.839	"

depth at station no. 1 (Table 2). The addition of $NiCl_2$ adjusted to molarities of 10^{-5}, 10^{-4}, and 10^{-3} was stimulatory generally to the microbial population producing an average increase of 111, 119, and 133% respectively, of the populations on the untreated marine nutrient agar. At a concentration of 5×10^{-3} and 10^{-2} M $NiCl_2$, the microbial populations developed at 5 and 0%, respectively, of the untreated medium as the chelating potential of the organic nutrients in the medium was exceeded. Individual depths showed population discrepancies from the average percentages but the trend of the effect was evident.

At station no. 2 (Table 3), the bacterial population was greater than at station

Table 2. Bacterial populations for 100 ml of seawater calculated from 25 ml samples collected from station no. 1 outside the Gulf Stream

Depth (meters)	Molarities of Added $NiCl_2$[a]										
	none	10^{-5}		10^{-4}		10^{-3}		5×10^{-3}		10^{-2}	
	No.	No.	%[b]	No.	%	No.	%	No.	%	No.	%
Surface	196	188	96	288	147	212	108	12	6	0	0
25	60	88	147	44	73	120	200	4	7	0	0
50	60	56	93	92	153	92	153	4	7	0	0
75	100	60	60	80	80	92	92	4	4	0	0
100	48	92	192	56	117	68	142	0	0	0	0
200	60	48	80	88	147	60	100	4	7	0	0
Average	87	89	111	108	119	107	133	5	5	0	0

[a] Culture grown on marine nutrient agar medium containing different concentrations of $NiCl_2$ and incubated at 15 C until maximum population was attained.

[b] Relative percentage bacterial populations on nickel-containing medium were calculated against the control.

Table 3. Bacterial populations for 100 ml of seawater calculated from 25 ml samples collected from station no. 2 outside the Gulf Stream

Depth (meters)	Molarities of Added $NiCl_2$[a]										
	None	10^{-5}		10^{-4}		10^{-3}		5×10^{-3}		10^{-2}	
	No.	No.	%[b]	No.	%	No.	%	No.	%	No.	%
Surface	484	496	102	512	106	792	163	44	9	0	0
25	588	452	77	420	71	572	97	28	5	0	0
50	532	428	80	280	53	356	67	16	3	0	0
100	1,064	1,040	98	1,112	105	1,108	104	96	9	0	0
200	440	456	104	568	129	480	109	8	2	0	0
Average	622	575	92	578	93	662	108	38	6	0	0

[a] Culture grown on marine nutrient agar medium containing different concentrations of $NiCl_2$ as incubated at 15 C until maximum population was attained.

[b] Relative percentage bacterial populations on nickel-containing medium were calculated against the control.

no. 1 (Table 2). The addition of $NiCl_2$ which was adjusted to molarities of 10^{-5}, 10^{-4}, and 10^{-3} caused 92, 93, and 108% of the bacterial populations on the average to develop as in the untreated marine nutrient agar. The microbial populations dropped to 6 and 0% on the average on marine nutrient agar medium containing 5×10^{-3} and 10^{-2} M $NiCl_2$, respectively. Again, there was an average stimulation of the microbial population where 10^{-3} M $NiCl_2$ was present in the marine nutrient agar medium.

The effect of $NiCl_2$ added to the marine nutrient agar medium showed a similar effect on the five other Gulf Stream stations (Tables 4–8). At station no. 3 (Table 4), an average stimulation of 121% of bacterial populations at 10^{-3} M $NiCl_2$ occurred, compared to the untreated marine nutrient agar medium. At station no. 4 (Table 5), the bacterial colonies developing from a 25 ml sample

Table 4. Bacterial populations per 100 ml of seawater calculated from 25 ml samples collected from station no. 3 inside the Gulf Stream

Depth (meters)	None No.	10^{-5} No.	10^{-5} %[b]	10^{-4} No.	10^{-4} %	10^{-3} No.	10^{-3} %	5×10^{-3} No.	5×10^{-3} %	10^{-2} No.	10^{-2} %
Surface	140	84	60	124	89	144	103	16	11	12	9
25	108	156	144	116	107	152	141	16	15	0	0
50	328	256	78	248	76	360	110	24	7	0	0
75	56	52	93	56	100	72	129	4	7	0	0
Average	158	137	94	136	93	182	121	15	10	3	2

Molarities of Added $NiCl_2$[a]

[a] Culture grown on marine nutrient agar medium containing different concentrations of $NiCl_2$ as incubated at 15 C until maximum population was attained.

[b] Relative percentage bacterial populations on nickel-containing medium were calculated against the control.

Table 5. Bacterial populations per 100 ml of seawater calculated from 25 ml samples collected from station no. 4 outside the Gulf Stream

Depth (meters)	None No.	10^{-5} No.	10^{-5} %[b]	10^{-4} No.	10^{-4} %	10^{-3} No.	10^{-3} %	5×10^{-3} No.	5×10^{-3} %	10^{-2} No.	10^{-2} %
Surface	880	1,260	143	1,236	140	1,180	134	72	8	12	1
25	4	8		32		24		8		0	
50	4	20		16		16		20		0	
75	4	8		8		4		0		0	
100	0	8		4		12		0		0	
200	0	4		0		0		0		0	

Molarities of Added $NiCl_2$[a]

[a] Culture grown on marine nutrient agar medium containing different concentrations of $NiCl_2$ as incubated at 15 C until maximum population was attained.

[b] Relative percentage bacterial populations on nickel-containing medium were calculated against the control.

were too low to have meaning except at the surface. The general trends were observed. Low bacterial populations were observed at station no. 5 (Table 6). The bacterial numbers rose on stations no. 6 and 7 (Tables 7 and 8). The bacterial stimulation by 10^{-3} M $NiCl_2$ was observed on these stations and 33 and 22% of the bacterial colonies appeared at 5×10^{-3} M $NiCl_2$ on station no. 6 and 7, respectively, compared to the untreated control.

From all of these Gulf Stream stations, there was a tendency for 10^{-5} and 10^{-4} M $NiCl_2$ added to the marine nutrient agar medium to have little consistent stimulatory or inhibitory effect. Definite stimulation appeared in the marine nutrient agar medium containing 10^{-3} M $NiCl_2$ on open ocean stations. Marked reduction in the microbial population occurred at 5×10^{-3} M $NiCl_2$. Twenty-

Table 6. Bacterial populations for 100 ml of seawater calculated from 25 ml samples collected from station no. 5 inside the Gulf Stream

Depth (meters)	Molarities of Added $NiCl_2$[a]										
	None	10^{-5}		10^{-4}		10^{-3}		5×10^{-3}		10^{-2}	
	No.	No.	%[b]	No.	%	No.	%	No.	%	No.	%
Surface	12	8		36		24		0		0	
25	0	8		8		20		8		0	
50	25	136		92		48		84		0	
100	180	144		148		100		72		0	
200	120	156		148		144		68		0	

[a] Culture grown on marine nutrient agar medium containing different concentrations of $NiCl_2$ as incubated at 15 C until maximum population was attained.

[b] Relative percentage bacterial populations on nickel-containing medium were calculated against the control.

Table 7. Bacterial populations for 100 ml of seawater calculated from 25 ml samples collected from station no. 6 inside the Gulf Stream

Depth (meters)	Molarities of Added $NiCl_2$[a]										
	None	10^{-5}		10^{-4}		10^{-3}		5×10^{-3}		10^{-2}	
	No.	No.	%[b]	No.	%	No.	%	No.	%	No.	%
Surface	224	184	82	172	77	212	95	78	34	0	
25	196	212	108	192	98	240	122	48	25	0	
50	200	180	90	196	98	204	102	78	39	0	
75	256	292	114	232	91	348	136	108	42	0	
100	192	152	79	112	58	220	115	52	27	0	
Average	213	204	95	181	84	245	114	73	33	0	

[a] Culture grown on marine nutrient agar medium containing different concentrations of $NiCl_2$ as incubated at 15 C until maximum population was attained.

[b] Relative percentage bacterial populations on nickel-containing medium were calculated against the control.

Table 8. Bacterial populations for 100 ml of seawater calculated from 25 ml samples collected from station no. 7 outside the Gulf Stream

| Depth (meters) | Molarities of Added NiCl$_2$[a] | | | | | | | | | | | |
| | None | 10^{-5} | | 10^{-4} | | 10^{-3} | | 5×10^{-3} | | 10^{-2} | |
	No.	No.	%[b]	No.	%	No.	%	No.	%	No.	%
Surface	448	544	121	444	99	552	123	144	32	4	1
25	168	224	133	156	93	256	152	8	5	0	
50	204	320	157	276	135	356	174	60	29	0	
75	144	156	108	124	86	172	119	56	39	0	
100	160	152	95	96	60	112	70	16	10	0	
200	220	196	89	180	82	212	96	32	15	0	
Average	224	265	117	212	93	276	122	52	22	0.6	

[a] Culture grown on marine nutrient agar medium containing different concentrations of NiCl$_2$ as incubated at 15 C until maximum population was attained.

[b] Relative percentage bacterial populations on nickel-containing medium were calculated against the control.

eight colonies developed from all of the Gulf Stream casts on marine nutrient agar medium containing 10^{-2} M NiCl$_2$.

Coastal Zone Area. Four stations were taken on 31 March 1971 in the Gulf of Maine between 4 and 8 mi from the mouth of the Piscataqua River which separates the states of Maine and New Hampshire. Hydrographic data for these Gulf of Maine stations indicated reduced temperature and salinity compared to the Gulf Stream area (Table 9). Generally, the bacterial populations rose by two orders of magnitude compared to the Gulf Stream stations (Table 10). Consistent stimulation from the addition of NiCl$_2$ to the marine nutrient agar occurred at 10^{-4} M in these samples, a full order of magnitude less nickel than stimulated the bacterial population in the Gulf Stream area. Marked bacterial toxicity with an average reduction to 35% of the untreated control was noted at

Table 9. Hydrographic data for Gulf of Maine stations on *Chase*-I cruise

Station No.	Date	Location	Depth (meters)	Temp. (C)	Salinity (ppt)
1	3/31/71	42°58.4′N	surface	3	31.7
		42°38.8′W	25	—	—
2	3/31/71	42°59.5′N	surface	3.25	32.4
		70°39.3′W	25	—	—
3	3/31/71	43°00.4′N	surface	3.5	31.3
		70°40′W	25	—	—
4	3/31/71	43°01.3′N	surface	3.5	31.9
		70°40.6′W	25	—	—

Table 10.   Bacterial populations per ml of seawater calculated from 1.0 ml samples collected from four stations at two depths in the Gulf of Maine within four miles of the mouth of the Piscataqua River

Station No.	Depth (meters)	Molarities of Added $NiCl_2$[a]										
		None	10^{-5}		10^{-4}		10^{-3}		5×10^{-3}		10^{-2}	
		No.	No.	%[b]	No.	%	No.	%	No.	%	No.	%
1	Surface	185	210	114	223	120	43	23	0	0	0	0
	25	170	108	64	200	118	53	31	1		0	0
2	Surface	280	219	78	297	106	107	38	0	0	0	0
	25	239	260	109	270	113	58	24	0	0	1	
3	Surface	332	334	101	350	105	186	56	1		0	0
	25	240	155	65	176	73	53	22	0	0	0	0
4	Surface	258	277	107	275	107	135	52	2	1	1	
	25	185	183	99	208	112	70	38	2	1	0	0
Average		236	218	92	250	107	88	35	0.75	0.3	0.25	0.1

[a]Culture growth on marine nutrient agar medium containing different concentrations of $NiCl_2$ as incubated at 4 C until maximum population was attained.

[b]Relative percentage bacterial populations on nickel-containing medium were calculated against the control.

10^{-3} M $NiCl_2$ in this water mass. At 5×10^{-3} M $NiCl_2$ less than 1% of the control colonies grew and at 10^{-2} M $NiCl_2$ only two colonies developed from all casts. Thus, the bacterial population from the coastal zone area was more sensitive to $NiCl_2$ than bacteria from the open ocean.

The Estuarine Area. On 11 April 1971 water was collected from four stations in the Great Bay–Little Bay estuarine complex which is supplied by the Piscataqua River in New Hampshire. Hydrographic data for the estuarine stations indicated that the salinity was drastically reduced at this time of year compared to the open ocean, as were the temperatures, which ranged from 5.5 to 8 C (Table 11). The bacterial population was an order of magnitude higher in the estuary compared to the coastal zone region (Table 12). None of the microbial populations from the estuarine areas showed consistent stimulation from the addition of $NiCl_2$ to the marine nutrient agar medium. The only $NiCl_2$ concentration that showed any stimulation was at 10^{-5} M in four samples. At 5×10^{-3} M $NiCl_2$, only 2% of the colonies developed as compared to the control medium. Not a single colony developed on medium containing 10^{-2} M $NiCl_2$ from the estuarine environment even though the overall numbers of microorganisms developing on these membrane filters were considerably greater (1,000x) than from the open ocean samples (Table 12). Thus, there appears to be a marked difference in the response of marine microorganisms obtained from the open ocean, the coastal zone, and the estuarine environment for the tolerance to concentration of $NiCl_2$ ions. Open ocean marine microbial populations showed evidence for

Table 11. Hydrographic data for Great Bay–Little Bay estuarine stations on *Chase*-II cruise

Station No.	Date	Location	Depth (meters)	Temp. (C)	Salinity (ppt)
1	4/12/71	43°07.5′N	surface	5.5	14.0
		70°51′W	10	–	14.5
2	4/12/71	43°07′N	surface	6.0	14.1
		70°52′W	10	–	13.8
3	4/12/71	43°06′N	surface	6.5	11.9
		70°51′W	10	–	15.2
4	4/12/71	43°05′N	surface	8.0	9.6
		70°52′W	10	–	12.4

Table 12. Bacterial populations per ml of seawater calculated from 0.1 ml samples collected from four stations at two depths in the Great Bay–Little Bay, New Hampshire, estuarine complex

Station No.	Depth (meters)	Molarities of Added $NiCl_2$[a]										
		None	10^{-5}		10^{-4}		10^{-3}		5×10^{-3}		10^{-2}	
		No.	No.	%[b]	No.	%	No.	%	No.	%	No.	%
1	Surface	4,790	1,330	28	1,210	25	1,860	39	30	1	0	0
	10	3,980	4,210	106	2,650	67	2,460	62	50	1	0	0
2	Surface	4,740	2,950	62	2,600	55	2,220	47	30	1	0	0
	10	2,360	2,550	108	4,340	184	2,650	112	80	3	0	0
3	Surface	3,960	2,710	68	2,070	52	2,830	71	50	1	0	0
	10	3,210	3,890	121	2,270	70	1,900	59	110	3	0	0
4	Surface	3,640	4,450	122	2,270	62	2,920	80	60	2	0	0
	10	4,320	850	20	2,110	49	2,300	53	110	3	0	0
Average		3,875	2,868	79	2,440	71	2,392	65	65	2	0	0

[a]Culture grown on nutrient agar medium containing different concentrations of $NiCl_2$ as incubated at 4 C until maximum population was attained.

[b]Relative percentage bacterial populations on nickel-containing medium were calculated against the control.

stimulation from the addition of 10^{-3} M $NiCl_2$ to the marine nutrient medium (Tables 2–8).

Discussion

The abundance of aerobic heterotrophic marine bacteria is approximately two orders of magnitude higher in coastal seawater than in open ocean seawater, and estuarine water samples are an order of magnitude higher than the bacteria in the coastal seawater. These bacterial densities reflect the relative amounts of organic matter available to microorganisms in each individual environment. Heavy metal tolerance, however, is greatest among the open ocean marine bacteria and the toxicity of nickel ions increases for the heterotrophic marine bacterial populations as samples are obtained near land. The low amounts of organic matter in the open ocean suggest less chelation of the metal ions in this environment. Thus, open ocean bacteria may have adapted mechanisms for resisting metal toxicities. The exact relationships involved need further investigation. The complexity of chelation by naturally occurring organic compounds in the sea has been discussed by Duursma (7).

The use of rich media, such as marine nutrient agar, for the development of colonies of marine bacteria caused a discrepancy in the concentrations of metal ions tolerated by marine bacteria experimentally as compared with those in the natural environment. Marine nutrient agar contains 8 g of organic nutrients per liter, whereas the concentration of organic matter in the open ocean ranges between 0.35 and 0.70 mg carbon per l (22). Thus, there is a difference of four orders of magnitude in the amount of carbon in the open ocean and in the nutrient agar used. The organic nutrients in these experiments were convenient and more than sufficient to develop any microbial colonies. The nickel concentration in the open ocean is approximately four orders of magnitude lower than the concentration that was stimulatory (10^{-3} M) in the marine nutrient agar medium. Thus, the ratio of organic matter and nickel remained equivalent. The amount of organic matter in the open ocean approaches threshold nutrient values for most marine bacteria (9). The delicate interrelationships between the concentrations of organic matter and trace elements need careful elucidation, but differences in response to these substances appear evident. One possible explanation is that the toxicity develops as the ratio of labile to nonlabile ions becomes greater in a particular medium (17).

The general phenomenon presented in this paper, i.e., developmental stimulation of marine bacteria by low levels of added nickel, has been previously noted (14). Trace elements such as nickel, which appear high on the avidity series, may displace from various ligands other more metabolically involved elements such as magnesium, calcium, or iron, which are lower on the avidity series (1). These labile elements may be utilized subsequently by the bacterial community, causing stimulation.

Open ocean bacteria examined in this study had a greater tolerance to nickel ions than the nearshore organisms of an earlier study conducted in Salem Harbor, Massachusetts (14). The relative amounts of organic matter used in the earlier study (2 g/l - modified 2216E) as compared to the present study (8 g/l - marine nutrient agar) are a partial explanation for the somewhat different nickel tolerance of bacteria in the nearshore seawater from these two sampling areas.

The "marineness" of a microorganism is perhaps not related as much to its biochemical or physiological characteristics as to its structural composition. There may then be a basic difference in the three-dimensional interaction of the various cell wall and membrane layers due to the structural or bridging influence of the major ions such as magnesium and calcium in the individual marine bacterial cell. The major ions with their lower avidity for ligands can be displaced by many of the trace elements. Such structural differences between marine and terrestrial bacteria may explain, in part, the morphological and ecological responses observed *(2, 3, 8, 14)*.

Marine bacteria have a marked ability to tolerate nickel and other heavy metal ions. The data continue to support the concept that this may be one of the distinguishing features of microorganisms from the sea.

Acknowledgments

The authors would like to thank the Woods Hole Oceanographic Institution for providing space on the *R. V. Knorr* during January, 1971. The help of Mr. James Maryanski in preparing for and participating in this investigation aboard the *R. V. Knorr* is gratefully acknowledged. This work was supported in part by a Predoctoral Fellowship, Grant No. GZ-2049 from the National Science Foundation.

Literature Cited

1. Bowen, H. J. M. 1966. *Trace elements in biochemistry.* New York: Academic Press Inc.
2. Cobet, A. B., C. Wirsen, and G. E. Jones. 1970. The effect of nickel on a marine bacterium, *Arthrobacter marinus* sp. nov. *J. Gen. Microbiol.* **62**:159–69.
3. Cobet, A. B., G. E. Jones, J. Albright, H. Simon, and C. Wirsen. 1971. The effect of nickel on a marine bacterium: Fine structure of *Arthrobacter marinus. J. Gen. Microbiol.* **66**:185–96.
4. Drapeau, G. R., and R. A. MacLeod. 1963. Sodium dependent active transport of α–aminoisobutyric acid into cells of a marine pseudomonad. *Biochem. Biophys. Res. Comm.* **12**:111–15.
5. Drapeau, G. R., and R. A. MacLeod. 1965. A role for inorganic ions in the maintenance of intracellular solute concentrations in a marine pseudomonad. *Nature* **206**:531.
6. Drapeau, G. R., T. I. Matula, and R. A. MacLeod. 1966. Nutrition and metabolism of marine bacteria. XV. Relation of sodium activated transport to sodium requirement of a marine pseudomonad for growth. *J. Bacteriol.* **92**:63–71.
7. Duursma, E. K. 1970. Organic chelation of ^{60}Co and ^{65}Zn by leucine in relation to sorption by sediments. In *Symposium on organic matter in natural waters,* ed. D. W. Hood, pp. 387–97. Institute of Marine Science, Occasional Publication No. 1. College, Alaska: University of Alaska.
8. Gonye, E. R., A. B. Cobet, and G. E. Jones. 1971. Abnormal morphogenesis of *Arthrobacter marinus* under heavy metal stress. *Bacteriol. Proc.,* p. 40.
9. Jannasch, H. W. 1967. Growth of marine bacteria at limiting concentrations of organic carbon in seawater. *Limnol. Oceanogr.* **12**:264–71.
10. Jones, G. E. 1964. Effect of chelating agents on the growth of *Escherichia coli* in seawater. *J. Bacteriol.* **87**:483–99.
11. Jones, G. E. 1967. Precipitates from autoclaved seawater. *Limnol. Oceanogr.* **12**:165–67.
12. Jones, G. E. 1967. Growth of *Escherichia coli* in heat- and copper-treated synthetic seawater. *Limnol. Oceanogr.* **12**:167–72.
13. Jones, G. E. 1968. In *Marine biology.* IV: *Unresolved problems in marine biology,* ed. C. H. Oppenheimer, New York: New York Academy of Sciences Interdisplinary Communications Program.
14. Jones, G. E. 1970. Metal organic complexes formed by marine bacteria. In *Symposium on organic matter in natural waters,* ed. D. W. Hood, pp. 301–319. Institute of Marine Science, Occasional Publication No. 1. College, Alaska: University of Alaska.

15. Jones, G. E. 1971. The fate of freshwater bacteria in the sea. *Devel. Ind. Microbiol.* **12**:141–51.

16. Lyman, J., and R. H. Fleming. 1940. Composition of sea water. *J. Mar. Res.* **3**:134–46.

17. McCarthy, L. R. 1971. The effect of copper on *Pseudomonas cuprodurans*, sp. nov. Ph.D. dissertation, Univ. of New Hampshire.

18. MacLeod, R. A. 1965. The question of the existence of specific marine bacteria. *Bacteriol. Rev.* **29**:9–24.

19. MacLeod, R. A., E. Onofrey, and E. Norris. 1954. Nutrition and metabolism of marine bacteria. I. Survey of nutritional requirements. *J. Bacteriol.* **68**:680–86.

20. MacLeod, R. A. and E. Onofrey. 1956. Nutrition and metabolism of marine bacteria. II. Observations on the relation of seawater to the growth of marine bacteria. *J. Bacteriol.* **71**:661–67.

21. MacLeod, R. A., C. A. Claridge, A. Hori, and J. F. Murray. 1958. A possible role of sodium in the metabolism of a marine bacterium. *Fed. Proc.* **17**:267.

22. Menzel, D. W., and J. H. Ryther. 1970. Distribution and cycling of organic matter in the oceans. In *Symposium on organic matter in natural waters*, ed. D. W. Hood, pp. 31–54. Institute of Marine Science, Occasional Publication No. 1. College, Alaska: University of Alaska.

23. Payne, W. J. 1958. Studies on bacterial utilization of uronic acids. III. Induction of oxidative enzymes in a marine isolate. *J. Bacteriol.* **76**:301–307.

24. Payne, W. J. 1960. Effects of sodium and potassium ions on growth and substrate penetration of a marine pseudomonad. *J. Bacteriol.* **80**:696–700.

25. Pratt, D., and F. C. Happold. 1960. Requirement for indole production by cells and extracts of a marine bacterium. *J. Bacteriol.* **80**:232–36.

26. Rhodes, M. E., and W. J. Payne. 1962. Further observations of effects of cations on enzyme induction in marine bacteria. *Antonie van Leeuwenhoek J. Microbiol. Serol.* **28**:302–14.

27. Rhodes, M. E., and W. J. Payne. 1967. Influence of sodium on synthesis of a substrate entry mechanism in a marine bacterium. *Proc. Soc. Exp. Biol. Med.* **124**:953–55.

28. Rhodes, M. E., and W. J. Payne. 1967. Influence of cations on spheroplasts of marine bacteria functioning as osmometers. *Appl. Microbiol.* **15**:537–42.

29. Scarpino, P. V., and D. Pramer. 1962. Evaluation of factors affecting the survival of *Escherichia coli* in seawater. VI. Cysteine. *Appl. Microbiol.* **10**:436–40.

30. Strickland, J. D. H., and T. R. Parsons. 1968. *A practical handbook of seawater analysis.* Bulletin 167. Fish. Res. Bd. Can. Ottawa.

31. Thompson, J., J. W. Costerton, and R. A. MacLeod. 1970. Potassium dependent plasmolysis of a marine pseudomonad plasmolyzed in a hypotonic solution. *J. Bacteriol.* **102**:843–54.

32. Tyler, M. E., M. C. Bielling, and D. B. Pratt. 1960. Mineral requirements and other characters of selected marine bacteria. *J. Gen. Microbiol.* **23**:153–61.

Comments

HOLM-HANSEN: I am sorry to be the first one to comment again, but I am very anxious to find out the nickel concentration in average seawater.

GONYE: The nickel concentration in open ocean seawater is around 3×10^{-8} molar (2 ppb). However, you have to take into account that in certain environments nickel can get much higher, as in the Red Sea where these values are around 10^{-5} molar.

HOLM-HANSEN: I will comment on two things. One concerns the possible artifacts caused by filtration. These questions have been raised by people like Hamilton and Watson in this meeting. When I look at your data, I just wonder if it is really significant when you postulate or claim an oligodynamic effect of nickel.

I wonder if this is really significant if you treated these numbers, or is it merely an interpretation on your part which would not hold up under close scrutiny.

The second thing I would like you to comment on is your data on toxicity. Your threshold for nickel toxicity is between 1×10^{-3} and 5×10^{-3} molar, but you tell me that nickel concentrations in average seawater is five magnitudes lower than that. I wonder if there is any real sense in asking the initial question you posed — i.e., how do marine forms of bacteria stand the high nickel and other heavy metals? It seems to me there is absolutely no problem. They are five orders of magnitude removed from any toxic level.

GONYE: These are excellent questions and ones which we ask ourselves. As far as the membrane filters are concerned, we have noted that if brands of filters are changed, differences are obtained. There may also be differences among filters from the same source. As far as the oligodynamic effect, it has been observed by other investigators using a similar technique.

With regard to the five orders of magnitude increase of nickel in the medium compared to seawater, we were using marine nutrient agar containing 5 g of peptone and 3 g of beef extract per liter. That is 8 g of readily available organic matter. The nickel was approximately 10^{-3} molar which would be 63 mg. The ratio is somewhere around 100:1, organic matter to nickel (125:1). In the ocean, the organic matter is approximately 0.5 mg per l as an average value. The amount of nickel is 10^{-8} molar. Once again, these organic matter to nickel ratios are close to 100:1 (250:1). So, we still feel that there is some relative significance to these types of data. Using conventional microbiological techniques, it is extremely difficult to obtain appreciable data on microbial populations functioning in the sea without raising the nutrient level.

WATSON: Have you examined any dumping area like Long Island Sound where National Lead Company is dumping huge amounts of iron? I do not know what the composition is, but I think you will have fairly high concentrations of nickel there, too. The other question is, what is the actual thing we are observing here? We are working with mercury and find that the toxic level for most organisms is 10^{-4} molar. We have gotten some organisms to grow down as high as 10^{-3} molar. It is hard to visualize why some organisms are more tolerant than others. I wonder if it is not in some way involved with their -SH groups. Has anybody ever looked at these organisms which are more tolerant in respect to their SH groups?

GONYE: In answer to the first question, we have not yet looked at any other polluted areas. The second question is really an extremely complicated one. Many of these heavy metal involvements may very well be membrane phenomena, since many of them are readily reversible. The membrane itself does not contain any great deal of SH groups. The proteinaceous material inside the cell and in the wall itself may certainly represent an area of interaction. It becomes a very difficult problem and we certainly do not have many of the answers for it.

G. JONES: Let me add to that. Spores of fungi and also yeast take up very large amounts of metal, more than any of the organisms that we have studied. They do have a high number of SH groups. I think there is a good correlation here.

GONYE: One other point is that when you say 10^{-3} molar mercury, you have to take into account the organic level that is involved. We find that if we raise the organic level, then we have to raise at the same time the amount of nickel it takes to become toxic. So one always has to consider the organic matter along with the concentration of the metal.

WATSON: I would like to comment on differences in the structure of the cell wall and membranes of marine and terrestrial bacteria. I think some people might jump to conclusions. I am not so sure that we have evidence for this. Certainly in the nitrifying bacteria and in the ammonia-oxidizing bacteria, we have very clear evidence that they have extra cell wall layers that are unique to the marine-nitrifying bacteria. But, we find exactly the same type of architecture in the freshwater photosynthetic organisms. I do not think Dr. Colwell has presented enough evidence to convince me that the plasma membrane is what is controlling rigidity of the organism that she has. I think there is an extra layer outside the plasma membrane controlling rigidity of the cell. I do not think we have enough evidence at this time, at least at the level of the electron microscope, that the cell walls and the membranes of marine bacteria are sufficiently different from other forms. I think I have looked at as many marine bacteria as anybody with an electron microscope, but this technique says nothing about the molecular structure. When you say "structure," you often infer things that we see with our naked eye.

COLWELL: I would like to comment on that, if I may. The work that we and Bob Murray have done, together with a paper by Bob MacLeod, suggests that the murein-containing layer is in juxtaposition to the cell membrane. It is not an outer wall.

WATSON: I do not think our evidence will support that statement. The outside surface may consist of a slime layer or extra outer cell wall layers. Underlying the outer surface is the typical double-track layer. Under this layer is a globular layer of proteins approximately 40 to 60 Å thick. Beneath this comes the peptidoglycan layer, which is about 10 to 15 angstroms thick. Below that, there is another globular layer, and then the plasma membrane. In no case that I know of in marine bacteria or any other bacteria is there any clear evidence that the peptidoglycan layer is right next to the plasma membrane. In fact, we even have cultured an organism in which the plasma membrane invaginates and partitions the cell. The peptidoglycan layer goes in to provide the internal rigidity. At the same time, the lower globular layer, which is right next to the plasma membrane, intrudes along with the peptidoglycan, possibly contributing to the rigidity of the cell. I question very seriously if it is only the peptidoglycan layer that is providing the support for the cell. We have overwhelming evidence to support this view. Have you seen our paper in the *Journal of Ultrastructure?* We can show you by negative staining, by freeze-etching, and by thin sections how this organism is put together.

COLWELL: Your evidence is good, but I am saying that you cannot take what you see in your organism and extrapolate to all the organisms in the sea. I am simply saying what MacLeod and others have put forward.

WATSON: I do not think they are saying that there is not something between the peptidoglycan layer and the plasma membrane. You only see a space there, and if you fix this properly, you will see that there is a protein layer. This is not the time to argue this point.

COLWELL: Your organism is not the same as that observed in terrestrial forms.

WATSON: In many cases, it is the same as the *Escherichia coli.*

GONYE: Actually, I would like to vindicate myself. I said previously that this was speculation. I said it in order to stimulate some discussion because it is the sort of thing we are thinking about. I can see that it worked. Dr. Watson, you know the state of the various wall layers as well as anybody. We know a considerable amount about each of the individual wall layers, however, where our knowledge is lacking is in the binding of one layer to the next and particularly in the three-dimensional interaction of the various wall layers. I feel that the true marine bacteria, those which have a dependence upon the major ions, might have evolved a different system of binding the various layers together. They might have developed a different system of binding the various tetrapeptide portions of the peptidoglycan and the lipopolysaccharide subunits. When the various trace elements are added, they might be replacing some of the major ions and the wall becomes unstable. The structure of the wall could provide it with some small mode of protection against some of the trace elements in that it can respond to them, whereas a terrestrial organism, strict terrestrial organisms, or freshwater organisms, do not have this option. It is pure speculation.

WATSON: At the molecular level, I think there is a great deal of evidence that the membranes of marine bacteria are quite different from those of the terrestrial bacteria. For example, it appears from our studies that the monovalent cations are needed to hold the membrane together. If we suspend organisms in sucrose or in divalent cations at the molarity found in seawater and leave out the monovalent cations, the internal membrane systems that we work with vascularize. It looks like they may need some monovalent cations to hold together. I think probably monovalent cations, if you want to really look at differences between a marine organism and a terrestrial organism on the structural level, are where you should be looking. I am not saying there are not differences as far as the binding properties of heavy metals. I suspect there is.

TODD: I was curious to see that you have some organisms that are capable of growing at 10^{-2} molar metal ion. Are these organisms *Arthrobacter?*

GONYE: We have looked at them; however, we do not have any classification for them. One is a gram-negative rod, which also plasmolyzes under heavy metal stress. They exhibit a plasmolysis effect in which the membrane pulls away from the wall proper.

The Oceanography of Block Island Sound
Part I—Sampling[1]

James E. Alexander, Rudolph Hollman, and Steven A. Fisher

In Dr. ZoBell's most informative talk, he mentioned (1) the need for long-term studies, (2) the need for models to be developed for predictive purposes, (3) the danger of making quick studies, (4) the search for representative figures, and (5) the vision of estuarine ecology as highly dynamic. These points are hardly debatable, yet I wonder how many of us have given them more than passing thought.

Block Island Sound is situated approximately between latitude 41°05' and 41°20'N, is bounded to the north by Rhode Island and Connecticut, to the west by the Race and Plum Gut, to the south by the Atlantic Ocean and to the east by Rhode Island Sound. Williams *(3)* reported its mean depth to be about 40 m and off Fishers Island the depth ranges up to 100 m.

Williams *(3)* also indicated that fairly large horizontal gradients of temperature and salinity were possible in the area, that the currents are primarily tidal, and that, with respect to the latter, the semi-diurnal component was the most important. Riley *(1)* showed that the waters of Long Island Sound flow along the surface past Montauk Point, while the more saline bottom water enters the sound from the southeast. These waters, he noted, are probably of Georges Bank and Gulf of Maine (and at times, Gulf Stream) origin. As noticed by Riley *(1)* and also observed during the course of our investigation, this leads to a rather confusing picture when the spatial and temporal distribution of phytoplankton and zooplankton is examined.

Since the tidal forces are so important in coastal oceanography and we recognized that we were working in a complex region, we developed a sampling program designed to collect data along a given transect over a lunar day. Figure 1 shows the location of our four sampling stations.

Figure 2 (a and b) depicts the temperature and salinity distribution along a transect between Montauk Point and Watch Hill, Rhode Island. This graph shows the influx of the cooler and more saline coastal waters into the region along the southern shore of Rhode Island. The lower salinities and temperatures to the south are characteristic of the waters of the Peconic Bay system and Long Island Sound.

It is apparent from an examination of these two figures that at no one location along this transect could the conditions be considered typical of the area as a

[1] Publication No. 001 of the New York Ocean Science Laboratory.

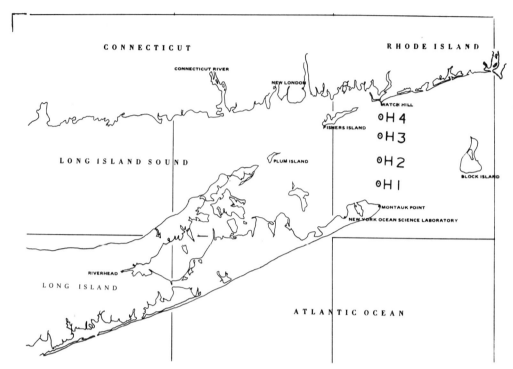

Fig. 1.   **Location of the sampling stations in Block Island Sound.**

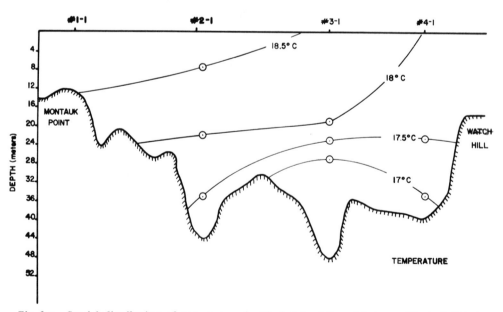

Fig. 2-a.   **Spatial distribution of temperature in Block Island Sound between Montauk Point, N.Y., and Watch Hill, R.I.**

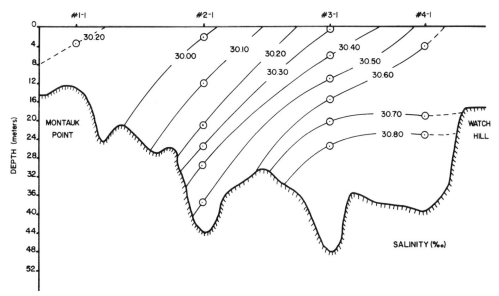

Fig. 2-b. Spatial distribution of salinity in Block Island Sound between Montauk Point, and Watch Hill.

whole; nor could the conditions present at a fixed point at a given instant be considered representative for that point for the entire day. For example, Figure 3 depicts the reactive phosphate variations occurring at such a station over a lunar day. This same variation with time was noted in the distribution of phytoplankton* (Fig. 4). These data were collected from our station H4, which is located immediately south of Watch Hill, Rhode Island. The symbols H4–1, H4–2, and H4–3 designate the particular hydrographic cast. The interval between each sampling period was approximately 3 hr. Since our ship was sampling this transect over a tidal period, we asked whether these differences were of a (1) navigational, (2) analytical, (3) tidal, or (4) biological nature. Our approach to the solution of these questions will constitute the remainder of this paper.

Methods

Navigational influences on the data were removed by anchoring the vessel with a minimum scope on the anchor chain. Our analytical data were well within the limits described by Strickland and Parsons *(2)* for each of the parameters described (except for iron). The latter was determined by atomic absorption spectrophotometry and our combined precision (both sampling and analytical) was ±4%.

We have, to date, completed two experiments at H4, the first of which was conducted in December 1970 and the second was completed in mid-May 1971. In each of these studies, samples were collected at hourly intervals from the water column. The parameters followed were temperature, salinity, oxygen,

* R. Nuzzi, personal communication.

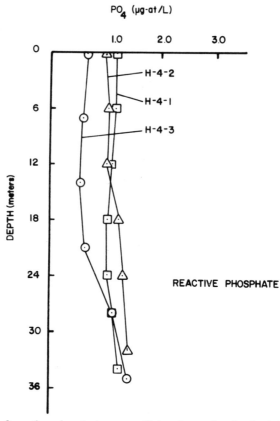

Fig. 3. Variation of reactive phosphate over a tidal cycle at a fixed point in Block Island Sound.

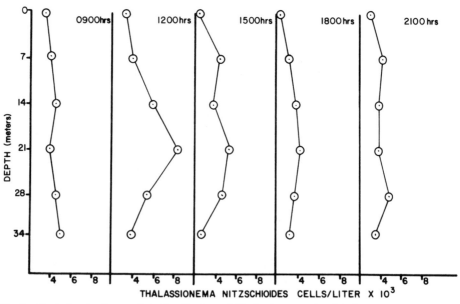

Fig. 4. Variation in the vertical distribution of *Thalassionema nitzschioides* over a tidal cycle at a fixed point in Long Island Sound.

phosphate (reactive, organic soluble, and particulate), nitrite, nitrate, silicate, chlorophyll, and iron.

Niskin bottles (5l) were used to collect the water samples and once on board each sampler was connected via Tygon tubing to an all-glass filtration apparatus, each of which was equipped with a bypass to allow for the removal of samples (prior to filtration) for the measurement of salinity and phytoplankton. Oxygen samples were removed from the samples as they came on board.

The scope of the May operation was broadened to include observations on not only the temporal effects but also the spatial fluctuations. This was accomplished by anchoring the *R. V. Kyma* at station H4 and surrounding her with four anchored buoys positioned at the four cardinal points of the compass, each of which was located one mile from her. These were sampled at hourly intervals—the times for which coincided approximately with the sampling activity on the *Kyma*. Two Boston Whalers were used to collect these samples. At each of these buoy stations, surface salinity and temperature were measured with a Beckman (RS-5) induction salinometer. Surface and bottom samples (collected with 2 l Niskin bottles) were returned to the support vessel for processing.

Results and Discussion

The results of the December study were interesting in that, with the exception of temperature, all of the parameters exhibited large and cyclic variations over the observational period. The temperature curves (Fig. 5) were characteristic of these waters during the winter; in that the bottom waters were warmer than the surface waters. The effect of the tidal influence was apparent in the variation of salinity over a lunar day at station H4 (Fig. 6, bottom). As might be expected, these influences were more noticeable in the surface waters. The bottom waters showed a higher salinity on the average, with a smaller range over the tidal cycle indicating a continuous source of water (which in this case would be coastal). The surface waters show a larger range over a tidal cycle, which would indicate a more turbulent source region—in this case, Long Island Sound.

The reactive phosphate (Fig. 6, top) also showed the influence of the tidal exchange, as well as an indication of a diurnal effect, demonstrated by variation in the amplitudes of the maxima, particularly the one occurring at approximately 0800. This diurnal effect was more evident in the surface water mass than in bottom waters.

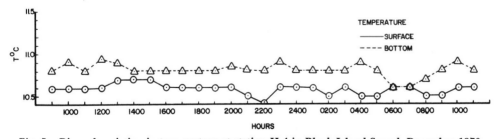

Fig. 5. **Diurnal variation in temperature at station H-4 in Block Island Sound, December 1970.**

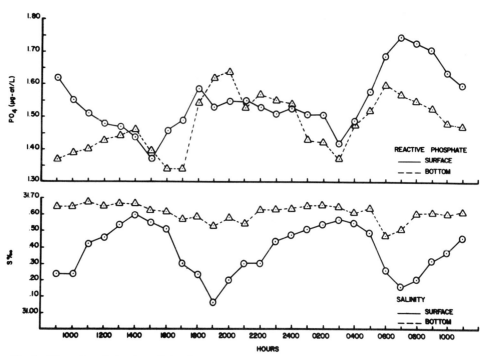

Fig. 6. **Diurnal variation in salinity (bottom) and reactive phosphate (top) at station H-4, December 1970.**

Figure 7 shows the variations in particulate phosphate and particulate iron over the same tidal period. The particulate phosphate distribution was difficult to interpret due to the apparent nontidal effects superimposed on the tidal components. Again, this is more evident in the surface waters. The effect of the tidal forces on particulate iron was particularly apparent and showed little evidence of other nontidal effects. The lag effect that was evident in these data, between the surface and bottom waters, we believe at this time to be attributable to the lag between the surface and bottom water movements.

From this series of observations, we concluded that at this time of the year the tidal influences were the dominant factors in determining the distribution of the parameters measured in this study. Accordingly, we adjusted our sampling frequency along this particular transect io a minimum of 3/tidal cycle to obtain more representative data.

Recognizing that biological factors would become of increasing importance as the waters became more stratified with the onset of vernal heating, we planned an experiment for mid-May, at which time biological production in the waters would have increased. In this experiment, we also planned to investigate on a small scale the spatial and temporal distribution of all of the above parameters.

Figure 8 shows the temporal distribution of the surface and bottom temperatures and salinities at station H-4 in mid-May. At this time of year the effects of vernal heating were apparent; that is, the surface waters were much warmer than the bottom waters and showed much larger tidal and nontidal excursions than was evident in December. The diurnal (heating) effect on the second or last day was less apparent than on the first. We attribute this to an overcast sky

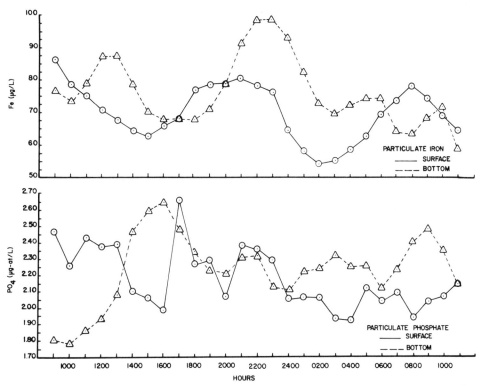

Fig. 7. Diurnal variation in particulate phosphate (bottom) and particulate iron (top) at station H-4, December 1970.

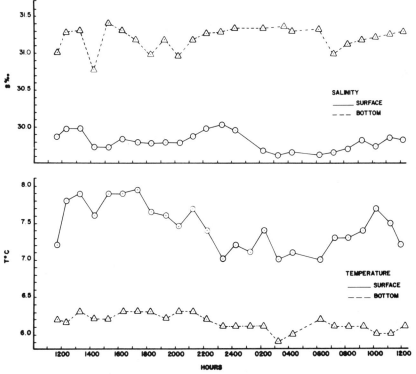

Fig. 8. Temporal distribution of temperature (bottom) and salinity (top) at station H-4, May, 1971.

and pouring rain. Only a small tidal periodicity in the salinity distribution was apparent at this time of the year. It should be noted, however, that the mean freshwater runoff for this period of time is approximately the same as for the December sampling period.

Figure 9 shows the spatial and temporal distribution of total soluble iron over a tidal cycle at station H-4. The tidal effects in each of these five stations is strong—especially on the ebb tide, for example, between 1700 and 2400 hr. Many of the higher frequency fluctuations were attributable to the geography of the area. For example, stations A and B in the lee of Fishers Island should, and did show the greatest amount of fluctuations.

As a result of the greater amount of higher frequency (other than tidal) fluctuations encountered, we have increased our sampling program to a minimum of five samples per transect. As a result of the variations seen in these data, both physical and biological in nature, we strongly urge that more attention be given to at least quasi-synoptic sampling over tidal periods.

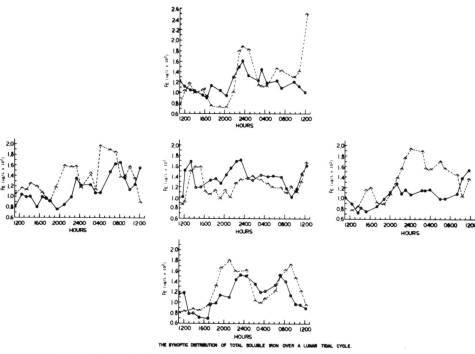

THE SYNOPTIC DISTRIBUTION OF TOTAL SOLUBLE IRON OVER A LUNAR TIDAL CYCLE.

Fig. 9. **Synoptic distribution of total soluble iron over a lunar tidal cycle.**

Literature Cited

1. Riley, G. A. 1952. Hydrographic and biological studies of Block Island Sound. *Bull. Bing. Oceanogr. Coll.* XIII (3).
2. Strickland, J. D. H., and T. R. Parsons. 1968. *A practical handbook of seawater analysis.* Bull. 197, Fish. Res. Bd. Can. Ottawa: Queen's Printer.
3. Williams, R. G. 1969. *Physical oceanography of Block Island Sound.* USL Rept. 966.

Comments

ZOBELL: What are your criteria for particulate versus reactive phosphate?

ALEXANDER: Particulate phosphate is that material retained by a Millipore filter (H.A.) or its equivalent. Many years ago, while I was in the tropics, I found the filtering rate to be very slow, and since I had to filter 8 ls of water, I shifted over to the Whatman (GFC) filter pad, which according to the company at that time, had a retention size of 0.49. I found it to be better than the Millipore 80% of the time and ever so much faster in filtering the water.

ZOBELL: What is the chemical composition of the filters?

ALEXANDER: It is glass.

ZOBELL: What was the chemical state of the phosphate? Is it organic or inorganic?

ALEXANDER: Particulate phosphate is anything that is retained by the filter pad and which, upon oxidation, will react with the reagents. This material will be both inorganic and organic in composition.

HOLM-HANSEN: I do not quite understand what you mean when you say particulate phosphates. Is this done according to the Strickland and Parsons Manual? Is this what is called total particulate phosphates? Did you follow the procedure of Strickland and Parsons?

ALEXANDER: Yes.

HOLM-HANSEN: I am a little skeptical in regard to what the manufacturer claims for the GFC filter. This filter was checked by Peter J. Williams in our lab. He found it to be effective between one and two microns, not 0.5.

ALEXANDER: Some years ago in Miami, I worked with Ferguson Wood, Kimbell, and Corcoran. We used pure cultures and checked the GFC against the Millipore. We were getting about 80% better recovery in terms of chlorophyll with the GFC. They work quite well for us.

HOLM-HANSEN: I know Peter used polystyrene spheres to check this. Below one micron, they just seem to pass right through and this coincides quite nicely with my biological data when I do ATP or some measurement with various membrane filters or with the GFC or some microfine glass filters. The GFC always seemed to let through considerably more material than, for example, the 0.5 membrane.

ALEXANDER: I do not know the Microfine filter.

HOLM-HANSEN: In regard to the GFC, did you have any problem during the digestion when you had a big glass fiber filter in there?

ALEXANDER: We had some problem with it. It fragments very much during the process and I have very great difficulty in removing such fragments in the centrifugation step. I filter it. We make sure the caps on the glass are quite loose to prevent bumping.

Asking Microorganisms to Classify Their Environment

E. H. Anthony and G. Vilks

In a series of papers published at the beginning of the last decade, Williams and his colleagues developed a method of numerical classification requiring only presence-or-absence data which they called association analysis *(2, 14, 15, 16, 17, 19)*. It was developed and tested in relation to problems in phytosociology, but it showed considerable promise for much wider application.

In general terms, and for ecological purposes, we suggest that association analysis is a technique for inducing certain members of a biota, through their association with one another, to provide us with a classification of their environment which may reveal discontinuities in factors controlling it. It may also provide an ecologically useful classification of the organisms concerned.

We admit that the very name, association analysis, appealed to us intuitively as ecologists, since it is a basic premise of our discipline that the ecosystem reflects the interactions of the members of its biota with one another and with the inanimate factors of their environment. These interactions tend to select certain species. The interspecific associations arising in this manner lend to the system much of the distinction it enjoys and enable us to distinguish a meadow from a bog or a seashore from its limnetic counterpart by casual observation.

Nevertheless, deductions from even casual observations are not independent of experience. Ecologists commonly face a problem of collecting information about both living and nonliving components of various ecosystems and then searching amid a confusing wealth of data for correlations that permit simpler descriptions. We can scarcely doubt the value of training in ecology for those undertaking such tasks, but even the well-trained ecologist is left with at least two important shortcomings.

First of all, training or experience is an open-ended process. The complete ecologist is never achieved. He needs a continuing diet of clues to nourish his concepts. Second, his perception is not only limited by lack of training or experience, but such training as he has received may shade his perception with a little cloud of preconceived notions. Hence, even well-trained ecologists may benefit from directing their attention somewhat independently of their personal biases.

The fact that association analysis required only presence-or-absence data appealed to something stronger than our intuition. Even brief reflection upon all of the tedious labor that has gone into counts of bacteria, counts of bottom fauna, counts of plankton, etc., leaves ample cause to wonder if useful informa-

269

tion thus obtained might have been forthcoming from a simpler floristic or faunistic approach.

The problems of where and at what to look most effectively loom large in most ecological surveys, but nowhere are they greater than in the marine environment, including, we insist, estuaries, where a formidable amount of survey work remains to be done. Ideally, we should like for a minimum of collecting and processing of data to provide us with a maximum exposure of factors influencing the system and with a minimum of distortion from subjective assessment.

Subjective assessment must rarely enjoy better advantage from good visibility than it finds in phytosociology. Hence, we regarded the comparative success of association analysis under those circumstances as a good recommendation for its application in areas less easily observed visually, such as the marine benthos. Initially, the senior author wished to induce bacteria to describe the microenvironment of estuarine sediments, but association analysis tends to be impractical where visual identification of the test organisms is not adequate.

The oceanographic institute of which we were then members had not been established as a discrete unit, but was diffused throughout the science faculty of the university in a deliberate attempt to break down interdisciplinary barriers. Thus, a channel was provided through which a geology student, Vilks, elected to work with a biologist, Anthony, and brought with him an intimate knowledge of foraminifera. It only remained for us to enlist some help from Professor Williams in order to proceed with the work reported here.

We have applied association analysis to aquatic ecology, using benthic foraminifera as test organisms *(10)*. Foraminifera offer certain advantages for this type of analysis in marine environments. They are ubiquitous, generally abundant, and consist of numerous species that may be readily identified. Our preliminary trial of this approach has been published recently *(10)*. Response to that paper, both before and after publication, prompts us to speak primarily about it here.

Association analysis may be applied in two different ways: as normal analysis and as inverse analysis. The results of those two analyses may be superimposed to provide further information about the study area.

Let us consider normal analysis first. In its ecological applications, normal analysis regards *quadrats* in, or samples of, an ecosystem as *entities* that are defined by underlying properties or *factors*. *Species* are thought to respond in varying degrees to those factors, hence species may be used as tests to identify *discontinuities* between *groups* of quadrats.

The first step in the analysis is sampling the estuary or other region of interest. We have sampled two regions: East Bay, MacKenzie King Island, which lies in the outer fringe of the Canadian Arctic, and Bras d'Or Lake, Cape Breton Island, Nova Scotia. It has commonly been convenient to use a systematic grid of sampling stations for this purpose, as we have done for Bras d'Or Lake.

East Bay (Fig. 1) was sampled at intervals that increased geometrically, proceeding on traverses running out over the ice from bases on the shore (50, 100, 200, 400 ft, etc., to 2 mi). Mr. Vilks had completed this work for the Geological Survey of Canada before either he or I had heard of association analysis *(9)*.

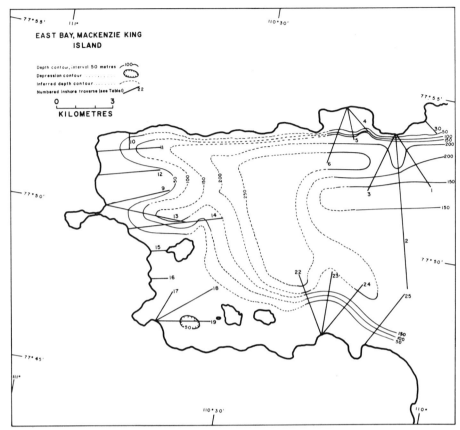

Fig. 1. Traverses along which samples were collected at specified intervals running out from shore. (Reprinted from Vilks, Anthony, and Williams *(10)* by permission of the National Research Council of Canada).

All foraminifera in 20 g of recent sediment from each station were identified, counted, and recorded in the original published account of Vilks' work *(9)*. Only the presence or absence of each species was entered in the analysis reported here.

A brief general description of the analytic method is provided here. The actual analysis was carried out by electronic digital computer at the Computer Research Section, C.S.I.R.O., Canberra, Australia, through the generosity of the Director, Mr. G. N. Lance. In lieu of the illustrations used in verbal presentation, and in the interest of conserving publication space, we provide a graphic summary of the procedure in Figure 2.

The survey of 75 stations collected 56 species. This provided a 75 × 56 "quadrats/species" matrix of presence-or-absence data. The degree of association between each pair of species for all of the quadrats was calculated as chi-squared from 2 × 2 contingency tables. The chi-squared values were entered in a species by species matrix. The scores in this matrix were standardized, i.e., transformed to zero mean and unit variance.

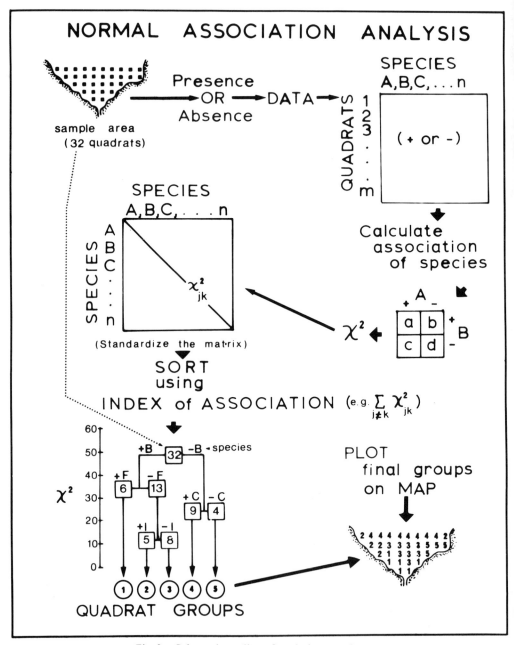

Fig. 2. **Schematic outline of analytic procedure.**

The next step is to select an appropriate *index* of association. In the present work, we used:

$$I = \sum_{j \neq k} \sqrt{(\chi^2_{jk}/n)} \equiv \sum_{j \neq k} |r_{jk}|$$

where n = number of species.

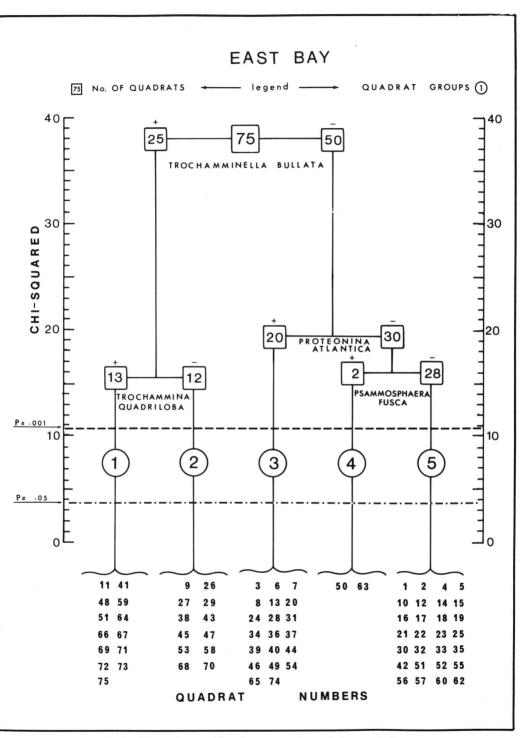

Fig. 3. **Hierarchy resulting from normal association analysis.**

This index was used to subdivide the quadrats into groups in a hierarchical fashion. Some rule is needed for deciding when to stop subdividing. We halted when chi-squared $= 3.84$ (for 1 degree of freedom, $P = 0.05$).

This strategy aims at maximum reduction of residual association at each subdivision. Components of the subdivision are commonly arranged in a dendogram (Fig. 3) to display the hierarchical nature of the classification and to relate it to levels of the index at which status in the hierarchy is established. Figure 3 is modified from the original version (Fig. 2, *10*) by omitting subdivisions below the level providing groups plotted here.

It should now be apparent why association analysis is classed as a divisive, monthetic type of numerical sorting procedure. We began with the entire population and progressively subdivided it, basing each subdivision on the presence or absence of one attribute (species).

The most abundant species in the East Bay sediment was *Spiroplectammina biformis (9)*. It did not play a significant role in this sort. The first sort is on *Trochamminella bullata*, which attracted no attention in the initial work *(9)*. Neither did *Trochammina quadriloba*, on which the next division on what we may call the positive side of the hierarchy is made. The next two divisions on the negative side are on *Proteonia atlantica* and *Psammosphaera fusca*, which are listed seventh and sixth, respectively, in order of abundance *(9)*. We mention these points simply because the numerically predominant species have loomed so large in most ecological surveys.

We now have the quadrats sorted into five groups, containing from two to twenty-eight quadrats, on a basis of the interspecific association of their foraminiferan fauna. To these final groups of quadrats, numbers or other symbols are assigned by which they may be plotted on the map of the study area. Note (Fig. 3) that group 1 is on the most positive side of the hierarchy; group 5 is on the most negative side.

The location of each quadrat is plotted on the map (Fig. 4), using the symbol of the final group with which it is associated. All that now remains for normal analysis is to try to make some ecological sense of the discontinuities revealed by the plot we have achieved in this manner.

Something of how that might proceed may be indicated by drawing attention to the distribution (Fig. 4) of quadrat groups 1 and 5. The former tend to be limited to the open sea; the latter tend to be crowded at the shore. From the hierarchy (Fig. 3), we know that group 1 has the richer fauna, hence group 5 may reflect a response of foraminifera to less stable nearshore conditions.

Turning to inverse analysis, we find, as the name implies, that species are classified on basis of the quadrats with which they are associated. Otherwise, the procedures are identical. Inverse analysis terminates (Fig. 5) with the species sorted into groups that may be designated with letters or other symbols. The fifty-six species of East Bay were sorted into six species-groups, A through F, running from three to twenty-five species in a group.

The original work *(9)* attempted to classify the area into depth zones on the basis of the species that were numerically predominant at various levels. *Spiroplectammina biformis* was thought to characterize a 0–90 m zone. In the present analysis, it is classed with six other species in group A, including the three other species that were considered to characterize three deeper zones.

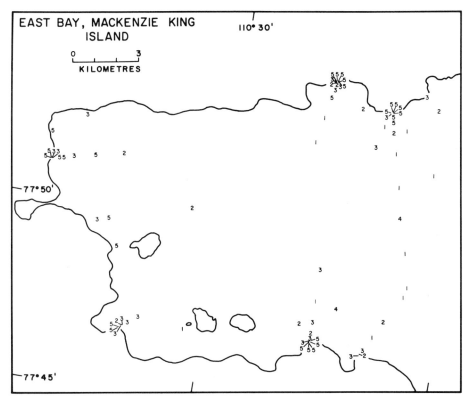

Fig. 4. **Distribution of final groups of the hierarchy of normal analysis. (Reprinted from Vilks, Anthony, and Williams *(10)* by permission of the National Research Council of Canada.)**

For those who envisage an experimental approach to understanding problems of marine ecology, we suggest the possibility that inverse association analysis may indicate groups of organisms particularly useful for comparative physiological ecology. Although both normal and inverse forms of association analysis may be ecologically meaningful as separate analyses, it may also be of interest to examine the extent to which species-groups are tied up with habitat discontinuities. Toward this end, results of the two analyses may be superimposed to form a "quadrat-groups" × "species-groups" table (Table 1). The significance of entries in the table may be checked in relation to expectations based upon size and nature of the entire population *(16)*.

For East Bay, Table 1 shows species-group A was significant over all quadrat-groups and it is the only one to reach significance in quadrat-group 5. Conversely, species-group D reached significance only in quadrat-group 1. These two species-groups emphasize the division of this system into two zones: near-shore shallow and offshore deep. Exclusion of other than A defines the former region; presence of D characterizes the latter.

Quadrats normally differ in the number of species they contain. The effects of this "richness" upon normal analysis may be measured as the mean number of species per quadrat for each quadrat-group. Species normally differ in the number of quadrats in which they occur. The effects of this "abundance" upon

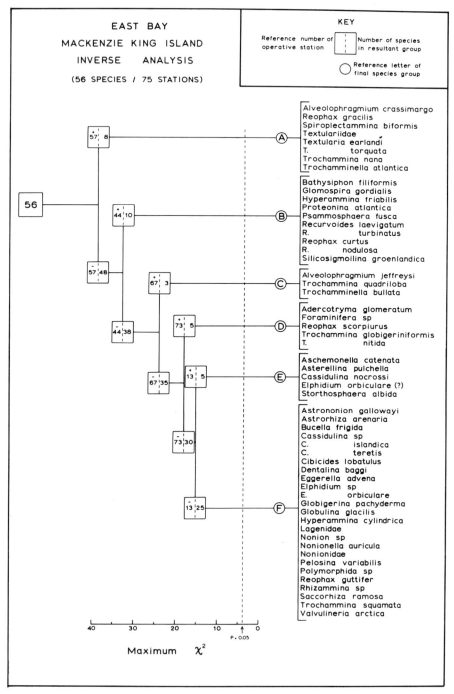

Fig. 5. Hierarchy resulting from inverse association analysis. (Reprinted from Vilks, Anthony, and Williams *(10)* by permission of the National Research Council of Canada).

Table 1. Comparison of individual species-entries per species group within each station subdivision[a]

Specific Groups		Station-Groups 1 (13)[b]	2 (12)	3 (20)	4 (2)	5 (28)
A	(8)	80.7[c]	83.3	71.3	50.0	43.3
B	(10)	62.3	50.8	39.0	40.0	4.3
C	(3)	89.7	52.8	3.3	33.3	3.6
D	(5)	41.5	10.0	6.0	–	4.3
E	(5)	15.4	13.3	13.0	–	0.7
F	(25)	5.5	1.7	2.6	2.0	2.0

[a]Reprinted from Vilks, Anthony, and Williams *(10)* by permission of the National Research Council of Canada.

[b]Bracketed figures represent numbers of stations and species in normal and inverse groups respectively.

[c]Expressed as a percentage of all possible entries within the doubly defined "cell."

inverse analysis may be measured as the mean number of quadrats per species for each species-group. Both of these measurements may be expressed (Table 2) either as absolute values or as ratios to the mean for the groups. We should point out that standardization eliminates effects of abundance from normal analysis and effects of richness from inverse analysis.

The original work *(9)* observed that the number of foraminifera per sample increased toward the shore, but the number of species per sample decreased. These observations are reflected in the present analysis. The distribution of quadrat-groups 1 and 5 are aligned with differences in faunal richness. Furthermore, the more marked effect of abundance upon inverse analysis (Table 2)

Table 2. The comparative effects of richness and abundance[a]

Normal Analysis Group	Richness	Richness Ratio	Inverse Analysis Group	Abundance	Abundance Ratio
1	19.62	1.63	A	47.88	2.61
2	14.92	1.24	B	24.00	1.31
3	11.30	0.94	C	20.33	1.11
4	9.50	0.79	D	9.00	0.49
5	4.75	0.40	E	6.40	0.35
			F	2.04	0.11
Mean	12.02			18.28	
Standard deviation	4.64			16.75	
Coefficient of variation	38.6%			91.7%	

[a]Reprinted from Vilks, Anthony, and Williams *(10)* by permission of the National Research Council of Canada.

can probably be attributed to the preponderance of samples near shore where the number of species per sample was lowest. In other words, the method of sampling has emphasized abundance through avoiding richness. Despite any limitations that may have been imposed by the sampling scheme used at East Bay, it seemed to us that association analysis had lived up to our expectations, and we proceeded to apply it to an estuarine environment, Bras d'Or Lake.

The use of numerical classification in ecology is not without its critics. We suggest it should not be adopted without weighing those criticisms. The work of Lange and colleagues (7) is particularly useful in this respect.

Williams and his colleagues proceeded from their simple but effective beginning with association analysis to a comparative study of numerical sorting procedures (1, 3, 4, 5, 6, 11, 13, 18). In a recent publication (8), they demonstrate very effectively the current state of their classificatory art through applying it to a tropical marine benthos. We propose to update our analytical procedures in light of their advances before publishing the results of the survey of Bras d'Or Lake. We find their approach all the more attractive because of their underlying philosophy that numerical classification is merely an hypothesis generating procedure (12, 20). They discard any notion that classifications are either *good* or *bad* in any absolute sense, preferring to judge any classification as to whether it proves useful, informative, or predictive.

We found our application of association analysis to East Bay data informative, but we face certain limitations on a thorough assessment of its usefulness. In any ecological work, the profits of an initial survey can best be realized and reinvested through a reexamination of the scene. This has not been possible for us in this instance.

Acknowledgment

We thank Mr. L. G. Richards of Audiovisual Services, University of Guelph, for advice and for preparation of graphics used in verbal presentation of our work.

Literature Cited

1. Lambert, J. M., and M. B. Dale. 1964. The use of statistics in phytosociology. In *Advances in ecological research,* vol. 2, ed. J. B. Cragg, pp. 55–99. New York: Academic Press Inc.
2. Lambert, J. M., and W. T. Williams. 1962. Multivariate methods in plant ecology. IV. Nodal analysis. *J. Ecol.* **50**:775–802.
3. Lambert, J. M., and W. T. Williams. 1966. Multivariate methods in plant ecology. VI. Comparison of information-analysis and association-analysis. *J. Ecol.* **54**:635–64.
4. Lance, G. N., and W. T. Williams. 1966. A generalized sorting strategy for computer classifications. *Nature* **212**:218.
5. Lance, G. N., and W. T. Williams. 1966. Computer programs for hierarchical polythetic classification ("similarity analyses"). *Comp. J.* **9**:60–64.
6. Lance, G. N., and W. T. Williams. 1967. General theory of classificatory sorting strategies. I. Hierarchical systems. *Comp. J.* **9**:373–80.
7. Lange, R. T., N. S. Stenhouse, and C. E. Offler. 1965. Experimental appraisal of certain procedures for the classification of data. *Aust. J. Biol. Sci.* **18**:1189–1205.

8. Stephenson, W., W. T. Williams, and G. N. Lance. 1970. The macrobenthos of Moreton Bay. *Ecol. Monogr.* **40:**459–94.
9. Vilks, G. 1964. Foraminiferal study of East Bay, MacKenzie King Island, District of Franklin. Geol. Surv. Can., Paper 64–53.
10. Vilks, G., E. H. Anthony, and W. T. Williams. 1970. Application of association-analysis to distribution studies of Recent Foraminifera. *Can. J. Earth Sci.* **7:**1462–69.
11. Watson, L., W. T. Williams, and G. N. Lance. 1966. Angiosperm taxonomy: A comarative study of some novel numerical techniques. *J. Linn. Soc. (Bot.)* **59:**491–501.
12. Williams, W. T. 1967. Numbers, taxonomy, and judgment. *Bot. Rev.* **38:**379–86.
13. Williams, W. T., and M. B. Dale. 1965. Fundamental problems in numerical taxonomy. In *Advances in botanical research*, vol. 2, ed. R. D. Preston, pp. 35–68. New York: Academic Press Inc.
14. Williams, W. T., and J. M. Lambert. 1959. Multivariate methods in plant ecology. I. Association-analysis in plant communities. *J. Ecol.* **47:**427–45.
15. Williams, W. T., and J. M. Lambert. 1960. Multivariate methods in plant ecology. II. The use of an electronic digital computer for association-analysis. *J. Ecol.* **48:**689–710.
16. Williams, W. T., and J. M. Lambert. 1961. Multivariate methods in plant ecology. III. Inverse association-analysis. *J. Ecol.* **49:**717–29.
17. Williams, W. T., and J. M. Lambert. 1961. Nodal analysis of associated populations. *Nature* **191:**202.
18. Williams, W. T., J. M. Lambert, and G. N. Lance. 1966. Multivariate methods in plant ecology. V. Similarity analyses and information-analysis. *J. Ecol.* **54:**427–46.
19. Williams, W. T., and G. N. Lance. 1958. Automatic subdivision of associated populations. *Nature* **182:**1755.
20. Williams, W. T., and G. N. Lance. 1965. Logic of computer-based intrinsic classifications. *Nature* **207:**159–61.

Comments

COULL: If you could get back to that environment, what parameters would you look at, from the environmental point of view, to correlate with the species distributions? What are you going to try and look at to correlate with this?

ANTHONY: That is a very good question because I expect you know what that region is like. I talked of the sampling as though it were something that we would not have done in that fashion had we started off from scratch. But in actual fact, that is a trivial question. It would have been difficult, perhaps impossible, to have imposed a grid on that area. They were going around on pack-ice. If you have ever been on pack-ice, you probably know better than I, what it is like. I have not been there, but it has been described to me quite accurately. If it were possible, I would like to compare these data with some fresh samples collected with a grid pattern to see how it supports what we have seen in the old samples. That is, I would be curious about the extent to which the sampling pattern has imposed something on our interpretations. One would like to look at the extent to which the area was eroded by the pack-ice. This seems to mean a number of visits. One would like to visit it at a time when the ice and things were in motion and have something real to look at on the surface.

COULL: Did you look at any of the standard species diversity analyses, such as, the things that have been done on deep sea forams? How does this correlate with your data? If indeed it does, then it seems to suggest the same thing. Others have suggested that as you go deeper, you find a more stable environment and therefore you increase the diversity, if you want to believe the stability hypothesis type thing. Now the other question, are these live forams or tests?

ANTHONY: Both tests and live forams are included. It is all that could be recognized as forams in what you might call recent sediments.

COULL: So that they could be palagic, and they could be benthic.

ANTHONY: These species were benthic organisms.

COULL: Did you look at other diversities of species?

ANTHONY: No, we did not.

Section Five
APPLIED ESTUARINE MICROBIOLOGY

Effect of Thermal Effluent on the Microbiology and Chemistry of the Connecticut River[1]

John D. Buck

As a consequence of population increase and technological advances, the subsequent demand for greater supplies of electricity has resulted in near-crisis situations in some locales. Utility companies have responded by the introduction of new power-generating facilities including nuclear-powered reactors. While these plants represent increased efficiency in electricity output and eliminate the air pollution aspect of conventional plants, considerable concern has been evoked over a new potential environmental threat—thermal pollution *(3, 4, 8, 13)*.

At present, few definitive and synoptic studies are available on power plant sites on a "before" and "after" basis to establish unequivocally the effect of thermal discharges on receiving waters although over 200 research programs are under way or proposed *(18)*. From a biological standpoint, most investigations concern the impact of heated discharges on macroorganisms, such as fish and benthic invertebrates *(11)*. The phytoplankton have received some attention with conflicting views on the response of primary producers in cooling waters *(7, 10)*. Other microorganisms, however, have been neglected. Accordingly, we chose to examine the effects, quantitatively and qualitatively, on a number of bacterial groups as well as phytoplankton and selected chemical parameters.

Materials and Methods

Sampling Area. Figure 1 shows the location of sampling stations near the site of the Connecticut Yankee Atomic Power Company's (CYAPC) nuclear-powered generating station on the Connecticut River at East Haddam, Connecticut. Additional information on other studies at the plant site is available *(9)*.

Station 1 (a control) was located approximately 100 m upstream from the intake on the same side of the river. Cooling water (883 cfs; 1,400 cu m/min) leaves the plant and travels through a 20 m wide canal parallel to the river and mixes with river water at a point 2,000 m downstream. Station 3 was established at the juncture of this effluent canal with the river; sampling was done 30 to 70 m out in the river depending on tide and extent of shoaling in the area but always in the area of thermal influence.

[1]Contribution No. 79 from the Marine Research Laboratory, University of Connecticut, Noank, Connecticut 06340.

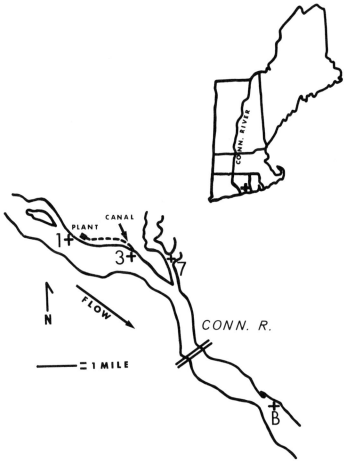

Fig. 1. **Sampling locations in the Connecticut River. Stations 1, 7, and B were controls; station 3 was in the area of thermal effluent.**

Initial plans called for discharging effluent into the Salmon River and station 7 was located there. Sampling at this site was discontinued in September of 1966 and begun at station B, 2.4 km below the mouth of the Salmon River. Both 7 and B represented downstream control areas and remained out of the thermal mixing zone.

Field Sampling. Our study began in September 1965 and terminated in September 1969 and afforded a "prethermal" period of about 27 months and a "thermal" duration of 21 months. Samplings were made every two or three weeks depending upon ice formation and weather conditions.

Table 1 lists the chemical and microbiological determinations which were made routinely. Salinity, temperature, conductivity, pH, and dissolved oxygen measurements were made *in situ* on "top" and "bottom" water samples (approximately 15 cm below the surface and 30 cm above the bottom, respectively). Samples for the other chemical determinations were collected at the same depths in

Van Dorn samples, dispensed into polyethylene bottles, and iced or frozen until processed in the laboratory within 12 hr of sampling.

Water for phytoplankton counts and identification was sampled in Van Dorn bottles and 1 l portions were preserved using a distilled water-ethanol-formalin mixture (6:3:1). For bacteriological analyses, top water samples were collected in sterile, screw-capped bottles, and Niskin samplers *(12)* were used to obtain bottom water. Bottles and Niskin bags were iced until laboratory examination within 2.5 hr. For enumeration of anaerobes, approximately 25 ml of water was transferred aseptically into sterile 1 oz prescription bottles containing a few ml of fluid thioglycollate medium (Difco). These bottles were capped loosely and placed in a desiccator which contained pyrogallic acid and 10% KOH. The desiccator was evacuated, and the chemicals were mixed to absorb residual oxygen. The desiccator was packed in ice until laboratory plating.

Table 1. Chemical and microbiological determinations

Chemical
  pH
  Temperature
  Salinity
  Conductivity
  Dissolved oxygen
  Nitrate-nitrogen
  Ammonia-nitrogen
  Organic nitrogen
  Total PO_4
  Ortho PO_4
Microbiological
  Phytoplankton
  Bacteria
    Total count; mesophiles; 30 C
    Thermophiles; 45 and/or 55 C
    Psychrophiles; 5 C
    Proteolytic
    Lipolytic
    Amylolytic
    Anaerobic
    Total and fecal coliforms
    Enterococci

Laboratory Procedures. In addition to *in situ* readings, dissolved oxygen values were obtained by the azide modification of the Winkler technique *(1)*. Orthophosphate (stannous chloride method), total phosphate, ammonia, and total-nitrogen (Kjeldahl distillation-titration) determinations were made according to recognized methods *(1)*. Nitrate levels were corrected for nitrite *(17)*.

For quantitative and qualitative phytoplankton determinations, 1 l volumes of water were concentrated by either sedimentation in graduated cylinders or continuous flow centrifugation. The Sedgwick-Rafter cell and Whipple disc technique were employed for counting and identifying algal cells. Some supplementary data were obtained using Foerster samplers (5) for phycoperiphyton and, on occasion, 15 min plankton tows using #20 mesh and 10 μm nannoplankton nets.

For the enumeration of bacteria, the spread-plate technique (2) was employed. Samples were diluted appropriately in buffered distilled water. All plates, unless otherwise specified, were incubated at 30 C. Total counts of heterotrophic, mesophilic bacteria were made after 48 hr on plate count agar (Difco) with 0.2% soluble starch. Flooding with iodine solution allowed detection of amylolytic organisms. "Psychrophilic" bacteria were enumerated on plate count agar incubated at 5 C after 14 days. Counts of "thermophilic" bacteria were made as above with incubation at 45 and/or 55 C and observed after 48 hr.

Lipolytic bacteria were detected on spirit blue fat agar (16) after incubation for 6 days. Proteolytic forms were counted on bromcresol purple milk agar and confirmed by flooding plates with 1 N HCl; the incubation period was 48 hr. Acid- and base-producing colonies were recorded prior to acid treatment. Plate count agar was used for anaerobes with plates incubated at 30 C for 5 to 7 days in Weiss-Spaulding anaerobe jars. The membrane filter technique was employed for enumeration of total and fecal coliforms and enterococci (1).

In addition to these quantitative studies, predominant colonies were isolated routinely from mesophile platings and identified as to genus (6, 15). Also, isolates of thermophilic and psychrophilic bacteria were incubated at 5, 25, 30, 35, 45, and 55 C for more proper evaluation of temperature optima.

Results

Water Chemistry. The temperature plume resulting from the introduction of heated water into the Connecticut River has been characterized (9). These data, together with our observations, confirm that only station 3 (outfall) was routinely under thermal influence. The other upstream and downstream stations were unaffected by the plume at the time of sampling; top and bottom temperatures at stations 1, 7, and B were within 1 degree F.

Figure 2 shows temperature distribution at the outfall area before and after plant operation. As mentioned above, these readings were taken at the sampling point at station 3 which varied with tide and degree of shoaling in the area. At the immediate juncture of the outfall canal and river, the depth was 6 to 8 ft and the temperature was vertically isothermal. Progressing out into the river, the warm water formed a layer (plume) which moved upstream or downstream depending on the tide (9).

One of the prime ecological considerations in the power plant controversy is the matter of dissolved oxygen (DO) since, other things being equal, oxygen is less soluble as the temperature of water increases. As expected, the lowest DO values were found during the summer months and reached prethermal lows of 2–3 μg/ml when water temperatures were above 80 F. Winter highs of over 12 μg/ml were characteristic. Comparison of prethermal data for all stations showed clearly a general horizontal and vertical uniformity. All of our readings were made between 1000 and 1300 hr and thus represent values approaching daily maxima. Predawn samplings may have shown greater extremes.

Comparison of data taken after plant operation showed that DO values were consistently higher at all stations during the thermal period and were more variable. Figure 3 shows our observations at the outfall station and is representative of all areas sampled. No uniform differences were noted between top and bottom waters. Aeration of the water as it leaves the plant may account for some higher values at station 3 but did not influence the other stations which showed similar trends. A possible algal effect is discussed below.

Nitrate values in the Connecticut River ranged from 0 to 2 $\mu g/ml$ (average <0.1 $\mu g/ml$) during both prethermal and thermal periods. Phosphate levels varied from 0 to 1.5 $\mu g/ml$ (average $<0.1\mu g/ml$). Both showed a general decrease during the warmer months with highest values recorded in the winter and probably reflected the influence of land drainage and algal activity. No thermal effect was noted in either nitrate or phosphate values nor was any difference seen among stations or between top and bottom waters.

Ammonia- and Kjeldahl-nitrogen determinations were made less frequently than biweekly. All levels observed were <1 $\mu g/ml$ and did not represent an

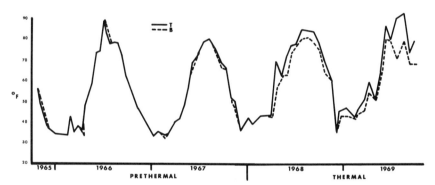

Fig. 2. Temperature (F) of top and bottom waters at station 3 (outfall).

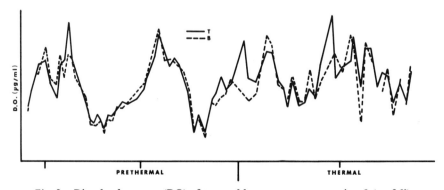

Fig. 3. Dissolved oxygen (DO) of top and bottom waters at station 3 (outfall).

unusual environmental quality. There was no suggestion of significant fluctuation from station to station or before and after plant operation.

The pH values were within the range of 6.2 to 8.4. The higher levels were recorded at station 7 and may have resulted from drainage from a local dump. Overall, no uniform pH fluctuations at any station were noted nor any correlation observed with thermal addition. All salinity levels were <2 mg/ml and were not considered to be of ecological importance in the stretch of the river studied.

Algae. Both quantitative and qualitative observations were analyzed statistically and several computerized routines were developed. These detailed data will be presented elsewhere. Figure 4 shows density graphs of the total phytoplankton counts at stations 1 (intake) and 3 (outfall) during the entire study. The general trends of increase and decrease at these two stations and those further downstream are almost identical; analysis of variance tests showed that there was no significant difference between prethermal and thermal data. Phytoplankton counts showed a definite increase in total numbers of all stations over the four-year study period. This trend is discussed below.

Numbers alone, however, form only a portion of the total effect since qualitative differences between stations were obvious. At station 3 (outfall) the predominant algae were members of Cyanophyta (blue-green), while diatoms were most prevalent at the other stations above and below the effluent canal. Analyses

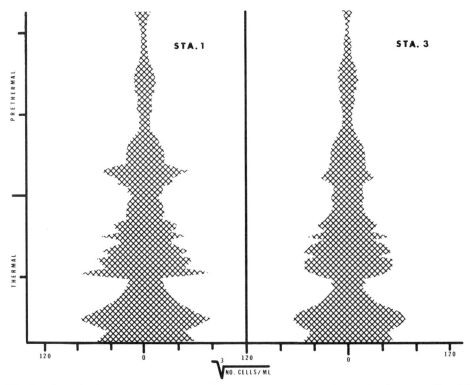

Fig. 4. Phytoplankton populations (cube root of total counts) at stations 1 (intake; control) and 3 (outfall).

of plankton tows and phycoperiphyton samples confirm these observations and others *(14)* of a shift from a diatom-dominated community to a blue-green dominated structure as a function of thermal addition. The influence was confined to the area around the outfall, since zones of effect and recovery were noted in the river progressing upward to station 1 and downstream to the lower stations.

Bacteria. It should be stressed that our studies considered only the heterotrophic, saprophytic forms which developed on the media employed; the autotrophs and pathogens were not included.

The presence of bacteria in rivers has been useful as a measure of specific pollutants (e.g., sewage) and as general contributors to oxygen consumption (BOD). It was our purpose here to focus attention both quantitatively and qualitatively on a variety of groups of bacteria as influenced by a single environmental additive and to evaluate the observations in view of the river ecosystem rather than isolated situations.

Mesophilic bacterial populations at stations 1 (intake; control) and 3 (outfall) varied randomly from 1,000 to 250,000/ml over the study period (Figs. 5 and 6).

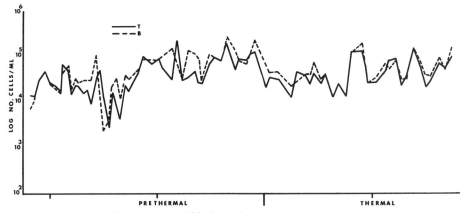

Fig. 5. **Mesophilic bacteria at station 1 (control).**

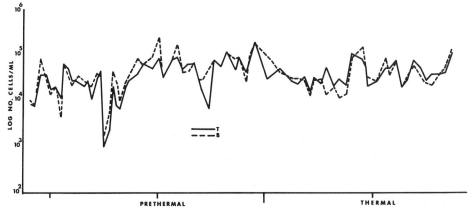

Fig. 6. **Mesophilic bacteria at station 3 (outfall).**

There was no pronounced variation between top and bottom waters or as a function of plant operation. Seasonal changes were minor.

Bacteria capable of growth at 5 C showed extreme reduction in numbers during the warmer months. This general pattern was evident throughout the four-year period and was reflected at all stations. Data for station 3 are shown in Figure 7. Since psychrophilic counts at the outfall area were not significantly different than those at the other stations, several comments may be made. Any obligate psychrophiles present may have had too short an exposure period to allow a lethal effect of thermal effluent and/or the zone of mixing at station 3 may have been confined spatially to the immediate juncture of the canal and the river. Our sampling site was not in close enough to detect changes in psychrophilic populations if in fact they were occurring. However, our top samples at station 3 were always under thermal influence. As shown below, our "psychrophilic" isolates were probably facultative forms.

Counts of bacteria capable of growth at 45 and/or 55 C ranged from <10 to 6,000/ml. Seasonal distribution paralleled that of the psychrophiles although the extremes in numbers were less. Figure 8 shows data obtained at station 3; graphs for the other stations were essentially superimposable on this. Microscopic examination of colonies from these plates revealed the common occurrence of *Bacillus* species. Since this genus is sporogenic and found in soil, we believe that our observations reflected the influence of higher winter precipitation and runoff. Our "thermophilic" isolates, like the "psychrophiles," were capable of growth over a wide range of temperature.

Amylolytic bacteria were present routinely and varied from 100 to 37,000/ml. No clear-cut distinctions were seen between seasons, stations, or plant operation. Lipolytic bacteria ranged from 90 to 5,000/ml and, as above, no obvious seasonal, thermal, or location effect was noted. Counts were lowest during the drought periods of 1966. Proteolytic and acid- and base-producing bacteria were present also throughout our study. The latter two showed summer minima. Proteolytic counts ranged from 300 to 8,100/ml while the other groups were present in smaller numbers, in the range of 10 to 1,800/ml. No differences were noted as a function of plant activity.

Counts of anaerobes varied randomly from a low of 250/ml to a maximum of 120,000/ml. There was no uniform correlation between these counts and DO levels in the river although the highest count recorded coincided with a period of low DO (<4 μg/ml) and high temperature (>80 F). Station 3 showed no anomalous counts as compared to the other sites during either prethermal or thermal periods.

Total coliform counts were below 5,000/100 ml on only six sampling dates which occurred during various seasons and during each of the four years and ranged from 700 to 728,000/100 ml. Lowest counts correlated with flood tide stages. A major source of domestic sewage is located approximately eight miles upstream; primary treatment effluent varies from 3 to 6 MGD. No thermal effect was observed in the distribution of total coliforms. Fecal coliform counts ranged from 86 to >13,000/100 ml. The lowest values were noted during summer and fall months and coincides with the chlorination periods of the above plant. Otherwise, no station of thermal influences were observed. Enterococcus

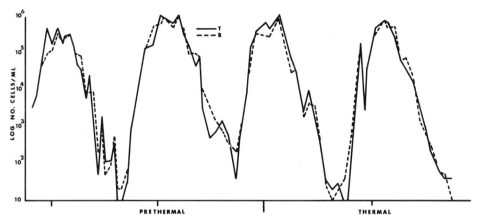

Fig. 7. Psychrophilic bacteria at station 3 (outfall).

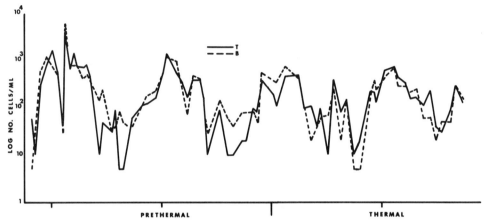

Fig. 8. Thermophilic bacteria at station 3 (outfall).

counts varied from 3 to >4,000/100 ml; minimal numbers were found in summer and fall months.

Considering quantitative aspects of the study, Table 2 shows the total numbers
and percentages of the various genera of bacteria isolated from mesophilic
platings at all stations during both prethermal and thermal periods. A general
decline in the Achromobacter-Alcaligenes and coryneform groups was noted,
along with an increase in the Enterobacteriaceae, *Pseudomonas* species, and the
Flavobacterium-Cytophaga groups. None of these figures, however, represented
gross quantitative or qualitative changes in the overall microflora of the river.
Nor were any drastic changes seen as a function of season. In comparing our
data with a previous study* during 1964–65 in the same area of the Connecticut
River, it was apparent that, while the same genera or groups of bacteria were
dominant and their frequencies similar, the actual percentages of the various
organisms found were inconsistent. No upward or downward trends in occur-

* Wolff, unpublished data.

Table 2. Bacteria isolated at all stations during prethermal and thermal periods

Genus or Group	Number and Percentage of Isolates			
	Prethermal (1965–67)		Thermal (1968–69)	
	No.	%	No.	%
Achromobacter-Alcaligenes group	53	37	37	27
Bacillus species	10	7	9	7
Chromobacterium species	0	0	1	1
Coryneforms	28	19	17	12
Enterobacteriaceae	8	6	22	16
Flavobacterium-Cytophaga group	13	9	21	15
Micrococci	2	1	1	1
Pseudomonas species	14	10	21	15
Spirillum species	0	0	0	0
Vibrio-Aeromonas group	10	7	9	7
Unidentified	7	5	0	0
Total number of isolates	145		138	

rence were evident from this 1964 study to the present work through 1969. Table 3 treats only station 3 and considers top and bottom samples separately since our sampling was in a thermally stratified area. Comparing these data with the overall picture (Table 2), some generalizations may be advanced: (1) Some differences were noted between 3T and 3B waters during the thermal period. The frequency of occurrence of the Achromobacter-Alcaligenes group increased while decreasing in the overall data: *Pseudomonas* species decreased at 3T during the thermal period but increased overall: and the Vibrio-Aeromonas group decreased as thermal effluent was added while remaining constant in the composite data. (2) The coryneforms increased at both 3T and 3B while de-

Table 3. Percentage distribution of bacterial groups at station 3 (outfall area)

Genus or Group	Prethermal				Thermal			
	No.	%	No.	%	No.	%	No.	%
	3T	3T	3B	3B	3T	3T	3B	3B
Achromobacter-Alcaligenes group	6	27	10	43	7	46	7	38
Bacillus species	4	18	3	13	0	0	1	6
Chromobacterium species	0	0	0	0	0	0	0	0
Coryneforms	2	9	2	9	3	20	4	22
Enterobacteriaceae	2	9	2	9	0	0	1	6
Flavobacterium-Cytophaga group	3	14	2	9	4	27	2	11
Micrococci	0	0	0	0	0	0	0	0
Pseudomonas species	3	14	1	4	1	7	1	6
Spirillum species	0	0	0	0	0	0	0	0
Vibrio-Aeromonas group	2	9	2	9	0	0	2	11
Unidentified	0	0	1	4	0	0	0	0
Total number of isolates	22		23		15		18	

creasing overall. (3) The Flavobacterium-Cytophaga complex increased at both station 3 and all stations. (4) The Enterobacteriaceae decreased in frequency at station 3 top and bottom but increased in frequency in all station data. These observations seem to indicate that thermal effluent may be selecting (stimulating? supporting?) the Achromobacter-Alcaligenes and coryneform bacteria while suppressing the Vibrio-Aeromonas group and the enteric bacteria.

During the course of our study, bacteria were isolated routinely from platings made at 5, 45, and 55 C. Once established in pure culture, these isolates were incubated at 5, 15, 25, 35, 45, and 55 C for up to 10 days with visible growth checked periodically. Of the "thermophilic" isolates, a total of 178 were cultured as above during the "before" portion of the study. Of these, a total of seven, or about 4%, grew only at 45 and/or 55 C and were properly classed as obligate thermophiles; one came from station 3. All others grew only at temperatures below 45 C or below 45 C as well as at 45 or 55 C and were considered facultative thermophiles or mesophiles. After plant operation, 7 of 133 isolates (5%) were shown to be obligate thermophiles and only one of these was isolated from station 3 (outfall).

Isolates from "psychrophilic" platings showed similar results. Prior to plant operation, 10% (10 of 102) of the isolates were obligate psychrophiles, i.e. grew only at 5 and/or 15 C. After CYAPC plant operation, 19 of 135 isolates (14%) showed a similar growth response.

Discussion

None of the chemical characteristics of the Connecticut River which were examined has showed any severe alteration as a direct consequence of thermal discharge from the CYAPC plant. The range of readings was similar for all parameters during prethermal and thermal periods of study, except for dissolved oxygen, nitrate, and phosphate levels which were uniformly somewhat higher at all stations after plant operation.

The first two years of our study (prethermal) were characterized by unusually low precipitation levels which resulted in the lowest annual mean stream flows in over 15 years. The thermal portion included two years of increased rainfall and subsequent runoff. Thus, the inorganic nutrient level rise noted may be attributed to these phenomena.

Since DO levels in the Connecticut River in the study area are not routinely so low as to be considered limiting to the biota, the increase in oxygen content near the plant outfall cannot be expected to result in significant stimulation of existing or "new" populations. Numbers of phytoplankton have increased also; this may be a result of higher nitrate and phosphate concentrations and could account for the increased daytime DO levels recorded. However, thermal addition may also be a factor in localized algal stimulation together with rising nutrient levels. If true, predawn DO levels could be depressed to critical levels.

A more tangible effect of thermal effluent was shown in qualitative plankton studies. A distinct shift from diatoms to blue-green algae was noted in both free-floating and attached plankton in the mixing zone near the outfall. If these changes continue or intensify, an alteration in the basic aquatic food chain may occur in this area, since blue-green algae are generally less beneficial than diatoms as a food source for herbivorous forms. A potential ultimate loss in trophic level efficiencies and diminished food stocks for fish is thus possible.

Some changes were noted in groups of bacteria studied. Quantitatively, bacterial counts did not show any distinct alteration at any station although there were indications of generic shifts among dominant members of the river microflora in the immediate outfall area. However, we cannot predict at this time whether the changes noted will be ultimately detrimental to the natural mineralization and other activities occurring in the river near the plant.

While we cannot conclude that the thermal discharge from the Connecticut Yankee nuclear plant has grossly affected the general ecology of the Connecticut River as of this date, it represents an environmental additive the ultimate effect of which remains unknown. We emphasize that our studies, following initiation of plant activity, lasted only 21 months and therefore must be considered to be of a short-term value.

We have no information on effects of fluctuating plant activity or daily chlorination on the parameters studied herein. It is likely that periods of low power output and subsequent decreases in effluent temperature will occur as a result of refueling, maintenance, or other reasons. Resumption of full power output and rising discharge temperatures may stimulate changes not noted in our study.

Acknowledgments

This investigation was supported by Contract No. 14–12–177 from the United States Department of the Interior, Federal Water Pollution Control Administration. I am particularly indebted to Julia S. Rankin and John W. Foerster for their invaluable assistance and expertise during portions of the study.

Literature Cited

1. American Public Health Assoc. Inc. 1965. *Standard methods for the examination of water and wastewater.* 12th ed. Albany: Boyd Printing Co.
2. Buck, J. D., and R. C. Cleverdon. 1960. The spread plate as a method for the enumeration of marine bacteria. *Limnol. Oceanogr.* **5:**78–80.
3. Cairns, J., Jr. 1971. Thermal pollution—a cause for concern. *J. Wat. Poll. Cont. Fed.* **43:**55–66.
4. Clark, J. R. 1969. Thermal pollution and aquatic life. *Sci. Amer.* **220:**19–27.
5. Foerster, J. W. 1969. A phyco-periphyton collector. *Turtox News* **47:**82–84.
6. Hendrie, M. S., and J. M. Shewan. 1966. The identification of certain *Pseudomonas* species. In *Identification methods for microbiologists, Part A,* ed. B. M. Gibbs and F. A. Skinner, pp. 1–7. New York: Academic Press Inc.
7. Hirayama, K., and R. Hirano. 1970. Influences of high temperature and residual chlorine on marine phytoplankton. *Mar. Biol.* **7:**205–13.
8. Krenkel, P. A., and F. L. Parker, eds. 1969. *Biological aspects of thermal pollution.* Nashville: Vanderbilt Univ. Press.
9. Merriman, D. 1970. The calefaction of a river. *Sci. Amer.* **222:**42–52.
10. Morgan, R. P., and R. G. Stross. 1969. Destruction of phytoplankton in the cooling water supply of a steam electric station. *Ches. Sci.* **10:**165–71.
11. Naylor, E. 1965. Effects of heated effluents upon marine and estuarine organisms. *Adv. Mar. Biol.* **3:**63–103.

12. Niskin, S. J. 1962. A water sampler for microbiological studies. *Deep-Sea Res.* **9**:501–503.
13. Parker, F. L., and P. A. Krenkel, eds. 1969. *Engineering aspects of thermal pollution.* Nashville: Vanderbilt Univ. Press.
14. Patrick, R. 1969. Some effects of temperature on freshwater algae. In *Biological aspects of thermal pollution*, ed. P. A. Krenkel and F. L. Parker. Nashville: Vanderbilt Univ. Press.
15. Shewan, J. M., G. Hobbs, and W. Hodgkiss. 1960. A determinative scheme for the identification of certain gram-negative bacteria, with special reference to the Pseudomonadaceae. *J. Appl. Bacteriol.* **23**:379–90.
16. Starr, M. P. 1941. Spirit blue agar: A medium for the detection of lipolytic micro-organisms. *Science* **93**:333–34.
17. Strickland, J. D. H. and T. R. Parsons. 1960. *A manual of sea water analysis.* Bull. No. 125. Ottawa: Fisheries Research Board of Canada.
18. Ulrikson, G. V., and W. G. Stockdale. 1971. Survey of thermal research programs sponsored by federal, state, and private agencies (1970). Oak Ridge, Tennessee: Oak Ridge National Laboratory.

Comments

STEVENSON: What is the microbiology of the canal?

BUCK: I do not know. We had a couple of occasions to sample the water immediately going into the plant and immediately coming out. The residence time of a particle of water in the plant itself is somewhere between 10 and 15 sec, I am told by the engineers. The bacterial counts going in and coming out were reasonably agreeable. We have done no studies in the canal itself. This was important and we would like to have done it. However, our goal at this point was the river ecosystem as a whole. I am sure there are some things going on in the canal. As far as the river itself is concerned, apart from this general mixing area, there seems to be very little effect. I will not comment on the invertebrates because I do not think our work was detailed enough to make any general comments. There were some changes in the phytoplankton. I do not know whether John Foerster would care to say anything about it or not. There was a definite shift to things other than diatoms and this may be significant in terms of food chains.

SMITH: Did you do any efficiency studies of intake and outflow related to chlorophyll and productivity?

BUCK: No, we did not.

HOLM-HANSEN: What physical or chemical parameter do you hold responsible for the rise of blue-greens?

BUCK: Temperature is perhaps the most significant. There are many of the greens and blue-greens which are thermal-tolerant and I think they are probably being stimulated by the higher temperatures. I am not sure that there is any chemical parameter which by itself controls their growth. I think the temperature is the main thing.

HOLM-HANSEN: What was the maximum temperature in the stream to which the phytoplankton would be exposed?

BUCK: In summer months, 102 to 103 F.

HOLM-HANSEN: One comment. When recording phytoplankton populations for ecological work like this, I consider it practically worthless to report cell numbers unless reporting average cell volume at the same time. I would much rather see total phytoplankton biomass unless you couple each cell number with a size distribution. You cannot say anything from biomass, which is the important factor ecologically. What is the phytoplankton doing as a group, not the total numbers? This comment does not hold for bacterial numbers, of course, but it does for phytoplankton, in which there is such an enormous range in size.

FOERSTER: I basically disagree with this. I think you have to take a look at numbers simply because of the stimulation possibility by a number of factors like size, volume, and amount of chlorophyll. I think we get erroneous chlorophyll data simply because people fail to take a look at the numbers of cells and/or the numbers of types of organisms.

HOLM-HANSEN: One correction to that. If there is any faulty chlorophyll due to cell size, all it does is reflect sloppy technique, because the chlorophyll is absolutely independent of cell number, cell size, cell phylum, and anything else. If you know how to extract chlorophyll, you get the answer regardless of what kind of plant you have—freshwater or marine, higher or lower plant.

BUCK: John Foerster, as a matter of fact, has quite a bit of statistical data on phytoplankton populations that I did not present here. I have a copy of the report in which some of these data are. I think it may clarify some of these questions. Again, John did most of the plankton work, but I will stand up for it and say that his overall findings are significant.

ALEXANDER: What kind of contribution does the flow of the canal make to the total flow of the river?

BUCK: I am told that the plant withdraws about 10% of the river volume and then returns it.

ALEXANDER: How far does the plume go down river?

BUCK: It varies. It goes up or down about a mile, depending on the tide. We have about a 2 ft tide even though it is 16 mi up river from Long Island Sound. It will push the plume up or down about a mile. One consideration, which we did not touch on and perhaps should have and will again if we have the opportunity, is to study the plankters that get trapped in this plume and move up and down with it. This is a very distinct possibility. Over a period of time there could be some thermal acclimation and some other forms may appear.

ALEXANDER: Once the canal enters the river, you have a surface plume that is only a couple of feet deep and extends up or down.

BUCK: Yes, that is right.

ALEXANDER: How much of the surface of the river, then, will that cover?

BUCK: The plume will go almost across the river, which is 200 or 300 yd across at that point. It will extend almost across the river on a flood tide and extend upstream. Now in an ebbing tide, when it comes out of the canal itself, it generally hugs that bank and streams down along that side. It moves up and down the river with the tide. There is about a one degree recirculation. Occasionally, there is some water which is raised one degree Fahrenheit which is recirculated through the plant but this does not bother the engineers at all. Incidentally, these are amazing plants. We spent 24 hr taking samples on one occasion. During the middle of the night, there are four men at the plant. This greatly surprised me. I knew it was automated, but I did not know it was quite this efficient.

KEIFER: If you are trying to find a change in the planktonic community because of a short-term temperature change, should you not have picked a station farther down the river? You might not expect a dramatic change in the community structure when you have a station planted very close to the outfall. What would be the residence time of a cell in warm water? That is basic to the question.

BUCK: The station furthest downstream was about 3 mi from the outfall. We detected nothing at all at that point. In other words, the diatoms were back, quantitatively and qualitatively. The genera, species, and numbers were pretty much what they were up above the plant. That is as far down as we went, about 3 mi. There was no change seen there at all.

ALEXANDER: I would rather have seen you present your oxygen data as percentage saturation. I think you would have seen quite a change. I really do not agree with just presenting dissolved oxygen in this case. It was very misleading. It just can not be the same when the physical characteristics have changed. I suggest you recalculate that.

BUCK: The reason it was done this way was so it was comparable with other data that were available at the time.

ALEXANDER: True, but it is so misleading to do it that way. That does not make it right. Second, I would like to know more about the location of your stations. Please correct me if I am wrong in my impression that you had a single station here and another one here, and so forth, right on down the river.

BUCK: This is correct.

ALEXANDER: You just finished saying that, based on the type of tide, either a big plume or one that bent and went on down occurred. If your stations are placed appropriately, you could present data that are biased. You could miss the whole ballgame. I would suggest that the stations should have been transect-type stations.

BUCK: There are transect-type data.

ALEXANDER: This was not brought out.

BUCK: I know this; I am quite aware of it.

ALEXANDER: In my final point, I have to agree with Dr. Holm-Hansen that community structure would have been much more valuable for us to see. Second, when he brings up the chlorophyll, I have to agree with him and would have requested that you talk about the chlorophyll to phytoplankton ratios. We might have seen a different picture.

BUCK: We have community structure information in the report which is available. Our stations were established with the personnel and the money that we had, with the thought in mind that the Environmental Protection Agency wanted to know what changes occurred in the river. We have considerable data which we got on our own and apart from our contract. It was a matter of having the people and money to do it. I quite agree that other stations were desirable, and perhaps John Foerster would concur with some of the phytoplankton comments. We do have some community structure, and I believe that, overall, we have at least begun to show some environmental effects of a nuclear power station in one river.

Ecological Evaluation of an Estuary Receiving Treated Sewage Effluent Prior to a Major Flow Increase[1]

E. Charles Pilcher and John D. Buck

The potential high productivity of estuaries has been well documented in the literature; however, the ability of estuaries to maintain this position in the delicate balance of aquatic ecology is threatened *(3, 5, 6)*. Closing of estuarine waters to swimming and shellfish harvesting, as a result of increased pollution levels, is commonplace along much of the coastal areas of the United States. With prudent use, however, coastal waters can continue to serve multiuse functions, including assimilation of treated sanitary wastes, and still maintain an acceptable aesthetic quality. The literature contains many reports on the effects of marine sewage outfalls (e.g., *4, 7, 8*). The study reported here represents a broad ecological approach to the evaluation of a specific case of potential significance to an expanding urban area.

The town of Groton, Conn., presently operates a secondary sewage treatment plant, with chlorination, that produces an average effluent flow of 0.4 million gallons per day (MGD) (1514 cu m/day). Discharge from the plant enters a small brook which flows into the head of Mumford Cove, approximately 0.4 km away. Within three years, expanded sewerage facilities and a new plant with an eventual output of 5 MGD (18,925 cu m/day) will be constructed on the same site. Although Mumford Cove has been receiving treated effluent for some 25 years, this estuary has not undergone obvious deterioration and continues to support swimming, fishing, and boating activities for a growing year-round and summer resort population on one side of the cove. A large, undeveloped state park is located on the opposite shore. Consequently, considerable concern exists over the ability of this valuable resource to maintain its multiuse status, if final plans include proposal of a greater than tenfold increase in effluent within the cove.

An ecological evaluation of present conditions and possible environmental alterations of suggested outfall sites are being considered. This paper represents a preliminary report of nine months of sampling in a program that will continue for another year.

[1] Contribution No. 80 from the Marine Research Laboratory, University of Connecticut, Noank, Connecticut, 06340.

Materials and Methods

   Palmer Cove, adjacent to Mumford Cove, is one of the few remaining areas along the Connecticut coast open for the taking of shellfish and receives no gross input of organic material. Thus, it was chosen as a control estuary for comparative purposes. Both bodies of water, in Palmer and Mumford Coves (Fig. 1), are shallow, except for a 3 m deep channel dredged through Mumford Cove, well mixed, and dominated by salt water. Fresh water enters both coves through small brooks at the upper reach. Both freshwater streams pass over small dams before entering the coves. Three stations were established for sampling in each cove during the period of study (1, M, and 4 in Mumford; 5, P, and 8 in Palmer). All water samples were taken at mid-depth. Data reported herein, unless stated otherwise, were collected from September 1970 through May 1971.

Chemical and Physical Methods. Field measurements that were made included salinity, pH, dissolved oxygen concentration (DO), and temperature. Laboratory analyses included dissolved oxygen (azide modification of the Winkler method), standard five-day biological oxygen demand (BOD), orthophosphate concentra-

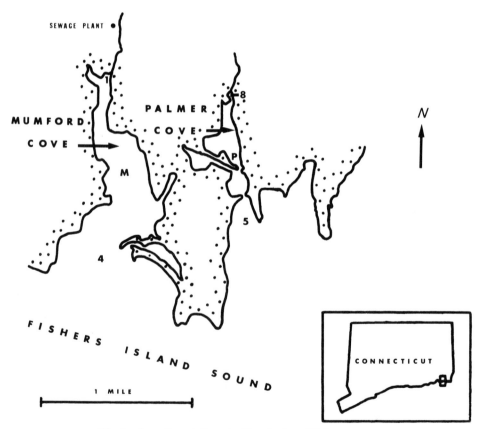

Fig. 1. **Sampling stations in Mumford and Palmer coves.**

tion (stannous chloride method), and nitrate-nitrogen concentration (UV spectrophotometric method). All procedures were those generally recommended *(1)*.

Bacteriological Methods. Samples were collected in sterile screw-cap bottles, iced, and processed immediately upon return to the laboratory. Total aerobic, heterotrophic counts were made on marine agar (MA; Difco) and plate count agar (PCA; Difco) prepared with distilled water. The spread plate technique *(2)* was utilized with incubation at 2, 25, and 35 C for PCA and 2, 20, and 25 C for MA. Total and fecal coliforms and fecal streptococci were enumerated by the membrane filter method *(1)*.

Phytoplankton Sampling. One-liter samples were collected in Van Dorn bottles and preserved with a distilled water, ethanol, and formalin solution (6:3:1). Samples were concentrated by continuous flow centrifugation and counted by means of a Sedgwick-Rafter cell. Identification was limited to four major groupings (green, diatoms, blue-greens, and flagellates), with dominant organisms identified to genus.

Benthic Invertebrate Sampling. Samples were collected with a Peterson grab, measured for volume, washed through 5.66, 2.0, and 1.0 mm mesh screens, and preserved with buffered formalin for subsequent identification.

Results

Chemical and Physical. Extreme salinity fluctuations occurred at the uppermost reaches in both coves, and ranged from 0 to 29 mg/ml, depending primarily upon tidal conditions and freshwater runoff. At the midcove stations (M and P), the salinity range was 24–29 mg/ml. The outer stations (4 and 5) remained relatively constant at 28–29 mg/ml. No DO values of <5 μg/ml were observed through the early summer of 1971 in either cove. One extensive daytime sampling during May, in Mumford Cove, showed DO values at all portions of the cove at all depths in the range of 9–10 μg/ml. However, more recent continuous 24 hr monitoring of DO over a several day period in both coves during July showed distinct diurnal changes at the upper stations. Early morning values of <3 μg/ml to late afternoon readings of >10 μg/ml were recorded. In addition, daytime values over a four-week period at these stations have been found to lie within the same range and were attributed to extensive "blooms" of macroalgae and subsequent decomposition. *Ulva* was the dominant algal form in Mumford Cove while *Enteromorpha* was more abundant in Palmer Cove. Since stations 1 and 8 are influenced most by salinity and temperature fluctuations, these factors assume significance in the interpretation of the DO levels observed.

BOD values ranged from <1 μg/ml in both coves to a maximum of 4.5 in Mumford Cove and 3.5 in Palmer Cove. Nitrate and phosphate levels are summarized in Table 1. Station 1 (effluent outfall area) revealed the highest values for both parameters.

Table 1. Phosphate and nitrate data collected over a nine-month period, September 1970 through May 1971, in Mumford Cove and Palmer Cove, Connecticut

Station	Orthophosphate		Nitrate-Nitrogen	
	Range	Mean	Range	Mean
Mumford Cove				
1	0.18–3.50	0.64	0.39–4.00	1.35
M	0–0.29	0.07	0.12–0.95	0.68
4	0–0.14	0.06	0.13–0.98	0.69
Palmer Cove				
5	0–0.08	0.03	0.02–0.91	0.62
P	0–0.08	0.03	0.11–0.97	0.59
8	0–0.04	0.01	0.11–1.68	0.81

Bacteriology. Total bacterial counts on MA at the upper (1 and 8) and outer (4 and 5) stations are shown in Figures 2 and 3. The most obvious feature was the extreme fluctuation in counts at the upper stations, which ranged from 300 to 70,000/ml. Data for both coves showed similar trends. Bacteria counts at the outer stations, which demonstrated a constant salinity, were more stable during the period of study. Figures 4 and 5 show counts on PCA. Maximum values at the upper stations differed from the maxima on MA by a factor of ten. Numbers of bacteria obtained in both coves were similar. At the outer stations (farther from shore), bacteria capable of growth on PCA were present, although these represented the greatest quantitative variability between the two bodies of water. In all cases, there was a seasonal selection of thermal types (9) which was most evident in the 2 C counts.

Considering the relative closeness of all sampling sites to land and the extensive coastal current patterns in the area, great significance cannot be attached to these data at present. More frequent samplings over a longer period of time are now being done for further evaluation of bacterial counts.

Total and fecal coliform and enterococcus counts showed great fluctuations in both coves (Table 2). Highest counts were obtained in water collected at both of the upper stations and, occasionally, at station 4; mid-cove stations and station 5 revealed routinely lower counts. These data indicated that factors other than the treatment plant were contributing to pollution in both coves. A large public bathing area near the mouth of Palmer Cove, a densely populated unsewered summer resort community, and the brooks themselves may be significant sources of enteric bacteria.

Phytoplankton. The highest phytoplankton counts (Fig. 6) were recorded at the upper stations, with greatest numbers found at the head of Palmer Cove. Both coves showed a significant increase in total counts during September and October. The most pronounced variation was evident in the upper cove areas, particularly station 8 where counts were 30,000/ml from pre-bloom levels of

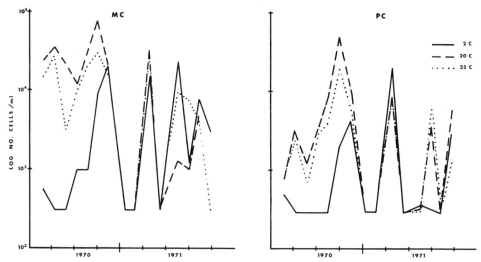

Fig. 2. **Bacterial counts on MA for samples collected at the upper stations (1 and 8).**

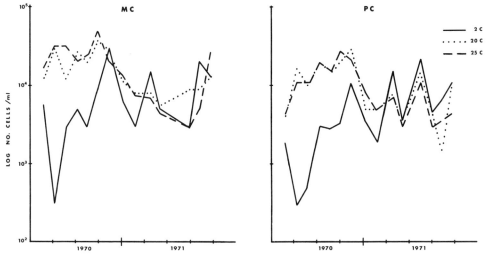

Fig. 3. **Bacterial counts on MA for samples collected at the outer stations (4 and 5).**

3,000/ml. Dominant organisms during this peak period were found to be several dinoflagellates, including a *Peridinium*-like form. Counts at station 1 increased to 7,000/ml. All other stations were below 4,000/ml and showed no significant abundance of dinoflagellates. Mid-cove (M and P) and lower stations (4 and 5) were found always to be diatom-dominated and the counts remained relatively constant, quantitatively and qualitatively. Species of *Chaetoceras, Skeletonema, Melosira,* and *Rhizosolenia* were most common. Counts for all stations were in the range of 2,000/ml during the winter months.

Thus far, the data indicate that the outer, constant salinity stations (4 and 5)

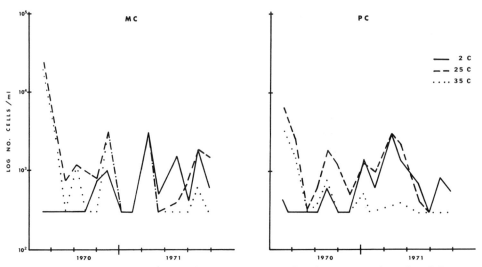

Fig. 4. **Bacterial counts on PCA for samples collected at the upper stations (1 and 8).**

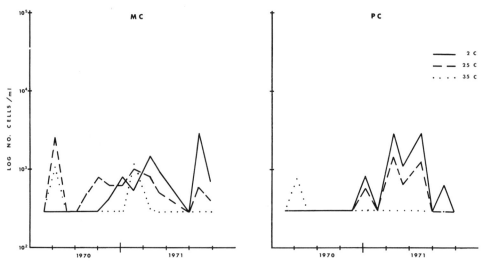

Fig. 5. **Bacterial counts on PCA for samples collected at the outer stations (4 and 5).**

are uninfluenced by chemical or biological changes within the coves. Mid-cove stations (M and P) have shown some changes but not to the extent observed at the upper stations (1 and 8). The latter showed consistently a variety of phytoplankters including diatoms; unicellular green algae occurred more frequently at the outfall station.

Benthic Invertebrates. No quantitative or qualitative data are available at this time. Detailed identifications of preserved material are in progress.

Table 2. Counts (/100ml) of pollution-indicator bacteria over a nine-month period in Mumford Cove and Palmer Cove

Station	Total Coliforms		Fecal Coliforms		Enterococci	
	Range	Mean	Range	Mean	Range	Mean
Mumford Cove						
1	<10->7,400	70	<1->390	36	1->1,000	16
M	2->1,220	295	<1-41	8	<1-229	23
4	12->10,000	1,817	<1-101	26	<1->200	30
Palmer Cove						
5	4->1,600	428	<1-46	17	<1->380	31
P	<2->656	168	<1-49	10	<1->712	68
8	70-7,100	2,720	<2-450	70	4-2,360	200

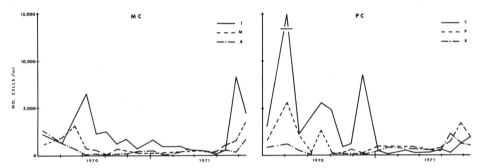

Fig. 6. **Total phytoplankton counts for samples collected at all stations.**

Discussion

Considering the various biological and chemical observations made throughout the nine-month sampling period, several generalizations are permitted. Overall, dissolved oxygen levels did not differ greatly between the two coves studied. Greatest daily fluctuations were found at the upper reaches of both coves, with narrower ranges of readings observed with seaward progression in sampling. At the higher salinities, dissolved oxygen fluctuations were principally influenced by seasonal temperatures. Increased nitrate-nitrogen and orthophosphate values at station 1 in Mumford Cove were the most notable effect of the presence of the treatment plant. Other chemical and biological observations were similar at corresponding stations within the two coves.

Salinity variations between stations will be of interest when interpreting bacterial populations. In the upper coves, abrupt environmental changes resulting from salinity changes and organic loading are major influences on counts recorded on the marine and freshwater media used in this study.

Phytoplankton counts were greatest at the upper reaches of both areas, with Palmer Cove supporting greater numbers during bloom periods. Diatoms were the dominant organisms at all stations, although station 1 (outfall) regularly showed varying numbers of flagellates and green algae to be present. Of potential importance is the occasional dominance of dinoflagellates known to be associated with certain toxic phenomena. Salinity and/or nutrient levels may influence these populations.

The warmest months of the year are yet to be evaluated and may represent the most critical period in these shallow, highly productive waters. Since DO saturation levels decrease with increased temperature and oxygen uptake is increased during the summer period, we will examine daily, and longer, variations in DO more critically.

In addition to routine sampling, further work will emphasize studies of daily bacterial fluctuations. Laboratory enrichment of raw water samples with nitrate, phosphate, and various increments of plant effluent to evaluate possible selection of algal groups by these increased loadings will be done. Also, studies of the natural input and loading sources to the freshwater creeks that empty into the coves must be undertaken.

Acknowledgments

This research was supported, in part, by funds provided by the U.S. Department of the Interior, as authorized under the Water Resources Research Act of 1964, Public Law 88–379 and administered through Dr. William C. Kennard, Director, Institute of Water Resources, University of Connecticut, and in part by the State of Connecticut Water Resources Commission.

We wish to express our appreciation to Mrs. John S. Rankin, Jr., for her invaluable contributions to both field and laboratory aspects of the study.

Literature Cited

1. American Public Health Association Inc. 1965. *Standard methods for the examination of water and wastewater.* 12th ed. Albany, N.Y.: Boyd Printing Co. Inc.
2. Buck, J. D., and R. C. Cleverdon. 1960. The spread plate as a method for the enumeration of marine bacteria. *Limnol. Oceanogr.* **5:**78–80.
3. Carpenter, J. H., D. W. Pritchard, and R. C. Whaley. 1969. Observations of eutrophication and nutrient cycles in some coastal plain estuaries. In *Eutrophication: Causes, consequences, correctives,* ed. G. A. Rohlich, pp. 210–221. Washington: Nat. Acad. Sci.
4. Chen, C. W. 1970. Effects of San Diego's wastewater discharge on the ocean environment. J. Wat. Poll. Cont. Fed. **42:**1458–67.
5. Cronin, L. E. 1970. The test of the estuaries. *Bioscience* **20:**395.
6. Lauff, G. H., ed. 1967. *Estuaries.* Washington: Amer. Assoc. Adv. Sci.
7. McNulty, J. K. 1970. Effects of abatement of domestic sewage pollution on the benthos, volumes of zooplankton, and the fouling organisms of Biscayne Bay, Florida. *Stud. Trop. Oceanogr.* No. 9. Coral Gables: Univ. of Miami Press.
8. Pike, E. B., and A. L. H. Gameson. 1969. *Effects of marine sewage disposal.* Ann. Conf. Isle of Man, England: Inst. Wat. Poll. Cont.
9. Sieburth, J. M. 1967. Seasonal selection of estuarine bacteria by water temperature. *J. Exp. Mar. Biol. Ecol.* **1:**98–121.

Comments

COULL: This is a question to Dr. Buck. I notice that you referred to benthic invertebrates. Not being a microbiologist, but being a benthic invertebrate ecologist, I am concerned. I just want to know what benthic invertebrates you were dealing with. Were these primarily macrofauna, the bigger things?

PILCHER: Actually, we do not have any data on this yet. We screen mud samples and the collected animals are preserved. These will be counted and identified at a later date.

COULL: This bothers me. My contribution to this symposium concerns the meiofauna. The work that is now being published shows that the bacteriologists have the right to claim most of the energetic role. We would like to say that the meiofauna are probably at least as significant as bacteria, if not more. In my review, you will see that there are only two studies in the entire United States dealing with the meiofauna. Investigators have neglected these animals and have gone on to macrofauna. I think this is an area where people who are trying to do an integrated study must place some attention. These organisms represent an important group.

BUCK: Can you put a size on the macrofauna?

COULL: Yes, I like to say any metazoan, which, of course, gets me away from dealing with ciliates and foraminifera. These include those animals that will pass through a 500 μm mesh, the animals that most people sift out. These animals, numerically, are 800 times more abundant per unit than are the macro-forms. I am simply making a plea for some attention to these organisms.

STEVENSON: Am I right in assuming that the coves you studied are affected by tides?

PILCHER: Definitely.

STEVENSON: Then, were your samples taken at the same tidal cycle, level, and so forth?

PILCHER: We have samples taken over tidal cycles. That is why we are setting up tidal gauges to correlate each sample of each definite tidal section.

STEVENSON: We attempted to run some seasonal data on bacterial population in a South Carolina estuary and the results showed fluctuations very much like your data. Then we found that the level of the tides had a definite effect on re-covery of the organisms. I suggest that working at one tide level, ebb or flow, will minimize the fluctuations in numbers.

BUCK: The major problem under investigation is where to locate a sewerage outfall when the plant starts pumping out five or six million gallons a day. Does the new outfall go where the old one is right now? Do you put it in the middle of the cove, or does somebody spend a million dollars to locate it in Long Island Sound? I think this is the problem they were faced with at Woods Hole. I am sure many other communities are faced with it too.

WATSON: We are not faced with it. It was defeated politically.

Bacterial Treatment of Oil Spills: Some Facts Considered

Paul A. LaRock and Merrily Severance

After his recent transatlantic voyage on the *Ra II*, Thor Heyerdahl commented that a continuous stretch of 1,400 mi was polluted by floating lumps of solidified asphaltlike oil *(10)*. Similarly, workers from the Woods Hole Oceanographic Institution reported that during sampling operations in the Mediterranean and eastern North Atlantic their surface sampling nets were commonly fouled by lumps of solidified crude oil. They found several kinds of organisms on or associated with the tarlike lumps. Tar was also found in the stomach of an epipelagic fish, the saury, demonstrating that the toxic material has been introduced into the oceanic food web *(12)*.

Although oil pollution is widespread and has been a problem for decades, most people relate oil pollution to massive releases such as the *Torrey Canyon* or the Santa Barbara blowout. The amount of oil spilled by the *Torrey Canyon*, however, was small compared to the annual releases to the sea by the world's merchant fleet. At the November 1970 NATO Oil Spills Conference in Brussels, Max Blumer estimated that the annual total influx of oil into the oceans is between 5 and 10 million tons. Tankers comprise only 4,000 of the world's merchant fleet of 48,000 ships of more than 100 gross tons, yet they account for about half of the total oil released to the oceans. All vessels accumulate oil in their bilges which is later released to the oceans, but tankers, because of the addition of seawater ballast to empty oil tanks, will release substantially larger quantities of oil. A tanker retains about 0.4% of the cargo as clingage on the insides of her tanks. When ballast water is added for the return trip, the residual oil becomes mixed with the water and is pumped from the tanks with the ballast water. With the adoption of the load-on-top procedure in 1961, oily ballast water is no longer pumped to the sea. Instead, residual oil is gradually separated and transferred to a storage tank. When the ship reaches its home port, the residual oil and water mixture either becomes a part of the next cargo or is pumped ashore for separation. An estimated 80% of the world tanker fleet has voluntarily adopted this procedure in an attempt to decrease the amount of oil released to the sea.

Further efforts to reduce day-to-day oil releases were proposed by Secretary of Transportation, John A. Volpe, at Brussels in November 1970. If adopted, such an agreement would prohibit the flushing of oily wastes into the sea, but would necessitate either construction of shore facilities to receive and separate oil and water wastes or design of tankers with separate ballast and oil tanks.

A major problem yet to be solved is the control of massive oil releases such as those associated with tanker collisions, groundings, and oil drilling accidents. With the closing of the Suez Canal, the Cape of Good Hope became a main route for tankers and was the scene of four accidents in the first nine months of 1968 *(6)*. Off the coast of England on 23 October 1970, the oil-laden *Pacific Glory* collided with a second tanker, the *Allegro*. With its cargo of 274,000 barrels of oil, the *Pacific Glory* brought forth uneasy memories of the *Torrey Canyon*. On 11 January 1971 the English Channel was the scene of yet another accident in which the *Texaco Caribbean* was cut in half by the *Paracas*. On the following day the German vessel *Brandenberg* struck the wreckage from the *Texaco Caribbean* and sank. Oil was not spilled because the main holds of the *Texaco Caribbean* were not pierced, but traffic conditions in the Channel — where 850 ships pass each day and vessels are advised, but not required, to keep to the right — are inviting disaster *(1)*. The early months of 1971 also witnessed several ship accidents off the coast of the United States. On January 19, two vessels of the Standard Oil Company of California fleet, the *Arizona* and the *Oregon Standard*, collided in the early morning hours beneath the Golden Gate bridge. Between 12,000 and 48,000 barrels of bunker oil were released into San Francisco Bay. Later in the same month, the *Esso Gettysburg* was grounded in New Haven, Connecticut, and spilled 8,300 barrels of heating oil.

Designing effective control measures for large oil spills requires some knowledge of the type of accident and the prevailing conditions at the time of the incident. In a recent report, the Dillingham Corporation, under contract to the American Petroleum Institute, analyzed thirty-eight major oil spills occurring between 1956 and 1969 in the United States *(8)*. The report examined the location of the incident, the type of petroleum product involved, the quantity of material released, the duration of the spillage, the prevailing weather conditions, and the amount and type of shore line affected. This report evaluated spills in which 2,000 barrels or more were released, or spills which received widespread publicity. Of the thirty-six incidents reported, 75% involved vessels; only 25% of the oil spills involved refineries, offshore drilling, or other incidents (Fig. 1). Twenty-five of the twenty-seven vessel incidents involved tankers. This fact assumes greater importance if one considers the increase in tanker size from the conventional 40,000-ton vessel to the new super tankers which may eventually reach 500,000 or a million tons dead weight.

The composition of the hydrocarbon material released in thirty-five incidents is summarized in Figure 2. The data indicate that the greatest volume of material spilled was crude oil and the least was residual oil, although crude and residual oils were involved in about the same number of accidents. From Figure 3 it can be seen that in twenty-three incidents the volume of the spill varied from 2,000 barrels to 700,000 barrels, with a median of approximately 25,000 barrels of oil. The distance from shore to the point of oil release was less than one mile from shore in 52% of all incidents and less than ten miles in 75% of all accidents (Fig. 4). The distance from spill to shore determines the time available for control measures. Assuming certain rates of oil slick drift, we can predict that in 75% of the cases, oil would reach shore in less than one day. Determination of the duration of a spill incident is subjective, but in 50% of all incidents the spill

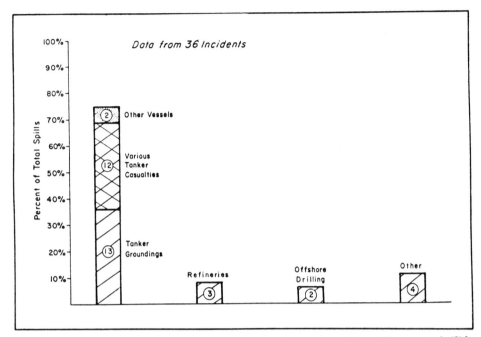

Fig. 1. Source of oil spill incidents. (Figures 1 through 9 reprinted from Gilmore et al. *(8)* by permission of the Dillingham Corporation.)

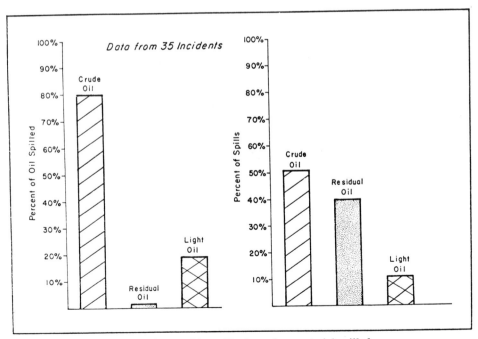

Fig. 2. Composition of hydrocarbon material spilled.

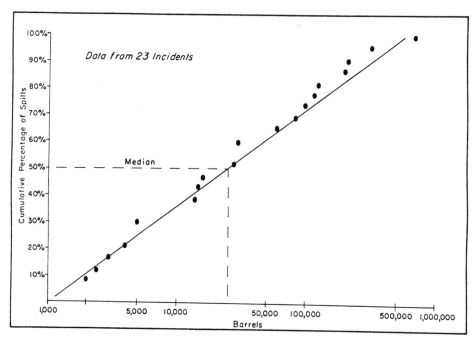

Fig. 3. **Volume of oil spilled.**

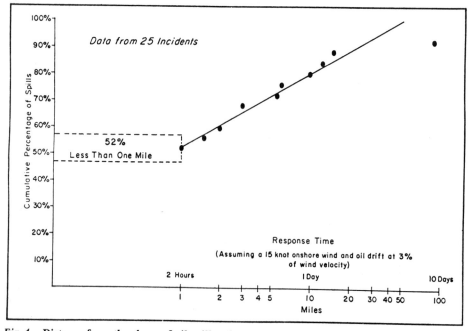

Fig. 4. **Distance from the shore of oil spill and response time available for shoreline protection.**

lasted for approximately 18 days (Fig. 5). The length of coastline affected by the oil released in 18 incidents, illustrated in Figure 6, indicates that in 50% of all accidents, slightly over 3 mi of coastline was coated with oil. The shoreline affected, in most cases, includes recreational areas, although industrial and residential coastlines will also be affected (Fig. 7). The weather conditions at the time of an oil spill will determine the preventive and protective measures taken. Data from eleven incidents (Fig. 8) indicate that the median wave height at the time of the accident was approximately 9 ft, and in 50% of all incidents the wind speed was 20 knots (Fig. 9). An oil slick moves at 3.3% of the prevailing wind velocity or at about ½ mph in a 20 knot wind.

From the Dillingham data, it can be predicted that 50% of future oil spills will occur within 10 mi of shore during strong winds and high seas. The source of the spill will most likely be a tanker; the product released will be either a crude or a residual oil; and the volume of the spill will probably be more than 25,000 barrels. The spill will last for an average of 18 days and will affect 3 to 4 mi of predominantly recreational shoreline. Weather conditions at the time of the accident will prevent the use of small boats or booms. With wind speeds of 20 knots, the oil will reach the shore in less than 24 hr after the incident. Given these conditions as the most probable under which a major spill will occur, it is easy to see that very little can be done to prevent the oil from reaching land by applying the current technology of oil spill cleanups. The increase in water traffic, the past history of tanker accidents, and the advent of the supertanker make more stringent navigational controls and reevaluation of tanker design necessary first steps in reducing the hazard of oil spills. Criteria for tanker construction have been summarized by Swift et al. *(18)*. Until engineering advances can be applied to increase the safety of marine operations, alternative measures are necessary to deal with oil spills on the sea. Hopefully, cleanup technology will not become so effective that we rely on cleanup devices rather than concentrating on spill prevention.

In the development of cleanup techniques, acceptable levels of residual oil in the oceans must be defined. According to Blumer *(3, 4, 5)*, carcinogenic components in oil will prove to be a hazard with increasing releases, and therefore, all spilled oil must be removed from the sea. Such a task is virtually impossible with present technology. Cleanup measures currently available include entraining devices (booms and pneumatic barriers), mechanical recovery devices, absorbing material, sinking materials, chemical treating agents or dispersants, combustion promoting materials, and solidifying or gelling agents.

Booms or entrainment devices are effective in calm waters but are ineffective in seas over 2 to 3 ft or in high winds. Skimming devices have a very low capacity and are ineffective in major spills. Absorbent materials have been used; straw has been the mainstay in the treatment of oil spills. Straw will absorb roughly eight to thirty times its weight in oil and can be burned, but it is slow to apply and must be harvested.

Sinking materials are apparently the most effective means presently available. Stearate-treated chalk was used in the *Torrey Canyon* episode to prevent oil from reaching the coast of France. Three thousand tons of chalk were used to sink about 30,000 tons of oil but more chalk would have been required to sink fresh

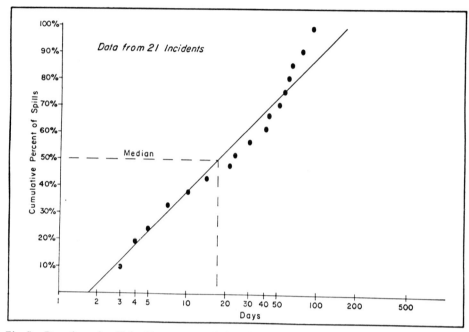

Fig. 5. **Duration of spill incident; reported duration of the spill will vary with reporter, but it does indicate that for several weeks afterward the spill will require cleanup operations.**

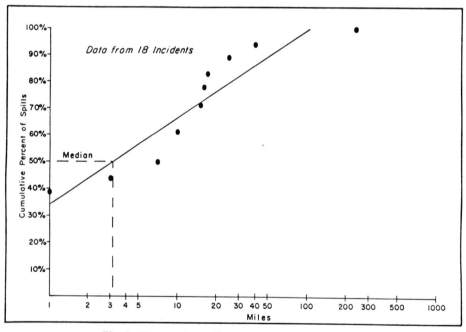

Fig. 6. **Extent of coastline contaminated by oil spills.**

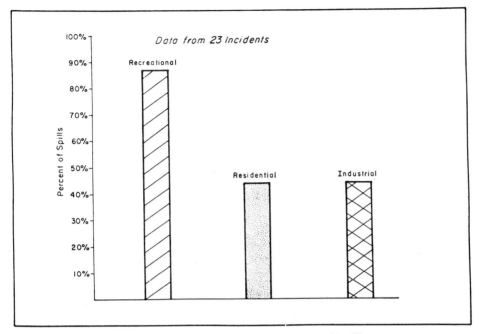

Fig. 7. **Utilization of coastline affected by oil spills.**

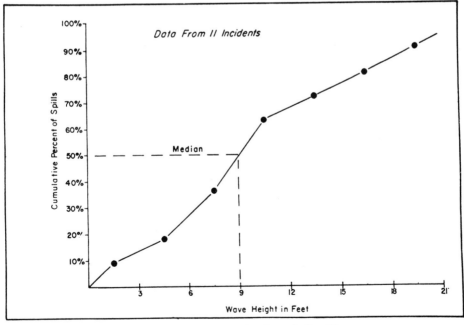

Fig. 8. **Wave height at the time of oil spills.**

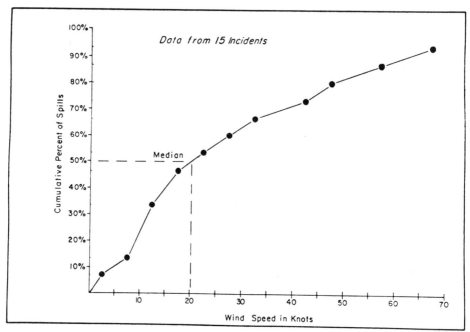

Fig. 9. **Wind velocity at the time of oil spills.**

oil *(16)*. Sinking removes oil from surface water and relocates it on the bottom where it is no longer an immediate problem, but its rate of degradation is greatly reduced. Sinking would also be objectionable in shallow waters or near inshore fisheries *(3, 4, 16, 20)*. Sinking agents are most successful when the oil is weathered; and if oil is to be dispersed by this method, it must have been exposed to the sun and winds to remove lighter fractions and increase its density or be aged by some artificial means.

Because of their apparent toxic effects, chemical agents have been used infrequently since the *Torrey Canyon* incident. Burning of the oil may be promoted by the addition of insulating material such as Cab-O-Sil, which prevents the heat of the flame from being conducted away from the oil. Burning requires the presence of the lighter fractions of oil to be effective, and should be attempted on freshly spilled oil rather than aged oil *(10)*. Solidifying agents change the state of the oil by gelling or polymerization. In this way, oil may be converted to a solid mass to prevent spreading. Gelling agents are not considered to have significant application for major oil spill control, and the availability of many of these chemicals is limited *(8)*.

One widely suggested, but as yet untried, method for cleaning up oil spills is that of bacterial seeding to degrade spilled oil. Since the bacteria are the ultimate degraders of oil released to the oceans, a logical approach would be to seed bacteria into a slick rather than wait for a natural population of hydrocarbon-oxidizers to develop. There are a number of drawbacks to such a scheme and these have been summarized by ZoBell *(21)*. Among the most important conditions that influence microbial oil degradation are the number of microorganisms present, concentration of nutrients, especially nitrogen and phosphorus compounds, chemical composition of the hydrocarbon materials spilled, availability of water or the dispersion of the oil in water, and temperature. Temperature can have a profound effect on microbial hydrocarbon degradation of oil in the ocean. One series of experiments indicated that the percentage of oil degraded per week was 50–80% at 25 C, 30–50% at 20 C, 20–30% at 10 C, with no apparent degradation at 4 C *(14)*. The average surface temperature of the Atlantic between 30 and 40°N is 20.4 C *(17)*. The effectiveness of bacterial seeding would vary by season and latitude. Oxygen availability in hydrocarbons is probably not limiting; air saturation level of oxygen in kerosene is 322 ppm and deoxygenated water takes up oxygen from kerosene in a matter of minutes *(11)*. Wave action at the time of a major spill can cause the bacteria to be dispersed from the oil-water interface, reducing the rate of microbial degradation of oil.

Miget et al. *(15)* attempted to evaluate feasibility of microbial degradation of an oil spill. Fifty cultures of oil–degrading bacteria were tested at 32 C in a phosphate and nitrate enriched seawater medium overlaid with crude oil. Results indicated that, under laboratory conditions, bacteria could degrade from 35 to 55% of the oxidizable crude oil within 60 hr. The Dillingham prediction that in 75% of all major spills the oil will be on the shore in less than 24 hr suggests that lack of time precludes the use of bacteria as a sole method of treating large–scale oil spills. In the absence of actual field data, projections of the efficiency of seeding cannot be made, but bacteria probably cannot be considered primary agents for oil spill cleanup if time is a critical factor.

Factors encouraging the use of bacteria in oil spill treatment are that bacteria can be applied in rough seas and that bacteria have been reported to degrade certain carcinogenic components found in crudes *(21)*. Seeded bacteria may also reach the shore with the crude oil and accelerate degradation on shore, even though rates of degradation on the water are ineffective as a control measure. However, if a bacterium or a group of bacteria can alter oil so that a spill is amenable to other control procedures, bacteria might be a tool in oil spill technology.

During an investigation supported by the National Aeronautics and Space Administration concerning microbial life in harsh environments, a bacterium was isolated which is different from most hydrocarbon bacteria in that the organism will grow only in the oil phase of an oil-water liquid medium rather than at the interface or in the aqueous phase. The organism requires water, but its development is confined to the hydrocarbon phase. This type of growth has been termed "oleophilic." The oil is visibly altered during the growth of the organism and eventually a population of bacteria in excess of 10^9 cells per ml develops. We have been studying this organism for other reasons but the unique pattern of growth exhibited by the bacterium suggests it may be useful for oil spill control. Limited experiments have been done with crude oil; hence, our discussion on this point is speculative.

Two oleophilic hydrocarbon-oxidizing cultures were isolated from oil-soaked soil. When grown in a two-phase liquid medium, the cultures developed first on the oil side of the interface and later dispersed throughout the oil. Three distinct layers are visible in a typical culture flask (Fig. 10). The uppermost

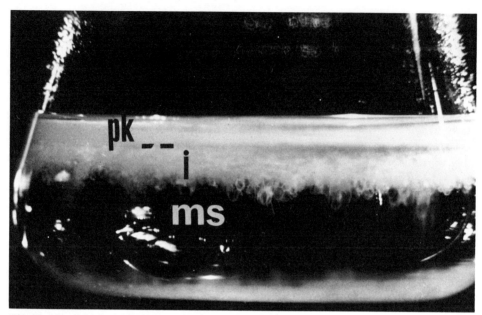

Fig. 10. **Growth of an oleophilic bacterium in a mineral salts and hydrocarbon medium: the three layers, from top to bottom, are light paraffin oil-kerosene, with bacterial growth (pk); frothy, interfacial growth (i); and aqueous mineral salts medium (ms).**

layer is the hydrocarbon phase, i.e., mineral oil and kerosene sterilized by Millipore filtration. The oil is turbid due to the extensive bacterial growth (Fig. 10). Just below the oil is a white frothy, interfacial layer. The interfacial layer resembles small spheres or bubbles penetrating the aqueous phase, the bottom layer in the flask. A second bacterial isolate exhibited a similar growth pattern in the earlier stages of growth, but later produced a pigment as well as large lumps or globules (Fig. 11). Figure 12 shows the contents of the flask when poured into a cylinder and allowed to stand. The three layers are observed clearly; the upper layer is the turbid oil phase containing bacteria; next the interfacial layer; and finally, the aqueous component. The significance of these three layers will be discussed later.

The growth of the isolates in the oil phase of two-phase hydrocarbon media at various pH values is illustrated in Figure 13. The media consisted of double strength modified Bushness and Haas (7) mineral salt solution overlaid with a sterile mineral oil and kerosene mixture. The medium was modified by replacing the phosphate salts of Bushness and Haas with an equivalent total molar phosphate concentration of Sorenson's buffer to achieve the desired pH values.

No growth was observed at pH 6.0, but the culture grew at pH values as low as 6.5. It is evident that pH values near or above neutrality do affect the growth rate of the bacteria in the oil phase. The increased growth rate at pH 8 suggests that this culture would grow well at the pH of seawater.

A long lag period of approximately 70 hr was consistently observed, but the length of the lag period does vary with inoculum size. Since the bacteria need water for growth, a possible explanation for the lag may involve the entrainment

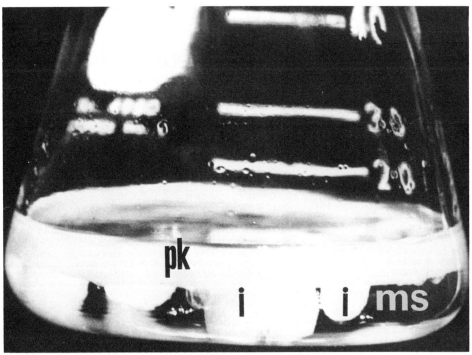

Fig. 11. **Typical growth of oleophilic *Brevibacterium* in a mineral salts and hydrocarbon medium: the three layers, from top to bottom, are light paraffin oil-kerosene with bacterial growth (pk); interfacial layer (i); and aqueous mineral salts medium (ms).**

of water in the oil. A minimal population size may be necessary to alter the oil so that the mineral salts and water become entrained. A part of this process may be the formation of bubblelike objects at the interface to increase interfacial surface area. The growth curves (Fig. 13) were obtained in the oil phase and indicate no changes for 70 hr, but the first evidence of growth occurred at the interface within 24 to 36 hr.

The interfacial layer is best examined microscopically using oil-soluble flaming red and water-soluble methylene blue dyes to stain the respective phases. In this procedure the dyes are added to the flasks; the flasks are shaken for about $\frac{1}{2}$ hr; a wet mount is prepared and examined with phase-contrast and bright-field optics. A stained interfacial sample is shown in Figure 14. The pink (P) and blue (B) areas, respectively, denote oil and water mixed together at the interface. The fiberlike structures are consistent features, and we feel this is related to the large globular structures observed in Figure 10. A higher power magnification of the central position of Figure 14 shows the presence of bacteria in the oil phase surrounded by the fiberlike material and water (Fig. 15). Due to the nature of the hydrocarbon, the presence of bacteria, and the extensive mixing of the materials, it has been impossible to separate and analyze the fiberlike objects. We have occasionally obtained slides in which the fibers appear broken and discontinuous, suggesting they are a solid, possibly a wax. The amount of the fibrous material appears to be a function of the age of the culture;

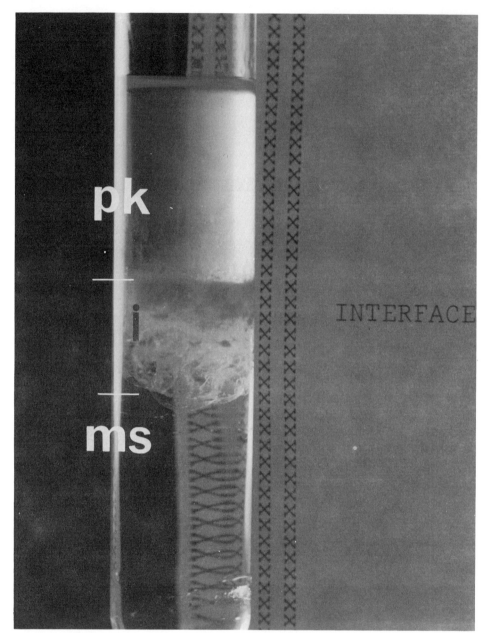

Fig. 12. **Contents of flask (Fig. 11) after poured into a narrow tube: the three layers are more easily distinguished.**

the older the culture, the more prevalent the fibers. In cultures of three or four days, fibers are seldom present.

Using the staining procedure on a five- or six-day old culture and examining it under low power (Fig. 16), a large number of bacteria and refractive bodies in the oil phase were observed, but the aqueous phase was essentially devoid of

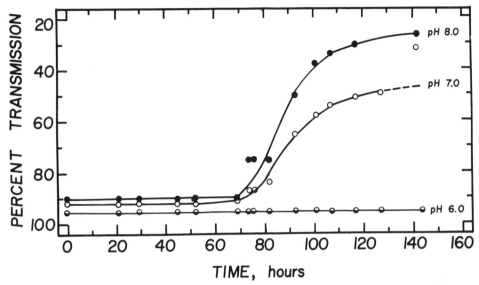

Fig. 13. Spectrophotometric growth curves of oleophilic organism in oil phase over aqueous media.

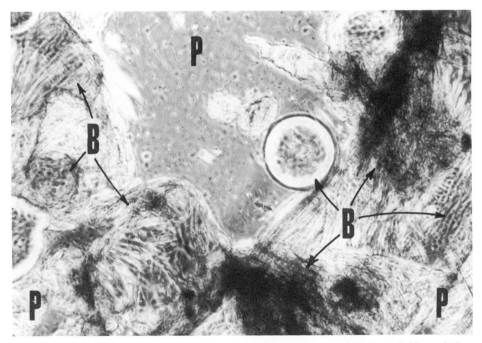

Fig. 14. An interfacial sample stained with oil-soluble flaming red and water-soluble methylene blue. Pink (P) and blue areas (B), respectively, denote oil and water mixed in the interface, x54. The fiberlike objects are consistent features at the interface and are possibly related to the globular structure of the interface seen in Figs. 10 and 11.

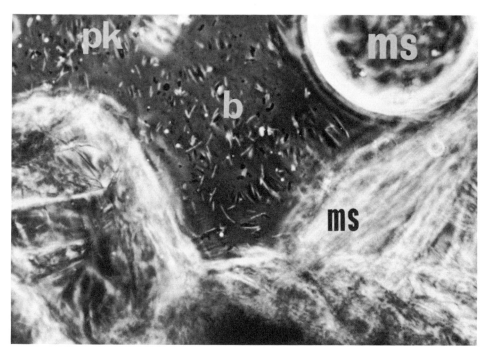

Fig. 15. An enlarged portion of the stained interfacial sample in Fig. 14 showing mixture of oil (pk) and mineral salts (ms), fibers, and bacteria (b) in the oil phase. Approximate magnification x325.

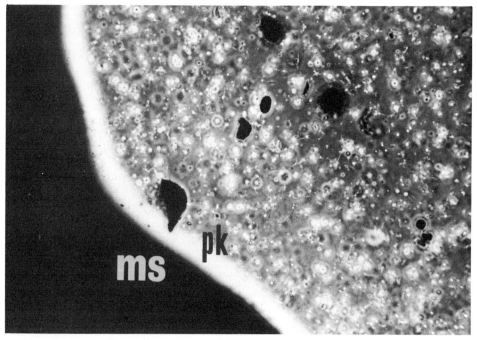

Fig. 16. Bacteria and refractive bodies present in oil phase of five- to six-day-old cultures. Symbols same as Fig. 15.

these objects. Sterile oil contained no such inclusions. By examining the oil phase under a higher magnification, we can see many bacteria that appear to be attached to the small refractive bodies (Fig. 17). In several staining experiments, refractive bodies, as well as the bacteria, pick up the methylene blue, suggesting that water may be present. A second bit of evidence suggesting that the refractive objects are composed of water is that when viewed with positive phase-contrast optics, objects with a refractive index lower than the background material appear brighter than the background (9). Since water has a lower refractive index than oil, any water present in the oil should appear as bright objects relative to the background such as seen in Figure 17. Definition of the nature of the refractive objects must await future isotope studies. The refractive objects may be water entrained by extensive interfacial growth of the organism. Further evidence for water entrainment is visible in Figure 18. Several large blue droplets of water surrounded by a matrix of oil can be seen. This sample, collected from the interface, indicated that water can be dispersed into the oil phase, which would be important in the acceleration of hydrocarbon degradation.

The bacterial culture studied in our laboratory has been identified as a *Brevibacterium* species. In an earlier review, Beerstecher (2) indicated that acid-fast bacteria penetrated the oil phase of liquid medium as an emulsion if the mixture was shaken. Our culture does not require emulsifying agents for its dispersal in oil, and in fact, all dilutions and suspensions must be made with oil. The only way we have been able to disperse the culture in water is to add an oil suspension to water containing 0.008% Triton X-100 and sonicate the mix-

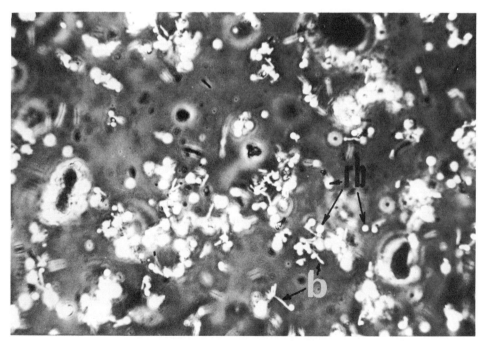

Fig. 17. **Refractive bodies (rb) apparently attached to bacteria (b) in oil phase of five- to six-day-old culture, x450.**

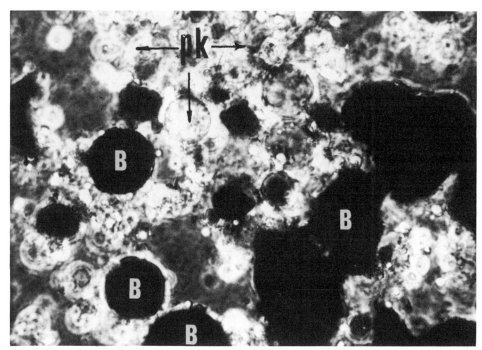

Fig. 18.   **Large blue droplets of water (B) surrounded by a matrix of oil (pk) from the oil phase of the oleophilic culture may be evidence of water entrainment.**

ture. This emulsion is stable for several hours, but when the emulsion breaks, the bacteria will float to the surface and leave the aqueous phase clear. If sufficient oil is present, the bacteria will grow at the surface of the water.

The features that render this oleophilic bacterium of possible use in oil spill treatment are its ability to remain in the oil phase and its ability to entrain water in the oil. Entrainment of water promotes a greater oil-water interfacial area and accelerates degradation by other nonspecific hydrocarbon bacteria.

Perhaps the most useful application of an oleophilic culture to oil spills is to age the oil. As discussed earlier, sinking of oil spills is probably the most effective means of dispersing oil from the surface. (See papers by Blumer *(3, 5)* for discussion of the merits of total removal vs. dispersion.) If the oil is to be sunk, its density must be increased. The lighter fractions can be removed by exposure to sunlight and wind or possibly by hydrocarbon–oxydizing bacteria. Kinney et al. *(13)* indicated that in 24 hr experiments, evaporation essentially removed the normal alkanes of C_{12} or shorter chain length from crude oils. Concentration of compounds of carbon length C_{14} and above remained the same.

Alkane degradation by the oleophilic culture was determined by growing the organism in a paraffin oil-kerosene mixture. The original oil phase and the oil after two weeks of bacterial growth were compared by gas chromatography. Measurements of peak areas were confined to the normal alkanes between C_{11} and C_{35}. The peak area for each alkane is expressed as a percentage of the total peak (area of *n*-alkanes between C_{11} and C_{35} (Table 1). It is evident that the cul-

Table 1. Change in *n*-alkane content of paraffin oil-kerosene mixture mediated by bacteria

Alkane Carbon Number	Peak Area – Percentage of Total		
	Sterile PK	Degraded PK	Net Change
11	1.29	2.68	1.29
12	7.55	9.45	1.90
13	11.41	10.20	−1.21
14	11.82	9.55	−2.27
15	9.80	5.50	−4.30
16	6.95	3.36	−3.59
17	7.50	Not Resolved	
18	5.42	2.37	−3.05
19	6.18	3.74	−2.44
20	4.59	3.40	−1.19
21	3.08	Not Resolved	
22	3.43	5.33	1.90
23	2.89	4.40	1.51
24	2.67	5.48	2.81
25	3.67	6.35	2.68
26	2.92	5.17	2.25
27	2.15	4.46	2.31
28	1.68	3.28	1.60
29	0.95	2.40	1.45
30	1.12	1.70	0.58
31	1.17	1.92	0.76
32	0.94	1.30	0.36
33	0.25	0.96	0.71
34	0.28	0.85	0.58
35	0.21	0.40	0.19

ture degrades alkanes between C_{13} and C_{20} and possibly up to C_{21}, leaving the heavier *n*-alkanes to make up an increased proportion of the crude. If the degradation of lighter hydrocarbons can be rapid enough, the oleophilic bacterium may then be a potential agent for aging oil spills. The slick could be sunk using chalk dust or sand. Both the bacteria and the sinking agent can be applied in rough seas, offering an advantage over current oil spill cleanup procedures. In this instance, the bacteria are not used as primary degrading agents, but rather to condition the oil for subsequent dispersal by physical means.

Because of the limited time available for the control or dispersal of oil spilled at sea, bacteria cannot be viewed as the principal mechanism for the erradication of a major oil spill before it reaches land. When used with another control technique such as sinking, bacteria may be helpful in the treatment of sea-borne oil slicks. Bacterial seeding may also have application in ameliorating the duration of spill effects on the coastline and the long–term effects of the oil, particularly the residual carcinogenic components in the ocean. Although the use of bacteria for spill cleanup has been discussed in the literature, there are no published data concerning the actual seeding of bacteria in an oil spill in the field. Laboratory data are valuable, but cannot substitute for field experience; until this evidence is in hand, the ideas presented in this paper and elsewhere must be considered speculative. The ultimate answer must be prevention of oil spills rather than their control.

Acknowledgment

A portion of the work reported in this paper was supported by Research Grant NGR 10–004–041 from the National Aeronautics and Space Administration.

Figures 1 through 9 were reprinted from reference *8* by permission of the Dillingham Corporation and the American Petroleum Institute.

Literature Cited

1. Anonymous. 1971. Revamp rules of the road for traffic in Channel? *Oceanology International*, January, p. 17.
2. Beerstecher, E. 1954. Petroleum microbiology. New York: Elsevier Publishing Co.
3. Blumer, M. 1969. Oil pollution of the ocean. In *Oil on the sea*, ed. D. P. Hoult, pp. 5–13. New York: Plenum Press.
4. Blumer, M., G. Souza, and J. Sass. 1970. Hydrocarbon pollution of edible shellfish by an oil spill. *Mar. Biol.* **5:**195–202.
5. Blumer, M., H. L. Sanders, J. F. Grassle, and G. R. Hampson. 1971. A small oil spill. *Environment* **13:**2–12.
6. Boyle, C. L. 1969. Oil pollution of the sea: Is the end in sight? *Biol. Conserv.* **1:**319–27.
7. Bushnell, L. D., and H. F. Haas. 1941. The utilization of certain hydrocarbons by micro-organisms. *J. Bacteriol.* **41:**653–73.
8. Gilmore, G. A., D. D. Smith, A. H. Rice, E. H. Shenton, and W. H. Moser. 1970. *Analysis of oil spills and control materials.* LaJolla, California: Dillingham Corporation.
9. Hartley, W. G. 1964. *How to use a microscope.* Garden City: Natural History Press.
10. Heyerdahl, T. 1971. Voyage of *Ra II. Nat. Geogr.* **139:**44–71.
11. Hill, E. C., D. A. Evans, and I. Davies. 1967. The growth and survival of micro-organisms in aviation kerosene. *J. Inst. Petrol.* **53:**280–84.
12. Horn, M. H., J. M. Teal, and R. H. Backus. 1970. Petroleum lumps on the surface of the sea. *Science* **168:**245–46.
13. Kinney, P. J., D. K. Button, and D. M. Schell. 1970. Kinetics of dissipation and biodegradation of crude oil in Alaska's Cook Inlet. In *Proc. Joint Conf. on Prevention and Control of Oil Spills, API and FWPCA, December 15–17, 1969*, pp. 333–340. New York: American Petroleum Institute.
14. Ludzack, F. L., and D. Kinkead. 1956. Persistence of oil wastes in polluted water under aerobic conditions. *Ind. Eng. Chem.* **48:**263–67.
15. Miget, R. J., C. H. Oppenheimer, H. I. Kator, and P. A. LaRock. 1970. Microbial degradation of normal paraffin hydrocarbons in crude oil. In *Proc. Joint Conf. on Prevention and Control of Oil Spills, API and FWPCA, December 15–17, 1969*, pp. 327–31. New York: American Petroleum Institute.
16. Smith, J. E., ed. 1968. *Torrey Canyon pollution and marine life.* Cambridge: Cambridge Univ. Press.
17. Sverdrup, H. V., M. W. Johnson, and R. H. Fleming. 1946. *The oceans; their physics, chemistry and general biology.* New York: Prentice-Hall Inc.
18. Swift, W. H., C. J. Touhill, W. L. Templeton, and D. P. Roseman. 1969. Oil spillage prevention, control, and restoration—the state of the art and research needs. Part 1. *J. Wat. Poll. Cont. Fed.* **41:**392–412.

19. Tully, P. R. 1969. Removal of floating oil slicks by the controlled combusion technique. In *Oil on the sea*, ed. D. H. Hoult, pp. 81–91. New York: Plenum Press.

20. ZoBell, C. E. 1962. The occurrence, effects, and fate of oil polluting the sea. *Adv. Wat. Poll. Res.* **3**:85–118.

21. ZoBell, C. E. 1970. Microbial modification of crude oil on the sea. In *Proc. Joint Conf. on Prevention and Control of Oil Spills, API and FWPCA, December 15–17, 1969*, pp. 317–26. New York: American Petroleum Institute.

Comments

QUESTION: You said before that it is possible to seed bacteria in case of an oil spill. How would you go about seeding in sufficient concentration to make some contribution?

LAROCK: The density of this bacterium is greater than that of the oil in which it grows but less than that of water. The organism can be centrifuged out of the oil at about 1,800 × g, yielding a very dense packing of cells that could then be used to seed. A recent paper presented at the ACS meeting proposed that the hydrocarbon bacteria be lyophilized. There is a firm in Alexandria, Virginia, which produces lyophilized bacterial preparations to be mixed with water, yielding a very dense suspension that can be distributed in very high concentrations.

QUESTION: What was your source material for isolation of the cultures?

LAROCK: We isolated both of the cultures from soil samples from fuel storage depots. The soil had been saturated with kerosene that leaked from tanks. The one shown in the first slide (Fig. 10) was a culture we lost. Certain growth factors are required which we are now in the process of determining. To maintain the second culture, we have to provide yeast extract in the medium.

QUESTION: How sensitive are the bacteria to fluctuations in temperature?

LAROCK: The cultures are normally grown at 35 C, but the organism grows quite well at 20 C. Frequently, we grow it at room temperature which is between 20 and 25 C.

QUESTION: Do you have any data on what short–chain end products are produced as a result of the degradation?

LAROCK: No, we are trying to determine that now.

QUESTION: What was the salt concentration in the medium?

LAROCK: The medium includes phosphates and nitrates. If the refractile bodies are water, and we feel they are, they may contain salts. Theoretically, one would have to add salts during seeding.

HOLM-HANSEN: What is the shortest generation time measured for the bacterium?

LAROCK: About 2 hr.

WATSON: Dr. Robert Mahoney from Skidmore has been working with us this last year on hydrocarbons. We can get a very good turbid culture in 18 hr. If we do a gas chromotographic analysis, about 90% of all the straight chain hydrocarbons are gone. But, even if we hold that same culture for a week, the aromatics are barely touched. As soon as the straight chain compounds are depleted, the culture or oil seems to emulsify and sink to the bottom. If you were to seed an oil spill with bacteria, I am afraid the first things that would go off would be the straight chains, and then the oil would emulsify and sink to the anaerobic zone and remain forever. It is a reducing atmosphere and the hydrocarbons just will not be decomposed down there at all.

LAROCK: That is right. This is actually the same point Dr. Blumer has made. This is why he feels that either burning or the total removal would be best. I am not advocating its use without some discriminating judgment.

WATSON: Actually, if you want to see oil naturally decomposed, the best place is right on the beach where we walk, which is rather obnoxious. I will make another comment about bacteria living in the oil phase. We commonly see this. It does seem that the bacteria orient themselves in the oil droplet very much like sulfur bacteria do on sulfur granules. This almost indicates that the heads of the bacteria are different from the rest of them. We get enriched cultures because oil acts as a selective medium. We see one organism that grows in long chains which are quite different from normal chains.

LAROCK: I have had a number of arguments with colleagues on the question of whether these organisms are common. Our isolate is different in some respects in that it grows only as single cells. There is no orientation at the interface, and we cannot pick it up with water. All our dilutions and culturing must be with oil. We cannot wet this organism unless we pick it up in oil and use Triton-X100 and sonicate a few minutes. This is the only way we can disperse the cells in water. We can obtain micrographs of the organism in oil droplets mixed in water but the organism does not grow in water. The organism will grow in brain-heart infusion broth. The organism grows in clumps that appear as little balls with a diameter of about an eighth of an inch. It is definitely hydrophobic.

WATSON: We have also asked the question, how is it possible that these bacteria can live in oil? We normally think of the hydrocarbons as capable of dissolving membranes. At least 50% of all the organisms we found growing in hydrocarbons have extra cell wall layers, probably composed of wax. This wax possibly protects the membrane systems. The oil may be digested by extracellular enzyme and hydrocarbons never get into the cells.

LAROCK: This may be. This organism has a very high total lipid content, about 34 to 36% of the dry weight of the cell.

G. JONES: What is the taxonomy of the organism?

LAROCK: We have identified it as a species of *Brevibacterium*.

Environmental Effects of Pulp Mill Wastes

M. L. Quammen, P. A. LaRock, and J. A. Calder

The Big Bend of the Florida panhandle is covered with pine forests owned by several paper-making firms. Pulp manufacture, a major industry in the northern portion of the state, contributes to the pollution of the Gulf of Mexico. Pulp manufacture requires large volumes of water as a transport vehicle for raw materials, for use in chemicals, and for washing and cooling. Typically, pulp mills in the area release between 30 and 60 million gallons/day (MGD) of wastes which has been treated by short-term lagooning.

The Florida Division of Air and Water Pollution recently issued citations for extensive water pollution to eight of the nine pulp mills in Florida. Under terms of pending legal action, the mills must initiate immediate construction of waste treatment facilities, even though the state of Florida has not decided upon standards for treated pulp mill wastes. Standards for treated pulp mill wastes must take into account the toxic and subtoxic effects of long-term releases. The state of Florida formerly consulted with regional and state health services to arrive at safe and reasonable levels of pulp mill waste. However, large coastal tourist industries, commercial fishing, and shell fishing industries are located in Florida. Thus, the potential effects of water pollution on these industries require that extreme care be taken in adopting standards for the release of wastes into the marine environment.

The quantities of water used in pulp manufacture vary with the pulping process used and the quality of the finished product. A bleached Kraft process may require up to 60 MGD per 1,000 tons of product, and discharge waste receiving little or no treatment.

Liquid effluent from a Kraft mill contains wood sugars, lignin, tannins, hemicellulose, tall oil, organic sulfur compounds, sulfides, and mercaptans (5). Chemical pulping operations generally have a pulp yield of about 50% (5) and produce a waste high in BOD and suspended solids. Suspended matter consists primarily of fibers, fiber debris, and coating material (5). Average waste characteristics for an unbleached Kraft mill effluent may be summarized as follows (5):

pH	8.2	
Total alkalinity	175	mg/l
Total solids	1,200	mg/l
Total suspended solids	150	mg/l
BOD$_5$	175	mg/l
Color	250	

329

Pulp mill wastes in a concentration of 1:5,000 (v/v) will adversely affect oyster populations (4). If the concentration of mill effluent is increased to 1:1,000 (v/v), deterioration of oysters is observed within a few days. Wolke (13) suggests pulp mill effluents in concentrations as low as 16 ppm will interrupt the reproductive cycle of the oyster, *Ostrea lurida*.

Galtsoff et al. (4) attribute the destruction of some Virginia oyster beds to pulp mill effluents. A similar series of events was observed in Apalachee Bay when the Kraft pulp mill at Perry, Florida, began releasing its untreated wastes to the Gulf of Mexico via the Fenholloway River. Dilution and dispersion of pulp mill effluents have been used as a method of waste disposal. Such practices are not to be condoned, but in areas of the country with large rivers to support such additions, noxious conditions might not develop. In the southeastern United States, however, many streams dry up in the summer months and thus cannot function in any capacity other than as an open channel to funnel mill wastes to another body of water. The existence of such a problem caused us to undertake this study.

Study Area

The pulp plant studied is located in Foley, Florida, which is about 50 mi west of Tallahassee in the Big Bend area of the state on the northern Gulf of Mexico. The Kraft pulp mill has a daily output of 1,000 tons of Kraft paper or bleached pulp and a normal effluent flow of between 50 and 55 MGD. Until recently, the method of waste treatment at this plant was lagooning the waste and releasing it to the Fenholloway River for dilution and dispersal.

The source of the Fenholloway River is San Pedro Bay, a covered swamp between 80 and 105 ft above the mean sea level. The paper company drained much of the swamp for tree farming by the construction of a system of canals to control the ground water and divert it to the Fenholloway and Econfina rivers. The flow in the canals and rivers provides a flushing mechanism for pulp mill wastes. The water from the canals is discharged into the Fenholloway in San Pedro Bay and again approximately 3 mi upstream from the mill at Foley (10).

The Fenholloway River, although small, is not sluggish. It has a mean velocity of approximately 1 ft/sec during periods of average flow, and water will travel the 25 mi to the tidal limits of its estuary in approximately 30 hr. The principal sources of fresh water to the Fenholloway are the swamp drainage canals and Spring Creek which flows into the river below the pulp mill. During periods of very high flow, two minor tributaries add fresh water to the river. The Fenholloway River drainage basin comprises an area of some 333 sq mi. However, before the drainage was controlled, the stream had been known to dry up in the summer months. Flow rates formerly averaged between 80 cu ft/sec in the winter to 2 cu ft/sec in the later summer months.* The tides affect only the lower 4 mi of the river, and estuarine salinities are normally found in the lower 2 mi.

The Gulf at the mouth of the Fenholloway is very shallow. The 6-ft and 12-ft

* Pulp mill staff, personal communication.

depth contours are located approximately 4 and 6 mi offshore, respectively (Fig. 1). The water depth at the mouth of the Fenholloway is 1 ft at mean low tide and an oyster bar beyond the river mouth serves to shunt the river flow either to the west or to the southeast, depending on wind conditions. The prevailing water flow is along the shore to the southeast and is indicated by the large, solid arrows in Figure 1. The open arrows indicate an occasional westerly movement. The dark portion along the shore is salt marsh. Landward of the marsh, there are hummocks and flatlands. Except for the pulp mill, the area is undeveloped. The bottom of the Gulf is covered with a dark, spongy material having the characteristic pulp mill odor and a fibrous texture.

Methods

Selected chemical and bacterial parameters were measured in the Fenholloway River above and below the pulp mill in February and November 1970. Figure 2 shows the sampling locations along the river. Station A was approximately 2.5 to 3 mi above the plant. Station B was just below effluent outflow. Station G was approximately 4 mi above the Gulf or just at the limit of tidal effects in the river. Water samples for all measurements were collected with a Hale water sampler *(12)* modified by LaRock *(6)*.

Analyses of river-water samples for phosphate, sulfate, and BOD were performed following standard methods *(1); in situ* temperature and dissolved oxygen (DO) were measured using a Yellow Springs Instrument Company Model 54 Dissolved Oxygen Meter, with thermistor. The pH of sediment samples was determined immediately after collecting by using an Orion Model

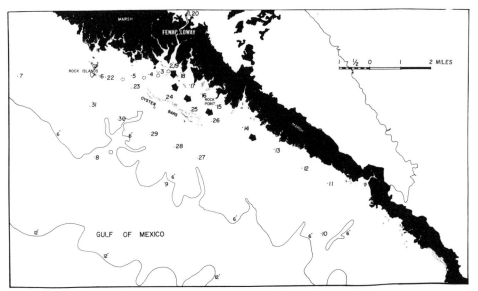

Fig. 1. **Gulf of Mexico at the mouth of the Fenholloway River; solid arrows indicate predominant direction of water flow and open arrows show a minor flow pattern; numbers indicate sampling stations.**

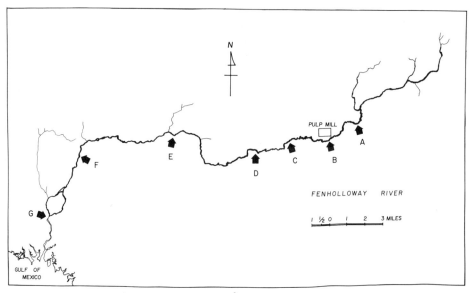

Fig. 2.   **Fenholloway River: location of pulp mill and sampling stations.**

401 Meter. Sulfide was determined using a specific ion electrode *(2);* estimates of sulfate-reducing and photosynthetic sulfide-oxidizing bacteria were made using a modification of the MPN technique using Postgate's medium *(9)* for sulfate reducers, and Larson's medium *(7)*, with a pH of 8.2 instead of 7.3, for sulfide oxidizers. Salinity was measured using an American Optical refractometer, model 10402.

The Gulf of Mexico in the vicinity of the Fenholloway River was surveyed twice in March 1971. Stations 1 to 20 (Fig. 1) were sampled 23 March 1971, and ATP content, total organic carbon, sulfide, and dissolved oxygen were measured. Stations 22 to 31 were sampled 5 March 1971 and were used for dissolved organic carbon (DOC) and isotope fractionation measurements. Only samples of surface water were collected on both days. The procedures for determination of these parameters have been published elsewhere *(3)*.

Total organic carbon was determined for samples collected on March 23 by using a Beckman Total Organic Carbon Analyzer. Analyses were made on unfiltered samples. Quadruplicate injections were made and the results averaged.

Measurements of inorganic carbon concentration (IOC) were made on samples collected in the river 2 mi above the mouth, at the mouth of the river, and on a transect running from the mouth of the river to the southeast (stations 2, 25, and 27), and on a transect running south from the mouth of the river (stations 2, 24, and 29). The method used in the determinations was that of Calder and Parker *(3)*.

The ATP content of the Gulf samples was assayed by (1) extracting the ATP from the sample, and (2) measuring the emission of light brought about by the reaction of ATP luciferin-luciferase according to the method of Strehler and Totter *(11)*. The assay procedure used in this work is summarized below.

Lyophilized extracts of firefly lanterns were obtained from Sigma Chemical Company (FLE-50) and were kept frozen until used. Each vial was reconstituted 24 hr before use and aged to reduce the background count rate of the preparation. Approximately forty assays could be performed per vial of FLE-50, and the enzyme preparation did not lose activity during the normal working day.

ATP standard curves were prepared for each enzyme batch. A standard solution, containing 2,000 mg/l ATP, was divided into 10 ml vials, sealed, and frozen until needed. One vial of the concentrated standard was thawed and diluted with Tris buffer (0.025 M, pH 7.75) to give final ATP concentrations of 20, 15, 10, 7, 5, 3, 2, 1, 0.7, 0.5, and 0.3 μg/l to prepare a standard curve. Each of these dilutions was assayed by adding 1 ml of the solution, 0.5 ml of Tris and 0.5 ml enzyme preparation to a scintillation vial and measuring the light emission. The resultant family of luminescent decay curves is depicted in Figure 3.

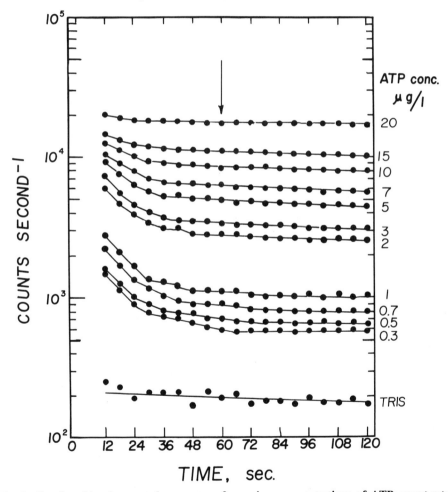

Fig. 3. **Family of luminescent decay curves for various concentrations of ATP; count rate at 60 sec (arrow) was used to prepare the standard curve given in Figure 4. (Initial rapid decrease in count rate results from phosphorescent decay of the caps of the liquid scintillation vials; the first 50 sec of data are routinely discarded.)**

A curve of best fit was drawn through the data points, and the count rate at 60 sec plotted against ATP concentration on a log-log scale to obtain the standard curve (Fig. 4).

Samples for ATP analysis were collected in 250 ml bottles held just below the water surface and capped with parafilm for return to the laboratory. A 200 ml aliquot was filtered through a 47 mm membrane filter (0.45 μm pore size). The filter was quickly plunged into screw cap tubes containing 10 ml of 0.025 M Tris buffer (pH 7.75) maintained at 100 C for 10 min in a 115 C oil bath. After ATP extraction, the samples were cooled in an ice bath, brought to a 10.0 ml volume, and assayed for ATP content.

The luminescent decay of the ATP-enzyme reaction mixture was measured using a Nuclear Chicago Model 4534 educational liquid scintillation counter interfaced with a Nuclear Data 110 multichannel analyzer. The scintillation counter was not equipped to run repeat counts and the multichannel analyzer was used to provide this function. The ND 110 was operated so that each of the

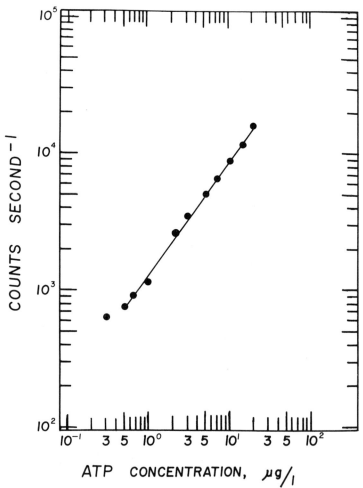

Fig. 4. Standard curve for ATP measurements (a curve must be prepared for each enzyme preparation.)

128 channels was open for data acquisition sequentially for a predetermined counting period over the range of 1 msec to 10 sec per channel.

The information was stored for display or print-out at the end of the decay period. The decay curves obtained using a 1 sec dwell period for ATP concentrations between 0.3 and 20 $\mu g/l$ are shown in Figure 3. The increased slope of the curves during the first 48 sec of the analysis is caused by the phosphorescent decay of the caps on the counting vials. All data taken during the first 50 sec of analysis were rejected.

Results and Discussion

Fenholloway River Survey. Results of chemical, bacterial, and physical determinations of Fenholloway River samples are summarized in Figures 5 and 6. The abscissa in each figure is the distance downstream from station A. The arrow at the 3 mi point indicates the location where the pulp mill effluent enters the river.

The phosphate concentration was highest at the point of effluent entry and decreased downstream with a slight increase at the lower end of the river (Fig. 5). A portion of the phosphate concentration may be due to the phosphoric acid added to the waste lagoon to accelerate biological degradation; however, the river flows through a limestone bed characterized by a high phosphate content.

The maximum sulfate concentrations were 60 ppm and 200 ppm in February and November, respectively (Fig. 5). The high sulfate burden is caused by the pulp-digesting chemicals present in the effluent. During both surveys total sulfide species reached a maximum of about 23 ppm, approximately 6 mi below the plant (Fig. 5).

At the point of effluent entry, the BOD reaches 100 ppm (Fig. 6). This gradually decreased as the waste flowed toward the Gulf of Mexico. Saville *(10)* suggested some waste degradation may occur and dilution may contribute to the decreases in the BOD. The DO measurements (Fig. 6) were made within 6 in of the water surface. At station B some dissolved oxygen was measured at the surface, but as the probe was lowered through the water column, no DO was detected.

Sulfate-reducing bacteria were detected above the mill, but their numbers increased by a thousandfold at station B (Fig. 6). The numbers remain relatively constant throughout the course of flow of the river to the Gulf. The photosynthetic sulfide-oxidizing bacteria were not detected above the plant, but gradually increased in number as the river flowed to the Gulf (Fig. 6).

The mill effluent is heated and causes an elevation in the river temperature (Fig. 5). Using the temperatures of the heated effluent, the river above the plant, and the combined flow below the mill, we estimated the waste volume contributed by the pulp mill accounted for 62% of the river flow in February 1970. Measurements in November indicated an even higher percentage, though runoff downstream may help to dilute the waste *(10)*. The pH of the river also changes at the point of effluent entry (Fig. 5). It is clear from our measurements of these physical and chemical parameters that the Fenholloway River takes on the characteristics of the mill waste.

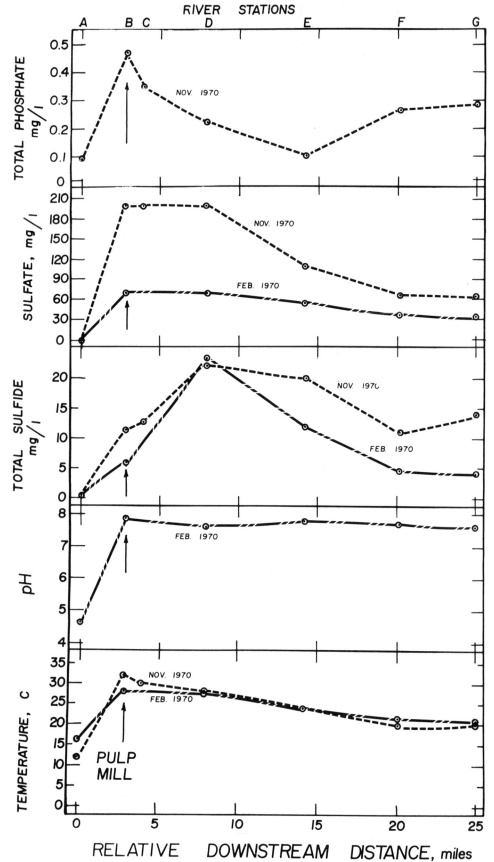

Fig. 5. Selected chemical and physical parameters measured in Fenholloway River in two separate surveys. (Sampling stations shown in Figure 2.)

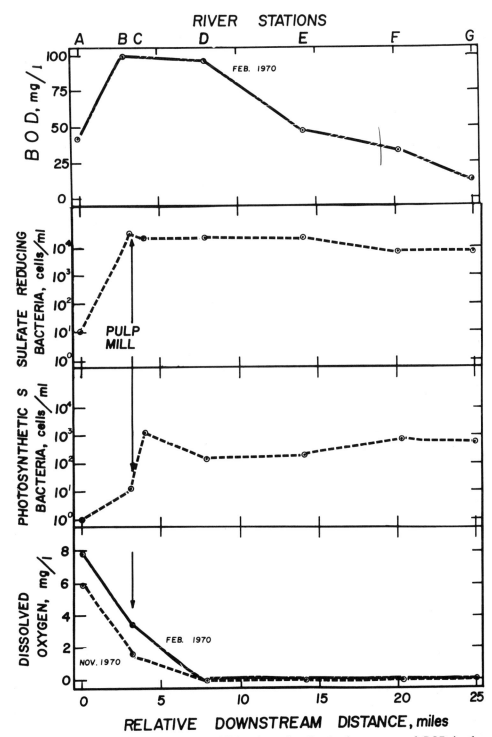

Fig. 6. Sulfate-reducing and sulfide-oxidizing bacteria, dissolved oxygen, and BOD in the Fenholloway River.

Gulf Water Survey. The Gulf of Mexico in the area receiving the flow of the Fenholloway was surveyed on 23 March 1971, and the data are presented in Table 1. The sampling stations shown in Figure 1 are numbered in a counter-clockwise fashion from the mouth of the Fenholloway River. Samples from stations 1, 2, 3, 4, and 5 were collected within an hour before high tide. All other stations were collected on an outgoing tide. At the time of the survey, the flow of Fenholloway waters was to the southeast (solid arrows).

The total organic carbon concentration (Table 1) was greatest in the river (station 1, 20, and 21) and decreased from the river mouth into the Gulf (stations 2 and 19). The carbon concentration decreased to the west (stations 3, 4, and 5), but was still higher than the offshore samples (stations 6, 7, 8, and 9). A steady decrease in organic carbon concentration was observed from the mouth of the river to the southeast (stations 19 to 11). This decrease could be due to dilution by mixing or to an oxidative process. Although stations 2 and 19 are in the same location at the mouth of the river, the difference in organic carbon concentration is a result of tidal differences. On an incoming tide, the flow in the river is composed of a downstream velocity component in the upper 1 to 2 ft and an upstream component between 2 ft and the bottom. An incoming tide promotes mixing and dilution of the mill wastes with the Gulf waters lower in organic carbon. The high total organic carbon values at stations 11 to 15 may be due to the presence of wood fibers or other refractory elements. This con-

Table 1. Selected chemical parameters and ATP concentrations for surface water samples collected 3/23/71

Station	Particulate and Dissolved Organic C mg/l	ATP, mμg/l	S^{-2} mg/l	Salinity	DO, mg/l
1	64	1,080	6.8	–	-0-
2	28	760	0.11	12.3	1.66
3	14	700	<0.1	18.1	3.71
4	11	740	<0.1	19.2	4.20
5	8	508	<0.1	20.4	4.77
6	6	620	<0.1	22.4	5.83
7	5	506	–	24.6	5.84
8	5	440	–	25.4	5.69
9	5	480	–	23.8	5.76
10	10	460	–	19.2	5.95
11	15	480	–	15.0	5.52
12	15	500	–	16.0	5.64
13	15	740	–	15.4	5.41
14	16	710	–	16.6	5.29
15	18	720	<0.1	14.4	3.87–4.22
16	25	2,000	<0.1	11.8	3.16
17	27	1,500	0.6	10.6	1.73
18	39	1,000	2.8	5.0	1.41
19	54	460	7.0	0.7	1.43
20	58	420	10.0	-0-	1.40
21	56	440	10.0	-0-	-0-

clusion is supported by the presence of a fibrous mat of pulp mill debris in the sediments.

The ATP values (Table 1) reflect the flow conditions prevailing in the Gulf on the day of sampling. At the mouth of the Fenholloway (station 19), the ATP concentration is low. ATP values for samples collected to the southeast, the direction of major river flow, increase markedly (1.0, 1.5, and 2.0 μg/l at stations 18, 17, and 16, respectively) for the first mile and then decrease for the next 4 mi. The ATP concentration 5 mi southeast of the river mouth (station 11) was comparable to those collected several miles offshore. The stations south of the oyster bar and several miles offshore receive the least amount of river flow, and these stations had the lowest and most consistent ATP values. Only a small part of the river flow moves west from the mouth and the ATP concentrations decrease rapidly in this direction.

Our analysis of the ATP data indicated a rapid increase in the total microbial biomass within 1 mi of the Fenholloway River mouth. This increase occurred in the direction of flow of the river water in the Gulf, with minor increases in the direction of minor flow. There was no increase in biomass south of the oyster bar. This illustrates the confinement of pulp mill wastes to a small inshore area. Slight mixing of river and Gulf waters at the river mouth created aerobic conditions as illustrated by the increase in dissolved oxygen and the decrease in sulfide (Table 1). The increase in biomass corresponded to these changes and to the presence of the relatively concentrated pulp wastes.

Results of geochemical studies supported the conclusions drawn from the ATP analyses. The values obtained for DOC are shown in Figure 7. The DOC

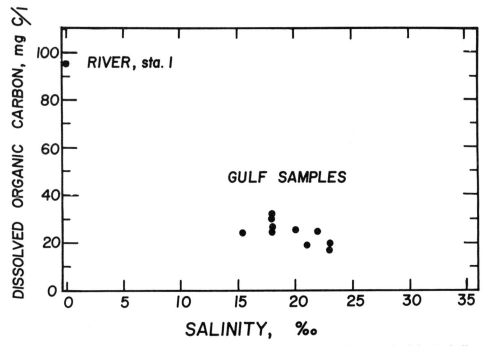

Fig. 7. **Dissolved organic carbon concentration in samples collected at mouth of the Fenholloway River along the salinity gradient into the Gulf of Mexico.**

concentration decreases as salinity increases from the river into the Gulf of Mexico. This may be a result of the waste mixing with seawater, or biological and/or chemical oxidation of the DOC.

To distinguish between the above processes, inorganic carbon (IOC) was studied. The data (Fig. 8) indicated a rapid production of IOC as one moves from the river to the Gulf, with a maximum production of IOC occurring at the mouth of the river (station 2) or to the southeast, in front of the oyster bars (station 25). The large input of CO_2 into the IOC pool can not be due to mixing but must result from an oxidative process. Therefore, the removal of DOC noted above is due in part to oxidation. The locations at which the maxi-

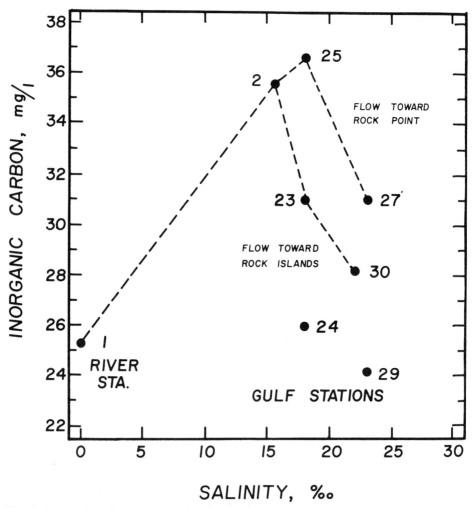

Fig. 8. Inorganic carbon concentration in samples collected at mouth of Fenholloway River along the salinity gradient into the Gulf of Mexico. Stations 2, 23, and 30 were sampled when the water current was to the west in the direction of minor flow; stations 25 and 27 were sampled when the water flow was to the southeast (Fig. 1). Data for stations 1, 24, and 29 did not vary appreciably with current movement.

mum input of IOC was observed (stations 2 and 25) correspond to locations between which the maximum ATP concentrations were found. This finding suggests the IOC production is caused by microbial oxidation of the DOC or pulp mill waste.

Isotope fractionation studies on Gulf and river water samples were performed and the results are illustrated in Figure 9. The value "$\delta^{13}C$" indicates the relative proportions of C^{12} to C^{13} in a sample and is calculated according to the formulation *(8)*:

$$\delta^{13}C = \frac{(^{13}C/^{12}C)\ \text{Sample} - (^{13}C/^{12}C)\ \text{Std.}}{(^{13}C/^{12}C)\ \text{Std.}} \times 1000$$

Bacterial oxidation can cause the $\delta^{13}C$ value of an organic substrate to vary by preferential consumption of the C^{12} isotope, thereby enriching the unoxidized material in C^{13}. The normal $\delta^{13}C$ of marine samples is approximately −20. More positive values indicate that the change in DOC concentration observed with salinity cannot be just a dilution factor but must be the result of a fractionation process, such as microbial growth leading to preferential consumption of the C^{12} isotope.

Assimilation of pulp mill wastes is confined to the area within 1 mi of the

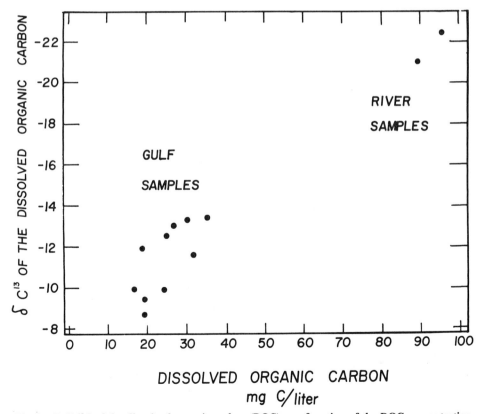

Fig. 9. Dell ^{13}C of the dissolved organic carbon (DOC) as a function of the DOC concentration.

mouth of the Fenholloway River near the oyster bars and is affected by prevailing wind conditions. Because of lack of mixing and the rapid increase in biomass, serious thought must be given to the flow patterns and the nature of the waste before the marine environment can be used as a dumping ground for inadequately treated wastes. The good agreement between the results of ATP analyses and the geochemical studies suggests these methods are well suited for quantitative and kinetic studies of estuarine pollution problems.

Literature Cited

1. American Public Health Ass. Inc. 1955. *Standard methods for the examination of water, sewage and industrial wastes*, 10th ed. Albany, N.Y.: Boyd Printing Co. Inc.
2. Anonymous. 1969. Determination of total sulfide. In *Water Applications Bulletin #12*. Cambridge, Massachusetts: The Orion Corporation.
3. Calder, J. A., and P. L. Parker. 1968. Stable carbon isotope ratios as indices of petrochemical pollution of aquatic systems. *Environ. Sci. Technol.* 2:535–39.
4. Galtsoff, P. S., W. A. Chipman, Jr., J. B. Engle, and H. N. Calderwood. 1947. Ecological and physiological studies of the effect of sulfate pulp mill wastes on oysters in the York River, Virginia. *Fishery Bulletin #43 of the Fish and Wildlife Service.* 51:59–186.
5. Gehm, H. W. 1953. Pulp, paper and paperboard. In *Industrial wastes*, ed. W. Rudolfs, pp. 194–231. New York: Reinhold Publishing Co.
6. LaRock, P. A. 1968. The bacterial oxidation of manganese in a fresh water lake. Ph.D. dissertation, Rensselaer Polytechnic Institute.
7. Larsen, H. 1952. On the culture and general physiology of green sulfur bacteria. *J. Bacteriol.* 64:187–96.
8. Parker, P. L., and J. A. Calder. 1970. Stable carbon isotope ratio variations in biological systems. In *Organic matter in natural waters*, ed. D. W. Hood, pp. 107–127. Institute of Marine Science, Occasional Publication No. 1. College, Alaska.
9. Postgate, J. R. 1966. Media for sulphur bacteria. *Lab. Pract.* 15:1239–44.
10. Saville, T. 1966. *A study of estuarine pollution problems on a small polluted estuary in Florida.* Bulletin Series 125 of the Florida Engineering and Industrial Experiment Station. Gainesville.
11. Strehler, B. L., and J. R. Totter. 1952. Firefly luminescence in the study of energy transfer mechanism. I. Substrate and enzyme determination. *Arch. Biochem. Biophys.* 40:28–41.
12. Welch, P. S. 1948. Limnological methods. New York: McGraw-Hill.
13. Woelke, C. E. 1959. Preliminary report on laboratory studies on the relationship between fresh sulphite waste liquor and the reproductive cycle of the Olympia oyster, *Ostrea lurida*. Washington State Department of Fisheries, Ms.

Comments

ALEXANDER: Was your data on surface water? Or, were profiles made?

QUAMMEN: This was just surface water data. The depth was 1 ft at low tide and about 4 ft at high tide. It is rather hard to do a profile.

ALEXANDER: Are there any data on iron? Are any of your colleagues measuring iron?

QUAMMEN: No, we are not.

ALEXANDER: It would be interesting to look at that. I think you would see a relationship between that and your phosphate, particularly where it is dumped. I think you would also find a relationship where the organic carbon decreased at about 19–20 ppt. It would have tied in very nicely. We saw the same effect in the Turner River in the Everglades.

ZOBELL: What remedial or corrective measures are being taken by the company?

QUAMMEN: Right now, none that I know of. Perhaps Dr. LaRock might have something more to add.

LAROCK: The company saw "the handwriting on the wall" and started adding treatment facilities. Settling bases and additional aeration equipment in the oxidation ponds have been added.

Methods and Techniques for the Isolation and Testing of Clostridia from the Estuarine Environment[1]

Jack R. Matches and John Liston

Bacteria in the marine and estuarine environments have been studied for a number of years. However, the sporing anaerobes in these environments received very little attention prior to the past eight or ten years. Nevertheless, marine sediments, as well as the intestinal contents of fish, shellfish, and other marine animals, have been known to contain both obligate and facultative anaerobes in large numbers. The early work on anaerobes in the sea is quite limited, and work in this general area prior to 1945 was reviewed by ZoBell (14), while more recent studies were reported by Kriss (5) and Oppenheimer (8). In recent years, most studies have been concerned with the incidence and characteristics of *Clostridium botulinum* type E, and an overall view of the research effort around the world has been presented at the *symposium on botulism* held in Moscow, U. S. S. R. (4).

The presence of anaerobic bacteria on fish, and particularly in fish intestine, has been noted by a number of investigators. Early reports include observations of anaerobes in herring and zooplankton by Obst (7), in salmon by Fellers (3), in haddock by Reed and Spence (9), and in fresh and decomposing fish by Schönberg (10). Shewan (11) reported the occurrence of specific clostridial species in haddock intestine. Shewan (12) has also summarized work by other European and Russian workers, indicating the occurrence of additional species in fish intestine. A few reports of the incidence of species in the sea have been published by Bonde (1), Smith (13), and Davies (2).

Obligate anaerobes important as pathogens causing disease or intoxication in man have been found in sediments and in marine animals. Such bacteria have been isolated from the alimentary canal of fish, with the highest numbers being found in feeding fish. The length of time these bacteria are retained in the stomach and gut of fish and their survival time are not well known. Fish feeding in a polluted area can pick up contamination on the surface of their gills or skin, as well as in the intestine. The transport of such bacteria by fish migrating into and out of polluted areas is poorly understood, and little is known of their transfer to man through the food chain or by other means. It is known that potentially pathogenic bacteria such as *Cl. perfringens* can gain entrance to marine and estuarine areas from sewage and land runoff and can contaminate fish used as food by man.

[1] Contribution No. 351, College of Fisheries, University of Washington, Seattle, Washington 98195.

345

Sampling

Our sampling area is Puget Sound (Fig. 1), an inland sea or estuarine area over 100 miles long, located in the state of Washington and bordered by a number of cities and towns. We are studying this environment to determine the total num-

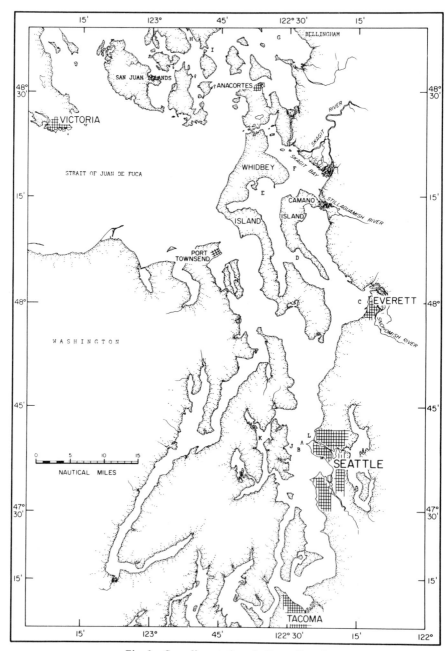

Fig. 1. **Sampling stations in Puget Sound.**

bers and species of *Clostridium* present. In addition, we are studying the growth capabilities and public health aspects of some of the Clostridia isolates.

Sampling stations used in these investigations were established as part of a regular sampling program within the College of Fisheries and were selected to represent different hydrographical conditions, depths, and proximity to populated and potentially polluted areas. The stations lie on a track line from West Point at Seattle (A and B, Fig. 1), north to Bellingham (G), with a loop into the San Juan Islands. A second track line of stations was established across Puget Sound to encompass hydrographic conditions ranging from deep marine to shallow freshwater areas. These areas are west, across Puget Sound from West Point (A and B) to Murden Cove (J), at Port Orchard (K), and also up the Lake Washington Ship Canal (L). Conditions in the sampling area are shown in Table 1.

Stations A and B at West Point are off a sewer outfall receiving between 1.5

Table 1. Puget Sound and other sampling stations

Station	Location	Depth (meters)	Temp. Range (C)	Type Bottom	Type Station
A	West Point	30	9–12	Sand-mud	Shallow, polluted
B	West Point	240	9–12	Brown mud	Deep marine, depositional channel
C	Everett Bay	100	10–13	Brown mud	Deep marine, polluted, depositional channel
D	N. Saratoga Pass	135	8–12	Brown mud	Deep marine, depositional channel
E	Penn Cove	63	7–12	Gray-green mud	Shallow marine embayment
F	Skagit Flats	25	8–12	Sand-gravel, plant debris	Estuarine, polluted depth
G	Bellingham Bay	27	9–14	Brown mud	Shallow marine, polluted, depositional embayment
H	East Sound	30	9–15	Gray-green mud, reduced H_2S	Shallow marine embayment
I	Lopez Island	40	9–14	Gray-green mud	Shallow marine embayment
J	Murden Cove	20–180	–	Sand-mud	Marine
K	Port Orchard	27	–	Brown mud	Marine
L	Lake Washington Ship Canal				
	Outer Buoy	13	wide	Sand	Estuarine
	R. R. Bridge	10	10–20	Sand-clay	Estuarine
	Above locks	10	9–21	Reduced mud	Fresh water
	Lake Union	8–10	9–21	Reduced mud	Fresh water

and 2.5 million gallons/day. Station J at Murden Cove is considered a clean area and was compared with stations A and B, considered polluted. These stations were sampled most often because of closeness to the University of Washington, requiring a sampling trip of one day or less. Stations in Northern Puget Sound required trips of 3 to 4 days' duration.

Samples of sediment were obtained at each of the stations by geological corer. Approximately equal amounts of sediment and water from the interface were removed and poured into sterile glass containers. On trips of several days' duration, the collected samples were run in the laboratory aboard the research vessel *M.V. Commando*, and on trips of one day's duration the samples were held refrigerated until return to the laboratory.

Isolation and Enumeration

The procedures used for isolation of organisms were either direct plating or an enrichment technique. In direct plating, 0.1 ml of sediment-water slurries were spread on the surface of either freshly poured or pre-reduced blood agar and incubated anaerobically. With the enrichment procedure, 1.0 ml of sediment-water slurry was inoculated into tubes of deoxygenated cooked meat medium and incubated. Fat from the fresh ground beef used in this medium solidified on the surface, forming an oxygen barrier. After turbidity and/or gas was evident, a drop of the broth from the medium was spread on the surface of blood agar. Colonies were picked from blood agar and restreaked on agar until pure. After purification, the colony and cellular morphology were recorded, and the organisms were subjected to a series of biochemical tests.

Quantitative counts were obtained by the direct plating method from blood agar plates. Duplicate plates were incubated both aerobically and anaerobically. The numbers of aerobic and facultative anaerobic organisms from 82 sediment samples are shown in Table 2. The numbers of aerobes and facultative anaerobes are surprisingly constant, with the average counts throughout the sampling stations and over a period of several years varying only about one log. Of more importance to this study are the obligate and facultative anaerobic counts obtained from 111 sediment samples (Table 3). Samples were collected from sediments over a period of three years, and during this time no pattern of seasonal variation could be detected. The numbers of organisms growing anaerobically were surprisingly constant in samples taken from Puget Sound sampling stations. Exceptions to this were the counts obtained at Bellingham Bay during the late summer and fall months, and the station in the Lake Washington Ship Canal above the locks. During late summer and fall, large quantities of waste water from fish, fruit, and vegetable processing plants are dumped into Bellingham Bay, carrying organic matter and soil washed from the produce, which may influence counts. The other area in the ship canal is a freshwater area, with an intrusion of saltwater.

The anaerobic count data are a composite of both obligate and facultative anaerobes. From these data, the numbers of obligate anaerobes in the population cannot be determined. To obtain numbers of obligate anaerobes, colonies were isolated by one of two methods from blood agar plates incubated anaerobi-

Table 2. Counts of aerobic and facultative anaerobic bacteria isolated from sediments

Station	Average Count $\times 10^4$	Range $\times 10^4$
West Point A	5.29	.52–51.0
West Point B	3.72	.45–9.5
Everett Bay C	10.97	.52–41.5
N. Saratoga D	4.05	.40–12.5
Penn Cove E	1.58	.65–1.09
Skagit Flats F	5.1	2.3–7.0
Bellingham Bay G	16.9	1.1–60.5
East Sound H	1.51	1.04–2.1
Lopez Sound I	1.21	1.1–1.3
Murden Cove J	1.32	.39–2.75
Port Orchard K	3.42	.217–6.3
Ship Canal L		
Outer buoy	1.96	.08–1.4
RR bridge	13.47	.104–50.0
Above locks	6.01	4.45–9.0
Lake Union	11.63	.25–23.0
Average	5.5	

Table 3. Counts of obligate and facultative anaerobic bacteria isolated from sediments

Station	Average Count $\times 10^4$	Range $\times 10^4$
West Point A	1.51	.05–6.3
West Point B	2.05	.08–5.9
Everett Bay C	4.1	1.4–9.7
N. Saratoga D	4.2	.42–11.7
Penn Cove E	2.5	.49–6.3
Skagit Flats F	3.3	1.9–4.4
Bellingham Bay G	23.49	2.3–103
East Sound H	4.29	.72–7.5
Lopez Sound I	1.9	1.1–2.7
Murden Cove J	.73	.21–1.05
Port Orchard K	3.56	.19–7.7
Ship Canal L		
Outer buoy	1.27	.13–3.5
RR bridge	1.36	.11–4.4
Above locks	18.2	8.4–28.0
Lake Union	4.5	.118–8.9
Average	5.12	

cally. Plates containing less than 100 well-isolated colonies were selected and every colony was picked, or a method of randomly selecting a proportion of the colonies was used. The isolated organisms were purified and then tested for both aerobic and anaerobic growth. The proportion of obligate anaerobes among the isolates is shown in Table 4. These data were collected from a number of sampling stations and over a period of several years. A total of 1,015 isolates were tested and 305, or 30%, were obligate anaerobic Clostridia. The percentage of obligate anaerobes isolated from the various sampling stations varied between 10 and 62%. This variation is not unlikely, since the sediment types and depth vary between stations.

Table 4. Proportion of obligate anaerobes among the obligate and facultative anaerobes isolated from sediment

Station	Depth (meters)	Numbers Isolated	Number Obligate Anaerobes	Percentage Obligate Anaerobes
Murden Cove	20–180	311	87	28
Mid-Sound	245	110	44	40
Port Orchard	27	16	10	62
West Point	76–180	328	139	42
Other Puget Sound Areas	25–135	250	25	10
Total		1,015	305	30

Distribution

The distribution of Clostridia in Puget Sound is important, and these organisms have been found at every sampling station in this investigation. To determine the species of *Clostridium* in Puget Sound sediments, the isolates were tested morphologically and biochemically, and the results were compared with data for known species reported in the literature, as well as with data collected in our laboratory using identified species, many of which were type cultures. Isolates from different areas within Puget Sound and from sediments collected at different times of the year fell into thirteen groups (Table 5). Toxicity tests

Table 5. *Clostridium* species isolated from sediments in Puget Sound

Species	Numbers	Percentage
Cl. perfringens	91	37.3
Cl. bifermentans	52	21.3
Cl. novyi	42	17.2
Cl. botulinum or sporogenes	12	4.9
Cl. baroti	7	2.9
Cl. subterminali	4	1.6
Cl. tonkinensis	2	.8
Cl. cadaveris (capitovali)	2	.8
Cl. difficile	2	.8
Cl. fallax	1	.4
Cl. butyricum	1	.4
Cl. mangenoti	1	.4
Unidentified	27	11.1
Total	244	

were not run; therefore, *Cl. novyi* A and B were combined as *Cl. novyi,* and *Cl. botulinum* and *Cl. sporogenes* were combined as *Cl. botulinum* or *sporogenes.*

The first four species make up 81% of the organisms isolated. *Cl. perfringens* has been reported to be more widely distributed than any other pathogenic bacterium, and its presence is therefore not unusual. In these studies, the greatest number of *Cl. perfringens* in sediments was collected from areas receiving domestic sewage, even though the organism was distributed throughout all sampling stations.

The next three groups, *Cl. bifermentans, novyi,* and *botulinum* or *sporogenes,* are also widely distributed and have been reported in marine sediments. The remaining eight organisms were found in marine sediments in low numbers. Three of these, *subterminali, cadaveris,* and *butyricum,* have recently been isolated from marine sediments by other workers. The remaining five species, composed of thirteen strains, were all isolated from sediments collected near the sewage outfall diffusion pipe. Several of these species were isolated only after enrichment in cooked meat medium for up to forty-five days. The significance of the extended incubation is not too well known; however, one can speculate that organisms present in very low numbers may be able to increase to a detectable level with extended incubation. It has been noted in our laboratory, by visual observation of colony types, that a change in population occurred during the enrichment periods. The colony types appeared to become more uniform during later isolations than during early isolations, indicating that a change was taking place.

Public Health Aspects

These quantitative and qualitative studies, which are still under way, have given some insight into the populations of *Clostridium* in the environment. Since we are in the Institute for Food Science and Technology, we are also interested in these organisms from other points of view. Some of the Clostridia are of public health significance in foods, and especially seafoods. Although *Cl. perfringens* has not been implicated in food poisoning outbreaks involving seafood as often as some other foods, the incidence of *Cl. perfringens* in foods reported in the literature is increasing. *Cl. perfringens* is a common organism in the gut of man and other warm-blooded animals. Our data and those of others show the prevalence of this organism in sewage which is being dumped into the environment. The possible cyclic transfer of organisms such as *Cl. perfringens* from man via sewage to the environment, to animals in the environment, and eventually back to man is of interest. This is important when one considers that our population is increasing, especially near estuarine areas, and the numbers of sewage treatment plants dumping into water are also increasing. As the proposed harvest from the seas increases, it is not unlikely that fish feeding in polluted areas will be caught and used as food by man.

Cl. perfringens have been isolated from sediments and from fish stomach and gut contents (Table 6). The greatest numbers of isolates have been from sediments and fish near the sewage disposal plant at West Point, station A. Fewer isolations were made from the other sampling stations, and none was detected

off the coast of Washington on the continental shelf. We cannot conclude that
Cl. perfringens are not present at these stations, but we use these data as an indication of the prevalence of the organism.

Table 6. Distribution of *Clostridium perfringens* isolated from the marine
environment

Sampling Station	Number of Isolates	Number of *Cl. Perfringens* Types	Percentage *Cl. Perfringens* Types
Sediments			
West Point, Station A	68	12	18
All other stations	17	1	6
Stations off Washington coast	142	0	0
Fish stomach and gut contents			
West Point, Station A	127	91	72
All other stations	73	39	53
Stations off Washington coast	15	0	0

Table 6 gives the percentage of *perfringens* in the total flora from sediments
and from fish. The percentage of *perfringens* in sediment from West Point and
other sampling stations are 18 and 6, respectively. Fish caught from these same
areas had a *Clostridium* flora in the gut and stomach contents composed of 72
and 53% *perfringens*. We have been unable to explain why *perfringens* makes up a
larger proportion of the gut population than of the sediment population from
the same areas. One possible explanation is that organisms or material used as
feed by fish may harbor *perfringens* in large numbers as a result of the normal
sequence of events in the food chain.

The question of how *Cl. perfringens* in the gut contents of fish can reach man
has been raised. We feel that this can happen during fishing and processing.
Most of the bottom fish on the West Coast are caught by otter trawl. While the
trawl net is operating, the lead line is dragging on the bottom, which disturbs
the sediment. The fish lying or feeding on the bottom are captured and collected
at the rear, or cod end of the net. As the number of fish accumulate in the cod
end of the net, pressure is exerted by the increasing weight of the fish, and this
pressure is most severe on the first fish caught. At the termination of the drag,
the net is retrieved and lifted aboard the vessel, with the captured fish as a mass
in the cod end of the net. The catch is dumped on the deck for sorting and
then stored in holding pens below deck with alternate layers of ice and fish.
During these fishing and unloading operations, the fish are subjected to pressure
sufficient to expel feces from the gut, contaminating the surface of the fish. The
organisms on the surface of the fish from feces can then come into contact with
the flesh during butchering and eventually reach man through ingestion of
the seafood.

An integral part of our Clostridia study is the uptake and retention of *Cl.
perfringens* by feeding fish. The feeding and uptake in the natural environment

are very difficult to study; therefore, a model system was set up in an aquarium. A test animal, the staghorn sculpin *(Leptocotlus armatus)* was selected because this animal is found in the same environment as most of the species of fish collected and is an aggressive and eager feeder. Several methods of inoculating *Cl. perfringens* into the gut of the animal were considered, such as force feeding, anesthetizing and feeding, and injection with a syringe. These methods, although effective, place the fish under undue stress. To eliminate stress, a method of natural feeding was devised. Small (5–10 cm) Coho salmon *(Oncorhynchus kisutch)* were obtained from a local hatchery. Small volumes of bacterial culture, containing known numbers of cells of *Cl. perfringens*, were injected into the gut area of the dead salmon. The inoculated salmon were dropped into an aquarium and the sculpins reacted by rapidly swimming to the surface to consume the inoculated bait. The fish were sacrificed after 1, 24, 96, 144, and 288 hr, the stomach and gut removed, and the numbers of *Cl. perfringens* determined (Table 7). In addition, total anaerobic and facultative anaerobic counts were made. The data show that after 24 hr the numbers of *perfringens* may have increased slightly. This is possible since the minimum growth temperatures of the organism are approximately 10–15 C, and the aquarium water temperature was 18.7 C. Beyond 24 hr, the counts decreased, due possibly to elimination and/or destruction of the bacteria. These studies will be repeated, using both vegetative cells and spores.

Table 7.   Bacteria recovered from staghorn sculpins

Time (hrs after feeding)	*Cl. Perfringens* Counts/g Gut and Contents	Anaerobes and Facultative Anaerobes Counts/g Gut and Contents
1	5.0×10^7	—
24	5.0×10^8	1.4×10^9
96	$<5.0 \times 10^3$	$<5.0 \times 10^3$
144	$<5.0 \times 10^1$	$<5.0 \times 10^1$
288	$<5.0 \times 10^1$	$<5.0 \times 10^1$

Control — 6.6×10^7/ml.

Our work has shown that *Cl. perfringens* is presented in the environment and in fish; likewise, the organism has been isolated from commercially prepared fillets. Man can consume foods contaminated with this ubiquitous organism, and from his excrement the organisms can be returned to the marine environment, completing the cyclic pattern of infection from the environment to man and eventually back to man again.

Low Temperature Studies

Clostridium species, normally considered to grow most rapidly at 30–37 C, have been isolated from the marine environment, areas which rarely warm to more than 5–15 C. Some of the isolates from sediment were tested for their

ability to grow at 5 and 10 C (Table 8). A total of 365 isolates from Puget Sound
and off the Washington coast were tested. Of this total, 288 were able to grow at
5 or 10 C. Of the 288, only 3 Puget Sound isolates were able to grow at 5 C during extended incubation. However, 285 (78%) were able to grow at 10 C.

Table 8. Ability of marine anaerobes* from sediment to grow at 5 and
10 C

| | No. Isolates Growing at | | Total No. |
Location	5 C	10 C	Tested
Puget Sound	3	150	224
Washington coast	0	135	141
Total	3	285	365

* All strains isolated initially at 30 C.

Little information is available in the literature on the low temperature growth
of *Clostridium*, except fairly recent studies with *Cl. botulinum*. Since these Clostridia were isolated from areas in the environment that rarely warm to more than
10–15 C, we were very interested in their ability to grow at these low temperatures. Measurements of minimum growth temperature were made, using a
temperature gradient incubator. In previous work with aerobes, we were able
to grow the organisms in small, short vials. However, with the anaerobes in the
vials, it was difficult to keep the medium reduced or deoxygenated during extended incubation, and the Clostridia grew very poorly. To solve this problem,
a new temperature gradient incubator was constructed (Figs. 2 and 3), using
16 × 120 mm screw cap tubes with a layer of oil over the medium.

The main part of the portable unit is a block of aluminum 13 × 36 × 5 in.
This unit has twelve rows of holes, with thirty-two holes per row. Twelve organisms can be placed on this unit, with thirty-two temperature variations for
each organism. One end of the block is cooled and the other end is warmed. A
constant temperature gradient is obtained over the block from one end to the
other, and the temperature can be adjusted for most temperature ranges. This
equipment has been used for minimum growth temperature determination,
growth temperature ranges, and the biochemical responses of the Clostridia
over temperature ranges.

The minimum growth temperature of 15 strains of *Cl. bifermentans* was tested
(Table 9). The first eleven organisms were obtained from Dr. B. Dowell at the
Communicable Disease Center, and the other four were strains isolated from
the marine environment. In these studies, we were interested in comparing
the range of minimum growth temperatures of strains of organisms from the
marine environment with strains from other sources. A three-degree range from
11.3 to 14.5 C was obtained for the fifteen strains tested during two experiments.
After incubation for seven days, the minimum growth temperature of many
of these strains from different sources was very similar. Further studies are
being carried out with strains of other *Clostridium* species and longer incubation
periods are being used.

The marine environment in Puget Sound is cool and contains a number of organisms that grow at low temperatures. In our study of *Clostridium* species tested for growth at 5 and 10 C, the data indicated that some organisms are capable of growth at temperatures commonly found in inshore waters. It was felt that in this environment organisms might be present which were both anaerobic and psychrophilic, and this was shown to be true. Psychrophilic Clostridia were isolated from sediments in Puget Sound *(6)*. The technique used to isolate these organisms was to inoculate samples of sediment slurry into prechilled media and to incubate at 0 to 5 C. After purification, organisms

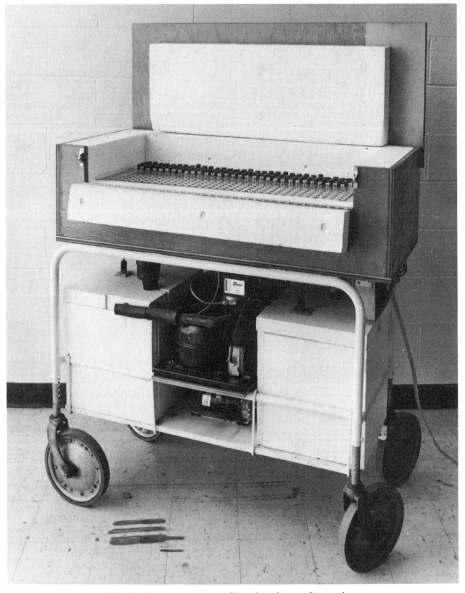

Fig. 2. **Temperature gradient incubator, front view.**

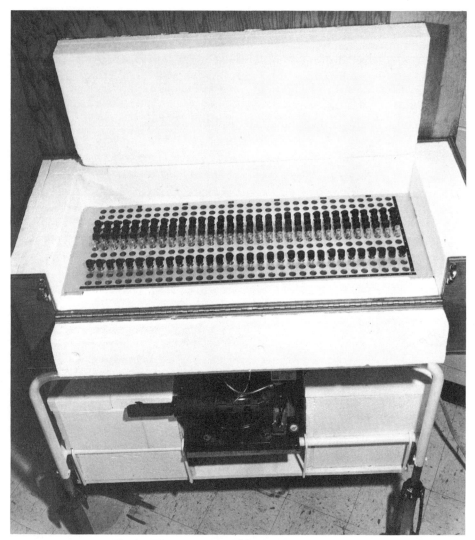

Fig. 3. **Temperature gradient incubator, top view.**

were tested for their ability to grow anaerobically over a temperature range, and those able to grow at 0 C but not at 30 C were selected as obligate psychrophilic anaerobes. Although the organisms have the general appearance of the Clostridia, spore production has not been demonstrated with all strains. A number of the isolates were obtained after heat treatment of sediments at 80 C for 10 min, which suggests that the organisms were originally present in the spore form.

Nine psychrophilic clostridial isolates were grown on the gradient block over a temperature range of −2 to 30 C (Table 10). In the table, the first column labeled "Growth" gives the temperature of the last tube to show growth, and

Table 9. Minimum growth temperatures of *Clostridium bifermentans*

Organism Number	Average Minimum Growth Temperature (C)
462	11.6
718A	11.6
4483	11.9
1668	11.9
5002	11.3
1628	12.6
409A	12.4
4511	11.7
1720A	12.5
4656	14.6
3023	11.3
75	12.2
1002	11.6
4	12.9
119	13.2

the next column labeled "No Growth" gives the temperature of the next warmest tube in which no growth was demonstrated. These data show that the maximum growth temperatures for organisms numbers 30 and 34 are 23.0 and 25.5 C, respectively, and the maximum growth temperatures for the remaining seven organisms are between 16 and 18 C. Two types of organisms are now distinguished as far as temperature is concerned: Type I organisms have a maximum growth temperature near 18 C, and Type II organisms with a maximum near 25 C. Both types show temperature optima near the midpoint of their growth range. The last three columns under "Growth Temperature" in Table 10 show some of these data. The minimum lag time is the temperature range over which

Table 10. Optimum and maximum growth temperatures (C) of anaerobic psychrophiles

Organism Number	Maximum Growth Temp.		Growth Temperature at		
	Growth	No Growth	Minimum Lag Time	Minimum Log Time	Maximum OD
1	17.5	18.6	6.5–13.9	10.4	10.4
2	18.3	19.5	11.9–14.2	10.7	10.7
5	17.0	18.2	8.7–9.8	9.8	9.8
19	16.3	17.5	10.6–14.0	8.3	11.7
30	23.0	24.3	12.7–17.5	11.6	11.6
34	25.5	26.5	not available	11.8	10.5
40	17.5	18.6	10.5–12.9	9.4	10.5
41	16.4	17.6	9.0–15.5	10.0	10–11.5
54	16.3	17.5	10.5–12.8	8.2	8.2

the delay between inoculation and the beginning of the logarithmic phase was shortest. The minimum log time is the optimum growth temperature. The maximum optical density (OD) is the temperature at which the highest population is achieved, as measured by OD readings.

It is difficult to explain the data obtained with some of these organisms. The temperature of most rapid growth is not necessarily the temperature at which the greatest population is obtained. In several cases the most rapid growth is obtained after a very long lag, but the greatest population is obtained at another higher or lower temperature, with a short lag. Examples of these are shown in Figs. 4 and 5 for Type I and Type II organisms respectively. The Type I organism number 5 shows most rapid growth at 3.9 C, but with the greatest population obtained at 9.8 C. The Type II strain number 30 shows most rapid growth at 19.8 C, with the greatest population obtained at 11.6 C.

One difficult feature of defining growth rates of psychrophiles is the long and variable lag which they show at different temperatures. These anaerobic psychrophiles were isolated at temperatures in the range of 0–5 C. The minimum growth temperatures of these organisms have not been determined because of the difficulty of working at the very low temperatures and the long incubation times required. We do feel, however, that these organisms may have minimum

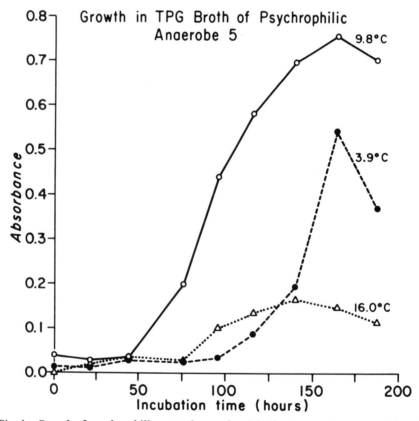

Fig. 4.   Growth of psychrophilic anaerobe number 5 in the temperature gradient incubator.

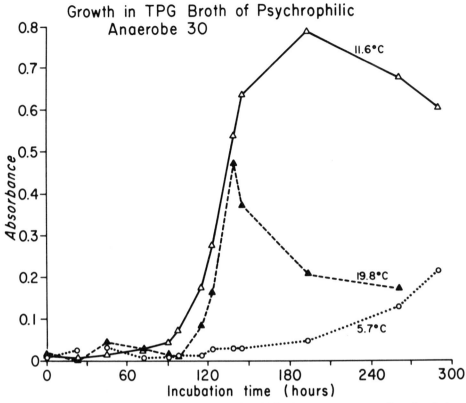

Fig. 5. **Growth of psychrophilic anaerobe number 30 in the temperature gradient incubator.**

growth temperatures much lower than we have measured. It is possible that at very low temperatures their lag periods are long and only several generations a year may be obtained. This can also occur in the case of mesophiles. If mesophilic Clostridia can grow very slowly at the temperatures prevailing in the marine environment, especially in the deeper waters, their presence and function in sediments would be easier to explain.

Pressure Studies

One other area we have investigated with the Clostridia is growth under pressure. *Cl. botulinum*, types A and E, were grown in pressure bombs for one week at room temperature, at pressures between 2,000 and 7,000 psi (Table 11). After incubation, the growth of the organisms was detected by visual observation of turbidity, as well as by increases in numbers of cells per ml. Visible growth was evident at pressures of 2,000 to 4,000 psi with both organisms. At pressures of 2,000 and 3,000 psi, increases in numbers of cells were obtained after seven days' incubation. At 4,000 psi the average number of cells recovered after seven days' incubation was only slightly higher than the inoculum in several cases and slightly lower in other cases. We have not repeated these studies as yet to determine if the maximum numbers of cells are obtained at some times earlier than seven days, with a reduction by the seventh day. This would, however, explain why turbidity was evident with only 10^4 and 10^5 cells per ml. Above 4,000 psi, the numbers of cells decreased by the end of the test period.

Table 11. Growth of *Clostridium botulinum* in pressure bombs at room temperature for 7 days

Initial Pressure (lb.)	Final Pressure (lb.)	Type A				Type E			
		Visible Growth		Cells/ml		Visible Growth		Cells/ml	
		#1	#2	#1	#2	#1	#2	#1	#2
2,000	2,000	+	+	1.6×10^5		+	+	3.9×10^6	2.1×10^6
3,000	3,000	+	+	5.6×10^5	1.0×10^5	+	+	2.9×10^5	3.3×10^5
4,000	4,000	+	+	1.3×10^4	1.5×10^5	+	+	2.9×10^4	1.1×10^4
5,000	5,000	−	−	1.3×10^3	1.7×10^3	−	−		
6,000	6,000	−	−	1.3×10^3	2.3×10^3	−	−	4.3×10^3	
7,000	7,000	−	−	2.2×10^3	1.6×10^3	−	−	2.7×10^3	1.9×10^4

Control   A $- 4.4 \times 10^4$ E $- 2.7 \times 10^4$.

Summary

We have been studying *Clostridium* species isolated from estuarine and marine areas. These organisms are widely distributed in the environment, in animals, and in sediments. Many of these organisms are pathogenic to man and animals and may cause food infections and food intoxications. *Cl. perfringens*, a food poisoning organism, can be found in man. The organism can be isolated from human waste entering the environment, from fish, and from fish fillets. As such, these organisms are potential hazards in food fish harvested from areas receiving contamination from man. The presence and growth of some of the Clostridia in the environment can be explained by the isolation of psychrophilic Clostridia. It is not clear what the other Clostridia are doing in the environment, especially in areas colder than their recognized minimum growth temperatures, but these organisms may be chance contaminants coming in from sewage, land runoff, and other sources of contamination. The sources of these organisms are being studied at the present time. Although the anaerobic Clostridia in the marine and estuarine environments have not been studied as much as aerobic organisms, the body of data is slowly increasing, and much work remains to be done with this interesting and important group of organisms.

Acknowledgment

The research was supported by PHS grants 5R01 EF00882 and 5R01 FD00292–01. The technical assistance of Mary Lou Holman, Nancy Debaste, and Valerie Fletcher is greatly appreciated.

Literature Cited

1. Bonde, G. J. 1968. Studies on the dispersion and disappearance phenomena of enteric bacteria in the marine environment. *Rev. Int. Oceanogr. Med., Tome* IX.
2. Davies, June A. 1969. Isolation and identification of *Clostridia* from North Sea sediments. *J. Appl. Bacteriol.* **32:**164–69.
3. Fellers, C. R. 1926. Bacteriological investigation of raw salmon spoilage. *Univ. Washington Publ. Fish.* **1:**157–88.
4. Ingram, M., and J. Roberts, eds. 1966. *Symposium on Botulism. Proceedings of the 5th International Symposium on Food Microbiology* (Moscow, July 1966). London: Chapman and Hall Ltd.
5. Kriss, A. E. 1963. *Marine microbiology.* Edinburgh: Oliver and Boyd.
6. Liston, J., M. Holman, and J. Matches. 1969. Psychrophilic Clostridia from marine sediments. *Bacteriol. Proc.,* p. 35.
7. Obst, M. 1919. A bacteriological study of sardines. *J. Infec. Dis.,* **24:**158–69.
8. Oppenheimer, C. H. 1963. *Symposium on marine microbiology.* Springfield, Ill.: Charles C Thomas.
9. Reed, G., and M. Spence. 1929. The intestinal and slime flora of the haddock—A preliminary report. *Contrib. Can. Biol. Fish.,* New Series **4:**257–64.
10. Schonberg, F. 1930. Über die Fischfaulnis und ihre bakteriologische Diagnose. *Berlin Tierarzt. Wochensch.* **46:**429–35.
11. Shewan, J. M. 1938. The strict anaerobes in the slime and intestines of the haddock (*Gadus aeglefinus*). *J. Bacteriol.* **35:**397–407.
12. Shewan, J. M. 1961. The microbiology of sea-water fish. In *Fish as food,* ed. G. Borgstrom, pp. 487–560. New York: Academic Press Inc.
13. Smith L. D. S. 1968. The clostridial flora of marine sediments from a production and from a non-production area. *Can. J. Microbiol.* **14:**1301–1304.
14. ZoBell, C. E. 1946. *Marine microbiology.* Waltham, Mass.: Chronica Botania Co.

Comments

SMITH: Have you done any work with shellfish?

MATCHES: We have attempted to isolate *Clostridium* from a number of shellfish samples with very poor results.

SMITH: Have you tried any of the concentrated forms of shellfish?

MATCHES: No, we have not.

COLWELL: What is your estimate of the ratio of anaerobes to the total bacterial population?

MATCHES: The obligate and facultative? We had between 10^4 and 10^5 in sediments.

COLWELL: What fraction of the total population would the anaerobes represent?

MATCHES: I do not know. We have not thought of it that way. We find approximately the same number of organisms growing aerobically as we do anaerobically. In both cases, the number is 10^3 to 10^4. The Clostridia reflect 30% of total obligate anaerobes.

COLWELL: Do you think gas chromotograph techniques for the quick identification of bacteria would work well for estuarine anaerobes?

MATCHES: Apparently, it works very nicely. I hope to spend some time learning this technique. We are getting set up to do this. Apparently, one can identify 95% of the cultures or at least put them into groups. Some of these groups are composed of maybe a dozen organisms. But at least it gets them in the right ball park.

ZOBELL: You mentioned the long generation time. Your charts indicated to me that the longest time you have used was fifteen days. In similar surveys that we have made with the same type of equipment, the temperature gradient box, we never incubated less than sixty days before we ended the experiment. We found so many examples of reproduction after 30, 40, and 50 days.

MATCHES: I would guess that there are others that reproduce after six months. We would like to look into this.

ZOBELL: Was fifteen days the longest that you used?

MATCHES: That was approximately the longest in those studies. Fifteen days' incubation was long enough to show the differences between pure cultures of strains of one species which we were working with in these studies. In other studies with aerobes, we incubated for thirty days or longer.

Section Six
PHYTOPLANKTON

Phytoplankton Cycle of Goose Creek, New York

Joseph M. Cassin and John J. A. McLaughlin

Goose Creek, New York (41°03′09″N, 72°25′23″W), is a small embayment receiving tidal water through a channel from Southold Bay (Fig. 1). The creek area is approximately 0.10 sq mi. Part of the shoreline borders high marsh of *Spartina* sp. and *Distichilis* sp. The remainder borders private dwellings and woodland. Depth averages 1.4 m with a mean tidal variation of approximately 0.7 m. This embayment is typical of those small bays and inlets of Long Island which nurture economically important shellfish and finfish, and are now subject to increasing ecological stresses, such as dredging and landfill.

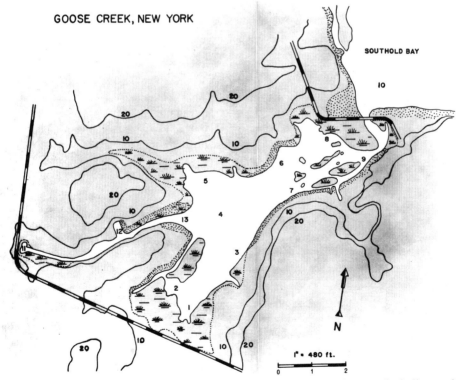

Fig. 1. Goose Creek, New York: stippled areas indicate sandy shore, hummocks indicate salt marsh, contour lines are in feet, and phytoplankton sampling stations were 2, 4, 6, 8, 10, 12, and 13.

Studies of physical, chemical, and biological parameters are not new to the waters of Long Island *(6, 29, 30)*, but little is known of the dynamics of these small embayments *(1)*. This paper reports phytoplankton methodology and dynamics from a synoptic study conducted at Goose Creek in 1966 and 1967. Traditional phytoplankton studies are limited to taxonomic considerations and seasonal distribution. This study has developed computer techniques permitting expansion of data to include major and minor variations which may exist within a given biotope, and to characterize dominant flora. Attempts were made to appraise sampling technology and such indices as frequency and diversity as evaluative tools in ecological management.

Materials and Methods

Thirteen stations were established for nutrient studies. Of these, seven were chosen for phytoplankton sampling based on water depth, current, and bottom sediment (Fig. 1). Stations 2, 4, 6, and 8 comprised Goose Creek proper; station 10 was in Southold Bay; stations 12 and 13 were in a canal dredged in the spring of 1965. Stations were sampled twice a week during the growing season and weekly during the remainder of the year. Samples were collected in 500 ml polyethylene bottles submerged to arm length. Tidal effects were minimized by sampling as close as possible to high tide. During the period from June 1966 to June 1967, a total of 378 samples were obtained and processed.

Samples were concentrated to a final volume of 10 ml by using continuous centrifugation *(2, 14)*. During times of high productivity, centrifugation was unnecessary. After microscopic examination to identify forms damaged by preservation, the concentrate was fixed in 3% neutral formalin. Counts were performed in a Palmer nannoplankton cell. Raw counts were adjusted to cells per liter using appropriate dilution and efficiency factors.

Numbers of zooplankton were recorded, with no attempt at speciation. Standing crop of phytoplankters was recorded and identified, as to genus or, when possible, as to species. Dimensions were measured by calibrated Whipple disc and volumes (in cubic microns) derived by assigning one of three geometrical forms (rectangle, cylinder, or sphere). Estimated biomass (mg/l) was derived from calculated cell volumes *(34, 35)*. Paasche *(22)* noted a 0.62 correlation between cell volume and productivity. The diversity index of Margalef *(18)*, which has had wide application in phytoplankton ecology, was employed in this study:

$$d = \frac{S-1}{\log_e N}$$

where S is the number of species, and N the total number of individuals.

The frequencies of major taxa were calculated for the creek, the bay, and the canal. Frequency distribution index is defined as the percentage of samples in which a particular taxon was present *(21)*.

Data Processing

Each taxon was assigned a four-digit identification number (Table 1). The first digit assigned the planktont taxonomic class; the second, a family; and the last

Table 1. Listing of phytoplankton coding data[a]

I. Bacillariophyceae (1100–2999)
  A. Centraceae:

Coscinodiscaceae	1100–1199
Actinodiscaceae	1200–1299
Rhizosoleniaceae	1300–1399
Chaetoceraceae	1400–1499
Biddulphiaceae	1500–1599

  B. Pinnatae:

Anaulaceae	1600–1699
Diatomaceae	1700–1799
Eutonaceae	1800–1899
Gomphonemaceae	1900–1999
Bacillariaceae	2000
Naviculaceae	2100–2199
Fragilariaceae	2300–2399
Nitzschiaceae	2400–2499
Cymbellaceae	2500–2599
Tabellariaceae	2600–2699
Meridionaceae	2700–2799
Achnanthaceae	2800–2899
Surirellaceae	2900–2999

II. Dinophyceae (3100–3199)

Prorocentraceae	3100–3199
Gymnodiniaceae	3200–3299
Glenodiniaceae	3300–3399
Goniaulaceae	3400–3499
Peridiniaceae	3500–3599
Dinophyciaceae	3600–3699
Polykrikaceae	3700–3799

III. Euglenophyceae (4100–4199)

Euglenaceae	4100–4199

IV. Chlorophyceae (5100–5399)

Polyblepharidaceae	5100–5199
Chlamydomonadaceae	5200–5299
Chlorococcaceae and Chlorellaceae	5300–5399

V. Myxophyceae	6000–6999
VI. Chrysophyceae	7100–7199
VIII. Zooplankton	9000
IX. Xanthophyceae	9100–9199
X. Silicoflagellates	9200–9299
XI. Mu-flagellates (unidentified ultra-plankton)	9300–9399

[a]The first digit assigns taxonomic class; the second, family; and the last two identify each taxon according to recorded sequences.

two identified each according to recorded sequence. Data were punched on standard IBM cards with an 80-column field. Field definition for raw data is listed in Table 2. The initial program provides standing crop of phytoplankton (cells/l), calculated biomass (mg/l), diversity index, and abundancy rank of individual phytoplanktons as percentage standing crop and biomass.

Table 2. Field definition for raw data processing

Column	Data	Example
1–6	Date	210666
7–8	Station	02
9–10	Sample	01
11–12	Depth	01
13–16	Time	0730
17–20	Taxon code	1601
21–28	Cells/l	00015210
29–34	Taxon volume	000246

Results

Continuous centrifugation proved a practical method for processing samples from small embayments. Net hauls, though valuable for qualitative work, did not retain "small forms" and were too easily clogged. Membrane filters (0.45 μm Millipore Corp.), while retaining most forms, damaged phytoflagellates. Clogging of the filter limited the sample volume.

The percentage phytoplankton recovered by continuous centrifugation was determined by efficiency tests based on chlorophyll *a (36, 37)*. The Kimball-Wood *(14)* and Foerst (Foerst Manufacturing Company, Chicago) apparatus yielded efficiencies of 85 and 90%, respectively. The recovery efficiency apparently was influenced by the type of organism present (Fig. 2). Seven unialgal cultures of liter volume were tested. *Eutreptia marina* showed the lowest percentage recovery in both apparati. The Foerst apparatus was most efficient in the recovery of diatoms. Damann *(7)* observed cell recoveries with the Foerst apparatus as 95% of the settling technique, but did not indicate whether all genera were removed with equal efficiency.

Annual Phytoplankton Cycle. The annual phytoplankton cycle of 1966–67 was triacmic with maxima occurring in spring, summer, and midwinter. Average standing crop for the period was 1.64×10^6 cells/l. Annual maxima (13×10^6 cells/l) occurred in June, while annual minimum of 8.56×10^4 cells/l occurred in October (Fig. 3). Biomass averaged 1.66 mg/l ranging from 12.4 mg/l in August to 0.19 mg/l in October (Fig. 4). Zooplankton usually increased following periods of phytoplankton maxima, and were particularly abundant during winter flowering to late spring (Fig. 5).

The phytoplankton community was monomictic during winter and early spring (Fig. 6). Diatoms, especially Centrales, dominated from October to March, while Chlorophyta dominated in May and June. Polymictic and pantomictic

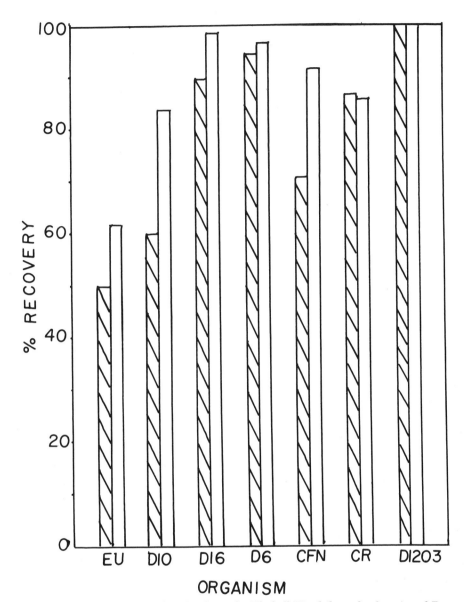

Fig. 2. **Differences of seven unialgal cultures by Kimball-Wood (barred column), and Foerster (open column) continuous centrifuges. Abbreviations: EU,** *Eutreptia marina***; D10, D16, D6, and CFN, pinnate diatoms; CR, unidentified chrysomonad; D1203,** *Gymnodinium* **sp.**

conditions prevailed during summer and fall. Dinophyceae were particularly abundant in late spring and summer. Cryptophyceae and Chrysophyceae were encountered mainly during spring and summer. Xanthophyceae were dominant flora for short periods during the summer. Unidentified μ-flagellates were confined to spring and summer months, and were minimal from October to March.

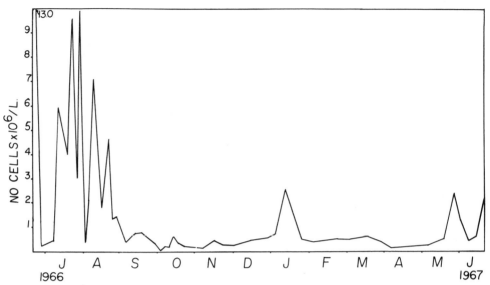

Fig. 3. **Annual phytoplankton cycle (1966–67) based on average of all stations sampled.**

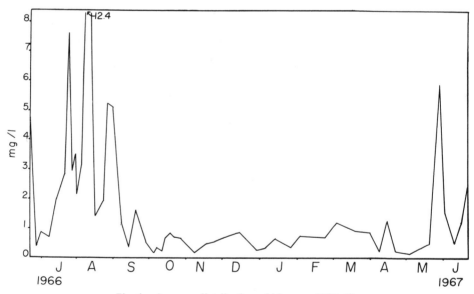

Fig. 4. **Average distribution of biomass 1966–67.**

Annual Cycle: Comparison of the Creek, Bay, and Canal. Definite variations were observed in the creek, bay, and canal. Average standing crop and biomass were highest in the canal, and lowest in the bay (Table 3). The summer creek maximum occurred in June and July with standing crop of 13.6 to 18.8×10^6 cells/l (Fig. 7). The summer maximum in the canal began in July and was of longer duration extending into August. Standing crop ranged from 10.8×10^6 to $33 \times$

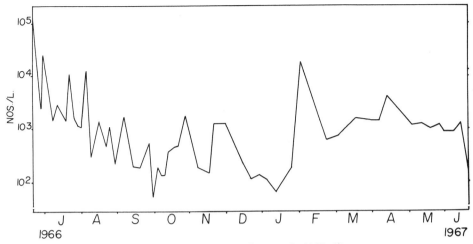

Fig. 5. **Annual zooplankton cycle 1966–67.**

10^6 cells/l. During this period biomass ranged from 13.3 to 14.3 mg/l (Fig. 8). The summer bay maxima occurred after those of the creek and the canal. Winter flowering was more pronounced in the bay (4.35 × 10^6 cells/l) and canal (3.0 × 10^6 cells/l) than the creek (1.56 × 10^6 cells/l). During postflowering period (February-March), the number of planktonts were at a minimum in the canal. Spring flowering was most prominent in the creek and the canal with standing crop of 2.3 and 3.23 × 10^6 cells/l and biomass of 2.2 and 20.5 mg/l. Spring and summer creek maxima were 2.5 and 1.9 × 10^5 cells/l, while winter maximum of 3.3 × 10^3 cells/l occurred in late March. Highest average standing crops occurred at stations 2 and 12 (Table 4). Highest average biomass and lowest diversity values occurred at stations 12 and 13.

A list of 175 species identified in the Goose Creek biotope, with the month of occurrence, and frequency distribution in the creek, bay, and canal, is given in the appendix. The monomictic condition which existed in the biotope from October through May was due to diatome dominance. The winter community was composed mainly of *Skeletonema costatum, Thalassiosira baltica, T. gravida, T. halina, Leptocylindrus* sp., *Rhizosolenia* sp., *Chaetoceros* sp., and *Asterionella japonica*. Diatom dominance was sporadic and short-lived in the canal. The bay and canal winter maxima were interrupted during January and February with the concomitant appearance of unidentified μ-flagellates. Dinophyceae and Chlorophyta were most abundant in the canal. Major dinoflagellates, *Cochlodinium catenatum* and *Gymnodinium* sp., were of the Gymnodiniaceae group. *Chlamydomonas* sp. and *Carteria* sp. dominated the Chlorophyta. Except for early spring 1966, few dinoflagellates or Chlorophyta appeared in the bay. Silicoflagellates, especially *Destephanus speculum*, and *Ebria tripartita* appeared most frequently in the bay and creek from January to June. *Chromulina pascheri, Crysocapsa pauldosa*, and *Pedinella* sp. were most frequent in the creek and canal. Cryptophyceae for the most part were confined to the creek. The Xanthophycean, *Olisthodiscus luteus*, was abundant in all areas in July and August, reaching bloom proportions in the canal.

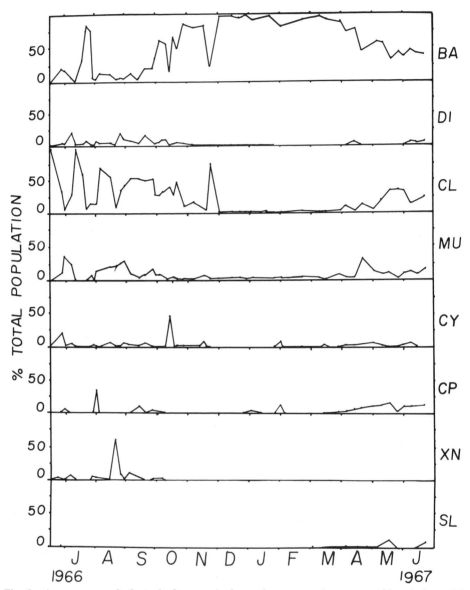

Fig. 6. **Average annual phytoplankton cycle for major taxonomic groups. Abbreviations: BA, Bacillariophyceae; DI, Dinophyceae; CL, Chlorophyta; MU, μ-flagellates; CY, Chrysophyceae; CP, Cryptophyceae; XN, Xanthophyceae; SL, Silicoflagellata.**

Discussion

The average standing crop of 1.64×10^6 cells/l for Goose Creek was lower than that reported by Smayda (*32*) and Pratt (*25, 26*) for Narragansett Bay, i.e., 6.7 and 5.54×10^6 cells/l, respectively. Conover (*6*) reported a mean standing crop for Long Island Sound (1952–53) as 2.38×10^6 cells/l. Bigelow et al. (*4*) gave a mean of 256×10^6 cells/l for the Gulf of Maine. The Goose Creek stand-

Table 3. Average standing crop, biomass, and diversity for creek, bay, and canal (1966–67)

Location	Standing Crop (cells/l)	Biomass (mg/l)	Diversity (bits/l)
Creek	1.46×10^6	1.67	1.82
Bay	1.17×10^6	0.90	1.78
Canal	2.43×10^6	2.10	1.30

ing crop was higher than that reported for Block Island Sound (29), and Vineyard Sound (16): 3.25×10^5 and 2×10^4 cells/l, respectively.

The phytoplankton cycle was unique in number and occurrence of maxima. Annual maxima occurred in the summer with minor peaks in January and May. A similar situation was reported for Great Pond, Massachusetts (12). Waters adjacent to Long Island are reported as having annual maxima in winter at times later than Goose Creek. Winter maxima of Block Island Sound (29) and Long Island Sound (6) occurred in February and March. April maxima were reported for Gulf of Maine (4, 8, 10), and Vineyard Sound (16). Fish (9) observed peaks in July and August, and an annual maximum in December for the waters adjacent to Woods Hole. In most cases, the annual minimum occurred in autumn.

Standing crop and biomass increased from the mouth of the embayment to the stations in the creek and the canal (Table 4). Sampling stations 2 and 12 had mean standing crops of 1.68 and 2.12 times greater than that of station 10. Other sampling stations, with the exception of 6, had values intermediate between those of Southold Bay and station 12.

The individuality of the sampling stations was revealed by the annual average data and dominant type flora. The average biomass for canal stations was higher and diversity indices lower, than any in the creek or the bay. A similar situation has been noted for Narragansett Bay (27) and Great Pond (12). Barlow et al. (3) attributed this to heavy "nutrient traps" in the embayment areas with the additional effect of little water exchange. Hair (11) has confirmed this for the canal area of Goose Creek.

Yearly productivity of the creek and canal were 97.1 and 158.0 g C/m² (11). Abundancy and frequency data attribute low diversity and high productivity in the canal to a dominant flora of *Eutreptia marina, Dunaliella* sp., Platymonads, Chlamydomonads, and Gymnodinioids. Ryther (31) and Lackey (15) noted that blooms in sections of Moriches and Great South Bay reflect high organic nutrients, temperature, and low salinity. Braarud (5) recorded Euglenophyceae and Dinoflagellata dominating waters of high organic content. Palmer (23) has shown Euglenophyta and Chlamydomonads to be among the most pollution-tolerant algae. Dinoflagellate blooms in Delaware Bay, New Jersey, were confined to shallow areas where the temperature reached 25 C. In the Goose Creek Canal area, temperature ranged from 20 to 25 C with salinity of 28‰ (11). Hulbert (13) attributed reduced diversity in estuaries to dominant forms which sink very slowly and are at an advantage over other species in monopolizing nutrient supply.

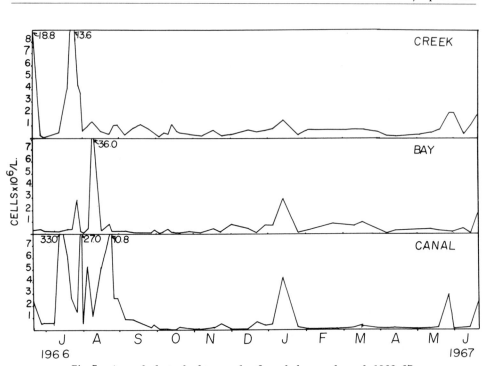

Fig. 7. **Annual phytoplankton cycle of creek, bay, and canal, 1966–67.**

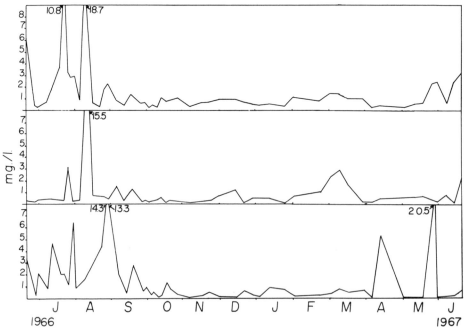

Fig. 8. **Annual distribution of biomass, 1966–67. (Upper, creek; middle, bay; lower, canal.)**

Table 4. Annual average phytoplankton data (1966–67)

Measurements	Station						
	2	4	6	8	10	12	13
Standing Crop (cells $\times$ 10⁶/l)	1.97	1.89	0.53	1.41	1.17	2.49	1.41
Biomass (mg/l)	1.00	1.22	1.79	1.22	0.93	2.42	2.24
Diversity Index (bits/l)	1.63	1.64	1.81	1.87	1.72	1.31	1.42

Correlation coefficients were calculated for numbers of zooplankton, phytoplankton standing crop, number of genera, biomass, and diversity with the following parameters: salinity, temperature, radiation, precipitation, wind, oxygen, phosphate (particulate, dissolved inorganic, and organic), nitrate, nitrite, silicate, and chlorophyll *a* (Table 5). Eight of the factors showed significant correlations.

Standing crop and biomass appeared to be dependent on temperature, radiation, dissolved inorganic phosphate, and silicate. The phosphorus cycle is typical of the North Atlantic region and is not considered to be a limiting factor in phytoplankton production *(1)*. Nitrate-nitrogen concentrations were minimal during summer (0.09 μg-at/l in the creek and 4.68 μg-at/l in the canal). Hair *(11)* suggests nitrate-nitrogen as a possible limiting factor. Correlation between standing crop and biomass reinforces the use of calculated volumes to estimate biomass. Zooplankton correlations with standing crop and biomass probably reflect influence of grazing on a seasonal cycle.

The diversity index adequately characterized individual stations within the Goose Creek biotope. Positive correlations were observed between biomass, genera, temperature, and phosphate. Steyaert *(33)* also indicated positive correlation of diversity with temperature, salinity, and primary nutrients.

Difference in standing crop, biomass, and diversity noted in all sectors of Goose Creek emphasizes the variety of microenvironments which exist within the estuarine biotope. Future studies should dissect these for presence of growth factors, preferential use of primary nutrients, trace element distribution, and exchange of growth factors between members of a plankton community *(17,28)*.

Summary

Computer technology was applied to systematize the processing of mass phytoplankton data. The phytoplankton dynamics of the Goose Creek biotope were found to be different from those of the adjoining waters of Long Island and Block Island sounds in standing crop and in time of plankton maxima. The annual phytoplankton cycle was triacmic, with three periods of phytoplankton abundance separated by a periods of phytoplankton paucity. Maxima occurred in the spring (May and June), summer (July and August), and winter (January and February). The spring maximum was characterized by unicellular, motile Chlorophyta, while dinoflagellates and *Olisthodiscus luteus* dominated the summer flora. *Skeletonema costatum* reached densities of 5.1 $\times$ 10⁵ cells/l during the winter maximum. The average phytoplankton standing crop (13 $\times$ 10⁶ cells/l)

Table 5. Correlation coefficients

	Standing Crop	Zooplankton	Biomass	Temperature	Radiation	Phosphate	Silicate	Chlorophyll a
Standing Crop		0.423	0.610	0.542	0.465		0.368	0.423
Biomass No.	0.610	0.389		0.482	0.389	0.373	0.350	0.250
Genera	0.361		0.549	0.451		0.506	0.329	
Diversity			0.391	0.422		0.521		

Validity 0.01 significance level. N = 53.

and biomass (12.4 mg/l) maxima occurred during the summer, while the minima occurred during autumn (8.56×10^5 cells/l; 0.18 mg/l, respectively). Phytoplankton distribution was monomictic during the winter, late spring, and summer. Diatoms dominated from October to March. Flagellates dominated from April to September.

Major differences in phytoplankton composition, abundance, and dynamics existed throughout the period of study in the creek, bay, and canal. The canal area had the highest average standing crop (2.44×10^6 cells/l) and biomass (2.10 mg/l). Standing crops and biomass increased with distance from the mouth of the Goose Creek embayment. Stations 2 and 12 had the highest average standing crops (1.97 and 2.49×10^6 cells/l, respectively), and biomass (2.00 and 2.42 mg/l, respectively). Chlorophyta and dinoflagellates were particularly indigenous to these areas during the spring and summer. Because of differences in standing crops, biomasses, and phytoplankton genera in various sectors, the biotope may be considered as a composite of microenvironments, each characterized by individual niche traits. The diversity index accompanied by concomitant biological data is a useful tool in predicting niche differences and characterizing succession of forms.

Acknowledgment

The authors wish to thank Dr. Malcolm E. Hair for permission to use Figure 1 and chemical data, and Dr. Patricia E. Cassin for helpful suggestions in the preparation of this manuscript. This work was supported in part by a grant from Arthur J. Schmitt Foundation to J. M. Cassin and Grant S-66 from Suffolk County, N.Y. to J. J. A. McLaughlin.

Literature Cited

1. Alexander, J. E. 1968. Water quality and growth of oysters, In *Proc. Conference Shellfish Culture*, pp. 33–37. Hauppauge, N.Y.: Regional Marine Resources Council.
2. American Public Health Association. 1965. *Standard methods for the examination of water and waste water.* 12th ed. Albany: Boyd Printing Co. Inc.
3. Barlow, J. P., C. J. Lorenzen, and R. T. Myren. 1963. Eutrophication of a tidal estuary. *Limnol. Oceanogr.* **8**:251–62.
4. Bigelow, H. B. 1926. Plankton of the off shore waters of the Gulf of Maine. *Bull. U.S. Bur. Fish.* (1924) **40**:1–500.
5. Braarud, T. 1945. A phytoplankton survey of the polluted waters of the inner Oslo Fjord. *Nordke Vindenskaps Akad. Oslo. Hualradets skrifter nr.* **28**:142.
6. Conover, S. 1952. Oceanography of Long Island Sound 1952–1954; Phytoplankton. *Bull. Bingham Oceanogr. Coll.* **15**:62–112.
7. Damann, K. E. 1950. A simplified plankton counting method. *Ill. Acad. Sci. Trans.* **43**:53–60.
8. Davidson, V. M., and A. G. Huntsman. 1926. The causation of diatom maxima: Passamaquoddy Bay, St. Croix River. *Trans. Roy. Soc. Can.* 5 (ser. 3) **20**:119–125.
9. Fish, C. J. 1962. Seasonal distribution of the plankton of the Woods Hole Region in relation to hydrography, temperature, density, and fishes. *Bull. U.S. Bur. Fish.* **41**:91–179.
10. Gran, H. H., and T. H. Braarud. 1935. A quantitative study of the phytoplankton

in the Bay of Fundy and the Gulf of Maine (including observations on hydrography, chemistry and turbidity). *J. Biol. Bd. Can.* **1**:279–467.

11. Hair, M. E. 1968. A study of the nutrient cycles in Goose Creek, New York 1966–67. Ph.D. dissertation, Fordham Univ.

12. Hulbert, E. M. 1956. The phytoplankton of Great Pond, Mass. *Biol. Bull.* **110**:157–68.

13. Hulbert, E. M. 1970. Competition for nutrients by marine phytoplankton in oceanic, coastal, and estuarine regions. *Ecology* **51**:475–84.

14. Kimball, J. F., and E. J. F. Wood. 1965. A simple centrifuge for phytoplankton studies. In *Contr. 000, Inst. Mar. Sci.*, Univ. of Miami, pp. 1–6.

15. Lackey, J. B. 1963. The microbiology of a Long Island Bay in summer of 1961. *Int. Rev. Ges. Hydro. Biol.* **48**:577–601.

16. Lillick, L. 1937. Seasonal studies of phytoplankton off Woods Hole. *Mass. Biol. Bull.* **73**:488–503.

17. Lucas, C. E. 1949. External metabolites and ecological adaptation. *Symp. Soc. Exp. Biol.* **3**:336–53.

18. Margalef, R. 1960. Temporal succession and spatial heterogeneity in phytoplankton. In *Perspectives in marine biology*, ed. A. A. Buzzati-Traverso, pp. 323–50. Berkeley: Univ. of California Press.

19. Martin, G. W. 1929. Dinoflagellates from marine and brackish waters of New Jersey. *Univ. Iowa Stud. Nat. Hist.* **12**:1–32.

20. Martin, G. W., and T. C. Nelson. 1929. Swarming of dinoflagellates in Delaware Bay, New Jersey. *Bot. Gaz.* **88**:218–24.

21. Odum, E. P., and H. T. Odum. 1959. *Fundamentals of ecology.* Philadelphia: W. B. Saunders Co.

22. Paasche, E. 1960. Phytoplankton distribution in the Norwegian Sea in June, 1954. Related to hydrography, and compared with primary production data. *Fisk. Skaaf.* **12**:5–51.

23. Palmer, C. M. 1969. A composite rating of algae tolerating organic pollution. *J. Phycol.* **5**:78–82.

24. Palmer, C. M., and T. E. Maloney. 1954. A new counting slide for nannoplankton. *Limnol. Oceanogr. Spec. pub. 21.*

25. Pratt, D. M. 1959. The phytoplankton of Narragansett Bay. *Limnol. Oceanogr.* **4**:425–40.

26. Pratt, D. M. 1965. The winter-spring diatom flowering in Narragansett Bay. *Limnol. Oceanogr.* **10**:173–84.

27. Pratt, D. M., and H. Berkowitz. 1958. Biological productivity project. *Ann. Prog. Rep., Narragansett Mar. Lab.*, pp. 1–8.

28. Provasoli, L., and I. J. Pintner. 1953. Ecological implications of *in vitro* nutritional requirements of algal flagellates. *Ann. N. Y. Acad. Sci.* **56**:839–51.

29. Riley, G. A. 1952. Phytoplankton of Block Island Sound, 1949. *Bull. Bingham Oceanogr. Coll.* **13**:40–64.

30. Riley, G. A. 1959. Oceanography of Long Island Sound, 1954–1955. *Bull. Bingham Oceanogr. Coll.* **17**:9–30.

31. Ryther, J. S. 1954. The ecology of phytoplankton blooms in Moriches Bay and Great South Bay, Long Island, New York: *Nannochloris, Stichococcus, Nitzschia. Biol. Bull.* **196**:198–209.

32. Smayda, T. J. 1957. Phytoplankton studies in lower Narragansett Bay. *Limnol. Oceanogr.* **2**:342–59.

33. Steyaert, M. 1966. Hydrologie et ecologie due phytoplankton de l'ocean. *Austral. Abstr. 2nd Int. Oceanogr. Cong.* (preprints). Moscow: Naura.

34. Strickland, J. D. H. 1966. Measuring the production of marine phytoplankton. *Bull. Fish. Res. Bd. Can.* **122**:172.

35. Willen, T. 1959. The phytoplankton of Gorvaln, a bay of Lake Malaren. *Oikos* **10**:241–74.

36. Wood, E. J. F. 1962. A method for phytoplankton study. *Limnol. Oceanogr.* **7**:32–35.

37. Wood, E. J. F., and E. F. Corcoran. 1966. Diurnal variation in phytoplankton. *Bull. Mar. Sci.* **16**:383–403.

Comments

HOLM-HANSEN: What is the advantage of centrifugation rather than the classical settling techniques?

CASSIN: First, I find it much more rapid. This is the real advantage of the system as it allows for immediate sample analysis. Second, sedimentation at times does not allow for recovery of many nannoplankton. Many organisms in this range remain suspended. We do apply an adjustment factor for the efficiency of the centrifuge when totaling the standing crop.

HOLM-HANSEN: Did you have any trouble resuspending the packed cells?

CASSIN: During times of high productivity, this can be a problem. For example, *Eutreptiella marina* will form a very lovely green pellet. The analyst must be aware of the sampling conditions and then choose the proper sampling method. During bloom conditions, raw counts should be done. At other times, centrifugation allows sufficient recovery of organisms from large volumes of water.

COLWELL: Have you tried to put bacteria into this program?

CASSIN: Studies on the distribution of coliform and heterotrophic bacteria were done by Dr. Robert Nuzzi of New York Ocean Sciences Laboratories, and Dr. J. J. A. McLaughlin. I would refer you to their dissertation abstracts as the most immediate source of this information. Some classical relationships between phytoplankton and bacterial total numbers and distribution were found.

COULL: The major criticism, I think, of Margalef's method has been that the number of species is affected by sample size. Did you notice a relationship between species diversity using this debatable index versus the total number of individuals in the sample? I would suspect that the index is greatly affected by sample size.

CASSIN: I would like to refer to the investigations of A. J. McErlean and J. A. Mihyrsky.* They were most interested in sample size factor and diversity in fish populations as related to thermal additions in the Patuxent River. A comparison of eight diversity indices led them to conclude that their values were independent of sample size. In using this particular index, it is the responsibility of the investigator to have strict control of sample size to assure maximal recovery. Proper sample size may be determined by chlorophyll analysis or counting genera recovered against sample volume. A point is reached at which no further recovery of phytoplanktonts occurs. We find that a volume between 500 to 1,000 ml gives us sufficient recovery.

WRIGHT: My question about diversity also relates to the importance of species. The Shannon-Weaver diversity index is much more sensitive to relative numbers of species or of individuals within species and also to the ecological importance of the species within the estuaries. I would suggest that if you use phytoplankton volume and plug this into your diversity index, the use of the Shannon-Weaver index would be more sensitive to the ecological importance of species rather than this index is to the occurrence of those numerous or rare forms.

CASSIN: I appreciate this suggestion. The use of abundance and frequency data in this study along with the diversity index did allow for assigning ecological importance to certain species.

* *Species diversity, species abundance of fish populations: an examination of various methods.* Contribution 368. Natural Resources Institute, University of Maryland.

Appendix

PHYTOPLANKTON DISTRIBUTION, GOOSE CREEK
1966-67

Taxa	Frequency Creek	Bay	Canal	J	F	M	A	M	J	J	A	S	O	N	D
I. Bacillariophyceae															
A. Centrales:															
1. Coscinodiscaceae:															
Coscinodiscus curvatulus Grunow	19.61	28.84	5.74	X	X	X	X	X		X	X	X	X	X	X
Coscinodiscus eccentricus Ehrenberg	17.24	23.07	8.52	X			X	X		X	X	X	X	X	
Coscinodiscus radiatus Ehrenberg			1.02					X							
Coscinodiscus sp.	2.41	9.61	2.01				X			X	X	X			
Cyclotella caspia Grunow	0.53		0.93							X	X				
Melosira nummuloides (Dillwyn) Agardh	20.43	23.07	70.25	X	X	X	X	X	X	X		X	X	X	X
Paralia sulcata (Ehrenberg) Gran	18.22	11.53	1.85	X	X	X		X		X	X	X	X	X	X
Skeletonema costatum (Greville) Cleve	77.78	11.58	70.25	X	X	X	X	X		X	X	X	X	X	X
Thalassiosira baltica (Grunow) Ostenfield	7.38	11.53	0.93	X	X	X	X	X		X	X	X			
Thalassiosira condensata Cleve		1.92	2.87							X	X	X			
Thalassiosira decipiens (Grunow) Jorgensen	0.51	1.92													X
Thalassiosira gravida Cleve	22.50	19.23	22.33	X	X	X	X	X	X						
Thalassiosira hyalina (Grunow) Gran	4.00	7.69	7.87	X	X	X									X
Thalassiosira nana Lohmann	3.80									X	X	X			
Thalassiosira nordenskioelrdii Cleve	22.50	19.23	22.33	X	X	X	X	X	X		X	X			X	X
2. Actinodiscaceae:															
Actinocyclus sp.	1.01									X	X				
3. Rhizosoleniaceae:															
Corethron criophilum Castracane	12.61	17.13	5.83							X	X	X	X		
Guinardia flaccida (Castracane) Peragallo	1.95		0.93	X	X						X	X			
Leptocylindrus danicus Cleve	8.77	1.92	2.87	X	X	X	X	X		X	X	X	X	X	
Leptocylindrus minimus Gran	3.84	3.84	1.02	X	X	X	X				X	X	X	X	
Rhizosolenia fragillissima Bergon	29.63	36.53	29.93	X	X	X	X	X							
Rhizosolenia hebeta form semispina (Hensen) Gran	.40	.11													
Rhizosolenia setigera Brightwell	25.38	38.46	17.79	X	X	X	X		X	X	X	X	X	X	X

Species				Occurrence
Rhizosolenia styliformis Brightwell	21.66	23.07	22.33	X X X X X X X X X X
Rhizosolenia sp.	5.04	8.65	5.43	X X X X X
4. Chaetoceraceae:				
Bacteriastrum sp.	1.92			X X X
Chaetoceros affinis Lauder	3.15			X X X X
Chaetoceros curvisetus Cleve	7.78	13.46	5.93	X X X X X X
Chaetoceros concavicornis Mangin	0.93			X
Chaetoceros danicus Cleve	6.67	1.92	3.06	X X X X X X
Chaetoceros decipiens Cleve	3.37	5.76	8.06	X X X X X X
Chaetoceros didymus Ehrenberg		1.92		
Chaetoceros gracilis Schutt	5.41	7.69	22.52	X X X X X
Chaetoceros lacinosus Schutt	1.92			X X X X
Chaetoceros lorenzianus Grunow	34.66	59.61	31.42	X X X X X X X X X X
Chaetoceros similis Cleve	30.83	42.30	38.09	X X X X X X X X X
Chaetoceros simplex Ostenfield	6.95	8.64	7.23	X X X X X X X
Chaetoceros socialis Lauder	1.02	1.42	1.02	X X X X X
Chaetoceros teres Cleve	6.75	3.84	3.79	X X X X X X X
Chaetoceros sp.	3.73	9.61	10.25	X X X X X X X X
5. Biddulphiaceae:				
Biddulphia aurita (Lyngbye) Brebisson and Godeyl	1.92			X X X
Biddulphia rhombus (Ehrenberg) Smith	1.48	13.47	0.93	X X X X X X X
Ditylum brightwellii (West) Grunow	25.00	40.39	16.58	X X X X X X X X X
Eucampia zoodiacus Ehrenberg	0.98	1.92	0.93	X X X X X
Lithodesmium undulatum Ehrenberg	0.98	1.80	0.93	X X X X X X X
B. Pennales:				
1. Anaulaceae:				
Anaulus mediterraneus Grunow	0.55			X
2. Bacillariaceae:				
Bacillaria paxillifer (Mueller) Hendey	3.86	13.46	0.93	X X X X X X
3. Naviculaceae:				
Amphiprora sp.	0.56			X X
Diploneis sp.	3.12	7.69	1.70	X X X X X X X
Gyrosigma acuminatum (Kutzing) Rabh	1.02	1.92		X X
Gyrosigma baltica (Ehrenberg) Cleve	12.40	11.53		X X X X X X X X X
Gyrosigma eximium (Thwaites) Boyer	2.66	2.96		X X

Taxa	Frequency Creek Bay	Canal	J	F	M	A	M	J	J	A	S	O	N	D
Gyrosigma distortum (W. Smith) Cleve	23.92	10.74	X	X	X		X	X		X	X	X	X	X
Gyrosigma faciola (Ehrenberg) Cleve	20.96	5.83	X	X	X	X		X	X	X	X	X	X	X
Gyrosigma sp.	2.98								X	X	X	X	X	X
Navicula humerosa Smith		1.02							X	X	X			
Navicula latissima Gregory	3.31	1.94	X			X	X	X		X		X	X	X
Navicula lyra Ehrenberg	2.28	0.93							X	X	X			
Navicula sp.	3.02	1.80	X	X	X	X	X	X	X	X	X	X	X	X
Pinnularia borealis Cleve	0.94								X	X	X	X		
Pleurosigma angulatum (de Brebisson) Van Heurk	4.72					X	X		X	X				
Pleurosigma elongatum W. Smith	33.53	18.63	X	X	X	X	X	X	X	X	X	X	X	X
Pleurosigma sp.	3.56	2.54							X		X		X	X
Tropodoneis vitrea (W. Smith) Cleve														
4. Fragillariaceae:														
Asterionella japonica Cleve	59.01	48.55	X	X	X	X	X	X		X	X	X	X	X
Fragilaria crotonis Kitton	56.96	12.05	X	X	X	X	X	X	X	X	X	X	X	X
Fragilaria sp.	2.14								X	X	X	X		
Synedra investiens W. Smith	0.56										X			
Synedra pulchella Kutzing	28.67	5.74	X	X	X	X	X	X	X	X	X	X		X
Synedra tabulata (Agardth) Kutzing	1.11								X					
Synedra tenera W. Smith	8.28	29.38	X	X	X	X	X	X	X	X	X	X		
Synedra ulna (Nitzch) Ehrenberg	2.12										X			
Synedra vaucheria Kutzing	3.70								X	X	X	X	X	
Tabellaria flocculosa (Roth) Kutzing	7.07								X	X	X			
Thalassiothrix frauenfeldi Grunow	0.97								X	X			X	X
Thalassionema nitzschiocoides Grunow	46.86	18.17	X	X	X	X	X	X	X	X	X	X	X	X
5. Nitzschiaceae:														
Cylindrotheca closterium (Ehrenberg) Reimann and Lewin	69.17	55.32	X	X	X	X	X	X	X	X	X	X	X	X
Cylindrotheca gracilis (de Brebisson) Grunow	1.11	1.92							X	X				
Hantzschia marina var *genuina* (Donkin) Grunow		5.76							X	X				
Nitzschia acuminata (W. Smith) Grunow	3.57								X	X	X		X	
Nitzschia affinis Grunow	2.06								X	X	X		X	

Species			
Nitzschia bilobata W. Smith	3.54		
Nitzschia longissima (Brebisson) Ralfe	60.98	57.69	55.32
Nitzschia sigmoidea (Ehrenberg) W. Smith	24.30	30.76	5.10
Nitzschia seriata Cleve	35.73	38.46	20.57
Nitzschia sp.	3.46	6.88	1.44
6. Cymbellaceae:			
Amphora aernicola (Grunow) Cleve		3.84	
Amphora sp.	2.01	1.92	2.40
Cymbella sp.	0.67	1.92	1.92
7. Tabellariaceae:			
Grammatophora marina (Lyngbye) Kutzing	40.25	48.07	5.10
Rhabdonema adriaticum Kutzing	24.05	21.15	5.93
Striatella unipunctata (Lyngbye) Agardh	3.50	5.76	
8. Meridionaceae:			
Licomorpha ehrenbergii (Kutzing) Grunow	16.29	17.30	0.93
Licomorpha lyngbyei (Kutzing) Grunow	2.08	3.84	0.93
9. Achnanthaceae:			
Achnanthes longipes Agardh	31.98	25.00	21.69
Cocconeis disculoides Hustedt	16.33	1.92	6.95
10. Surirellaceae:			
Camptylodiscus sp.	1.01		
Surirella sp.	1.53		
II. Dinophyceae			
A. Prorocentraceae:			
Exuviella marina Cenowski	4.09		7.84
Prorocentrum micans Ehrenberg	25.01	28.84	24.18
Prorocentrum minimum var *triangulatum* (Martin) Hulbert			
Prorocentrum scutellum Schroeder	18.09	26.92	13.90
B. Gymnodiniaceae:			
Amphidinium crassum Lohmann	0.51		0.93
Amphidinium sp.	1.32		
Cochlodinium catenatum Okamura	6.79	7.69	8.80
Gymnodinium aeruginosum Stein	4.21	3.84	5.83
Gymnodinium grammaticum Pouchet	5.06	1.92	5.00
Gymnodinium minor Lebour	3.10	7.69	

Taxa	Frequency			Month of Occurrence											
	Creek	Bay	Canal	J	F	M	A	M	J	J	A	S	O	N	D
Gymnodinium mirabile Penard	1.62	1.92	2.87									X			
Gymnodinium nelsoni Martin	13.87	5.76	14.45								X	X		X	X
Gymnodinium resplendens Hulbert	3.57	3.84	3.70								X	X		X	X
Gymnodinium simplex (Lohmann) Kofoid and Swezy	2.68		2.87								X	X			
Gymnodinium splendens Lebour	8.99	9.61	10.74								X	X		X	X
Gymnodinium subrufecens Martin	2.41	1.92	5.10								X	X			
Massartia asymmetrica (Massart) Schiller	5.12	15.38	4.62								X	X		X	
Massartia rotundata (Lohmann) Schiller	10.85	7.69	17.78								X	X		X	X
C. Glenodiniaceae:															
Glenodinium sp.	9.65	12.49	41.60	X	X	X		X	X		X	X	X	X	X
D. Goniaulacaceae:															
Goniaulax diegensis Kofoid	1.91										X	X			
Goniaulax digitale (Pouchet) Kofoid	8.62	1.92	6.02			X					X	X	X		
Goniaulax longispina Lebour			1.02									X			
Goniaulax monocantha Pavillard	0.94		1.02								X		X	X	
Goniaulax polygramma Stein	2.08	1.92	1.94								X	X	X		
Goniaulax scrippsae Kofoid	24.95	15.38	17.23	X	X		X	X		X	X	X	X		
Goniaulax spinifera (Claparede and Lachmann) Diesing	6.52	1.92	1.02					X	X		X	X	X		X
Goniaulax sp.	17.17	13.46	15.28				X	X		X	X	X	X		
E. Peridiniaceae:															
Diplopsalsis lenticula Bergh	2.00	15.38	7.68					X	X		X	X		X	X
Diplopsalsis minor Lebour	2.12								X		X	X	X		
Peridinium achromaticum Leuander	1.06	3.84		X					X				X		X
Peridinium balticum (Levander) Lemmermann	15.01	5.76	6.57	X	X	X		X	X		X	X	X		
Peridinium breve Paulsen	14.44	21.15	5.74	X	X	X	X		X	X		X	X	X	X
Peridinium claudicans Paulsen	3.45	15.38		X	X	X		X	X	X					
Peridinium marielebourae Paulsen	7.92	5.76	0.93				X	X		X		X		X	X
Peridinium minusculum Pavillard	17.90	17.30	17.51	X	X	X		X	X		X	X	X		
Peridinium ovatum (Bouchet) Schuett	0.47			X			X								
Peridinium pelliculum (Bergh) Schuett	14.44	15.38	10.74	X	X		X	X		X	X	X	X	X	X
Peridinium steinii Jorgensen	20.73	21.15	16.86	X	X		X	X		X	X	X	X	X	X
Peridinium triquetrum (Ehrenberg) Lebour	1.41		2.77								X	X		X	X

Species	% A	% B	% C	1	2	3	4	5	6	7	8
Peridinium sp.	1.70	1.92	1.17				X		X	X	X
Peridinopsis rotunda Lebour	4.12							X	X	X	X
F. Dinophyceae:											
Dinophysis acuminata Claparede and Lechmann	1.39	2.90	1.09	X	X	X	X	X	X	X	X
Dinophysis ovatum Schuett	1.20	0.59					X	X	X	X	X
G. Polykrikaceae:											
Polykrikos kofoidi Chatton	0.98	9.61	0.93	X	X	X			X	X	X
III. Euglenophyceae											
Euglena acus Ehrenberg	1.11		4.72	X	X	X			X	X	X
Eutreptiella marina De Cunha	47.61	26.92	69.04	X	X	X	X	X	X	X	X
Phacus sp.			0.93	X							
Trachelomonas sp.	6.21		10.84	X	X	X			X	X	X
IV. Chlorophyta											
A. Polyblepharidaceae:											
Dunaliella sp.	22.61	13.46	16.58	X	X	X	X		X	X	X
Pyramimonas sp.	13.80	3.84	8.42	X	X	X	X		X	X	X
B. Chlamydomonadaceae:											
Brachiomonas submarina Bohlin	1.91		14.35	X	X	X	X		X	X	X
Carteria sp.	30.68	25.00	38.92	X	X	X	X	X	X	X	X
Chlamydomonas sp.	32.48	11.53	29.83	X	X	X	X	X	X	X	X
C. Other Chlorophyta											
Actinastrum			3.90						X	X	X
Akistrodesmus sp.	0.56										X
Oedogonium sp.			1.02	X	X	X			X	X	X
Pandorina sp.	1.48	1.92	7.04	X	X	X	X				
Scendesmus sp.	3.81	3.84	3.89	X	X	X			X	X	X
V. Cyanophyceae											
Johnnes baptista sp.			1.02				X				
Ocillatoria sp.	4.54	1.92	0.93	X	X	X			X	X	X
Rivularia sp.	0.47	1.92								X	X
Spiralema major Kutzing	2.17	1.92		X	X	X	X		X	X	X
Trichodesmium sp.		1.92								X	
VI. Chrysophyceae											
Angulochrysis erratica Lakey	3.38	1.92	2.96					X	X		
Chromulina pascheri Hoefender	18.74	9.61	9.44	X	X	X	X	X	X	X	X
Chrysoamoeba sp.	3.40	3.84	7.68	X	X	X	X		X	X	X

Taxa	Frequency			Month of Occurrence											
	Creek	Bay	Canal	J	F	M	A	M	J	J	A	S	O	N	D
Chrysocapsa pauldosa Pascher	9.28	17.30	8.61				X		X	X	X		X		
Chrysococcus sp.	4.70		1.02				X								X
Dinobryon sp.			0.93								X				
Mallomonas sp.	3.34		1.85								X	X	X		
Monas sp.		1.92	1.02								X				
Pedinella sp.	5.52	7.96									X	X	X		
VII. Cryptophyceae															
Cryptomonas salina Wislouch	1.88						X								
Rhodomonas sp.	1.41			X	X	X					X	X	X		
VIII. Xanthophyceae															
Olisthodiscus luteus Carter	17.15	19.23	17.23	X	X	X	X	X	X		X	X	X	X	X
IX. Silicoflagellatae															
Destephanus speculum (Ehrenberg) Haeckel	5.10	19.23	13.90	X	X	X	X	X	X		X	X	X		X
Eboria tripartita (Schumann) Lemmermann	2.47	3.84		X	X	X		X	X		X	X			
X. Unidentified μ-flagellates	3.52	2.74	4.87	X	X	X	X	X	X	X	X	X	X	X	X

The Fate of Freshwater Algae Entering an Estuary[1]

John W. Foerster

Through the ages of recorded time, man has pondered the vast workings and intricate phenomena of the sea. He has recorded volumes on tides, fish migrations, chemical components, harvests, and so on. One feature often overlooked in past studies is the fate of the freshwater microorganisms, the plankton, as they descend a river, enter the ecotone of an estuary, and are funneled into the neritic ocean. In this paper, the survival of the freshwater algal flora as it enters a marine environment is explored. What happens to these algae? Do they survive the changing environment? If they survive, how does the changing environment affect their morphology? What are the environmental conditions that an alga must encounter in its passive transport to the sea?

In this study, these questions are analyzed in the framework of an environmental system (Fig. 1). The combination of a ninety-year-old reservoir, a stream, an estuary, and the neritic ocean was investigated. This combination of fresh and marine waters will be referred to as the Quiambog Environmental System. While literature is available on lakes, streams, estuaries, and oceans *(30, 31, 41, 48, 53)*, little work has been done to relate them as interdependent bodies of water in a single study. Further, little work to date appears to have been done to trace the algal flora through a system such as described in the following pages.

Many workers throughout the world have studied estuaries and attempted to delineate the inhabiting flora. Except for Drum and Webber *(13)* and Hendey *(25)*, little attempt has been made specifically to note the presence of freshwater algae in marine waters. Of the 151 species of algae listed by Drum and Webber *(13)* for a salt marsh in Massachusetts, 52 species were freshwater. In fact, they concurred with Hendey *(25)* that the presence of freshwater algae was indicative of brackish water.

Therefore, because the fate of freshwater algae has not been detailed in the context of a freshwater-marine system with an effort to study live algae, the following study is presented. The underlying premise is that some freshwater algae survive transport into an estuarine system. How they survive and where they are found will be reported.

[1] Publication No. 5 of the Goucher College Environmental Studies Program and No. 81 of the University of Connecticut Marine Research Laboratory.

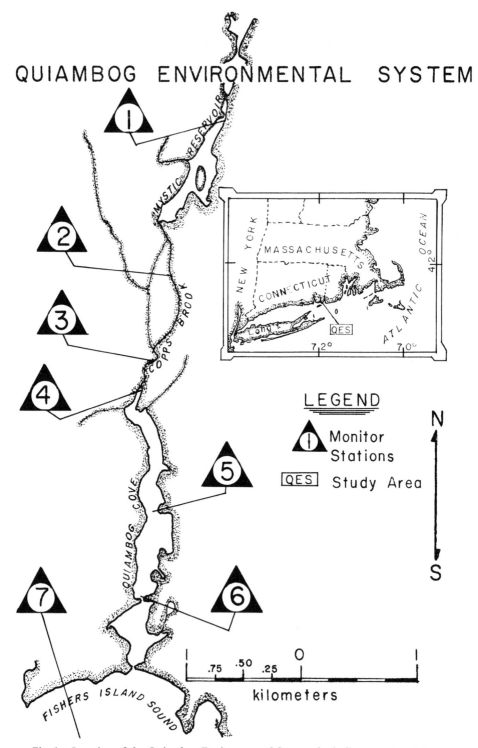

Fig. 1. Location of the Quiambog Environmental System, including monitor stations.

Methods

The study of four different habitats required various methods that would yield a large volume of information in a short period of time. All chemical-physical analyses used were performed according to those outlined in *Standard methods (1)* and in Strickland and Parsons (47), except that current speed was measured with a nonelectrical current meter (17), and the current pattern in the estuary was plotted using aerial photography (45), drogues, and fluorescein dye.

Of the ten chemical, eight physical, six meteorological, and five biological parameters measured, only orthophosphate, nitrate-nitrogen, bicarbonate, silicon, salinity, total dissolved solids, pH, current velocity, discharge, light penetration, primary production, and species composition will be presented here. The selection of these items is based on the standard statistical tests of analysis of variance, multiple regression, and correlation. All calculations were programed and computed on an I.B.M. 1130 computer. These chemical-physical factors showed the greatest correlation as controlling environmental factors and were most demonstrative of the survival of freshwater algae in the Quiambog Environmental System.

During the period of this study (December 1966 through August 1968) weekly cruises were made of six established monitor stations (Fig. 1). A seventh station was established in Fishers Island Sound and was sampled monthly. Further, monthly transects and transects during flood stages of the stream were completed. From all of this information, means have been calculated with the confidence that extensive sampling has given the means validity. It is believed that the means, as presented in the following work, will be adequate to describe and delineate the environmental changes faced by an alga as it transits the system. Therefore, it is not the purpose of this paper to detail the limnology-oceanography of the system, but to concentrate on the survival of the freshwater algae using pertinent environmental data.

Plankton samples were taken at the seven stations described, at the western tip of Fishers Island, in the Race upwelling off the western end of Fishers Island, in Long Island Sound south of Fishers Island, and southwest of Block Island, Rhode Island. In addition, plankton samples were obtained from the Connecticut River, Deep Creek Lake, Maryland, Saginaw Bay of Lake Huron, and from twenty-four stations on a transect from Bermuda to New York and along the Southern New England Coast to Woods Hole, Massachusetts. All plankton samples were taken using a 10 μm nannoplankton net (which was thoroughly washed between stations) and compared to centrifugation techniques *(1)* to determine if contamination from previous samplings occurred. Contamination was found not to be a problem.

Another device used to collect freshwater algae for experimentation was a phyco-periphyton collector *(18)*. Growths of algae that developed on the slides in the instrument were transferred to the estuary at varying salinities and the effect was noted.

Dialysis membranes were prepared containing 40 ml of concentrated Mystic Reservoir water (10 l concentrated to 1 l). These were affixed to a wooden-styrofoam float and incubated in the environmental system or used in the laboratory.

To obtain a measure of the phytoplankton primary production the light/dark bottle technique *(21)* was used on a weekly basis over an entire year (June 1967–June 1968). The differential oxygen analyses were performed on a colorimeter. Thus, a higher sensitivity was permitted in detection of differences between the light and dark bottles than could be obtained by titration *(46)*.

Finally, collections were periodically cultured using various types of media as described by Starr *(44)* and Chu *(9)*. Some of the media used was simply enriched Mystic Reservoir and Fishers Island Sound water (enriched with .01 mg/l PO_4, .10 mg/l NO_3, and 1.0 mg/l SiO_3).

Identification of algal organisms was done by phase microscopy and consultation of numerous literature sources *(2, 4, 6, 8, 11, 12, 14, 20, 22, 23, 24, 25, 26, 27, 28, 29, 33, 36, 37, 39, 40, 51, 52, 56, 57, 58)*. Neutral red dye was used to test live freshwater algae to determine if a pH change in the cytoplasm could be detected when the alga was exposed to seawater.

Results

The Quiambog Environmental System is characterized by mean salinities ranging from 0.03‰ to 35.0‰ (Fig. 2). It is composed of a ninety-year-old reservoir whose lower section has a mean depth of 3 m; a stream 3 km in length with a mean depth of 0.3 m; an estuary of over 8 km in length and a mean depth of 1 m, and Fishers Island Sound.

The system lies in a region that slopes to the south and is poorly drained. Geologically the region is underlain by Paleozoic igneous and metamorphic rocks *(5, 38)* and has been extensively glaciated during the Wisconsin Glacial Period *(38)*. Thornthwaite *(50)* characterizes the region as a B_3 humid climate with a moisture index of 60 to 80. Brooks and Deevy *(5)* report the area receives rainfall from 115 to 120 cm per annum, evaporation is 65 cm, and, along with Rumney *(42)*, they characterize the region as having a maritime climate.

Biologically the macrobiota shifts from a reservoir characterized by the plant *Anacharis canadensis* and sunfish *(Lepomis machrochirus)* to a stream with typical riffle fauna (stoneflies, caddis flies) and long strands of the alga *Stigeoclonium*.

The estuary harbors beds of eel grass *(Zostera marina)* from mid-estuary seaward, has the green alga *Enteromorpha entestinalis* throughout, and at the seaward end luxuriant growths of algae, *Fucus vesiculosus* and *Ascophylum nodosum*, cover the rocks while American eels, winter flounder, lobsters, and Quohog clams abound.

Figure 3 has been designed to demonstrate the estuarine physical components during high freshwater flow periods (Fig. 4) of the year. It is of importance to note that Quiambog Cove is a small, shallow, vertically mixed estuary. An organism entering the estuary has a mean residence time of 1.9 to 3.4 days, depending on the time of the month and the tides. The freshwater alga, whether plankton or periphyton, descends the stream and enters the head of the cove where a completely different physical and chemical regime is encountered. An alga must first go from a relatively nonshaded reservoir where currents are slight, to a rapidly flowing, shaded stream. It takes the passive organism on the average of 40 to 150 min to traverse the stream and enter the chemically dif-

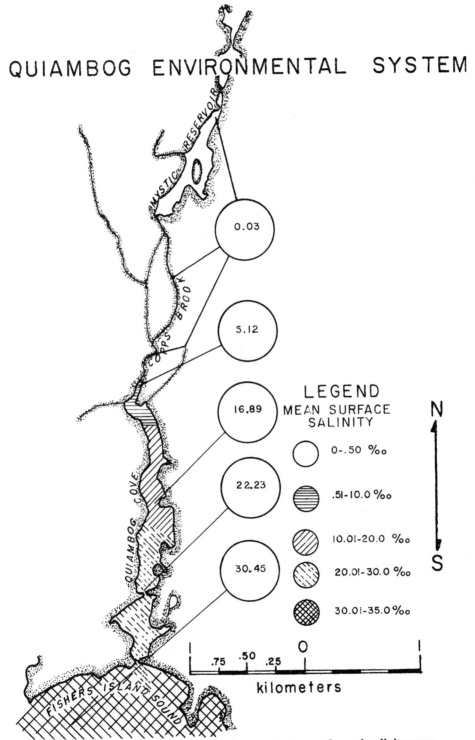

Fig. 2. **Mean surface salinities and approximate area for each salinity range.**

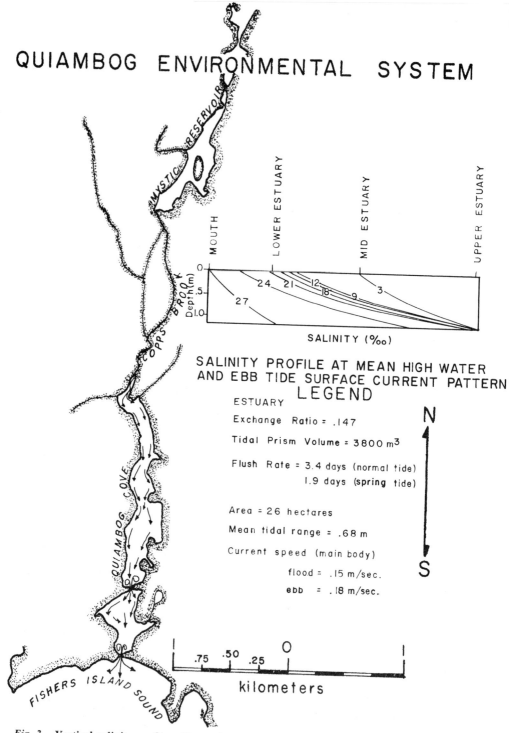

QUIAMBOG ENVIRONMENTAL SYSTEM

MYSTIC RESERVOIR

COPPS BROOK

QUIAMBOG COVE

FISHERS ISLAND SOUND

MOUTH | LOWER ESTUARY | MID ESTUARY | UPPER ESTUARY

Depth(m)
0
.5
1.0

24 21 12 18 9 3

27

SALINITY (‰)

SALINITY PROFILE AT MEAN HIGH WATER
AND EBB TIDE SURFACE CURRENT PATTERN
LEGEND

ESTUARY

Exchange Ratio = .147

Tidal Prism Volume = 3800 m^3

Flush Rate = 3.4 days (normal tide)
1.9 days (spring tide)

Area = 26 hectares

Mean tidal range = .68 m

Current speed (main body)

flood = .15 m/sec.

ebb = .18 m/sec.

N
S

.75 .50 .25 0

kilometers

Fig. 3. **Vertical salinity profile with pertinent morphometric data for Quiambog Cove during the spring of the year when stream flow is highest. The ebb tide current patterns give some idea as to the distribution of fresh water after it enters the estuary.**

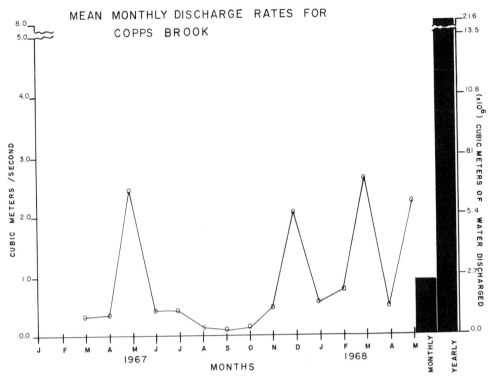

Fig. 4. **Mean discharges of fresh water from Copps Brook into Quiambog Cove. Seasonal peaks correspond to maximum periods of rainfall.**

ferent, nonshaded, less current-affected head of the estuary. Here at the head of the estuary an alga encounters changed physical-chemical conditions and competition for nutrients by a new set of organisms.

The pH (Fig. 5) increases seaward and the range becomes greater. Salinity (Fig. 2) increases 100-fold as do total solids (Fig. 6). The calcium-magnesium ratio reverses (Fig. 7) and the concentrations increase 50 to 100-fold respectively. During spring, summer, and early fall the stream is shaded. The light increases from the shaded stream to open estuary 75 to 92%. Light penetration is increased from the lake through the stream into the estuary and the sound as shown by the changes in the extinction coefficients (K, Fig. 8). This now means a thicker photosynthetic layer is encountered and thus greater competition for available nutrients. The stream and estuary are clear to the bottom.

Chemically, orthophosphate (Fig. 9) and bicarbonate (Fig. 10) increase while nitrate-nitrogen (Fig. 11) and silicon (Fig. 12) decrease. When these items are compared to primary production (Fig. 13), it would appear that nitrate-nitrogen and silicon show the effects of increased utilization seaward. Calculations of correlation coefficients for nitrate-nitrogen and orthophosphate with primary production and standing crop indicate that the increase of orthophosphate may stimulate nitrate-nitrogen utilization and in fact be a significant limiting factor in the system, more so than nitrate-nitrogen.

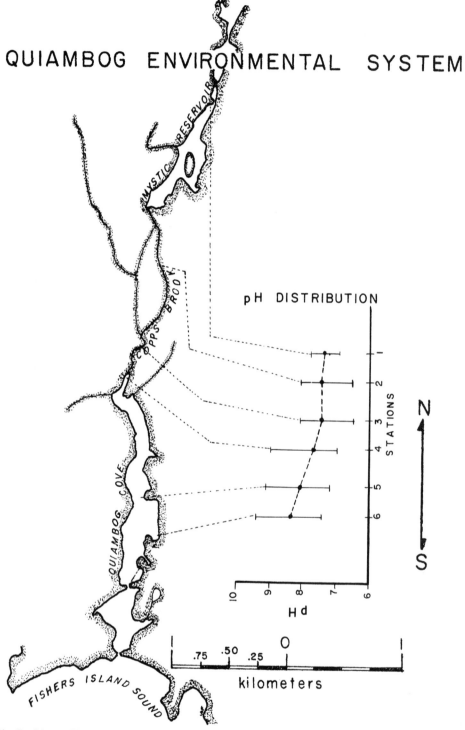

Fig. 5. **Mean pH values showing the increase in pH seaward with the overall range. Not only does the pH increase but the range of pH fluctuation becomes larger.**

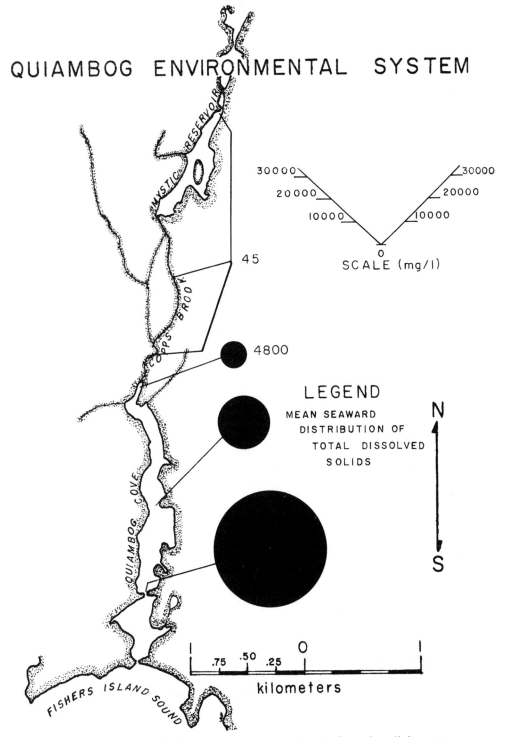

Fig. 6. **Total solids distribution increases seaward and reflects the salinity pattern.**

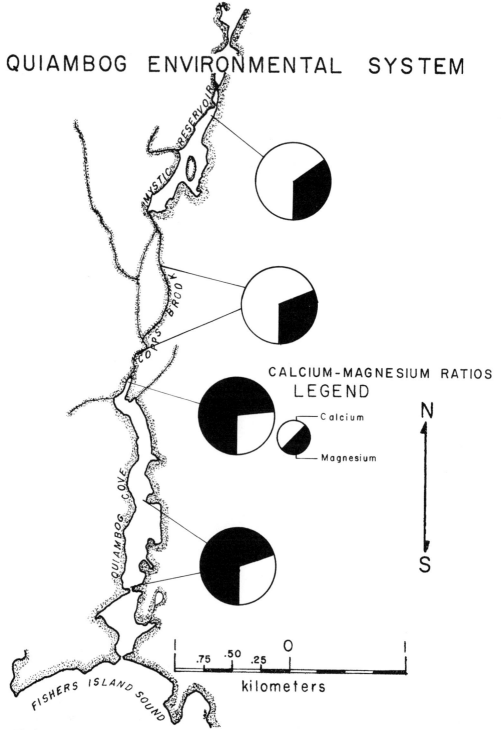

Fig. 7. **Calcium: magnesium ratios reverse as the saline segment of the Environmental System is encountered. Magnesium is the dominant cation in estuarine water while calcium is predominant in fresh water.**

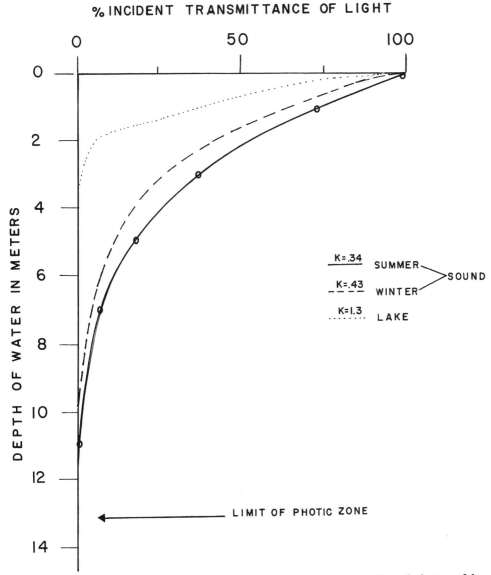

Fig. 8. Light extinction curves for the Quiambog Environmental System. Since the bottom of the stream and the estuary are visible, only the lake and sound are presented. The stream and estuary lie within the 50–100% incident transmittance region.

Looking at the nutrients and the phytoplankton in terms of gross primary production (Fig. 13), a peak of carbon fixation is reached at the head of the estuary. This would agree with the work of Flemer *(15)* for subestuaries on the Chesapeake Bay. In the Quiambog Environmental System the region of highest primary production corresponded to the region of mixed freshwater and marine phytoplankton—a high diversity. This was somewhat different from that reported by Ivanov *(34)* where blue-green algae dominated the upper regions of

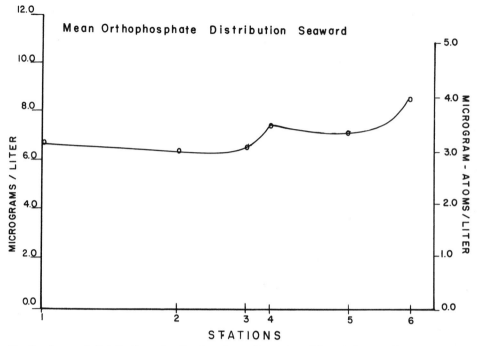

Fig. 9. Seaward distribution of orthophosphate. Data plotted in relation to distance between stations.

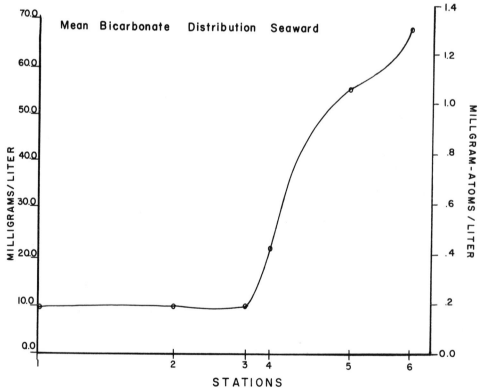

Fig. 10. Seaward distribution of bicarbonate. Data plotted in relation to distance between stations.

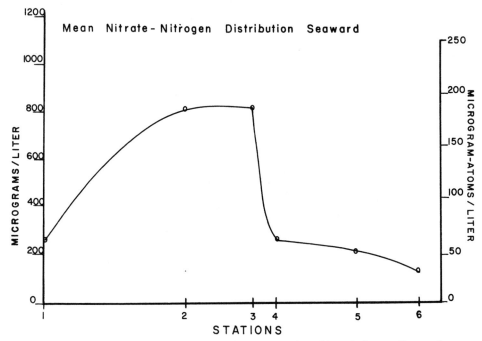

Fig. 11. Seaward distribution of nitrate-nitrogen. Data plotted in relation to distance between stations.

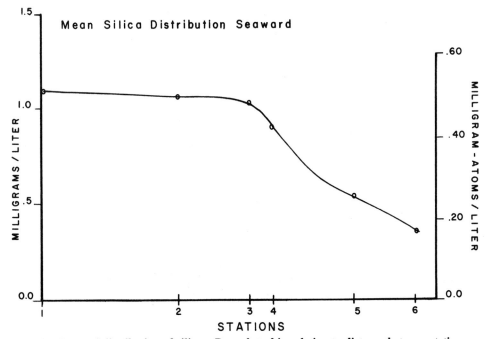

Fig. 12. Seaward distribution of silicon. Data plotted in relation to distance between stations.

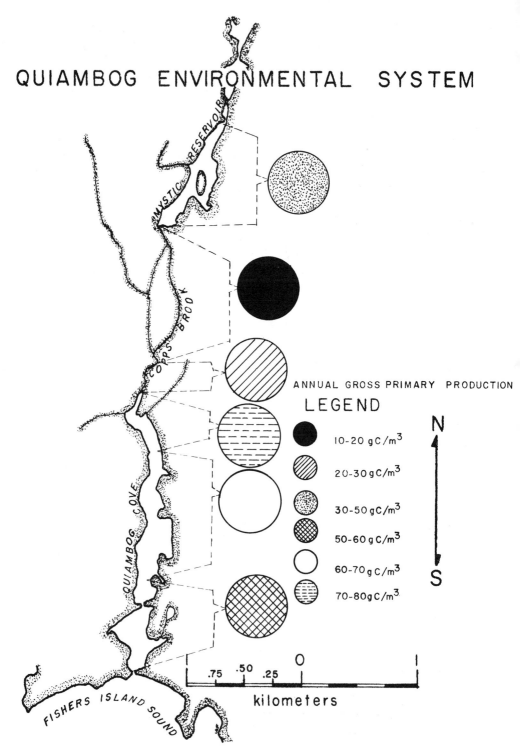

Fig. 13. The annual gross primary production for the Quiambog Environmental System. Highest production appears at the head of the estuary.

estuaries on the Black, Caspian, and Azov seas. In terms of survival and production, Blanc et al. *(3)* noted that in the estuarine dilution zone at the head of the Rhone River estuary, freshwater algae appeared "dead" or "almost dead" and this corresponded to a low chlorophyll content. Since this is a discussion of survival, no further detail will be reported concerning the overall ecological relationships except where necessary to indicate the conditions encountered by an alga.

Table 1 provides a list of freshwater algae found in the marine plankton collections. It further includes freshwater algae from the periphyton slides, and the dialysis experiments in the estuary. A question justly asked at this time is, "Are these organisms alive?" By use of neutral red dye the cell contents of the algal cells were checked both in culture experiments and in live collections. The pH difference between live and dead cells, young and senescent cells in culture was significant. However, in the cells sampled during the experiments and from the live collections no significant change in color rendition of the cytoplasm was noted even though the freshwater organism was subjected to seawater. Further, culturing of seawater samples containing freshwater algae in freshwater media (.03‰ salinity) showed that many of these forms could grow. Therefore, in

Table 1. Freshwater algal species found in the estuarine and offshore plankton and periphyton collections

Species	U.E.[a]	M.E.	L.E.	O.	C.[b]
Bacillariophyta					
Achnanthes					
brevipes Ag.	X	X	X		
deflexa Reim.	X				
hauckiana Grun.	X	X	X		
minutissima Kuetz.	X	X	X		
Amphiprora					
paludosa W. Sm.	X	X			
Amphora					
coffaeiformis (Ag.) Kuetz.	X	X	X		
ovalis (Breb.) Kuetz.					
Asterionella					
formosa Hassal.	X	X	X	X	
Cocconeis					
diminuta Pant.	X	X	X		
placentula Ehr.	X				
Coscinodiscus					
lineatus Ehr.	X				
Cyclotella					
bodanica Eulenst.	X				
glomerata Bachn.	X	X	X		X
menegheniana Kuetz.	X	X	X	X	
michiganiana Skv.	X	X	X		X
stelligra Cleve & Grun.	X				
Cymbella					
affinis Kuetz.	X	X			
cymbiformis (Kuetz.) Breb.	X				

Table 1. Freshwater algal species continued

Species	U.E.[a]	M.E.	L.E.	O.	C.[b]
Diatoma					
hiemale (Lyngb.) Heib.	X	X			
vulgare Bory	X	X			
Eunotia					
curvata (Kuetz.) Lager.	X	X			
elegans Oestrup	X				
naegelii Migula	X	X	X	X	
pectinalis					
var. *minor* (Kuetz.) Rabnh.	X	X	X		
var. *ventricosa* Grun.	X				
Fragilaria					
brevistriata Grun.	X	X	X		
constricta Ehr.	X				
crotonensis Kitton	X				
inflata Pant.	X				
intermedia (from Weber-52)	X	X			
pinnata Ehr.	X				
Frustulia					
rhomboides (Ehr.) deTone	X	X			
Gomphonema					
acuminatum Ehr.	X				
olivaceum Lyngb.	X	X	X		
parvulum Kuetz.	X				
Melosira					
ambigua (Grun.) Muller	X				
crenulata (Ehr.) Kuetz.	X				
granulata (Ehr.) Ralfs	X	X	X		
islandica Muller	X	X	X		
italica (Ehr.) Kuetz.	X				
juergensis Ag.	X	X	X		
varians Ag.	X	X			
Meridian					
circulare (Grev.) Ag.	X				
Navicula					
cryptocephala Kuetz.	X	X	X		
cuspidata Kuetz.	X				
gregaria Donk.	X				
radiosa					
var. *tenella* (Breb.) Cleve	X				
zanoni Hust.	X				
Nedium					
bisulcatum (Lager.) Cleve	X				
Nitzschia					
acicularis W. Sm.	X				
amphibia Grun.	X				
dissipata (Kuetz.) Grun.	X				
holsatica (from Weber-52)	X	X	X		
hungarica Grun.	X				

Table 1.   Freshwater algal species continued

Species	U.E.[a]	M.E.	L.E.	O.	C.[b]
lorenziana Grun.	X	X	X		
palea (Kuetz.) W. Sm.	X	X	X		
paradoxa (Gmel.) Grun.	X				
Ophephora					
martyri Heib.	X				
Pleurosigma					
delicatulum W. Sm.	X				
Rhabdonema					
adriaticum Kuetz.	X	X	X		
Rhopolodia					
gibba (Ehr.) Muller	X	X	X		
gibberula (Ehr.) Muller	X	X	X		
Stauroneis					
obtusa Lager.	X				
Surirella					
ovata Kuetz.	X				
pinnata W. Sm.	X				
Synedra					
actinastroides Lemm.	X				
nana Meister	X	X	X		
radians Kuetz.	X	X	X		
rumpens Kuetz.	X				
rumpens					
var. *familiaris* (Kuetz.) Hust.	X				
var. *scotica* Grun.	X	X	X		
tabulata (Ag.) Kuetz.	X				
ulna (Nitzsch.) Ehr.	X	X	X	X	
Tabellaria					
fenestrata (Lyngb.) Kuetz.	X	X	X	X	X
flocculosa (Roth) Kuetz.	X	X	X	X	X
Chlorophyta					
Arthrodesmus					
triangularis Larg.	X				
Bulbochaete					
sp.	X				
Chlamydomonas					
angulosa Dillw.	X	X			X
Cylindrocapsa					
conferta W. West	X				
Dichotomosiphon					
tuberosa (Braun) Ernst	X	X	X		
Oedogonium					
sp.	X				
Oocystis					
lacustris chodat					X
parvula W. & G. S. West	X				X
sp.					X

Table 1.   Freshwater algal species continued

Species	U.E.[a]	M.E.	L.E.	O.	C.[b]
Pediastrum					
duplex					
var. *clathratrum* (Braun) La Lager.	X				
var. *gracilimum* W. & G. S. West	X	X			X
var. *reticulum* Larg.	X				
Protococcus					
viridis Ag.	X	X	X	X	X
Scenedesmus					
incrassatulus Bohlin	X				
obliquus (Turp.) Kuetz.	X	X	X		X
quadricauda (Turp.) Breb.	X	X	X		
Staurastrum					
connatum (Lund) Roy & Bisset	X				
natator (?) W. West	X				
vestitum					
var. *subanatinium* W. & G. S. West	X				
Ulothrix					
sp.					X
subtilissima Rabhn.	X				
Chrysophyta					
Dinobryon					
bavaricum Imhof	X				
setularia Ehr.	X	X	X		
Mallomonas					
acaroides Perty	X				
alpina Pascher & Ruttner	X	X	X		
caudata Iwanoff	X				
producta (Zach.) Iwanoff	X				
Synura					
uvella Ehr.	X				
Cyanophyta					
Anabaena					
sp.	X	X			
Chroococcus					
dispersus (Keissler) Lemm.	X	X	X		
dispersus					
var. *minor* G. M. Sm.	X	X	X		
turgidus (Kuetz.) Naeg.	X	X	X		
Cylindrospermum					
catenatum Ralfs	X				
Dactylococcopsis					
smithii R. & F. Chodat	X	X			
Dermocarpa					
prasina (Reinsch) Bornet & Thuret	X	X			
Hapalosiphon					
intricatus W. & G. S. West	X	X	X		

Table 1. Freshwater algal species continued

Species	U.E.[a]	M.E.	L.E.	O.	C.[b]
Lyngbya					
ochrea (Kuetz.) Thuret	X	X	X		
Microcystis					
aeruginosa Kuetz.	X	X			
Nostoc					
spongiaeforme (?) Ag.	X				
verrucosum Vaucher	X				
Oscillatoria					
amphibia Ag.	X	X			
angustissima W. & G. S. West	X				
formosa Bory	X	X	X		
lacustris (Kleb.) Geitler	X	X	X		
tenuis Ag.	X	X	X		
Phormidium					
retzii (Ag.) Gom.	X				
tenue (Menegh.) Gom.	X	X	X		
Spirulina					
subsalsa Oerstd.	X	X	X		
Synechococcus					
aeruginosa Naeg.	X	X	X		
Euglenophyta					
Trachelomonas					
charkowiensis Swirenko	X				

[a] Abbreviations used: U.E. = Upper Estuary; M.E. = Middle Estuary; L.E. = Lower Estuary; O. = Oceanic; C. = Cultures.
[b] Freshwater species found in seawater cultures (32–36‰).

contrast to the work of Drum and Webber *(13)*, Bursa *(7)*, and Hendey *(25)*, the data presented here are based on living algae and the assessment of vitality in seawater of various salt concentrations. These workers reported organisms from preserved samples.

Table 2 provides a listing of freshwater algae found when samples of seawater from near Block Island, Rhode Island, and around Fishers Island, New York, were cultured. Some of the forms listed, i.e. *Tabellaria fenestrata, Cyclotella michiganiana, Asterionella formosa,* and *Protococcus viridis,* were found in the plankton tows.

Several types of experiments were designed using multialgal cultures subjected to laboratory culture tube and plate analyses, and dialysis in the field and laboratory. In the following discussion several representatives of the more commonly known algae have been selected to illustrate survival and to suggest morphological changes, with possible mechanisms to explain the changes.

The alga *Scenedesmus obliquus* was found in the Mystic Reservoir water during March and April and was subjected to field and laboratory experimentation. Figure 14 shows what happened when this alga was subjected to laboratory dialysis. In this experiment and other similar experiments, four aquaria with

Table 2. Culture results from subjecting samples of marine plankton tows to freshwater media

Species[a]	F.I.S.[b]	L.I.S.	B.I.S.	C.S.	O.O.
Bacillariophyta					
Achnanthes					
lanceolata					
var. *dubia* Grun.	X				
minutissima Kuetz.		X		X	
Amphora					
coffaeiformis (Ag.) Kuetz.	X			X	
Asterionella					
formosa Hassal.	X		X		
Cocconeis					
fluviatilis Wallace	X				
Coscinodiscus					
lineatus Ehr.	X				
Cyclotella					
glomerata Bachn.	X				
menegheniana Kuetz.	X				
michiganiana Skv.	X		X	X	
stillegra Cleve & Grun.	X		X	X	
Cymbella					
turgida Greg.		X			
Diatoma					
hiemale (Lyngb.) Lager.	X	X			
Eunotia					
curvata (Kuetz.) Lager.		X			
pectinalis					
var. *minor* (Kuetz.) Rabhn.		X			
Fragilaria					
brevistriata Grun.		X			
Gomphonema					
parvulum Kuetz.		X			
Melosira					
crenulata (?) (Ehr.) Kuetz.		X			
Navicula					
sp.				X	
Nitzschia					
palea (Kuetz.) W. Sm.		X			
Stephanodiscus					
invisitatus Weber	X	X	X	X	
Synedra					
tabulata (?) (Ag.) Kuetz.				X	X
ulna (Nitzsch.) Kuetz.		X			
Tabellaria					
fenestrata (Lynogb.) Kuetz.	X	X	X		
Chlorophyta					
Chlamydomonas					
sp.	X				

Table 2. Culture results continued

Species[a]	F.I.S.[b]	L.I.S.	B.I.S.	C.S.	O.O.
Chlorella					
vulgaris Beyer.				X	
Chlorella-like	X	X	X	X	X
Cosmarium-like				X	X
Oocystis					
sp.				X	X
Pediastrum					
boryanum (Turp.) Menegh.	X		X		
Protococcus					
viridis Ag.					X
Scenedesmus					
dimorphus (?) (Turp.) Kuetz.	X			X	
Selanstrum					
westii G. M. Smith	X			X	
Tetraedron					
planctonicum (?) G. M. Smith	X			X	
Westella					
linearis G. M. Smith				X	
Cyanophyta					
Anabaena					
circinalis (?) Rabhn.				X	
Chroococcus					
dispersus (Keissler.) Lemm.	X		X	X	
dispersus					
var. *minor* G. M. Smith	X		X	X	
turgidus (Kuetz.) Naeg.	X		X	X	
Microcystis					
aeruginosa Kuetz.				X	

[a] Best success was obtained using Mystic Reservoir water and soil-water extract media.

[b] Abbreviations used: F.I.S. = Fishers Island Sound; L.I.S. = Long Island Sound; B.I.S. = Block Island Sound; C.S. = Continental Shelf; O.O. = Open Ocean.

salinity ranges of (1) lake water, 0.03‰, (2) 1–5‰, (3) 10–15‰ and (4) 25–30‰ were illuminated (250 ft-c continuous light) and with temperatures simulating ambient conditions in the system. Salinities were prepared by combining filtered lake and sound water in the proportions providing the desired salinities. Later these experiments were repeated using a 12/12 light cycle, with similar results. Dialysis membranes containing 40 ml sample of *S. obliquus* were suspended in the aquaria. There were fourteen membranes to each aquarium and two membranes were sacrificed each day for seven days. Similar experiments were performed simultaneously in the field.

In the field, the membranes were placed in the lake and in the upper, middle, and lower estuary. Also, plankton samples were collected and analyzed. Effects of salinity (Fig. 14) appeared to be production of a monstrosity that developed

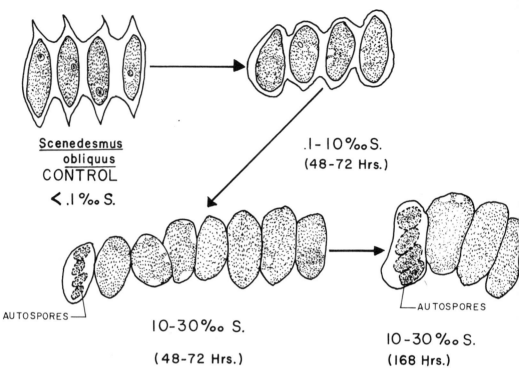

Fig. 14. **Morphological changes of *Scenedesmus obliquus* in a series of laboratory dialysis experiments using various salinities. These were experiments on plankton samples with an attempt to simulate the natural conditions in the System.**

reproductive cells (autospores) but did not release them. The chloroplast became swollen and the pyrenoid disappeared. When these results were compared to Wetherell's *(54, 55)* respiration experiments with *S. obliquus,* similar trends were noted.

Figure 15 provides a series of photomicrographs made from the plankton samples and field dialysis experiments. They depict a similar change in *Scenedesmus* as shown in Figure 14. Other organisms at other times of the year under the same kinds of conditions also showed salinity responses but did not "die" as Blanc et al. *(3)* found. Organisms like *Chlamydomonas angulosa* and *Chlorella vulgaris* swelled and the chloroplast became misshapen when subjected to increasing salinity. Above 10‰ *Chlamydomas* formed a palmelloid stage. *Protococcus viridis* (Phytoconnis) appeared undisturbed, while *Oocystis parvula* formed a thickened mother-cell wall. In other algal groups, *Euglena sanguiena* appeared destroyed after 10‰ while *Tabellaria fenestrata,* *Synedra ulna,* and *Navicula sp.* appeared to have larger and fewer chloroplasts than the freshwater controls.

Chroococcus dispersus seemed unaffected while the cell size of *Oscillatoria tenuis* became smaller and mucilage was more apparent. *Dinobryon setularia* was found in the estuarine dialysis at salinities up to 15‰. Under a very slight salinity increase, the colony of *Synura uvella* dissociated. *Dichotomosiphon tuberosa,* a green filamentous periphyton alga, became increasingly prostrate and the size decreased as the salinity increased. This organism was dominant on the periphyton

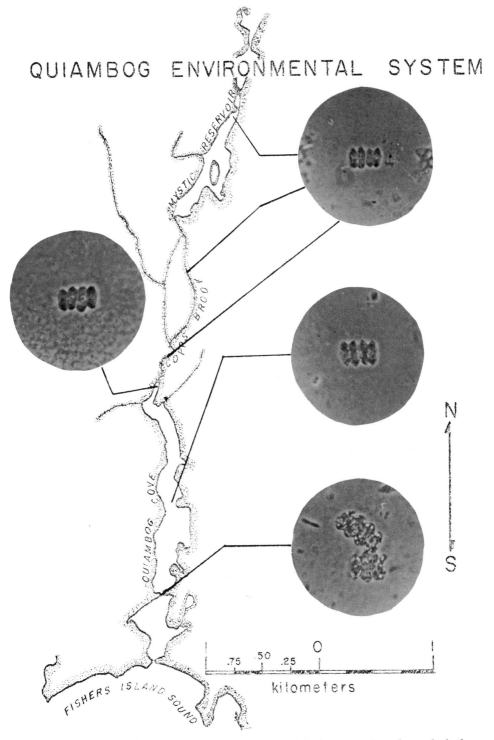

Fig. 15. Photomicrographs of *S. obliquus* as it appeared in field dialysis and actual plankton collections. It is related to the position in the System.

slides incubated in Copps Brook and transferred to the estuary. In the literature, Collins *(10)*, Hylander *(32)*, and Taylor *(49)* discussed this organism as consisting of either two species or a species in their collections. The second species was from upper estuarine areas. In the more saline environment of the estuary, the thinner, more prostrate filament was derived from the thicker, tufted form found in the stream. The data indicate a single species, *Dichotomosiphon tuberosa*.

Quantitative data for *T. fenestrata* is presented in Figure 16. *T. fenestrata* was a very common organism in the freshwater of the area. It is noted that as the salinity increased the number of cells decreased. The data were obtained from a series of dialysis experiments as described above. Only the cells containing chloroplasts were counted, and the chloroplasts became larger and less numerous. With these figures and tables a case is presented to show that some of the freshwater algae survive. Is this an anomaly of nature, a phenomenon associated with this system?

In an attempt to show that this hypothesis has validity, plankton samples from the Connecticut River, Connecticut, Deep Creek Lake, Maryland, and Saginaw Bay of Lake Huron were collected. Subsamples of the plankton tows made in these areas were inoculated into the following freshwater media: enriched Mystic Reservoir water, Chu 10, Bristols, Cyanophycean, soil water extract, and enriched seawater from Fishers Island Sound. After six months of incubation at 20 C±1 C, 250 ft-c of light,on a 12/12 light cycle, growth developed in

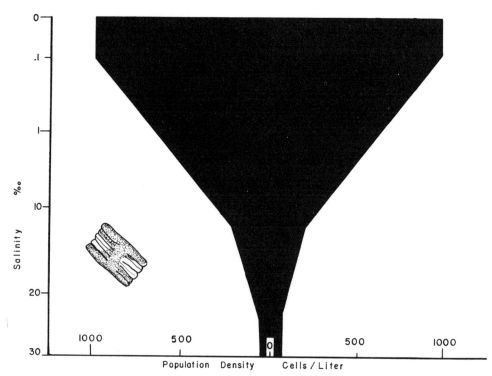

Fig. 16. **Mean results of several field and laboratory dialysis experiments on a population of the diatom *Tabellaria fenestrata*. The surviving cells have an enlarged chloroplast.**

the seawater tubes. The organisms that survived the rigors of seawater culture are listed in Table 3. Figure 17 gives an example of an organism that survived culturing in seawater and changed when morphology transferred into soil water extract medium. The organism is probably *Oocyctis lacustris*, although,

Table 3. Freshwater algae (from various environments) that survived culturing in increased salinity

Species	Quiambog System 1[a]	2	3	Conn. River 1	2	3	Deep Creek L. 3	Lake Huron 3
Bacillariophyta								
Asterionella								
formosa Hassal.				X				
Cyclotella								
glomerata Buchn.				X	X			
menegheniana Kuetz.				X	X			
michiganiana Skv.				X	X			
Eunotia								
curvata (Kuetz.) Largest				X	X			
Fragilaria								
crotonensis Kitton.				X	X			
Melosira								
crenulata (Ehr.) Kuetz.				X	X			
islandica Mueller							X	X
italica (Ehr.) Kuetz.				X	X	X		
Nitzschia								
amphibia Grun.				X	X			
dissipata (Kuetz.) Grun.				X	X			
palea (Kuetz.) W. Sm.				X	X			
Synedra								
acus								
var. *radians* (Kuetz.) Hust.				X	X	X		
ulna (Nitzsch.) Ehr.				X				
Tabellaria								
fenestrata (Lyngb.) Kuetz.			X				X	
Chlorophyta								
Botryococcus								
sudeticus (?) Lemm.				X		X		
Chlamydomonas								
angulosa Dillw.	X	X	X					
Chlorella								
vulgaris Beyer.	X	X	X				X	
Oocystis								
lacustris (?) Chodat								X
sp.							X	
Pediastrum								
boryanum (Turp.) Menegh.				X				
duplex Meyen				X				
duplex								

Table 3. Freshwater algae from various environments continued

Species	Quiambog System 1[a]	2	3	Conn. River 1	2	3	Deep Creek L. 3	Lake Huron 3
var. *gracilimum* W. & G. S. West				X	X	X		
Protococcus								
viridus Ag.				X	X	X		
Scenedesmus								
dimorphus (Turp.) Kuetz.				X	X	X		
quadricauda (Turp.) Breb.	X	X	X	X	X	X		
Selanstrum								
Westii G. M. Sm.				X	X			
Cyanophyta								
Chroococcus								
dispersus (Keissler) Lemm.	X	X	X	X	X	X		
Dactylocnccopsis								
fasicularis Lemm.				X	X	X		
Microcystis								
aeruginosa Kuetz.				X	X	X		
Oscillatoria								
lacustris Kleb.) Geit.						X		
limnetica Lemm.				X				
tenuis Ag.				X				
Synechococcus								
aeruginosa Naeg.							X	
Euglenophyta								
Euglena								
sanguinea Ehr.	X	X						

[a]Numbers used: 1 = upper estuary; 2 = mid-estuary; 3 = lower estuary (usually corresponds to a salinity over 30‰).

only *O. parva* was found in the original sample *(19)*. Numbers 2, 3, and 4 (Fig. 17) were found in the 33.8‰ salinity medium. Soon after transfer to 0.03‰ salinity soil water extract (1 drop culture in 10 ml of medium), number 2 broke open and variously appendaged forms developed (1, 5, 6, 7, 8, 9). The result was a confusing morphology in terms of taxonomy, but a key to a possible mechanism for adaptation and survival is the heavy cell wall.

The next question that can be asked, is, "Even if the freshwater organism makes it through the estuary will it survive and develop in the open sea?" This assumes that the alga is not selectively grazed and can compete well with the salinity-adapted forms.

In an attempt to gather some information in this area, twenty-four samples were taken along a transect from Bermuda to New York to Woods Hole, Massachusetts, during cruise 19–5 of the R. V. *Knorr* (April 1971). Into each of several tubes containing 10 ml freshwater media (Chu 10, Bristols, soil water extract, enriched pond water, and artificial seawater) was placed a drop of

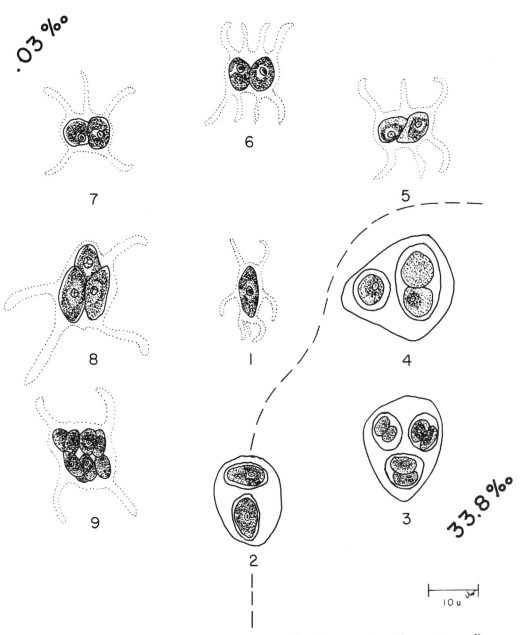

.03%₀₀

33.8%₀₀

Fig. 17.   No. 1–9. *Oocystis lacustris* (?). No. 2–4. *Tetraedron*-like stages found in seawater medium culture (33.8%₀₀ salinity). No. 2 and 5–9. Various stages found in soil-water extract medium culture one month after transfer from seawater medium to soil-water extract medium (.03%₀₀ salinity). The dashed line separates the morphological forms found in seawater and soil-water extract medium. No. 2 is found in both media. In soil-water extract, No. 2 breaks open to release daughter cells. (Reprinted from Foerster *(19)* by permission of the Torrey Botanical Club.)

sample from the plankton tow. After incubation on board the ship at 22 C±3 C at 100 ft-c of continuous illumination the tubes were transferred to a culture unit with 250 ft-c of illumination, a 12/12 light cycle, and a temperature of 20 C±1 C. Within four weeks, growth was noted in all tubes of the soil water and enriched pond water from the western edge of the Gulf Stream shoreward. Figure 18 shows some of the forms that developed in culture. Colonial green cells with the same morphology were noted in the original plankton tow. Continuing shoreward into the New York Bight area, at the head of the Hudson Canyon, at the sludge dumping grounds near the canyon, and along the coast, algae presently described as *Oocystis*-type cells, *Chroococcus*-like, *Chlorella*-type cells, and a *Synedra*-like diatom have been found. At the time of this writing the analyses on the cultures have just begun. Suffice it to mention that the freshwater organisms appeared to be present in the surface meter of seawater.

Discussion

What has been shown is the capacity for many types of so-called freshwater algae to survive transport into seawater. Some of the morphological changes, great or small, were noted. Two questions can be asked: (1) If an alga survives, what is its role in the economy of the estuary and open ocean? and (2) If an alga survives, why is it not prevalent or at least easily recognized in plankton samples?

At this time there is no answer to the first question. At the head of an estuary, freshwater algae appear to contribute 50 to 80% of the total biomass, which may amount to over 50% of the total carbon fixation. As the stream volume is spread more thinly seaward over the estuary, the chance of finding freshwater algae becomes slim.

Considering that of all the water on earth, 97% is represented by seawater. Of the 3% remaining, only 3% is not ice. Therefore, only about .09% of all the water in the world is freshwater runoff, eventually reaching the oceans. It may be reasonable to assume that the total contribution of freshwater forms to the economy in the oceans has not reached a point of genuine significance. However, since morphological changes are apparent when algae move into varying salinities, there may be forms which are unrecognizable as algae and may form what is now considered the important organic detritus segment of the ecosystem. The amorphous development of *Chlorella* and *Scenedesmus* suggests this possibility to be a valid one. Further, the possibility that polymorphism exists when organisms have been exposed to changes in salinity has been documented for *Cyclotella mehegheniana* by Schulz and Trainor (43), and for *Oocystis lacutris* and *Oscillatoria limnosa* by Foerster (19, 16).

In answer to the second question, if one knows freshwater forms are present in samples, then an amorphous *Scenedesmus*, the very thin *Oscillatoria*, or the larger form of *Cyclotella* can be recognized. More work must be done, and is being done, on this problem, for it has a very practical aspect. Is a body of water undergoing eutrophication actually losing some of its important microflora or are the microflora adapting through morphological changes which are an outward manifestation of internal physiological responses? Further, are

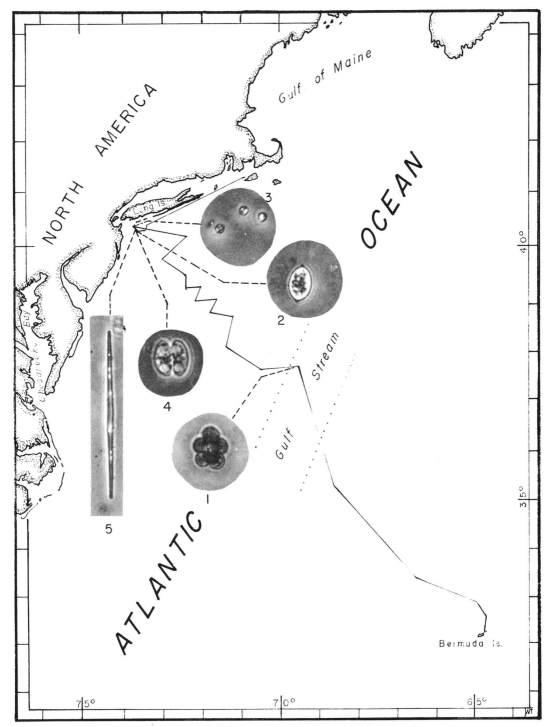

Fig. 18. Photomicrographs of some algae recovered from the surface meter of water. Portions of plankton samples were placed in freshwater media (Chu, o, soil-water extract, Bristols, enriched pond water). After 4 wks of growth, several organisms resembling freshwater algae grew. The algae (1-*Protococcus*-like; 2-*Oocystis*-like; 3-*Chlorella*-like; 4-*Chroococcus*-like; 5-*Synedra*-like) are related to areas from which each was recovered.

these organisms lying dormant, encysted, in spores, etc. until the appropriate conditions are present? In this study, it has been possible to determine that even under drastic change in environmental conditions in a natural system or in the artificality of a test tube, survival by some algae is possible. Therefore, it can be said at this point that some of the freshwater algae survive the rigors of a changing environment in moving from fresh water to the sea.

Summary

This study has determined that a freshwater alga can survive its transport from fresh water into seawater. By exploring various environments from the complicated Quiambog Environmental System to the Great Lakes to the Gulf Stream, various freshwater algae have been found that can survive the chemical-physical forces of seawater. We can now ask, "What is a marine microalga?"

Polymorphism was present and occurred naturally. This was further demonstrated in culture and dialysis experiments. Some of the data has suggested that freshwater algae formed part of the important organic detritus of an estuary.

Obviously, many questions have not been answered, and many more now can be asked. For centuries, algal investigators have considered that two separate microalgal floras exist, one in seawater, the other in fresh water. They have admitted on occasion (24, 39, 58) that some freshwater forms were eurybiont and could be found in estuaries. The existence of freshwater algae in the surface meter of water sampled in the deep sea has not been reported previously. Therefore, it has been shown that some freshwater algae survive the rigors of a changing environment and further questions may be asked on their role in the economy of a saline environment, speciation concepts, and survival capabilities under stress.

Acknowledgments

This investigation has been supported at various stages by an NSF Grant G.B. 4306, a Sigma Xi Grant-in-aide of Research, a Goucher College Faculty Research Grant, an NSF-sponsored cruise aboard the R. V. *Knorr* from the Woods Hole Oceanographic Institution, and the United States Coast Guard (Air Branch). Deep appreciation is extended to Dr. F. R. Trainor and J. D. Buck for their interest and criticism of this research.

Literature Cited

1. American Public Health Association. 1965. *Standard methods for the examination of water and waste water.* 12th ed. Albany: Boyd Printing Co. Inc.
2. Bigelow, H. B. 1926. *Plankton of the offshore waters of the Gulf of Maine.* Department of Commerce, Bureau of Fisheries. Doc. No. 968. Washington, D.C.
3. Blanc, F., M. Leveau, and K. H. Szekielda. 1969. Effects Eutrophiques an debouche d'un granda flerve (Grand Rhone). *Mar. Biol.* **3:**233–42.
4. Boyer, C. S. 1916. *The Diatomaceae of Philadelphia and vicinity.* Philadelphia: J. B. Lippincott Co.

5. Brooks, J. L., and E. S. Deevey, Jr. 1966. New England. In *Limnology of North America*, ed. D. G. Frey, Madison: Univ. of Wisconsin Press.

6. Brunel, J. 1962. *Le phytoplancton de la baie des chaleurs.* No. 91. Contributions du Ministere de la Chasse et des Percheries. Montreal.

7. Bursa, A. 1961. Phytoplankton of the *Calanus* expeditions in Hudson Bay, 1963 and 1964. *J. Fish. Res. Bd. Can.* **18**:51–83.

8. Carter, N. 1938. New or interesting algae from brackish water. *Arch. Protistenkonde* **90**:1–68.

9. Chu, S. P. 1942. The influence of the mineral composition of the medium on the growth of planktonic algae. I. Methods and culture media. *J. Ecol.* **30**:284–325.

10. Collins, R. S. 1909. The green algae of North America. *Tufts College Stud.* **2**:1–480.

11. Davidson, V. M. 1931. The planktonic diatoms of Hudson Bay. *Contrib. Can. Biol. Fish. N. S.* **6**:495–509.

12. Desikachary, T. V. 1959. *Cyanophyta.* New Delhi: Indian Council of Agricultural Research.

13. Drum, R. W., and E. Webber. 1966. Diatoms from a Massachusetts salt marsh. *Bot. Mar.* **9**:70–77.

14. Elmer, C. J. 1921. The diatoms of Nebraska. *Univ. Nebraska Stud.* **21**:22–215.

15. Flemer, D. A. 1970. Primary production in the Chesapeake Bay. *Chesapeake Sci.* **11**:117–29.

16. Foerster, J. W. 1964. The use of calcium and magnesium hardness ions to stimulate the sheath formation in *Oscillatoria limnosa* (Roth) C. A. Agardh. *Trans. Amer. Microsc. Soc.* **83**:420–27.

17. Foerster, J. W. 1968. A portable non-electric current meter. *Chesapeake Sci.* **9**:52–55.

18. Foerster, J. W. 1969. A phyco-periphyton collector. *Turtox News* **47**:82–84.

19. Foerster, J. W. 1971. Environmentally induced morphological changes in *Oocystis lacustris* Chodat (Chlorophyta). *Bull. Torrey Bot. Club* **98**:225–27.

20. Fritz, N. 1921. Plankton diatoms, their distribution and bathymetric range in St. Andrews waters. *Contrib. Can. Biol.* **1918–1920**:49–62.

21. Gaarder, T., and H. H. Gran. 1927. Investigations of the production of plankton in the Oslo Fjord. *Rapp. Conserv. Explor. Mer.* **42**:1–48.

22. Graham, H. W. 1942. *Studies in the morphology, taxonomy and ecology of the peridiniales.* Publ. 542. Washington, D.C.: Carnegie Institution of Washington.

23. Graham, H. W., and N. Bronikovsky. 1944. *The genus Ceratium in the Pacific and North Atlantic Oceans.* Publ. 565. Washington, D.C.: Carnegie Institution of Washington.

24. Griffith, R. E. 1961. *Phytoplankton of the Chesapeake Bay.* Hood College Monograph No. 1. Frederick, Md.

25. Hendey, N. I. 1964. *An introductory account of the smaller algae of British coastal waters. Part V: Bacillariophycea (Diatoms).* London: H. M. S. Office.

26. Heurck, H. van. 1896. *A treatise on the Diatomaceae.* London: Wm. Wesley and Sons.

27. Hohn, M. H., and J. Hellerman. 1963. The taxonomy and structure of diatom populations from three eastern North American rivers using three sampling methods. *Trans. Amer. Microsc. Soc.* **82**:250–329.

28. Hustedt, F. 1955. Marine littoral diatoms of Beaufort, North Carolina. Marine Station Publ. No. 6. Durham: Duke University Press.

29. Hustedt, F. 1965. *Kieselalgen.* Stuttgart: Kosmos-Verlag.

30. Hutchinson, G. E. 1958. *A treatise on limnology. Vol. 1. Geography, physics, and chemistry.* New York: Wiley.

31. Hutchinson, G. E. 1967. *A treatise on limnology. Vol. II. Introduction to lake biology and the limnoplankton.* New York: Wiley.

32. Hylander, J. C. 1928. *The algae of Connecticut.* State of Connecticut Geological and Natural History Survey Bull. No. 42.

33. Ireneae-Marie, Frere. 1939. Flore desmidiale de la region de Montreal. Des freres de l'instruction chretienne, Laprairie.

34. Ivanov, A. A. 1965. Krooprosu o Tipologii fitoplanktona priust evykh raionov cher-

nogo, azovskogo, i Raspiiskogo morei. In *Materialy Zakavkozskoi konferentsii po sporovym rasteniyam*, pp. 29–31. Baku: Akad. Nank. Azerb. USSR.

35. Lauf, G., ed. 1967. *Estuaries.* Washington, D.C.: A. A. A. S.
36. Lebour, M. V. 1925. The dinoflagellates of Northern Seas. *Plymouth Mar. Biol. Ass. U.K.* **13**:1–250.
37. Lebour, M. V. 1930. The plankton diatoms of Northern Seas. *Ray Soc. Publ.* **116**:1–244.
33. Martin, L. H. 1925. *The geology of the Stonington region, Connecticut.* State of Connecticut Geological and Natural History Survey. Bull. No. 33.
39. Patrick, R. and C. W. Reimer. 1966. *The diatoms of the United States* Vol. I. Monog. 13. Philadelphia: Acad. Nat. Sci.
40. Prescott, G. W. 1962. *Algae of the western Great Lakes area.* Dubuque: Wm. C. Brown Co.
41. Reid, G. K. 1961. *Ecology of inland waters and estuaries.* New York: Reinhold Publishing Co.
42. Rumney, G. R. 1968. *Climatology and the world climates.* New York: Macmillan Co.
43. Schultz, M. E., and R. R. Trainor. 1968. Production of male gametes and auxospores in the centric diatoms *Cyclotella menegheniana* and *C. cryptica. J. Phy. Soc.* **4**:85–88.
44. Starr, R. C. 1960. The culture collection of algae at Indiana University. *Amer. J. Bot.* **47**:67–86.
45. Strandberg, C. H. 1967. *Aerial discovery manual.* New York: Wiley.
46. Strickland, J. D. H. 1966. *Measuring the production of marine phytoplankton.* Fish. Res. Bd. Can. Bull. No. 122. Ottawa: Queens Printer.
47. Strickland, J. D. H., and T. R. Parsons. 1968. *A practical handbook of seawater analysis.* Fish. Res. Bd. Can. Ottawa: Queens Printer.
48. Sverdrup, H. V., M. W. Johnson, and R. H. Fleming. 1942. The oceans, their physics, chemistry and general biology. Englewood Cliffs, N.J.: Prentice-Hall Inc.
49. Taylor, W. R. 1953. The relation of *Dichotomosiphon pusillus* to the agal genus *Boodleopsis. Pap. Mich. Acad. Sci. Arts and Lett.* **38**:97.
50. Thornthwaite, C. W. 1948. An approach toward a rational classification of climate. *Geogr. Rev.* **38**:55–94.
51. Tiffany, L. H., and M. E. Britton. 1952. The algae of Illinois. Chicago: Univ. of Chicago Press.
52. Weber, C. I. 1966. *A guide to the common diatoms at water pollution surveillance system stations.* Cincinnati: U.S. Dept. Federal Water Pollution Control Association.
53. Welch, P. S. 1952. *Limnology.* 2nd ed. New York: McGraw-Hill.
54. Wetherell, D. F. 1961. Culture of freshwater algae in enriched natural sea water. *Physiol. Plant.* **14**:1–6.
55. Wetherell, D. F. 1963. Osmotic equilibrium and growth of *Scenedesmus obliquus* in saline media. *Physiol. Plant.* **16**:82–91.
56. Whitford, L. A., and G. J. Schumacher. 1969. *A manual of fresh-water algae in North Carolina.* North Carolina Agr. Exp. Sta. Tech. Bull. No. 188.
57. Wolle, F. 1890. *Diatomaceae of North America.* Bethlehem, Pa.: The Comenius Press.
58. Wood, R. D., and J. Lutes. 1967. *Guide to the phytoplankton of Narragansett Bay, Rhode Island.* Kingston: Stella's Printing.

Comments

G. JONES: Have you given any thought to the mechanisms that may be affecting these cells as they go from one environment to another? What factors in the environment seem to affect these changes most?

FOERSTER: This is one item we are beginning to focus on. We are in the process of trying to get *Oocystis* bacteria-free in an attempt to do nutritional studies. It has been somewhat refractory.

LITCHFIELD: When *Scenedesmus* has been exposed to high salinities, it has a polymorphic structure. Can you then take the organism back into a freshwater medium and restore the "normal morphology?"

FOERSTER: This was not done in this particular study. I did notice that *Scenedesmus* would show up when cultures of Fishers Island Sound plankton were placed in freshwater media. In light of some recent work by Frank Trainor at the University of Connecticut, it is interesting that a lot of the projections, spines, and so forth are determined by environmental nutritional levels. I did some work a number of years ago with *Oscillatoria*. I subjected it to a calcium-magnesium hardness up to 500 mg/l and then brought it back down. In increasing hardness the size and shape changed, but it recovered its original shape as the hardness concentrations were reduced.

COULL: Would you accept the term "ecological race," as some people have proposed, for the same species of blue-greens?

FOERSTER: We find the term "ecological races" used with fish and other plants and animals. I am not sure. I would tend to keep it simplified. In sitting around and nit-picking about whether this is a certain exact species of blue-green or green algae because it has X projections, I am not sure it is all that important. At least I am not convinced it is, especially with polymorphism showing up and being more recognizable in culture work and reported more often in the literature. We used to talk about polymorphism between different kinds of culture media. Many researchers thought we were not doing the experiments correctly. But, it exists, and now I am beginning to doubt some of the species criteria.

COULL: So do you reject the concept?

FOERSTER: Yes, at the present time.

HOLM-HANSEN: On the basis of all your experience, can you generalize on the ability of procaryotic cells to survive salinity changes as contrasted to that of eucaryotic cells?

FOERSTER: The procaryotic cell would seem to have a better chance. I found that some organisms like *Oscillatoria* were well represented in the stream and estuary. Other types of algae (eucaryotic), TABELLARIA, etc., would be spotty. I am not quite sure. I think the organisms that are the simplest may have a better chance. If it survives, it must have some sort of mechanism. I would postulate that it be genetically based.

HOLM-HANSEN: Have you ever seen morphological changes other than the cell wall characteristics on blue-greens?

FOERSTER: The only changes I noted were increased numbers of necridia, the dying cells that break colonies of filamentous blue-green algae apart. Matrix formation, cell attenuation, shrinking of cells, or in some cases swelling, were some of the other changes noted. It depended on the organism.

ANTHONY: I would like to reinforce the last question. Although we have heard a lot about temperature parameters in these discussions, we have not heard much about freezing-thawing. I would like to direct the same question to the freezing-thawing of organisms and whether or not the procaryotes are also more resistant in this respect. I might just say briefly why I asked this. We attempted to investigate the possibility that the freezing of sediment cores might be a satisfactory way of preserving them to return them to the laboratory for subsequent investigation of bacterial content. As one of the referees pointed out when we attempted to publish this, the results seem to suggest that it would be best actually to freeze cores from estuarine environments before you start investigating. I was not quite so surprised at that because I think perhaps the sediment might offer some protection to the cells for freezing purposes. What did surprise me, when we looked at the flora on marine pebbles, was that the same thing turned out to be true. You could hold the pebbles for a period of several weeks, in a deep freeze without any apparent change in the number of bacteria on the surface as determined by our methods.

The *In Vivo* Measurement of Chlorophyll by Fluorometry

Dale A. Kiefer

The measurement of the size and distribution of phytoplankton populations is necessary for a general description of a planktonic community. Knowledge of phytoplankton standing crop and rate of primary production is the basis for a trophic-dynamic description of an aquatic ecosystem (5), while patterns of phytoplankton distribution may be closely related to important physical, chemical, or biological gradients within the system (4).

Of the methods available for estimating phytoplankton standing crop, the determination of chlorophyll concentration continues to be most favored (10). The ease and speed of chlorophyll determination and its specificity with regard to the phytoplankton often outweigh the difficulty in relating chlorophyll concentration to the ecological parameters of total cell volume or carbon content. Since chlorophyll sensitizes plant conversion of light energy to chemical energy, it is more closely correlated with primary production than these other biomass indicators. Thus, Ryther and Yentsch (9) among others have successfully combined light intensity and chlorophyll data to estimate primary production.

Chlorophyll *a*, *b*, and *c* concentrations from laboratory cultures and field samples are routinely determined by pigment extraction in an organic solvent, followed by trichromatic measurement with a spectrophotometer (8). Spectrophotometric analysis was used exclusively until Yentsch and Menzgel (13) introduced a more sensitive fluorometric technique for measuring chlorophyll *a* concentration. In this method, an algal pigment extract is exposed to a blue light source (maximum transmission of the primary filter is about 430 nm). The amount of red fluorescence (emission peak is about 676 nm) emitted by chlorophyll *a* is measured and is proportional to concentration over a wide range. Since extracted chlorophyll *b* and *c* also have blue absorption bands and red emission bands at about 660 nm and 642 nm, respectively, care must be taken to choose a secondary filter which eliminates contaminating fluorescence of these two accessory chlorophylls (3).

The fluorometric determination of extracted chlorophyll *a* is over two orders of magnitude more sensitive than spectrophotometric measurements. The determination is also relatively insensitive to turbidity, and allows one to distinguish chlorophyll *a* from its major degradation product, phaeophytin *a*.

A logical extension of the fluorometric technique was its application to *in vivo* measurements of chlorophyll *a*. This was first done by Lorenzen (6), who

421

found a linear relationship between the red fluorescence (emission peak between 670 and 680 nm) of seawater and its chlorophyll *a* concentration. Since its introduction, the *in vivo* measurement of chlorophyll has been used by relatively few workers.

In this paper, I will discuss the methods and instrumentation of *in vivo* measurements as well as the advantages and difficulties in such continuous chlorophyll measurements. In particular, I will show how we have adopted *in vivo* fluorescence measurements to vertical profiling in lakes and oceans, and how such profiling yields more detailed information about the size and distribution of the phytoplankton. I will also discuss the variability of *in vivo* chlorophyll fluorescence and two important factors which cause such variability. Hopefully, such a discussion will help the participants in this symposium to decide upon the applicability of such measurements to estuarine phytoplankton studies.

Materials and Methods

Measurements of *in vivo* chlorophyll use the same fluorometer (Turner 110 or 111) and optical system as used for *in vitro* determinations. The fluorometer is equipped with a flow-through cuvette, and the voltage output of the fluorometer is recorded on a time-base, chart recorder. Water is continuously supplied to the fluorometer from an on-deck or submersible pump. The water which passes through the fluorometer can be sampled discretely for microscopic examination and for direct determination of chlorophyll *a* concentration upon extraction *(12)*. The trace of *in vivo* fluorescence can be standardized to units of chlorophyll *a* concentration by measuring the *in vivo* fluorescence and extracted chlorophyll *a* concentration for discrete samples taken during the profile. The ratio of $\frac{in\ vivo\ \text{fluorescence}}{\text{chlorophyll } a}$ for the discrete samples can then be divided into the trace of *in vivo* fluorescence values yielding a standardized profile of *in vivo* chlorophyll *a*.

Although the *in vivo* measurement of chlorophyll has been applied mainly to horizontal transects, we have adapted the method to studies of phytoplankton vertical distribution. Vertical profiles are obtained by lowering a hose and submersible pump at a constant rate through the water column. The time at which the pump reaches a given depth is recorded with a depth transducer or by hydrowire readings. After corrections for delay time in the hose, one then knows the fluorescence of water at any depth. Each profile is standardized with four or five discrete chlorophyll *a* samples.

Results of Field Investigations

Figure 1 shows an up-down fluorescence profile and discrete chlorophyll *a* concentrations for a station 14 mi off the San Diego coast. The small differences between the up and down profiles are ascribed to drifting of the ship combined with horizontal patchiness of the phytoplankton. Each of these profiles required about 20 min. Figure 2 shows a chlorophyll and phytoplankton cell volume profile from Lake Tahoe, California. Here a continuous profile was

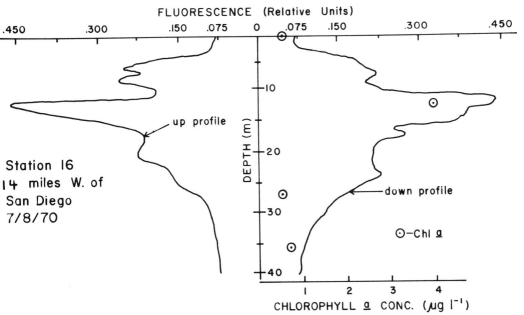

FLUORESCENCE (Relative Units)

Station 16
14 miles W. of
San Diego
7/8/70

CHLOROPHYLL a CONC. (μg l⁻¹)

Fig. 1. Fluorescence trace and discrete chlorophyll *a* concentrations for up–down profiles made off San Diego. Discrete chlorophyll *a* concentrations were determined by sampling the seawater outflow of the fluorometer, extracting algal pigments in 90% acetone, and measuring extracted chlorophyll *a* by fluorometry.

made to a depth of 100 m; below 100 m, discrete samples were taken. The Tahoe profile shows a good correlation between chlorophyll *a* concentration and total phytoplankton cell volume.

We have made over 70 such profiles, and they are of interest for three reasons. (1) The profiles allow a more accurate determination of phytoplankton standing crop of chlorophyll *(11)*. This increased accuracy is, of course, largest when phytoplankton are layered or patchy in distribution. (2) Since the profiles yield distribution information quickly, sampling can be adjusted in the field in order to exploit this information. For example, one can investigate species differences and physiological differences between layers or depth intervals within a profile. (3) Patterns of vertical distribution may help one to determine the factors which determine phytoplankton standing crop. For example, phytoplankton growth and respiration, grazing, and vertical sinking of cells can be thought to control both total standing crop and the vertical distribution of phytoplankton. In addition, it would be interesting to know the extent to which algal distribution patterns determine the structure and function of the entire planktonic community.

While *in vivo* chlorophyll measurements supply much information about the size and location of phytoplankton populations, the measurements are limited in precision and accuracy. Figure 3 shows a plot of *in vivo* fluorescence versus chlorophyll *a* concentration for vertical profiles made in California coastal waters; the scatter about the regression indicates that *in vivo* fluorescence per

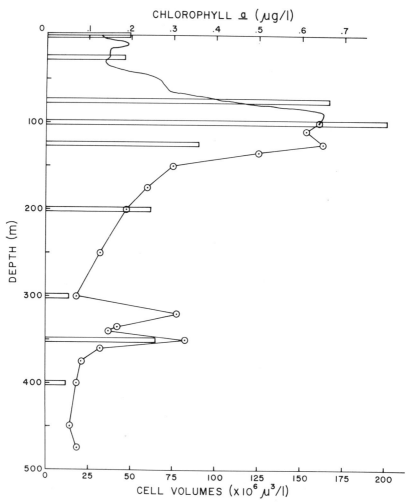

Fig. 2. **Vertical distribution of phytoplankton volumes and chlorophyll concentrations in Lake Tahoe, 30 September 1970. The horizontal bars show the total phytoplankton cellular volume per liter at discrete depths. The continuous line shows the concentration of chlorophyll** *a*. **From 0 to 100 m concentrations were continuously measured with an** *in vivo* **fluorescence profile calibrated with discrete samples. Below 100 m concentrations were determined directly by extraction of discrete samples.**

unit chlorophyll *a* is variable. Both Lorenzen *(6)* and Flemer *(2)* found a variable yield in fluorescence for horizontal profiles at the sea surface.

   In many vertical profiles, two–to threefold changes in the *in vivo* fluorescence per unit chlorophyll *a* have been observed. Most often, profiles taken during the day show low fluorescence per unit chlorophyll *a* at the surface, with a rapid increase with depth. During the night there is no such increase. Figure 4 shows changes in the ratio with changes in depth for a Lake Tahoe profile; the increase in chlorophyll fluorescence with depth is easily seen as a 2½-fold increase.

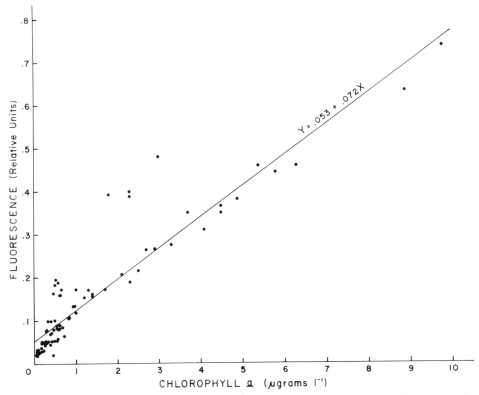

Fig. 3. Plot of *in vivo* fluorescence versus chlorophyll *a* concentrations for discrete samples taken from pump during vertical profiling in San Diego coastal waters (26 May to 4 June 1970). Chlorophyll *a* concentration and *in vivo* fluorescence were measured for discrete samples taken from the outflow of the fluorometer.

Similar, but less predictable, changes have been observed in California coastal waters and in the Gulf of California.

Since the decrease in fluorescence per unit chlorophyll is found near the surface and only during the day, a 24-hr study of sea surface fluorescence was made at a station in the Gulf of California. *In vivo* fluorescence and chlorophyll concentration were measured at regular intervals. The results of this work are shown in Fig. 5, and indicate that the higher ambient light intensities of midday inhibit phytoplankton fluorescence. Fluorescence per unit chlorophyll *a* is minimal near 1300 and maximal at night; the range is threefold. I have made similar observations along the California coast and in the central Pacific. It is likely that such periodicity in surface waters is caused by large changes in ambient light intensity (see laboratory results). Although the mechanism of fluorescence changes is unknown at present, their importance in interpreting and calibrating fluorescence profiles is obvious; one expects large changes in fluorescence per unit chlorophyll *a* for phytoplankton exposed to different ambient light intensities.

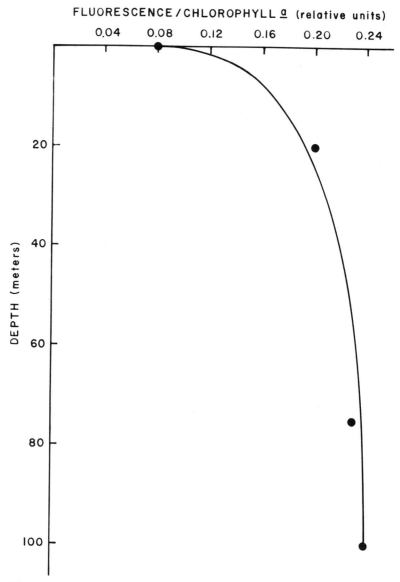

FLUORESCENCE / CHLOROPHYLL $\underline{a}$ (relative units)

Fig. 4. Fluorescence/chlorophyll *a* versus depth at station 26 on Lake Tahoe, near midday, 30 September 1970.

Results of Laboratory Investigations

In addition to field research, the *in vivo* fluorescence and chlorophyll *a* concentration in batch cultures of marine phytoplankton have been measured. The algae were grown under constant illumination and temperature, and they were limited in growth by 25 μM nitrate. The results of this work are summarized in Tables 1 and 2. The first table shows that ambient light intensity but not temperature affects fluorescence in the diatom, *Skeletonema costatum*.

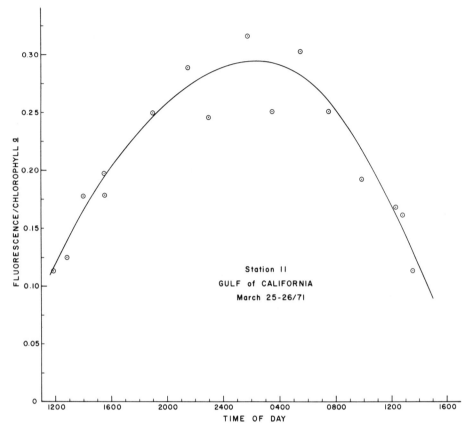

Fig. 5. *In vivo* fluorescence per unit chlorophyll *a* changes over 24 hr for surface water in the Gulf of California. *In vivo* fluorescence and chlorophyll *a* concentration were measured at regular intervals (25–26 March 1971) during pumping of surface seawater.

High light intensity (.12 langleys/min) caused a twofold inhibition of fluorescence, while a 6 C change in temperature had no effect. McCarthy and Strickland *(7)* have observed a similar inhibition of fluorescence by high light intensities. They also noted that the illumination effect is rapid; *in vivo* fluorescence increases or decreases within 10 to 15 min after changes in light intensity. It should be noted that these laboratory observations are in agreement with fluorescence changes in the field.

Table 1. Measurements of fluorescence/chlorophyll *a* for *Skeletonema costatum* grown under constant illumination and temperature

Temp C	Light Intensity (lg/min)			
	.0014	.010	.030	.13
20	.080	.122	.112	.066
14	.080	.104	.104	.066

Table 2 shows fluorescence per unit chlorophyll *a* for different phytoplankton species grown under constant conditions. Measurements were recorded for both exponential and stationary growth phases, but in most cases there was little difference between the phases. The ratios of *in vivo* fluorescence/chlorophyll *a* are approximately the same except for a significantly higher value for *Coccolithus huxleyi*.

Table 2. Measurements of fluorescence/chlorophyll *a* for single species cultures

Division	Species[a]	Mean $\frac{\text{Fluorescence}}{\text{Chlorophylla}}$ (Exponential Phase)	Mean $\frac{\text{Fluorescence}}{\text{Chlorophylla}}$ (Stationary Phase)
Bacillariophyceae	*Skeletonema costatum*	.072	.080
Bacillariophyceae	*Cyclotella nana*	.077	.11
Bacillariophyceae	*Ditylum brightwelli*	.068	.073
Chlorophyceae	*Dunaliella tertiolecta*	.053	.058
Haptophyceae	*Coccolithus huxleyi*	.15	.20
Haptophyceae	*Monochrysis lutheri*	.085	.16
Dinophyceae	[b]*Ceratium dens + C. Furca*	.074	—
Dinophyceae	*Cachonina niei*	.069	—

[a] Grown under continuous illumination (0.11–0.14 lg/1 min) and constant temperature (20 C).

[b] Measured with flow-through system at sea during a dinoflagellate bloom.

In order to investigate the reason for a higher fluorescence ratio in *C. huxleyi*, a comparison of the pigment composition of *C. huxleyi* and *S. costatum* was undertaken. It is well known that accessory pigments sensitize both the light reactions of photosynthesis and chlorophyll *a* fluorescence (1). The sensitization of chlorophyll *a* fluorescence by chloroplast pigments such as chlorophyll *b* and *c* and the xanthophylls has a high efficiency and explains the absence of *in vivo* fluorescence by pigments other than chlorophyll *a*. Since the fluorometer used in my study supplies a broad blue excitation band, both chlorophyll *a* and the blue-absorbing accessory pigments are excited. Such nonspecific excitation in cells of differing pigment composition will yield different *in vivo* fluorescence/chlorophyll *a* values.

Figure 6 shows the fluorescence excitation spectra (emission recorded at 678 nm) for *Skeletonema costatum* and *Coccolithus huxleyi*. The spectra are distinct, indicating a much higher proportion of fucoxanthin to chlorophyll *a* in *C. huxleyi* than in *S. costatum*. Pigment extracts of the two cultures likewise showed *C. huxleyi* enriched with fucoxanthin, when compared to *S. costatum*. Both the *in vivo* excitation spectra and the absorption spectra of extracted pigments showed a twofold increase in fucoxanthin/chlorophyll *a* in *C. huxleyi* compared with *S. costatum*. Thus, differences in fluorescence per unit chlorophyll *a* for the two species can be explained by differences in pigment composition. It is interesting to note that under similar culture conditions *C. huxleyi* also has a twofold larger assimilation number (photosynthetic rate/chlorophyll *a*) than *S. costatum*.

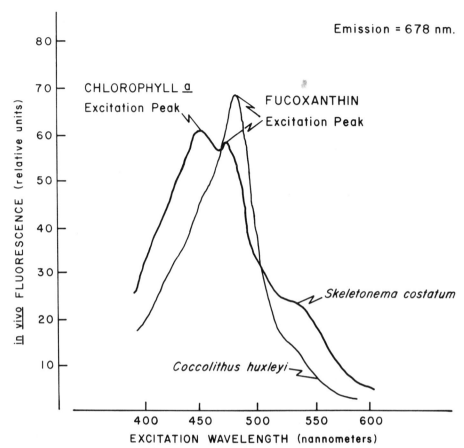

Fig. 6. **Fluorescence excitation spectra for cell suspensions of *Skeletonema costatum* and *Coccolithus huxluyi*. The spectra were made on an Aminco-Bowman scanning spectrofluorometer with the emission recorded at 678 nm. The spectra are uncorrected for photomultiplier sensitivity changes and the small differences in wave length excitation intensity.**

Discussion

The *in vivo* measurement of chlorophyll *a* is a powerful tool to be used in analyzing the planktonic community. Vertical profiles of chlorophyll concentration yield a fine resolution of phytoplankton distribution and a more accurate determination of total algal standing crop. The data supplied by the profiles also allow ecologists to modify sampling programs in the field. These features of the *in vivo* measurements are most valuable when studying ecosystems where temporal and spatial inhomogeneities are large. Such is the case for estuaries.

The major limitation of the *in vivo* measurement of chlorophyll is, of course, the variability of *in vivo* chlorophyll fluorescence. *In vivo* fluorescence per unit chlorophyll can change with species, as in the case of *C. huxleyi* and *S. costatum*, where there are differences in pigment composition. There may also be cases where morphological and physiological differences cause variations between the

chlorophyll fluorescence of species. Besides differences, *in vivo* fluorescence per
unit chlorophyll depends upon ambient light intensity; high intensities cause
an inhibition of fluorescence. The effect of ambient light intensity explains both
diurnal changes in fluorescence at the sea surface and changes within the water
column. If these factors are considered when using fluorometry, the precision
and accuracy of the method will be increased.

Acknowledgments

This research was sponsored by the International Women's Fishing Association and the Marine Life Research Group, Scripps Institution of Oceanography.
I would also like to thank Steve Chaiken for help with pigment analysis.

Literature Cited

1. Duysens, L. N. M. 1952. Transfer of excitation energy in photosynthesis. Thesis,
   University of Utrecht.
2. Flemer, D. A. 1969. Continuous measurement of *in vivo* chlorophyll of a dinoflagellate bloom in Chesapeake Bay. *Ches. Sci.* **10**:99–103.
3. Holm-Hansen, O., C. J. Lorenzen, R. W. Holmes, and J. D. H. Strickland. 1965.
   Fluorometric determination of chlorophyll. *J. Conseil, Conseil Perm. Intern. Exploration Mer.* **30**:3–15.
4. Hutchinson, G. E. 1953. The concept of pattern in ecology. *Proc. Nat. Acad. Sci.*
   105:1–12.
5. Lindeman, R. L. 1942. The trophic-dynamic aspect of ecology. *Ecology* **23**:399–418.
6. Lorenzen, C. J. 1966. A method for the continuous measurement of *in vivo* chlorophyll concentration. *Deep-Sea Res.* **13**:223–27.
7. McCarthy, J. J., and J. D. H. Strickland. 1969. *In vivo* chlorophyll *a* fluorescence.
   In *Progress Report, Research on the Marine Food Chain,* pp. 57–59. La Jollo, California:
   Univ. of California, Institute of Marine Resources.
8. Richards, F. A., and T. G. Thompson. 1952. The estimation and characterization
   of phytoplankton populations by pigment analysis. II. A spectrophotometric
   method for the estimation of plankton pigments. *J. Mar. Res.* **11**:156–72.
9. Ryther, J. H., and C. S. Yentsch. 1957. The estimation of phytoplankton production
   in the oceans from chlorophyll and light data. *Limnol. Oceanogr.* **2**:281–86.
10. Strickland, J. D. H. 1966. *Measuring the production of marine phytoplankton.* 2nd ed.
    Fish. Res. Bd. Can. Bull. 122. Ottawa: Queens Printer.
11. Strickland, J. D. H. 1968. A comparison of profiles of nutrient and chlorophyll con-
    centration taken from discrete depths and by continuous recordings. *Limnol.
    Oceanogr.* **13**:388–91.
12. Strickland, J. D. H., and T. R. Parsons. 1968. *A practical handbook of seawater analysis.*
    3rd rev. ed. Fish. Res. Bd. Can. Bull. 167. Ottawa: Queens Printer.
13. Yentsch, C. S., and D. W. Menzel. 1963. A method for the determination of phyto-
    plankton chlorophyll and phaeophytin by fluorescence. *Deep-Sea Res.* **10**:221–31.

Section Seven
FUNGI

Effects of Environmental Stress
on Aquatic Yeast Populations

Donald G. Ahearn

The presence of yeasts in waters has been associated with organic content *(8, 13, 15, 20, 21)*, but factors related to the survival and distribution of individual species in aquatic habitats have not been clearly defined. This report examines the incidence and distribution of yeasts in waters of the southeastern United States and relates physiological properties of selected species to their ecological occurrence.

Materials and Methods

Yeasts were collected from numerous waters of the southeastern United States including Biscayne Bay, Florida, the Florida Everglades, rivers and lakes of southern Florida, the Gulf of Mexico, and spartina marshes of Jekyll Island, Georgia, and Barataria Bay, Louisiana. Details of collection and isolation procedures and methods of isolation and identification have been published *(3, 14, 19)*. This report reviews these earlier studies and includes additional data in the same areas. Cell densities are given as colony-forming units (cfu) per volume of water. The procedures for determination of hydrocarbon and surfactant utilization are essentially as previously described *(2, 16, 19)*. Cells for respiration studies were grown with agitation for 48 hr at room temperature in Yeast Nitrogen Base *(22)* with 2% (v/v) hydrocarbon or 0.5% glucose. The cells were centrifuged and washed twice in cold phosphate buffer. Rates of oxygen uptake in the presence of hydrocarbons (4% v/v in 0.05 M phosphate buffer, pH 5.8) were determined by conventional Warburg manometric procedures with air as the gaseous phase at 30 C. Crude oil was "weathered" in phosphate buffer for 12 hr prior to the respiration studies.

Results and Discussion

The incidence of the more common species of yeasts in waters is presented in Table 1. All the urban sites were influenced by effluents from sewage treatment installations. The most frequently encountered species in the urban sites, *Candida tropicalis, C. krusei, C. guilliermondii,* and *C. parapsilosis,* were all in high densities in sewage. In general, fermentative species, e.g. *C. tropicalis* and *C. guilliermondii,* occurred in significant densities (>100 cfu/100 ml) only in waters with

Table 1. Incidence of common yeasts in waters of the Southeastern United States

Species	Fresh		Estuarine	
	Rural	Urban	Rural	Urban
Candida guilliermondii	5[a]	20	2	16
Candida krusei	18	32	9	29
Candida parapsilosis	5	20	3	19
Candida reukaufii	21	10	8	4
Candida tropicalis	4	32	3	43
Cryptococcus albidus	20	16	22	19
Cryptococcus laurentii	54	30	41	43
Debaryomyces hansenii	32	26	25	23
Kluyveromyces sp.	3	2	20	6
Pichia sp.	18	10	21	11
Rhodotorula glutinis	31	30	29	31
Rhodotorula rubra	24	23	28	22
Sporobolomyces sp.	60	20	18	5
Total stations	83	79	101	66

[a] Incidence of occurrence (%) in waters from different stations.

a BOD in excess of 2 mg/l. The continued presence of these yeasts in high numbers is suggestive of eutrophication. Species of *Pichia* and *Kluyveromyces* occurred in high densities in marshland sediments and in association with plants, but they were generally in low concentrations in adjacent waters. Oxidative species of *Rhodotorula*, *Cryptococcus*, and *Debaryomyces* (including *Torulopsis* sp.) occurred in all waters and sporadically in densities to about 10^3 cfu/100 ml. Quantitative analysis of collections from all areas (Table 2) supports the report of van Uden *(20)* that yeasts frequently occur in high densities in rivers and lakes and are generally in low concentrations in ocean waters. Blooms of yeasts, however, may occur in all habitats *(2, 8, 13)*.

Environmental stress, particularly chronic enrichment with available carbon,

Table 2. Summary of data on yeasts collected from aquatic habitats in the Southeastern United States[a]

Type of Data	Habitat and Salinity Range					
	Fresh 0–9‰		Estuarine 10–40‰		Marine 34–36‰	
	Rivers	Lakes	Urban	Rural	Coastal	Offshore
No. of stations	73	106	66	101	105	230
No. of stations with yeast	68	83	58	91	96	169
Maximum no. cfu/l	5.2×10^9	9×10^3	5×10^4	1.6×10^4	1.8×10^4	9.2×10^3
Average No. species/l	5.6	3.2	4.3	3.1	2.8	2.4
Average No. cfu/l	592	221	246	132	62	21

[a] Samples from different stations.

directly influences species development and cell densities in aquatic habitats. The cell maxima of yeasts in rivers (Table 2) consisted mainly of strongly fermentative species of *Saccharomyces*. These occurred in sugarcane field drainage areas in waters with a BOD of 90 to 120 mg/l. Twenty miles downstream in waters with a BOD of 5 mg/l, the species of *Saccharomyces* were absent and a more typical oxidative flora was present.

In northern Biscayne Bay, Florida, the predominant species appeared to have been introduced and sustained by chronic urban pollution. There were, however, distinct differences between the species in raw and treated sewage and those in the bay (Table 3). Since high densities of yeasts were frequently encountered in sewage surface foams containing detergents, the effect of linear (LAS) and branch chained (ABS) alkylbenzene sulfonates on yeast growth was examined. The minimal inhibitory concentrations (MIC) of detergents for representative yeasts are given in Table 4. The yeasts tolerated surfactant concentrations exceeding at least five to ten times greater than levels reported for sewage and streams *(4, 17, 18)*. Nevertheless, the amounts of detergents in sewage and natural waters usually are calculated by concentrating large volumes of water samples. Since detergent molecules are adsorbed on organic molecules and detritus in the water (very frequently the microecosystems supporting microbial development), the concentrations of detergents affecting specific

Table 3. Common yeasts in raw and treated sewage, Miami, Florida (1965)[a]

Species	Raw	Treated
Candida albicans	++	—
Candida tropicalis	+++	+++
Candida krusei	+	+++
Candida guilliermondii	+	++
Candida parapsilosis	+++	++
Trichosporon sp.	+	+++
Torulopsis glabrata	+	+
No. samples	43	38

[a] Incidence, −10%; +50%; ++50–75%; +++75–100%.

Table 4. Range of minimal inhibitor concentrations of detergents for yeasts[a]

Species	No. Isolates	LAS (mg/l)	ABS (mg/l)
Candida krusei	14	50 to 180	200 to 400
Candida tropicalis	9	20 to 55	15 to 25
Candida albicans	5	20 to 25	5 to 10
Candida parapsilosis	5	20 to 25	10
Rhodotorula glutinis	5	45 to 60	10 to 25
Sporobolomyces roseus	2	10 to 15	3 to 5

[a] A yeast nitrogen base broth with 0.5% added glucose was used; plates were examined after incubation at 25 C for 7 days.

microhabitats probably well exceed reported values. The marked resistance of *C. krusei* to high levels of detergents and its ability to degrade the detergent molecule *(16)* may partially explain its distribution patterns in aquatic habitats. The association of *C. krusel* with sewage and sewage effluents has been noted previously *(2, 7, 12)*.

Although *Candida albicans* occurs in high densities in human excrement *(1, 5)*, it is not commonly found in natural waters or treated sewage effluents. Hence, its isolation from water has been associated with recreational bathing or recent contamination with raw sewage *(2, 6)*. Several studies have demonstrated that pure cultures of *C. albicans* survive and maintain their pathogenicity for prolonged periods in fresh or seawater *(10, 11)*.

Our preliminary work indicated that the survival period of *C. albicans* in water may be significantly reduced in the presence of other microorganisms (Table 5). When held in water for 4 hr at 22 to 25 C, pure cultures of *C. albicans* showed restricted budding, and analysis of the water indicated an increased absorbance at 260–280 nm and a positive ninhydrin reaction. After 10 to 12 hr, budding was initiated. Isolates of *Rhodotorula* from aquatic habitats expressed a similar phenomenon, but resumed budding within 4 to 6 hr under similar conditions. It appeared that *C. albicans* may undergo a "shock excretion" of essential nitrogenous compounds when suspended in water. In the mixed culture systems of nature, strains of *C. albicans* appeared to be unable to compete effectively for the uptake of essential compounds. The general inability of *C. albicans* to exist without its animal host makes this yeast useful as an indicator organism.

Table 5. Survival of *Candida albicans* in pure and mixed culture in fresh and marine waters[a]

	Lake Water				Coastal Water			
Time in Days	0	3	6	12	0	3	6	12
Pure								
Candida albicans	195[b]	93	82	40	184	198	185	142
Rhodotorula rubra	168	148	151	140	176	180	141	150
Combination								
Candida albicans	181	42	0	2	169	110	112	48
Rhodotorula rubra	164	181	162	151	162	308	250	190

[a] Cells were grown in Yeast Nitrogen Base broth (Difco), with 0.01% added glucose, for 72 hr at 25 C and with constant agitation. The yeast were washed once with the water to be used in the experiment (either lake or coastal water that had been filter-sterilized) and diluted to the indicated cell densities. The final volume in each flask was 200 ml. The yeast suspensions were incubated at 25 C with constant agitation.
[b] cfu/ml, test repeated twice with different water samples with similar results.

Candida tropicalis occurred commonly in northern Biscayne Bay and in process waters of an asphalt refinery in Georgia *(19)*, but it was not isolated in rural regions from waters with an average BOD of <3 mg/l. Strains of *C. tropicalis* are known to utilize hydrocarbons, which constitute a significant carbon source in sewage and harbor waters. Of forty-six cultures from Biscayne Bay, all utilized hexadecane as a carbon source. Five randomly selected isolates assimilated alkanes from C_{10} to C_{18} and grew slowly on crude oil. However, *C. tropicalis* was not found in high densities among occasional oil slicks in the harbors. Unlike randomly selected isolates of *C. tropicalis* from sewage and from Biscayne Bay, cultures from an asphalt refinery *(19)* grew rapidly on crude oil without a lag phase and apparently utilized hydrocarbons with constitutive enzyme systems (Table 6).

Table 6. Oxygen consumption by *Candida tropicalis* in the presence of hydrocarbons

Strain[a]	Growth Substrate	Glucose	Kerosene	Light Gas Oil	Endogenous
R12	Crude	44[b]	48	56	22
R12	Glucose	58	43	51	23
R25	Crude	49	45	59	18
R25	Glucose	53	46	64	20
B8	Crude	41	38	40	19
B8	Glucose	53	32	29	24
S43	Crude	66	51	53	20
S43	Glucose	74	28	36	21

[a] R strains isolated from refinery; B strains isolated from Biscayne Bay; S strains isolated from sewage.

[b] Microliters of oxygen consumed per hour per mg dry weight of cells.

Catastrophic organic stress from oil spills may appreciably affect yeast populations. Samples taken from oil slicks originating from a spewing well in the Gulf of Mexico showed larger populations of rhodotorulas (0–10 cfu/100 ml to 100–2,000 cfu/100 ml). However, the blooms were short-lived, and within 4 to 6 wks, regardless of the presence of oil, cell densities decreased. In laboratory cultures, nitrogen and vitamin enrichment was required for noticeable and sustained oil utilization in seawater by representative isolates.

At estuarine marsh sites in Barataria Bay, demonstrating characteristically high yeast populations *(14)*, yeast densities of plots soaked with oil gradually increased (from 9,000 cfu/g to 18,000 cfu/g wet sediment) over a three-month period. During this time, the predominant species shifted from *Pichia spartinae* and *Kluyveromyces drosophilarum* to *Rhodotorula* sp., *Aureobasidium* sp., and *Trichosporon cutaneum*. The latter species utilized hydrocarbons, but growth was slow. Only in the third month did an actively hydrocarbonoclastic yeast, *P. ohmeri*, become established in the test plot. Of nine isolates examined, only a homothallic ascosporogenous isolate demonstrated constitutive enzyme activity.

Possession of a constitutive enzyme system for the utilization of an available carbon source and resistance to detergents were found to correlate with the aquatic distributions of strains of *C. tropicalis* and *C. krusei*, respectively. In preliminary studies, a general correlation between BOD values and yeast bionomics was observed. It is recognized, however, that yeast distributions are controlled by interaction of a variety of multiple factors. For example, temperatures greater than 35 C selectively affected yeasts present in paper mill effluents *(12)* and asphalt refinery treatment waters *(19)*. Low water temperatures (lower than the body temperature of the human host) were probably responsible, in part, for the poor survival of *C. albicans* observed in aquatic habitats. Since only a few yeasts, mainly oxidative species, appeared to occur normally in oligiotrophic waters, the presence and persistence of strongly fermentative species suggested a chronic or catastrophic organic enrichment. Crucial examination of the metabolic properties of these latter yeasts may permit their use as specific indicator organisms of water quality.

Acknowledgments

This research was supported in part by Office of Naval Research Contract N00014-71-0145 (NR 133-041).

Literature Cited

1. Ahearn, D. G., J. Jannach, and F. J. Roth, Jr. 1966. Speciation and densities of yeasts in human urine specimens. *Sabouraudia* **5**:110–19.
2. Ahearn, D. G., S. P. Meyers, and P. G. Standard. 1971. The role of yeasts in the decomposition of oils in marine environments. *Develop. Indust. Microbiol.* **12**:126–34.
3. Ahearn, D. G., F. J. Roth, Jr., and S. P. Meyers. 1968. Ecology and characterization of yeasts from aquatic regions of South Florida. *Mar. Biol.* **1**:291–308.
4. Brenner, T. E. 1968. The impact of biodegradable surfactants on water quality. *J. Amer. Oil Chem. Soc.* **45**:433–36.
5. Cohen, R., F. J. Roth, E. Delgado, D. G. Ahearn, and M. H. Kalser. 1969. Fungal flora of the normal human small and large intestine. *N. Eng. J. Med.* **280**:638–41.
6. Cook, W. L. 1970. Effects of pollution on the seasonal population of yeasts in Lake Champlain. In *Recent trends in yeast research*, ed. D. G. Ahearn, pp. 107–112. Spectrum, Monograph Series in the Arts and Sciences. Atlanta: Georgia State Univ.
7. Cooke, W. B., and G. S. Matsuura. 1962. A study of yeast populations in a waste stabilization pond system. *Protoplasma* **57**:163–87.
8. Fell, J. W. 1967. Distribution of yeasts in the Indian Ocean. *Bull. Mar. Sci.* **17**:454–70.
9. Fell, J. W., D. G. Ahearn, S. P. Meyers, and F. J. Roth, Jr. 1960. Isolation of yeasts from Biscayne Bay, Florida, and adjacent benthic areas. *Limnol. Oceanogr.* **5**:366–71.
10. Fell, J. W., and S. A. Meyer. 1967. Systematics of yeast species in the *Candida parapsilosis* group. *Mycopath. Mycol. Appl.* **32**:177–93.
11. Madri, P. A. 1966. Factors influencing growth and morphology of *Candida albicans* in a marine environment. *Bot. Mar.* **11**:31–35.
12. Meyers, S. P., D. G. Ahearn, and W. L. Cook. 1970. Mycological studies of Lake Champlain. *Mycolgia* **62**:505–15.
13. Meyers, S. P., D. G. Ahearn, W. Gunkel, and F. J. Roth, Jr. 1967. Yeasts from the North Sea. *Mar. Biol.* **1**:118–23.
14. Meyers, S. P., D. G. Ahearn, and P. C. Miles. 1971. Characterization of yeasts in Barataria Bay. *Louisiana State Univ. Coast. Stud. Bull.* **6**:7–15.
15. Spencer, J. F. T., P. A. J. Gorin, and N. R. Gardner. 1970. Yeasts isolated from the South Saskatchewan, a polluted river. *Can. J. Microbiol.* **16**:1051–57.
16. Standard, P. G., and D. G. Ahearn. 1970. Effects of alkylbenzene sulfonates on yeasts. *Appl. Microbiol.* **20**:646–48.
17. Sullivan, W. T., and Swisher, R. D. 1969. MBAS and LAS surfactants in the Illinois River, 1968. *Environ. Sci. Tech.* **3**:481–83.
18. Sweeney, W. A. 1966. Note on straight-chain ABS removal by adsorption during activated sludge treatment. *J. Wat. Pol. Cont. Fed.* **38**:1023–25.
19. Turner, W. E., and D. G. Ahearn. 1970. Ecology and physiology of yeasts of an

asphalt refinery and its watershed. In *Recent trends in yeast research*, ed. D. G. Ahearn, pp. 113–23. Spectrum, Monograph Series in the Arts and Sciences. Atlanta: Georgia State Univ.

20. Uden, N. van. 1967. Occurrence and origin of yeasts in estuaries. In *Estuaries*, ed. G. H. Lauf, pp. 306–10. Washington, D.C.: AAAS.
21. Uden, N. van, and J. W. Fell. 1968. Marine yeasts. *Adv. Microbiol. Sea* **1**:167–201.
22. Wickerham, L. J. 1951. Taxonomy of yeasts. *Tech. Bull. U.S. Dep. Agr.* **1029**:1–55.

Comments

FELL: You are using galvanized drums for your test plots. Do you think there is any effect from heavy metals leaching out of the drums?

AHEARN: Possibly. However, the yeast populations within these drums are quite high, up to 90,000 cells or colony-forming units per wet gram of sediment.

FELL: *Candida tropicalis* is probably a good pollution indicator. We find it in polluted areas. Do you think it has a specific role in the environment or does it simply demonstrate good survival?

AHEARN: *Candida tropicalis* is one of the strains used for the production of single cell protein from crude oil. It is a good indicator of organic enrichment in urban areas. We have not found it in the *Spartina* marshes or rural regions of the Louisiana delta. We hope to put it into that region in one of our control plots in the near future to see if it will survive. It probably plays a role in the degradation of complex organics in sewage and urban waters but it does not survive in sea-water to any great extent.

ZOBELL: I found this tremendously exciting, and I find it very difficult to refrain from making many comments that are mostly complimentary to Dr. Ahearn for a wonderful presentation. It is a very important, highly significant bit of work. Numerous questions come to mind, most of them personal and professional, but I shall restrict myself to one question on technique. In your test using 4% volume to volume of different kinds of hydrocarbons, were these shaken cultures? I ask this question because most of these organisms were probably aerobic. Oil can prevent penetration of oxygen into the medium if a heavier asphalt oil is used. There is less tendency for oxygen to go through such heavy oil than the lighter oils such as kerosene. Very often investigators making these observations come up with the conclusions that the more viscous oils are not attacked whereas the lighter ones are. Frequently, it is merely a matter of getting oxygen to the cells.

AHEARN: We used both static and agitated cultures, and have compared agitated and standing cultures. Agitation yielded better growth. This may not be so much related to the oxygen as to "bathing" cells with enough utilizable nitrogen.

Variations in Soil Fungus Populations in a South Carolina Salt Marsh[1]

G. T. Cowley

Since the publication of the Bayliss-Elliot research *(2)* on the fungi of the Dovey salt marsh soils, little new data have appeared in the literature which would contribute to the understanding of salt marsh soil fungus populations. The work done to date has concentrated primarily on taxonomy of isolates rather than population structure.

Terrestrial soil mycologists began in the late 1940s to emphasize differences in soil fungus populations in habitats differing in vegetational cover and in various physical and chemical factors *(3, 5, 6, 7, 8, 10, 11, 13, 14, 15)*. Vegetational zonation within a salt marsh is rather obvious even to the untrained observer. Thus, the salt marsh environment lends itself well to a study of soil fungus population zonation. However, tidal flooding and its accompanying effects on the physical and chemical factors within the marsh tend to complicate interpretation of results.

This study was an attempt to assess variations in soil fungus populations and to prepare groundwork for further studies on evaluation of effects of environmental variables. The role of saprophytic fungi in salt marsh soil environments was considered. Although many salt marshes are available for study in South Carolina, the Hobcaw Barony salt marsh near Georgetown, South Carolina, was selected because it is apparently less influenced by domestic and industrial activities than most others in the state. In addition, its associated estuary system is under intensive investigation by a number of biologists.

Materials and Methods

The area selected for study is a 50 yd wide strip within the salt marsh on the Hobcaw Barony. The strip extends for approximately 200 yd, from the edge of a tidal creek to the base of a pine stand. Four vegetational regions are present: a tall *Spartina* region at the low end of the study area, a dwarf *Spartina* region, a *Salicornia* region, and a *Juncus* region at the base of the pine stand. In addition to these vegetational regions, a zone covered by *Spartina* debris normally lies between the *Salicornia* and the *Juncus* regions. However, the debris is subject to movement to higher regions by excessively high tides. Such displacement took place between the two sampling periods of this study.

[1] Contribution No. 34, of the Belle W. Baruch Coastal Research Institute.

441

In November 1970 and March 1971, two sets of six soil samples (ca. 20 g) were collected in each of seven sampling zones, designated as follows: (1) TS1 = low end of the tall *Spartina* region, (2) TS2 = midpoint of the tall *Spartina* region, (3) TS3 = high end of the tall *Spartina* region, (4) DS = dwarf *Spartina*, (5) Sa = *Salicornia*, (6) SD = *Spartina* debris, and (7) J = *Juncus*. Each sample was collected from the top 2 in of soil with a spatula rinsed in 65% alcohol between each sampling. Samples were placed in sterile plastic bags, refrigerated in a portable ice chest, and transported to the laboratory for processing on the same day.

Processing consisted of measurement of pH, determination of moisture content, and enumeration and isolation of fungi. The soil pH from the wet end of the sample strip was measured directly with a Coleman Metrion III pH meter, while samples from the dry end of the strip were moistened with distilled water before measurements were made. Soil moisture content was determined by weighing a portion of each sample in a tared vessel, followed by drying at 105 C and reweighing. Moisture content was calculated as percentage of soil dry weight:

$$\frac{(\text{fresh weight} - \text{dry weight})}{(\text{dry weight})}.$$

Enumeration and isolation of fungi were accomplished by suspending 10 g of each soil sample in 90 ml of sterile distilled water and preparing further dilutions to 1:100 and 1:1,000. One ml of each dilution was pipetted onto, and distributed over, the surfaces of four petri dishes of Martin's medium *(1)* and incubated at room temperature (ca 27 C). Martin's medium was selected since more colonies and species developed on it than on any of several other media tested. After incubation for five to seven days, colonies were counted and total propagule populations per gram fresh weight and dry weight of soil were calculated. Thirty random isolates from each sample were transferred to tube slants of malt extract agar *(12)*, incubated for seven to fourteen days, and sorted into groups of visibly distinct entities.

Distributional data were recorded as the number of isolates of each species per sample and the number of samples showing occurrence of each species within each set of six samples. Indices of similarity, based on frequency of occurrence (number of samples of occurrence in each set of six samples) and on presence or absence of species within a population derived from six samples were calculated for all possible pairs of populations. The formula used for calculating the index of similarity was

$$\frac{2W}{A + B} \times 100,$$

where A is the total of the measures of all species in one population, B is the similar total for a second population, and W is the total of the lower measures of those species appearing in both populations *(4)*. A half matrix of all indices was constructed for each set of calculations. Such matrices, while showing relationships between all populations, does not show a simultaneous relationship between all populations. To accomplish simultaneous demonstration of relationships, three-dimensional ordinations of populations were constructed by the method of Bray and Curtis *(4)* as modified by Loucks *(9)*. By this ordination method, each population was projected into three-dimensional space and the placement of each population in relation to all others was a reflection of its relationship with respect to the others. End (reference) populations for the first *(X)* axis were selected by first totaling the indices of similarity in each column of the matrix. The population with the lowest average similarity to all other populations (the column with the lowest total) was selected as a reference population. The second reference population was that which was least similar to the first reference population. The interpopulation distance between each two populations was calculated (100–index of similarity). To avoid the problem of working with percentages, interpopulation distances were subjected to an angle transformation by determining arcsines *(9)*. The transformed inter-population distance between the two reference populations was the number of units separating them on the first axis. Each other population was placed on the first axis using the formula,

$$\frac{b^2 + a^2 - c^2}{2a},$$

where *a* is the transformed interpopulation distance between reference populations, *b* is the transformed interpopulation distance between the population to be placed and the first reference population, and *c* is the transformed distance between the population to be placed and the second reference population.

Reference populations for the second *(Y)* axis were selected by subtracting the difference between ordination points on the *X* axis from the transformed interpopulation difference for all pairs of populations. The pair yielding the highest total was selected as *Y* axis reference populations. They were separated from each other on the *Y* axis by the transformed interpopulation difference. Positions of other populations on the *Y* axis were determined in the same manner as for the *X* axis. Reference populations for the third *(Z)* axis were selected similarly, except differences between ordination points on both axes were subtracted from interpopulation distances. Once ordinations were plotted, not only could closely related populations be detected by clustering of points, but also distribution of individual species could be plotted in relation to population differences.

Results

Visual observations indicated that the soils in the TS zones were composed primarily of silt, in the DS zone, of silt mixed with sand, and in the remaining zones, primarily sand. The DS zone was further characterized by a dense mat of roots immediately beneath the soil surface. Average soil pH range was from 6.2 (SD zone) to 6.8 (TS zones). Between these extremes, average pH values were 6.3 (Sa zone), 6.5 (J zone), and 6.6 (DS zone). Moisture content at the time of sampling in November 1970 ranged from 194% in the TS zones and 129% in the DS zone to 45, 40, and 30% in the SD, Sa, and J zones. Soil moisture was considerably higher (up to 279%) in the low end of the marsh during the March 1971 sampling.

Total populations per gram fresh weight and dry weight of soil were highest in the SD zone during both sampling periods (Table 1). Although it was not measured, the dark color of the soil in this zone suggested it contained a higher organic matter content than surrounding soils. November TS and DS populations, per gram of fresh soil, were much lower than those in the SD and J zones. However, TS and DS zone populations per gram dry soil were near those of the J zone. In March, populations per gram dry soil in the wet end of the marsh were much lower than either the SD or J zones. Sa zone populations per gram fresh and dry soil were lowest during both sample periods.

Table 1. Average number of propagules $\times 10^3$ per gram of fresh soil and per gram of dry soil in each sampling zone (in November 1970 and March 1971)

Weight	Zone[a]						
	TS1	TS2	TS3	DS	Sa	SD	J
Nov. fresh wt.	9.0	12.6	18.5	17.4	8.6	67.0	33.9
dry wt.	26.5	35.0	48.1	39.9	12.0	97.6	44.2
Mar. fresh wt.	2.0	1.6	1.3	3.0	1.7	30.6	21.4
dry wt.	6.9	6.1	4.0	6.9	2.4	44.6	27.9

[a] TS1 = tall *Spartina* region at the low end of the study area.
TS2 = midpoint of the tall *Spartina* region.
TS3 = high end of the tall *Spartina* region.
DS = dwarf *Spartina* region.
Sa = *Salicornia* region.
SD = *Spartina* debris (between the *Salicornia* and *Juncus* regions)
J = Juncus region at the base of the pine stand.

The difference between SD and J zone populations was less in March than in November. However, the position of the SD zone had changed between sampling periods. Excessively high tides during a December storm moved the *Spartina* debris from its usual position at the base of the J zone to near the high end of the J zone.

Although the number of detectable propagules was highest in the SD zone, the total number of species and the number of abundant species (isolated from six or more samples) was lowest (Table 2). The number of abundant species in the J zone was also relatively low.

Table 2. Total number of species and number of abundant species (isolated from 6 or more samples) isolated from each zone

Species	Zone[a]						
	TS1	TS2	TS3	DS	Sa	SD	J
Nov. Total Species	55	51	51	62	66	22	55
Abundant Species	30	27	27	30	32	17	17
Mar. Total Species	73	63	60	50	45	14	57
Abundant Species	27	27	27	27	21	9	19

[a] Same as Table 1.

Discussion

Indices of similarity based on frequency of isolation and on presence or absence of abundant species within populations in November (Tables 3 and 4) and in March (Tables 5 and 6) clearly demonstrated that populations from vegetationally similar sample areas were more similar to each other than to populations from soils inhabited by different vegetational types. Although indices of similarity indicate relationships between pairs of populations, they cannot show interrelationships among all populations simultaneously. The ordination procedure used, in part, overcomes this problem by placing each stand at a point in space relative to all other stands. Ordination points along each of three axes are shown in Table 7. The points were plotted and are shown in Figures 1 and 2. In the figures, points representing similar vegetational regions are encircled to show clustering of populations. Figure 3 illustrates the distribution of four representative species in relation to population positions.

There are many methods and media for isolation of fungi from soil populations; apparently none result in isolation of all forms present. The method and medium used in this study, while not one of those used by most marine mycologists, did allow for isolation of a wide variety of species (over 200 in this study) and a somewhat quantitative treatment of distributional data.

While some abundant species were isolated from only one or two vegetational regions, many were more widespread. For example, in March, thirteen of the abundant species were isolated from only the TS and DS zones and three were restricted to the higher regions of the marsh. Of the remaining twenty-six abundant species, eleven were isolated from both ends of the sample area. The majority of the widespread species, however, did appear in greater abundance in certain vegetational regions than in others. The differences in population composition between vegetational regions were not surprising, in the light of distributional patterns observed in terrestrial soils. However, whether population differences observed in this salt marsh were a result only of vegetation influences is doubtful. Differences in soil pH were not great enough to be considered as a strong influence, but other factors could well have been important. For example, populations from the TS and DS zones were quite similar, as indicated by indices of similarity and relative positions or ordination plots. These similarities could result from similar vegetational types or from a high silt content on both soils. The high silt content, in addition to resulting in soil texture similarities, also resulted in a higher moisture content than for the other soils.

The duration and frequency of tidal flooding are obvious factors to be considered. The TS and DS zones are regularly flooded with each high tide, while the Sa and SD zones may not be flooded as often; the J zone is normally flooded only during excessively high tides, such as during the December storm, and then only for short times. Another factor with a potentially strong effect on fungus populations is temperature. Cold winter waters could well be the factor causing the very low populations occurring in March in the low end of the marsh.

In treatment of the distributional data, only species isolated from six or more soil samples were used. The reasons for doing this were to eliminate the effects

Table 3. Indices of similarity for November 1970 populations based on frequency of abundant species (lower left) and angle transformations of interstand differences (upper right)

	TS1a[a]	TS1b	TS2a	TS2b	TS3a	TS3b	DSa	DSb	Saa	Sab	SDa	SDb	Ja	Jb
TS1a	—	17.7	27.3	31.7	25.0	24.3	26.8	29.0	38.0	37.5	56.2	54.7	55.8	53.7
TS1b	69.5	—	24.8	30.5	28.3	29.7	34.0	32.2	36.0	38.5	51.3	50.2	59.8	54.3
TS2a	54.0	51.8	—	24.5	32.0	27.3	35.5	28.3	40.3	41.5	50.2	49.2	58.0	46.5
TS2b	47.6	48.6	58.5	—	34.7	31.7	45.5	45.3	52.3	46.2	67.4	75.0	75.5	65.0
TS3a	57.8	52.6	46.9	43.1	—	17.5	36.2	29.2	42.2	36.2	50.3	59.3	47.8	40.5
TS3b	58.8	50.5	54.0	47.6	69.9	—	32.0	33.0	35.2	37.5	50.8	57.3	51.0	51.3
DSa	57.9	44.2	42.0	28.6	41.0	47.1	—	17.8	27.5	31.8	43.7	46.0	48.8	43.0
DSb	51.5	46.8	52.5	28.9	51.2	45.5	69.3	—	31.0	28.7	39.0	45.0	48.7	37.2
Saa	38.4	41.2	35.3	20.9	32.9	42.3	53.8	48.5	—	22.0	40.0	38.0	43.3	38.2
Sab	39.2	37.8	33.7	27.8	41.0	39.2	47.4	52.1	62.6	—	41.8	35.2	45.3	37.5
SDa	16.9	21.9	23.2	7.5	23.1	22.5	31.0	37.1	30.1	33.3	—	19.5	46.0	43.3
SDb	18.4	23.2	24.3	3.4	14.0	15.8	28.1	29.3	38.5	42.3	66.7	—	39.5	40.0
Ja	17.3	13.5	15.2	3.2	25.8	22.2	24.7	25.0	31.3	28.9	28.0	36.4	—	22.3
Jb	19.5	18.7	27.5	9.4	28.6	22.0	31.7	39.5	38.0	39.0	31.4	35.7	62.3	—

[a] a and b designate populations from different sets of six soil samples within each zone.

Table 4. Indices for similarity of November 1970 populations, based on presence or absence of abundant species (lower left) and angle transformations of interstand differences (upper right)

	TS1a[a]	TS1b	TS2a	TS2b	TS3a	TS3b	DSa	DSb	Saa	Sab	SDa	SDb	Ja	Jb
TS1a	—	17.7	20.3	25.8	19.0	17.0	24.7	25.3	29.2	23.0	49.3	41.8	51.0	46.8
TS1b	69.6	—	17.7	29.2	20.5	19.5	28.0	25.3	22.8	23.0	39.5	37.7	45.3	46.8
TS2a	65.2	69.6	—	19.5	20.5	19.5	22.8	17.8	29.2	27.3	39.5	37.7	45.3	32.3
TS2b	56.4	51.3	66.7	—	19.5	15.5	31.7	31.7	38.2	39.8	57.8	62.5	59.5	53.2
TS3a	67.4	65.1	65.1	66.7	—	11.5	31.5	25.8	34.3	26.2	47.2	54.8	41.2	39.5
TS3b	70.8	66.7	66.7	73.2	80.0	—	26.8	29.3	24.3	24.7	45.5	47.5	47.5	48.0
DSa	58.3	53.1	61.2	47.6	47.8	54.9	—	8.8	20.2	17.8	33.8	32.5	39.8	36.8
DSb	57.1	57.1	69.4	47.6	56.5	51.0	84.6	—	20.2	20.3	33.8	32.5	39.8	30.0
Saa	49.0	61.2	49.0	38.1	43.5	58.8	65.4	65.4	—	17.8	37.7	29.2	32.5	30.0
Sab	60.9	60.9	52.2	35.9	55.8	58.3	69.4	65.3	69.4	—	39.5	30.0	37.7	23.8
SDa	24.2	36.4	36.4	15.4	26.7	28.6	44.4	44.4	38.9	36.4	—	28.5	34.3	35.7
SDb	33.3	38.9	38.9	6.9	18.2	26.3	46.2	46.2	51.3	50.0	52.2	—	27.5	28.8
Ja	22.2	27.8	27.8	13.8	30.3	26.3	35.9	35.9	46.2	38.9	43.5	53.8	—	15.0
Jb	27.0	27.0	43.5	20.0	36.4	25.6	40.0	50.0	50.0	59.5	41.7	51.9	74.1	—

[a] a and b designate populations from different sets of six soil samples within each zone.

Table 5. Indices of similarity for March 1971 populations, based on frequency of abundant species (lower left) and angle transformations of interstand differences (upper right)

	TS1a[a]	TS1b	TS2a	TS2b	TS3a	TS3b	DSa	DSb	Saa	Sab	SDa	SDb	Ja	Jb
TS1a	—	16.5	25.8	24.8	23.0	28.2	34.5	36.7	42.3	39.0	61.3	58.8	36.8	50.0
TS1b	71.5	—	26.8	27.2	22.7	25.3	28.3	36.2	48.0	48.7	68.8	70.2	55.3	54.7
TS2a	56.5	54.9	—	21.8	22.7	30.7	31.7	38.5	45.3	43.5	75.2	65.2	57.3	53.8
TS2b	58.1	54.3	62.9	—	20.3	22.3	36.3	35.3	43.7	38.0	64.5	61.5	43.5	44.7
TS3a	60.9	61.5	61.4	65.2	—	16.0	40.5	40.2	45.3	41.5	60.3	54.7	52.2	48.8
TS3b	52.8	57.1	49.0	62.1	72.5	—	39.8	38.0	42.0	42.0	57.2	52.8	41.8	44.5
DSa	43.3	52.6	47.4	40.8	35.1	36.0	—	18.2	42.0	40.7	70.5	76.7	56.7	49.2
DSb	40.4	40.9	37.8	42.2	35.6	38.5	68.7	—	39.0	42.2	65.0	70.0	57.8	60.3
Saa	32.6	25.6	28.9	31.0	28.9	33.0	37.0	32.9	—	16.2	68.3	75.2	45.8	38.2
Sab	37.0	25.0	31.2	38.5	33.8	33.0	34.9	32.9	72.2	—	53.2	54.7	43.8	40.2
SDa	12.3	6.8	3.3	9.7	13.1	16.0	5.7	9.4	7.1	20.0	—	24.8	36.3	37.5
SDb	14.5	5.9	9.2	12.1	18.5	20.3	2.7	6.0	3.3	18.5	57.9	—	33.2	42.8
Ja	30.0	17.7	15.8	31.2	21.1	33.3	16.5	15.4	28.2	30.8	40.8	45.3	—	24.2
Jb	23.4	18.4	19.2	29.7	24.7	29.9	24.4	13.2	38.2	35.5	39.1	32.0	59.0	—

[a] a and b designate populations from different sets of six soil samples within each zone.

Table 6. Indices of similarity for March 1971 populations, based on presence or absence of abundant species (lower left) and angle transformations of interstand differences (upper right)

	TS1a[a]	TS1b	TS2a	TS2b	TS3a	TS3b	DSa	DSb	Saa	Sab	SDa	SDb	Ja	Jb
TS1a	–	13.2	21.5	22.3	23.8	23.8	27.8	28.0	36.7	34.2	51.8	46.4	37.7	36.7
TS1b	77.3	–	24.5	19.0	19.0	18.3	22.3	26.2	48.1	49.2	59.0	59.5	46.2	36.7
TS2a	63.4	58.5	–	16.7	25.5	26.2	21.8	30.0	37.7	39.2	67.0	57.8	49.3	46.2
TS2b	62.2	66.7	71.4	–	17.7	17.0	23.0	30.0	36.8	31.7	59.5	53.2	37.8	31.7
TS3a	59.6	66.7	57.1	69.6	–	9.7	31.5	33.0	40.5	38.2	52.5	41.8	42.5	40.5
TD3b	59.6	68.1	54.5	70.8	83.3	–	30.0	31.5	38.2	36.2	47.8	43.3	32.5	32.7
DSa	53.3	62.2	61.9	60.9	47.8	50.0	–	13.2	23.7	24.5	59.5	68.8	51.7	33.3
DSb	51.2	55.8	50.0	50.0	45.5	47.8	77.3	–	24.8	26.7	51.0	59.0	50.5	52.2
Saa	35.9	25.6	38.9	40.0	35.0	38.1	60.0	57.9	–	9.7	55.7	66.5	33.2	28.2
Sab	48.9	24.4	36.8	47.6	38.1	40.9	57.1	55.0	83.3	–	42.8	43.8	26.2	30.0
SDa	21.4	14.3	8.0	13.8	20.7	25.8	13.8	22.2	17.4	32.0	–	22.7	23.7	37.3
SDb	27.6	13.8	15.4	20.0	33.3	31.3	6.7	14.3	8.3	30.8	61.5	–	25.5	41.8
Ja	38.9	27.8	24.2	38.7	32.4	46.2	21.6	22.9	45.2	54.5	60.0	57.1	–	13.0
Jb	35.9	35.9	27.8	47.5	35.0	42.9	45.0	21.1	52.9	50.0	34.8	33.3	77.4	–

[a] a and b designate populations from different sets of six soil samples within each zone.

Table 7. Ordination points on each axis, based on the angle transformation of inter-stand distances in Tables 3 through 6

	November						March					
	Frequency, Axis			Pres. or Abs., Axis			Frequency, Axis			Pres. or Abs., Axis		
Zone	X	Y	Z	X	Y	Z	X	Y	Z	X	Y	Z
TS1a	30.8	46.4	34.1	27.6	25.3	61.7	30.6	29.2	33.9	35.4	32.9	39.7
TS1b	27.9	39.7	39.7	36.8	22.0	52.2	18.4	21.0	37.0	23.3	21.0	38.8
TS2a	26.4	48.2	37.6	32.7	27.6	52.2	24.2	31.1	33.6	24.4	36.5	33.3
TS2b	7.0	62.5	43.7	14.0	24.3	26.0	29.3	32.8	39.2	38.7	34.8	44.2
TS3a	37.6	63.1	44.9	26.4	33.2	36.1	36.5	29.0	37.9	39.9	31.8	39.8
TS3b	34.2	53.1	38.7	27.1	24.8	46.5	37.5	33.3	42.4	38.8	33.4	44.4
DSa	42.7	48.3	29.0	41.7	39.3	51.2	7.0	37.5	36.6	11.0	44.5	38.0
DSb	42.6	53.0	36.9	41.7	42.6	51.2	15.6	41.1	23.9	21.4	45.8	19.0
Saa	50.4	44.8	52.4	43.2	41.4	59.7	18.9	69.0	31.2	17.3	69.1	43.4
Sab	45.3	42.9	60.8	43.2	45.0	61.3	36.6	67.0	26.0	35.9	48.5	43.3
SDa	61.1	26.3	47.5	71.8	48.4	35.5	68.2	45.7	63.8	67.4	49.0	56.6
SDb	71.7	25.0	58.5	76.5	51.7	64.0	83.7	53.8	60.5	79.8	35.9	61.7
Ja	82.5	58.3	50.0	63.4	64.9	57.1	59.1	62.7	58.9	62.3	56.5	67.5
Jb	70.5	65.0	51.9	56.4	68.8	55.8	49.2	69.0	66.2	40.7	51.9	71.2

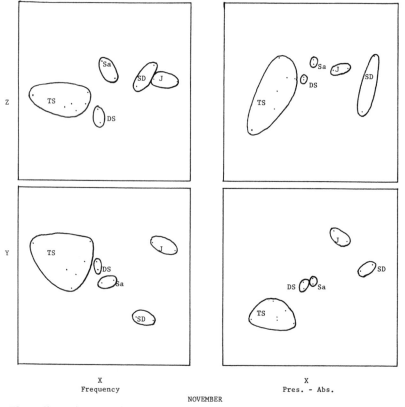

NOVEMBER

Fig. 1. **Three-dimensional ordinations of November 1970 populations based on frequency and presence or absence of abundant species.**

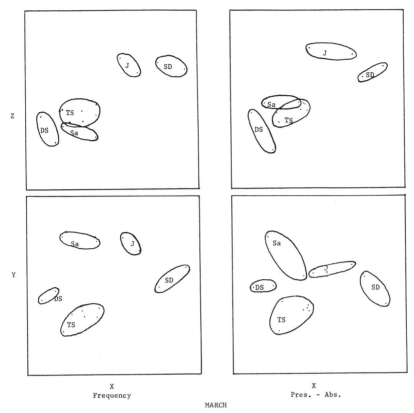

Fig. 2. **Three-dimensional ordinations of March 1971 populations based on frequency and presence or absence of abundant species.**

of rare species and to reduce the effects of species which appeared only because of the presence of spores carried in by tidal waters.

The low populations in the Sa zone were not unexpected. Vegetation in this zone is more sparse than in any other vegetational region. Thus, it may be assumed that less organic matter is available for fungal use. High populations in the SD zone may well be a function of abundant organic matter leached from the dead *Spartina* parts, or they may be a reflection of large spore populations reaching the soil from fungi inhabiting the dead *Spartina*. This latter possibility is being investigated by comparing populations on the dead *Spartina* and in the soil.

Our unpublished studies concerning terrestrial systems have shown little relationship between litter and soil populations. However, the addition of tidal waters in the salt marsh system may well result in a washing of spores from the dead plant parts into the soil. Regardless of which explanation is used, *Spartina* debris does have a strong influence on populations in the soil. This influence is demonstrated by the change from a *Juncus* population to a *Spartina* debris population after the debris had been moved into the *Juncus* zone. The population changes after movement of the *Spartina* debris into the rarely flooded *Juncus* zone suggest that tidal flooding has little influence on the SD populations.

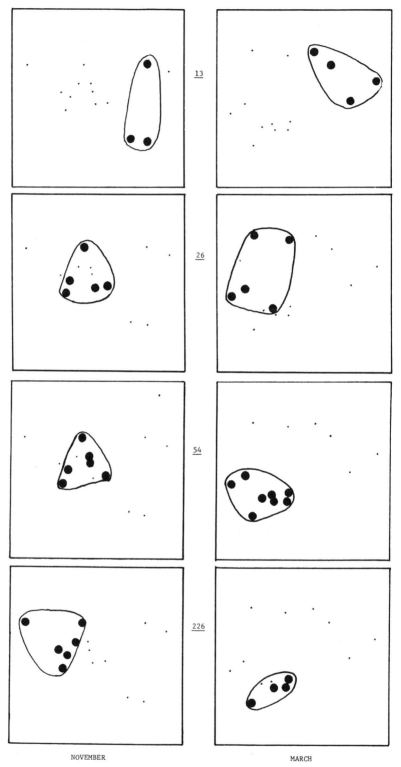

Fig. 3. Location of four representative species in November and March two-dimensional (X and Y axes) ordinations. Species are designated by culture collection number only.

While the study reported here has been preliminary in nature, it has demonstrated an apparent zonation of fungal populations. In the future, emphasis will be placed on factors influencing distribution, the ecological role of soil and debris fungi within the salt marsh system, and the taxonomy of the fungi.

Literature Cited

1. Allen, O. N. 1957. *A laboratory manual for soil microbiology.* Minneapolis: Burgess Publishing Co.
2. Bayliss-Elliot, J. S. 1930. The soil fungi of the Dovey salt marsh. *Ann. Appl. Biol.* **17:**284–305.
3. Benson, G. L., and G. T. Cowley. 1967. Distribution of soil microfungi in a South Carolina sand hill. *Ass. SE Biol. Bull.* **14:**23.
4. Bray, J. R., and J. T. Curtis. 1957. An ordination of upland forest communities in southern Wisconsin. *Ecol. Monogr.* **25:**325–49.
5. Burgess, A. 1949. Soil fungi and humus decomposition. *7th Pacif. Sci. Soc. Congr. N. Z.* **7:**47–53.
6. Christensen, M., W. F. Whittingham, and R. O. Novak. 1962. The soil microfungi of wet-mesic forests in southern Wisconsin. *Mycologia* **54:**374–88.
7. Cowley, G. T., and W. F. Whittingham. 1961. The effect of tannin on the growth of selected soil microfungi in culture. *Mycologia* **53:**539–42.
8. Cowley, G. T. 1964. A soil microfungal population analysis of fifteen stands on the Savannah River Plant. *Ass. SE Biol. Bull.* **11:**41.
9. Loucks, O. L. 1962. Ordinating forest communities by means of environmental scalars and phytosociological indices. *Ecol. Monogr.* **32:**137–66.
10. Miller, J. H., J. E. Giddens, and A. H. Foster. 1957. A survey of fungi from forest and cultivated soils. *Mycologia* **49:**779–808.
11. Orpurt, P. A., and J. T. Curtis. 1957. Soil microfungi in relation to the prairie continuum in Wisconsin. *Ecology* **38:**628–37.
12. Raper, K. B., and C. Thom. 1949. *A manual of the Penicillia.* Baltimore: The Williams and Wilkins Co.
13. Tresner, H. D., M. P. Backus, and J. T. Curtis, 1954. Soil microfungi in relation to the hardwood forest continuum in southern Wisconsin. *Mycologia* **46:**314–33.
14. Warcup, J. H. 1951. The ecology of soil fungi. *Trans. Brit. Mycol. Soc.* **34:**376–97.
15. Ward, J. E., and G. T. Cowley. 1971. Relationship of soil microfungal populations to tree populations. *Ass. SE Biol. Bull.* **18:**61.

Comments

ANTHONY: I am rather curious about the choice of ordination procedures, since there is really sharp discontinuity in the zones that you chose and not merely subtle frequency differences or gradations. That makes it interesting in that you have chosen ordination rather than a sort of classification procedure. One can not help but wonder whether looking at these groups perhaps by information statistic might be rather revealing.

COWLEY: I am sure that there are other procedures to be used. The differentiation between vegetational zones is rather sharp. However, when we look at fungal populations, it is not as sharp. There is more of a continuum type of distribution where one group of organisms or an organism might appear in abundance in the low end of the marsh and diminish or disappear toward the high end. Others will peak in the middle or toward the high end of the marsh. So there is a continuum as far as the fungal populations are concerned.

ANTHONY: I am interested in the actual index of similarity. Is this something that is common to fungal type sociology or something that you developed? Could you tell us a little about this?

COWLEY: This is a process that was called a coefficiency similarity at one time. It was developed in the 1920s and was picked up again in the 1940s and 1950s by a group of higher plant ecologists at the University of Wisconsin. Having been trained at the University of Wisconsin, I grew up with the system. I and one or two others who did some work in Wisconsin picked it up and applied it to fungal populations. I am a little bit lazy and find it is relatively simple to use. Actually, it is a percentage similarity coefficient, and seems to be a fairly useful measure.

ANTHONY: Do you have canned programs for this or have you written your own?

COWLEY: I do not have programs for the ordination. I have tried to write one and have it worked out so I can do one axis at a time. However, I have to stop and select new reference stands for each axis. I am not much of a computer programmer, and consequently I have done the ordination calculations by hand. When I work with 90 and 100 different populations, I will have to develop a program. One of the other members of our department and I did sit down a few years ago and work out a program for calculating the index of similarity. Originally it was done in connection with a rain forest irradiation project in Puerto Rico. Working with 120 different populations, I have found it necessary to prepare a program for that project. What would have taken me about a month to do by hand, took eight minutes using the computer.

ANTHONY: Thank you, I find your work fascinating.

TODD: In cooperation with Bob Rinehold at Sapplo Island, we are investigating the fungal populations associated with *Spartina* grass. That is another advantage of these meetings. We can get together and find out what others are doing. We have not looked at soils, but in about twelve weeks of study we have not been able to detect any fungi on free standing *Spartina* in the field. Likewise, we have not observed fungi on *Spartina* in litter bag samples in the tidal water. I will be very interested in your future results.

COWLEY: I will be very interested in following your work, too, because this is one of the kinds of things we do intend to get into.

Fungi Associated with the Degradation of Mangrove (*Rhizophora mangle* L.) Leaves in South Florida[1]

Jack W. Fell and I. M. Master

In a series of studies of the mangrove *(Rhizophora mangle* L.) forests in Everglades National Park, Florida, Odum and Heald *(15, 22, 23)* estimated that the annual contribution of organic debris to the estuarine ecosystem, through leaf fall, is approximately three tons (dry wt) per acre. These authors reported that the fallen leaves are converted by microbial activity to detrital particles and that the latter support large populations of detrital consumers such as amphipods, nematodes, polychaetes, small crabs, shrimps, and fishes. In turn these consumers are a primary source of food for game and commercial fishes, such as tarpon, snook, grey snapper, sheepshead, and spotted seatrout.

Odum and Heald observed that during the degradation of the mangrove leaf there is a significant increase in the relative amount of protein. Kaushik and Hynes *(17)* also observed a similar increase in protein during degradation of elm *(Ulmus americana)* leaves in freshwater streams and attributed this increase to the associated fungi and not to the bacteria. Odum and Heald conjectured that the fungal protein might be an important food source for the detrital consumers.

As a continuation of Odum and Heald's observations, we initiated a program to ascertain the role of fungi in this degradation system. In the initial phases of this study, we examined the decaying leaves to determine which fungi were present during the various stages of the process. In the present communication, we will list those fungi that we have identified, discuss our observations on the sequences of infestations, and generally characterize the leaf fungal community.

To our knowledge, there have not been any extensive studies of the mangrove leaf mycoflora. The few available reports (Table 1) have been taxonomically oriented, either to present descriptions of new species or to examine the habitat range of a particular fungal group.

Methods

Fungal communities in decaying submerged leaves were compared from different locations in south Florida: Turkey Point, 21 nautical miles south of

[1]Contribution No. 1635 from the Rosenstiel School of Marine and Atmospheric Science, University of Miami, Miami, Florida 33149.

Table 1. Fungi reported from mangrove *(Rhizophora mangle)* leaves

Species	Location	Reference
Hyphomycetes		
Cercospora rhizophorae	* Miami, Fla.	7
Creager	* Throughout host range in Fla.	19
Sphaeropsidales		
Ascochytella rhizophoropsis Ciferri and Fragoso	* Dominican Republic	6
Deptothyrium rhizophorae Fragoso and Ciferri	Dominican Republic (dry leaves)	9
Phomopsis rhizophorae Batista et Maia	* Recife, Brazil	5
Melanconiales		
Pestalotia disseminata Thuem	Little River, Fla.	14
P. longi-aristata Maubl.	Coconut Grove, Fla.	14
P. versicolor Speg.	Bermuda	14
P. zahlbruckneriana P. Henn	Miami, Fla.	14
Ascomycetes		
Anthostomella rhizomorphae (Ktz.) Berl. and Vogl.	* Puerto Rico	26
A. rhizophorae Vizioli	* Bermuda	28
Physalosphora rhizophorae Batista and Maia	* Recife, Brazil	5
Physalosporopsis rhizophoricola Batista and Maia	* Recife, Brazil	4

* Report specified that the observations were from living leaves.

Miami on Biscayne Bay (April 1969); the Florida west coast at Flamingo on Florida Bay (May 1969) and at Demi-john and Pavillion Keys in the 10,000 Islands (March 1970); and Bear Cut in Biscayne Bay, a tidal cut near Miami (March, May, and July 1970). The Bear Cut leaves were collected with a hand net from the tidal waters. Leaves at the other sites were collected from the sediments. The duration of submergence was estimated from the color of the leaves and the state of decay. This study also served as a comparison of the mycoflora of the leaves retained in mesh bags in the succession study.

Fungal succession in the decaying leaves was examined at four locations in the Biscayne Bay area: Matheson Hammock in Coral Gables, just south of Miami (March–April 1969); Turkey Point, in south Biscayne Bay (three different time periods: September–December 1969, March–April 1970, and July–September 1970, and at three stations: the vegetation line, three-fourths mile inland, and one mile inland); Mangrove Point, also in south Biscayne Bay (August–September 1970); and Card Sound, adjacent to the southern end of Biscayne Bay (July–September 1970). Water temperatures and salinities were recorded at the time of collection.

For these fungal succession studies, collections were made of dead yellow leaves that easily detached from the tree and were considered ready to fall.

The leaves were placed in nylon mesh bags with a pore size of 2.5 mm². The bags, with fifty leaves/bag, were tied to mangrove roots so that they were continuously submerged and lay on the bottom in the leaf litter. Leaves were removed from the nylon bags weekly or biweekly for periods up to three months. At each sampling period, two leaves were removed from each bag and placed in sterile plastic containers for transportation to the laboratory.

The leaves were rinsed with sterile distilled water to remove surface debris. Four discs were cut from each leaf with a sterile 4 mm (I.D.) cork borer. The leaf discs were placed on plates of corn meal agar (Difco) prepared with 15‰ seawater and 0.02% Chloromycetin (Parke-Davis) and incubated at 25 C. During the July–August 1970 collections, fungal development on corn meal agar was compared with a mangrove leaf extract agar consisting of 1% powdered mangrove leaves, 2% agar, 15‰ seawater and 0.02% chloromycetin.

The remainder of the leaf was submerged for 5 min in $HgCl_2$ (1:10,000) in 5% ethanol-distilled water for surface sterilization *(16)*. The leaf was then rinsed four times with sterile seawater to remove the mercuric chloride ($HgCl_2$). Four leaf discs were plated as before and incubated at 25 C. The plates were examined twice a week for one month. These techniques were specifically designed to isolate filamentous fungi; other microbes such as bacteria, yeasts, and some of the lower fungi were excluded by this procedure.

In addition to the isolation of Phycomycetes by the leaf disc technique, some very preliminary studies were undertaken to determine if other Phycomycetes were present in the leaves and surrounding waters of the mangrove habitat. Phycomycetes were baited from two leaves from Bear Cut and four leaves from the 10,000 Islands, using cellophane coated with egg albumin *(10)*. Isolation was on the Fuller et al. *(10)* modification of Vishniac's gelatin hydrolysate agar. Intertidal waters at Turkey Point were sampled on a falling tide in February 1970. Eighteen liters were concentrated *(11)* to 400 ml with a continuous flow centrifuge (27,000 G at 5 C) and 0.2 ml of both the concentrated and unconcentrated water was plated on the Fuller et al. *(11)* medium, Emerson's YpSs medium (Difco), and incubated at 25 C. Yeast colonies also developed on these two media. The colonies were enumerated as either white or red yeasts. Further identifications were not made.

Transverse sections of leaves were prepared at a thickness of 25 μ with an A-O Spencer Freezing Microtome. Identifications of the fungi followed the classical methods *(2, 3, 12, 13, 18, 25)*. Because of the difficulties and time involved in identifying the fungi to species, it was decided for the purposes of this program to limit the majority of the identifications to the genus.

Results and Discussion

The rates of fungal infestation and degradation varied with individual leaves. Generally, fungi developed on and in the leaves within the first week of submergence in the water; small invertebrates colonized the leaf surface and, on occasion, the animals could be found in the internal layers. After approximately six weeks, the leaves were fragile and the epidermal layers could be easily torn apart. At this time, a large variety of small invertebrates, particularly ciliates, foraminifera, flatworms, nematodes, polychaetes, and copepods were found in the internal leaf layers. At the shore line stations, where some wave agitation occurred, the leaves became detrital particles after ten weeks, whereas in protected inshore areas the leaves were still intact after one year. It is possible that the nylon bags gave some protection to the leaves and retarded the rate of detrital formation, as noted by Wiegert and Evans (29).

Transverse sections of several living green leaves did not indicate the presence of fungal hyphae; however, as indicated in Table 1, the leaves can be parasitized by several species of fungi. In transverse sections of leaves submerged for one week, encysted Phycomycete zoospores were found attached to the leaf epidermis (Fig. 1) and the rhizoids penetrated into the hypodermis. After 4 to 5 weeks of submergence, all of the layers were infested with fungal hyphae (as represented in Figure 2). With the exception of the fungi that produce large fruiting bodies, it was not possible to identify the specific genera in these sections, although in many cases *Thraustochytrium* and *Phytophthora* were recognizable.

Fifty-three genera of fungi were isolated from the decaying leaves (Tables 2, 3, and 4) including forty-three Deuteromycetes, three Ascomycetes, five Phycomycetes, one Myxomycete, and one Actinomycete. It was not always possible to identify accurately all of the fungi growing on the leaves. This was due to the intermixing of populations 'and to the similar morphological characteris-

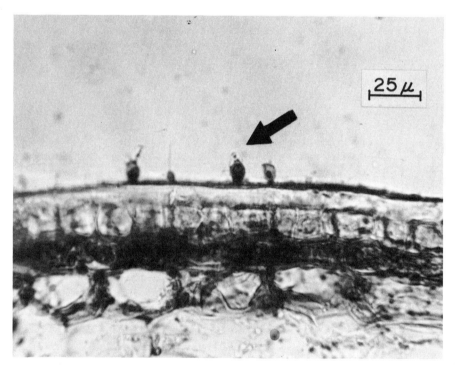

Fig. 1. **Encysted Phycomycete zoospores (indicated by arrow) attached to leaf epidermis of** *Rhizophora mangle* **L; transverse section from a leaf collected from tidal waters of Bear Cut, Miami, Florida.**

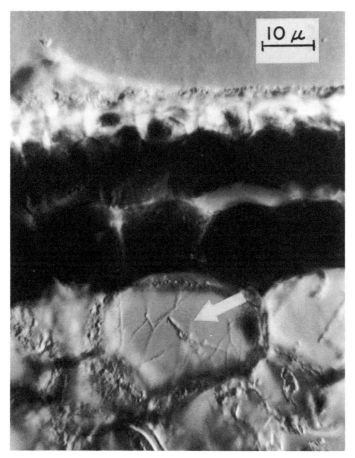

Fig. 2. **Fungal hyphae (indicated by arrow) in hypodermis (water storage cells) of a *Rhizophora mangle* L. leaf; transverse section from a leaf maintained in a mesh bag and submerged in 10 ft of water for seven weeks.**

tics of some of the immature stages. This resulted in the listing of numerous unidentified forms and the grouping of *Cephalosporium-Fusarium-Verticillium*, *Curvularia-Dendryphiella*, and *Cirrenalia-Zalerion*. In some cases, we were able to determine the species of some of the genera (Table 5).

Comparative studies of the incubated discs on corn meal agar and mangrove leaf extract agar resulted in the same generic composition on the two media. The advantage of mangrove leaf extract agar was that the fungal growth was considerably slower on that medium and, in the absence of heavy fungal overgrowth, the fruiting bodies of *Gloeosporium*, *Phyllosticta*, and the unidentified 3-celled Ascomycete were able to develop.

The sequence of fungal infestation is presented in Table 2. The fungi inhabiting the senescent leaves, prior to leaf fall, were restricted to a few genera. *Pestalotia* and *Phyllosticta* were readily discernible due to their large fruiting bodies that were often visible as small dark spots erumpent through the upper epidermis of the leaf. As indicated by the results of the HgCl$_2$ washings, some

Table 2. Genera identified from fungal succession studies of *Rhizophora mangle* leaves (Biscayne Bay, Florida) and their percentage frequency of occurrence from the leaf discs[a]

Genus	Days of Submergence													
	0		7		14		21		28		35		35+	
	B[b]	A[c]	B	A	B	A	B	A	B	A	B	A	B	A
Hyphomycetes														
Alternaria	10[d]	2	6	1	5	1	3	11					3	
Aspergillus			4	8		1	6	11	9	9	25	16	3	
Aureobasidium	6												1	
Cephalosporium														
Fusarium														
Verticillium	4	4	46	25	34	19	53	33	31	28	53	38	74	38
Cladosporium	37	8	19	2	11	3	3				3		19	
Curvularia														
Dendryphiella			5	2	2	1	11	14	16	3		3	3	
Cylindrocephalum	6	10	6	11	5	3								3
Dictyosporium					3	8	6							4
Fusarium			14	5	5	6	3						15	4
Gliocladium			1	1				22	28	13	13		4	
Myrothecium	4						3						1	1
Nigrospora	23	12	1		2				3					1
Penicillium			14	15	8	3	42	39	28	16	19	9	25	15
Stachybotrys				1								13		7
Trichoderma			18	11	9	10	39	44	19	34	50	34	31	28
Virgaria				1									3	1
Zygosporium	4	2	2	2		1							10	1
Sphaeropsidales														
Phoma			5	4	1							13	3	
Phyllosticta	13	13	1											
Melanconiales														
Gloeosporium			6	4			6	3	9	3	9		1	
Pestalotia	50	42	37	19	18	15	19	3		13	9	6	28	6
Ascomycetes														
Lulworthia					2	6				3	13	25	12	13
"3-celled Ascomycete"			4	6							3			
Phycomycetes														
Phytophthora			23	23	11	15		8						4
Rhizophydium[e]			10	8	14	10							1	
Unidentified	29	25	13	25	17	19	14	19	41	50	22	6	1	1
Sterile		27	1			1								
Total number of sample discs	52	52	84	84	88	88	36	36	32	32	32	32	68	68

[a] Genera found in 3 leaf discs or less are listed in Table 3.
[b] B   Before surface sterilization.
[c] A   After sterilization.
[d] Percentage frequency of occurrence.
[e] Reclassified as *Phytophthora vesicula* subsequent to this writing.

Table 3. Genera found infrequently (in 3 leaf discs or less) from fungal succession studies of *Rhizophora mangle* leaves (Biscayne Bay, Florida)

Hyphomycetes:	Sphaeropsidales:
Beltrania	*Phomopsis*
Botryosporium	Melanconiales:
Cercospora	*Colletotrichum*
Cirrenalia-Zalerion	Ascomycetes:
Cladobotryum	*Chaetomium*
Cylindrocladium	Phycomycetes:
Geotrichum	*Mucor*
Helminthosporium	
Idriella	
Isaria	
Monocillium	
Pithomyces	
Scolecobasidium	
Sporotrichum	
Stemphylium	
Varicosporina	

of the other fungi, particularly *Aureobasidium* and *Myrothecium*, were adventitious or leaf surface colonizers. Many of the fungi, e.g., *Cladosporium*, *Cylindrocephalum*, and *Nigrospora*, penetrated into the internal layers of the leaf and were isolated following the washing procedure.

The fungi inhabiting the senescent leaves varied in their ability to survive throughout the degradation process. *Phyllosticta* disappeared after the first week in the water and the remaining genera diminished in prevalence at differing rates. *Pestalotia* was the most prevalent genus and was found throughout the decay process.

A variety of fungi invaded the decaying leaves. Two of the more abundant initial invaders were the Phycomycetes, *Phytophthora* and *Rhizophydium*. Their occurrence appeared to be dependent on season and age of the leaf. The infestations were limited to spring and summer with the greatest abundance during the first two to three weeks of leaf decay. Often, under these conditions, the Phycomycetes were the only fungi isolated from the leaves. Other primary invaders were *Aspergillus*, *Penicillium*, *Trichoderma*, and members of the *Cephalosporium-Fusarium-Verticillium* and the *Curvularia-Dendryphiella* complexes. The only significant secondary invader that appeared after the second week of submergence was *Lulworthia*, which is a cellulolytic marine Ascomycete.

Confinement of the leaves to the mesh bags during the succession study did not appear to have any significant effect on the composition of the mycoflora that inhabited the leaves. The populations of the leaves collected from the sediments and tidal waters (Table 4) appear to be essentially the same as those from the mesh bags (Table 2). Most of the prevalent genera, particularly *Cephalosporium-Fusarium-Verticillium*, *Cladosporium*, *Curvularia-Dendryphiella*, *Pestalotia*, *Phytophthora*, and *Rhizophydium*, have widespread geographical distributions in south Florida. The apparent differences in distribution, as seen in Table 4, can probably be attributed to collection errors.

Table 4. Fungi from submerged leaves collected at different geographical locations in south Florida

Genus	Estimated Days of Submergence[a]					
	0–15 Days		7–25 Days		15+ Days	
	B[b]	A[c]	B	A	B	A
Hyphomycetes						
Alternaria	2[d]		3	2 3		
Aspergillus		1			1	
Aureobasidium	2				2	2
Beltrania	1					
Cephalosporium-Fusarium-Verticillium	1 2	1 2	1 2 3	2 3	1 2	1 2
Cercospora	1	1				
Cladosporium	1 2	3	1 3	2 3	1 2	1 2
Cirrenalia-Zalerion			1		2	2
Curvularia-Dendryphiella	1 2	1 2	1 2 3	2	1	
Cylindrocephalum	1				1	
Cylindrocladium					1	1
Dictyosporium	·			2		2
Epicoccum			3			
Fusarium	2				1	
Helminthosporium		1				
Idriella	1		3	3	1	1
Hyphomycetes						
Monocillium			3			
Myrothecium					2	
Nigrospora				3		
Penicillium	1		1 3		2	1
Trichoderma	2 3	2	3	3		2
Zygosporium			3	3		
Sphaeropsidales						
Phoma	1 2		1			
Phomopsis	1					
Phyllosticta	1	1				
Melanconiales						
Colletotrichum	1				1	
Gloeosporium	2	2				
Pestalotia	1 3	2 3	1 2 3	3	1 2	
Ascomycetes						
Lulworthia			2	2	2	2
"3-celled Ascomycete"	1					
Phycomycetes						
Phytophthora	1 2 3	1 2 3	1 2 3	1 2 3		
Rhizophydium	1 2 3	1 2 3	1 3	1 2 3		

[a] Time was estimated from color changes on leaves maintained in nylon bags.
[b] B   Before surface sterilization.
[c] A   After sterilization.
[d] 1. Bear Cut (14 leaves), 2. Turkey Point (14 leaves), 3. Florida West Coast (6 leaves).

Table 5. Genera identified to species

Hyphomycetes:
Cercospora rhizophorae Creager
Cirrenalia macrocephala (Kohlm.) Meyers and Moore
Dictyosporium pelagicum Hughes
Idriella lunata Nelson and Wilhelm
Monocillium indicum Saksena
Varicosporina ramulosa Meyers and Kohlmeyer
Zalerion varium Anastasiou
Zygosporium masonii Hughes
Phycomycetes:
Phytophthora vesicula Anastasiou and Churchland
Ascomycetes:
Lulworthia floridana Meyers
Lulworthia grandispora Meyers
Lulworthia medusa var. *biscaynia* Meyers

The Phycomycetes are possibly one of the most important fungal groups associated with leaf degradation. In addition to *Phytophthora* and *Rhizophydium*, isolated by the leaf disc plating techniques, the genera *Schizochytrium* and *Thraustochytrium* were abundant in leaves sampled by the baiting technique. Phycomycetes were also prevalent in the one study of waters in the intertidal mangrove region. Colony counts of *Schizochytrium* averaged 5–10 colonies/ml of seawater and *Thraustochytrium*, 1/ml. In comparison, yeast colony counts in the same water averaged 5–10/ml for white yeasts and 1/ml for red yeasts.

The majority of the genera isolated from mangrove leaves were ubiquitous saprophytes often associated with the breakdown of plant material. Many of the genera have been isolated from a variety of substrates including terrestrial soils (3); terrestrial leaf litter in terrestrial soils (16), in fresh water (17), and in seawater (1); the marine grass *Thalassia* (20); mangrove muds (24, 27); salt marshes (8), etc. In contrast, the fungal populations reported from mangrove wood (18) are quite distinct from the populations from leaves. Only *Phoma* and *Lulworthia* were found in both substrates. Newell (21) has discussed the similarities and differences in fungal communities from seedlings and leaves. It would appear that the most significant genera identified from leaves that were not observed during seedling degradation are the Phycomycetes, *Phytophthora* and *Rhizophydium*.

In conclusion, it appears that a wide range of fungal genera, including Phycomycetes, Deuteromycetes, and Ascomycetes, are associated with the degradation of the mangrove leaves. Considerable research is required to determine the specific role of these organisms in the food web, although it has been postulated that they are important in the production of protein as a food source for small fishes and invertebrates. It is extremely important, particularly at this time, to understand the dynamics of this ecosystem and the effects of alterations in environmental conditions. The mangroves of south Florida are being subjected to a variety of abnormal conditions, including effluents from domestic and industrial sources; therefore, it is necessary to know how these alterations will affect the entire food web. More important, the value of the mangrove system to the marine environment must be determined, as the mangroves are rapidly disappearing to urbanization.

Acknowledgments

This research was supported by the National Institutes of Health Grant No. HEW 5R01 FD–00031, and the Atomic Energy Commission Grant No. AT–(40–1)3801. The authors gratefully acknowledge Dr. Eric Heald, Tela Railroad Co., Tela, Honduras, and Dr. William Odum, University of Virginia, for their suggestions and discussions during the study and for allowing us to review their results on mangrove detrital systems prior to the publication of their manuscripts.

Literature Cited

1. Anastasiou, C. J., and L. M. Churchland. 1969. Fungi on decaying leaves in marine habitats. *Can. J. Bot.* **47:**251–57.
2. Barnett, H. L. 1960. *Illustrated genera of imperfect fungi.* Minneapolis: Burgess Pub. Co.
3. Barron, G. L. 1968. *The genera of hyphomycetes from soil.* Baltimore: Williams & Wilkins Co.
4. Batista, A. C., H. S. Maia, and A. F. Vital. 1955. Ascomycetidae aliquot novarum. *An. Soc. Biol. Pernambuco* **13:**72–86.
5. Batista, A. C., A. F. Vital, H. S. Maia, and I. H. Lima. 1955. Coletanea de novas especies de fungos. *An. Soc. Biol. Pernambuco* **13:**187–224.
6. Ciferri, R., and R. G. Fragoso. 1927. Hongos parasitos y saprofitos de la Republica Dominicana. *Boln. R. Soc. est. Hist. nat.* **27:**68–81.
7. Creager, D. P. 1962. A new *Cercospora* of *Rhizophora mangle. Mycologia* **54:**536–39.
8. Dickinson, C. H. 1965. The mycoflora associated with *Halimione portulacoides.* III. Fungi on green and moribund leaves. *Trans. Brit. Mycol. Soc.* **48:**603–10.
9. Fragoso, R. G., and R. Ciferri. 1928. Hongos parasitos y saprofitos de la Republica Dominicana (16a Series). *Publnes Estac. agron. Moca (Ser. B., Botanica)* **13:**1–17.
10. Fuller, M. S., B. E. Fowles, and D. J. McLaughlin. 1964. Isolation and pure culture study of marine Phycomycetes. *Mycologia* **56:**745–56.
11. Fuller, M. S., and R. O. Poyton. 1964. A new technique for the isolation of aquatic fungi. *Bioscience* **14:**45–46.
12. Grove, W. B. 1935 (Reprint 1967). *British stem- and leaf-fungi (Coelomycetes). Vol. I. Sphaeropsidales.* New Rochelle, N.Y.: Cambridge Univ. Press.
13. Grove, W. B. 1937 (Reprint 1967). *British stem- and leaf-fungi (Coelomycetes). Vol. II. Sphaeropsidales and Melanconiales.* New Rochelle, N.Y.: Cambridge Univ. Press.
14. Guba, E. F. 1961. *Monograph of* Monochaetia *and* Pestalotia. Cambridge: Harvard Univ. Press.
15. Heald, E. J., and W. E. Odum. 1972. The role of detritus in a South Florida estuary. II. Production and breakdown of vascular plant material. *Ecology* (in press).
16. Hering, T. F. 1965. Succession of fungi in the litter of a lake district oakwood. *Trans. Brit. Mycol. Soc.* **48:**391–408.
17. Kaushik, N. K., and H. B. N. Hynes. 1968. Experimental study on the role of autumn-shed leaves in aquatic environments. *J. Ecol.* **56:**229–43.
18. Kohlmeyer, J. 1969. Ecological notes on fungi in mangrove forests. *Trans. Brit. Mycol. Soc.* **53:**237–50.
19. McMillan, R. T., Jr. 1964. Studies of a recently described *Cercospora* on *Rhizophora mangle. Plant Disease Reporter* **48:**909–11.
20. Meyers, S. P., P. A. Orpurt, J. Simms, and L. L. Boral. 1965. Thalassiomycetes. VII.

Observations on fungal infestation of turtle grass, *Thalassia testudinum. Konig. Bull. Mar. Sci.* **15**:548–64.

21. Newell, S. Y. 1972. Succession and role of fungi in the degradation of red mangrove seedlings. In *Belle W. Baruch Library in Marine Science, Vol. I. Estuarine Microbial Ecology*, ed. L. H. Stevenson and R. R. Colwell. Columbia: Univ. of South Carolina Press. (This volume).
22. Odum, W. E., and E. J. Heald. 1972. The role of detritus in a South Florida estuary. I. Stomach analysis of the heterotrophic community. *Bull. Mar. Sci.* (in press).
23. Odum, W. E., and E. J. Heald. 1972. The role of detritus in a South Florida estuary. III. Pathways of energy flow. *Ecology* (in press).
24. Rai, J. N., J. P. Tewari, and K. G. Mukerji. 1969. Mycoflora of mangrove mud. *Mycopath. Mycol. Appl.* **38**:17–31.
25. Sparrow, F. K., Jr. 1960. *Aquatic Phycomycetes.* Ann Arbor: Univ. of Michigan Press.
26. Stevens, F. L. 1920. New or noteworthy Porto Rican fungi. *Bot. Gaz.* **70**:399–402.
27. Swart, H. J. 1958. An investigation of the mycoflora in the soil of some mangrove swamps. *Acta. Bot. Neerl.* **7**:741–68.
28. Vizioli, J. 1923. Some Pyrenomycetes of Bermuda. *Mycologia* **15**:107–19.
29. Wiegert, R. G., and F. E. Evans. 1964. Primary production and the disappearance of dead vegetation on an old field in southeastern Michigan. *Ecology* **45**:49–63.

Comments

TODD: Do you have any evidence that the fungi that degrade these leaves are a segment or a portion of the residual population on the leaf? Or are they part of the marine population?

FELL: The majority of the residual populations diminish in abundance after the leaf falls into the water. The leaves are then invaded by a variety of saprophytes that continue the degradation process. This entire process is probably a combined effect of the residual and the marine-occurring fungi. Most of these fungi are also found in terrestrial sources; possibly the only organisms that we identified that can be considered true marine fungi are *Cirrenalia macrocephala, Dictyosporium pelagicum, Varicosporina ramulosa, Zalerion varium, Phytophthora vesticula,* and the species of *Lulworthia.*

TODD: In addition to the litter bag studies, have you thought about incorporating a litter bag consisting of a semipermeable membrane? With this you could draw nutrients from the marine environment without the introduction of additional organisms. With this, one could see if the original organisms promote degradation.

FELL: That is a good idea.

LITCHFIELD: Can you estimate the survival time for the leaves in the mangrove swamp with your bag studies? Have you any idea how long it takes for complete degradation to occur?

FELL: This depends a lot on where the bags are placed. If they are down near the shoreline, they will break up very rapidly, within ten weeks. Back further in the mangroves where there is no wave agitation, it takes considerably longer, possibly up to a year.

ZOBELL: What are the mesh bags made of? Do you find that they undergo degradation?

FELL: The bags are made of nylon. They do not undergo degradation in the time period of our study.

Succession and Role of Fungi in the Degradation of Red Mangrove Seedlings[1]

Steven Y. Newell

As outlined by Fell and Master in this volume, the red mangrove, *Rhizophora mangle*, is an extremely important contributor to the productivity of south Florida estuarine waters; a constant flow of energy passes from the red mangrove plants to the estuarine ecosystem via the vast amounts of organic matter cast off by the plants. Microbial mediators convert this detritus into nutrient matter utilizable by small macroorganisms at the base of the food web. In the case of *Rhizophora* leaf and twig materials, this conversion is well documented *(9, 10, 24)*. The production of the viviparous seedlings and the contribution of this production to the estuarine ecosystem as nutritive biomass have been ignored.

However, combining Heald's *(9)* data for mean density of mature *Rhizophora* trees per unit area (0.96/m²), Davis's *(4)* data for production of seedlings per adult tree per producing season (300), and data on average seedling mass (6.5 g) gathered during the present study, one arrives at a figure for seedling biomass production per unit area (7.9 metric ton/acre/summer season) which is twice that for leaf and twig debris (3.6 metric ton/acre/yr). The great majority of the seedlings fail to take root *(5)*, and the number of seedlings set adrift in southeastern Florida coastal waters has been estimated to run into the millions *(29)*. Assuredly, there is room for error in the foregoing estimates, but they certainly suggest that cast-off mangrove seedlings form a significant contribution to the estuarine productivity of south Florida. The present report deals with the study of the fungal populations involved in the energy turnover of the seedling segment of *Rhizophora* production. At the time of this report, the study was in its sixth month of operation, and the results reported below were compiled over the first five months.

Materials and Methods

Over 1,000 mature, healthy seedlings were picked from parent *Rhizophora mangle* trees in late October 1970. The criteria for maturity were the appearance of the fused cotyledonary collar below the fruit and the ease of separation of the seedling from the cotyledons, indicating the onset of natural abcission *(1, 2)*. No seedlings were used which were shorter than 10 cm long, since seedlings less

[1] Contribution No. 1637 from the Rosenstiel School of Marine and Atmospheric Science, University of Miami, Miami, Florida 33149.

467

than 5 cm long were found by La Rue and Muzik *(20)* to be usable to root readily.

The experimental seedlings were divided into three sets. One set was weighed (wet) and marked by attaching plastic numbered tags to the seedlings with rubber bands (Fig. 1). One set was not treated, and one set was given uniform artificial injuries at the level of the radicle (Fig. 2). ("Radicle" here refers to the lowest 4–5 cm of the radicular end *(2)* of the seedling, which is covered by brown phellem or cork tissue.) These injuries were administered by removing a 2.0 cm² (surface area) disc of cortex plus epidermis tissue with a cork borer. The injured set was prepared because it was discovered during collection that many of the seedlings are naturally injured before or shortly after their fall from the trees. The agents of these injuries, which occur primarily at the level of the radicle, include rodents, crabs, insects, and perhaps other animals. Injuries of the type administered to the experimental seedlings do not kill seedlings; La Rue and Muzik *(20)* found that cutting 5 cm lengths from the radicles of seedlings did not prevent them from rooting.

Station sites for the experiment were chosen so that a range of hydrographic conditions characteristic of coastal waters of southeastern Florida would be represented (Table 1). Station 7 exhibited the daily passage of a salinity wedge back and forth across it; stations 1 and 4 were midestuary stations with stable salinity characteristics; station 2 was located within a tidal mangrove stream

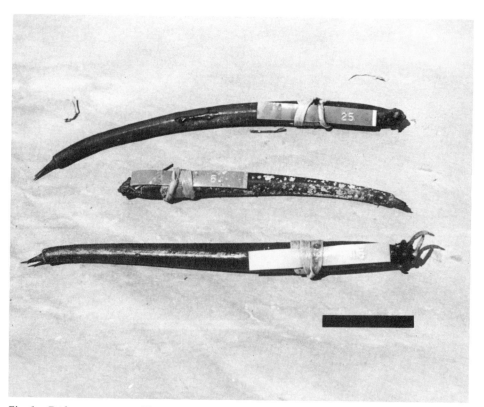

Fig. 1. **Red mangrove seedlings with plastic numbered tags attached by rubber bands (bar at lower right represents 5 cm).**

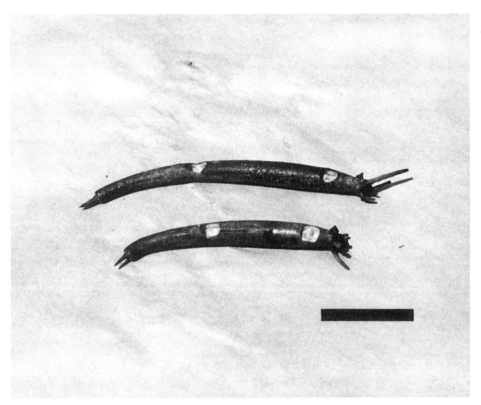

Fig. 2. **Red mangrove seedlings with cork-borer discs of epidermis plus cortex tissue removed at the mid-radicle and mid-hypocotyl levels. Discs such as these were removed from the radicles of one set of seedlings as artificial injuries before they were placed in the field (bar at lower right represents 5 cm).**

Table 1. Hydrographic data and distance from the red mangrove shoreline for the experiment stations

Station Number	Distance from Mangrove Trees	Depth, Mean Low Water	S‰,r[a]	S‰,m[b]	T,r[c]	T,m[d]
1	230 m	2 m	28–40	32.2	23–26	25.2
2	adjacent	½ m	30–43	36.7	20–30	25.5
3	1,800 m	3 m	35–36	35.5	21–24	22.5
4	2,000 m	3 m	35–40	37.8	18–28	22.8
5	3,000 m	6 m	36–38	37.0	20–25	22.6
6	adjacent	2 m	10–32	20.5	12–28	21.5
7	adjacent	½ m	2–51	26.3	15–31	24.2
8	adjacent	0[e]	22–47	35.0	16–29	24.3

[a] S‰,r = the range of salinities measured at subsampling times.

[b] S‰,m = the mean salinity.

[c] T,r = the range of water temperature in degrees C.

[d] T,m = the mean water temperature in degrees C.

[e] This station was located at the level of highest spring tides, and was immersed only at highest tides or under storm conditions. Hydrography is for the water 1 m away at mean low water.

having stable salinity characteristics; station 5 was located outside the estuary, in the barrier reef lagoon east of the chain of Florida keys. Stations 1, 2, 4, 5, and 7 contributed the bulk of the data compiled.

Nylon mesh bags (mesh size 4 mm × 4 mm) were used to contain the seedlings (63–80 per bag) at the station sites. The bags were large enough that excessive packing of the seedlings was avoided. Three bags of seedlings, one for each of the experimental sets, were tied with nylon cord to anchors or mangrove roots at each station. The bags were submerged at all times (except at station 8). After one month in the water, no test seedlings were noted which were of density less than seawater. The seedlings were allotted to stations at random, so that no station would have a preponderance of seedlings from any one of the parent trees.

Subsamples of six seedlings were taken monthly from each bag of injured and uninjured seedlings and returned to the laboratory in presterilized plastic bags (Nasco Whirl-Pak). The seedlings were treated by a combination of sampling-observation methods derived principally from the fungal succession studies of Kendrick and Burges (17) and Hering (11), with modifications and additions as described below.

Cork borer discs of epidermis plus cortex tissue were taken from three levels (mid-radicle—2.0 cm² disc; mid-hypocotyl—2.0 cm² disc; upper hypocotyl—1.1 cm² disc) on two of the six seedlings from each bag. These discs were given a preliminary wash (5 sec shake) in glass vials of 11 ml sterile seawater, then washed thoroughly (1 min shake) in a second vial of 11 ml sterile seawater. The discs were taken from the final wash and placed in a 1:10,000 $HgCl_2$ solution in 5% ethanol for surface sterilizing (1 min shake). After rinsing (1 min shaking each) in three changes of sterile deionized water, the discs were placed on plates of a mangrove seedling agar medium. This medium consisted of 4% whole, finely ground, dried mangrove seedlings plus 2% agar (Difco) in 15‰ seawater. In addition to the epidermis plus cortex discs, a 5 mm length of 0.6 mm diameter cylindrical core was taken by cork borer from the pith tissue at upper radicle level and placed on the mangrove medium plates. These disc plates were incubated at 25 C and observed periodically for development of fungi.

The final seawater wash from the above disc treatments was used to enumerate surface-inhabiting filamentous fungi, yeasts, and bacteria. First, it was serially diluted to 1:10,000. The final wash and dilution waters were then aseptically filtered through cellulose ester membrane filters (Millipore Corp.) of pore size 0.45 μm, as is commonly done in the study of marine yeasts (6). The filters were transferred to plates of a general heterotroph isolation medium (2.3% Difco Nutrient Agar, 2.0% glucose, 0.1% yeast extract, 0.2% Difco agar, in 15‰ seawater). These plates were incubated at 12 C, in order to retard overgrowth of the fungus-yeast plates by filamentous fungi. Two sets of the isolation medium were used, one set with 0.02% Chloromycetin (Parke-Davis; added before autoclaving) for observation of fungi, and one set without the antibiotic for enumeration of bacteria. All filamentous fungi appearing on the wash-water plates were transferred to mangrove media plates and incubated at 25 C for identification.

Two seedlings from each bag were incubated in presterilized damp chambers according to the methods described by Meyers and Reynolds *(23)* and Johnson, Ferchau, and Gold *(13)*. These seedlings were not treated prior to incubation. Fruiting fungi were recorded after thirty days of incubation at 25 C.

The final two seedlings of each subsample were subjected to cork borer discing, but in this case the discs were taken from mid-radicle and mid-hypocotyl level, washed briefly (10 sec) in 10 ml of sterile seawater, and placed in covered glass dishes (5 cm diameter) of filter-sterilized 15‰ seawater containing 0.5% each of penicillin G and streptomycin sulfate (Nutritional Biochemicals). The dishes were then baited with autoclave-sterilized, dried, ground seedling powder and incubated at 25 C. They were observed periodically for development of lower fungi.

In addition to the incubation observations, determinations of dry weight loss were made monthly on two seedlings from each of the bags of preweighed and marked seedlings. Original dry weight was determined by performing a regression of original wet weight on original dry weight with a sample of thirty seedlings. The regression was linear and significant at the 99.9% level of confidence.

Cross sections (25 μm thickness) were prepared monthly from seedlings from each subsample on a freezing microtome at the level of the upper radicle. They were stained for fungi with trypan blue in lactophenol.

In reporting of occurrence data below, two kinds of frequency figures are used. For individual species, frequency of occurrence is used. This means number of occurrences per possibilities of occurrence, expressed as a percentage. One possibility of occurrence was allotted for each series of wash plates, each disc plate, and each damp chamber seedling examined. For fungal groups, relative occurrence is used. This means the percentage of all observations of fungi for which the members of each group accounted.

Results and Discussion

Fifty-three species of fungi were recorded from the seedlings (Table 2). Twenty-seven were Hyphomycetes, fourteen Sphaeropsidales, six Ascomycetes, two Oömycetes, and one each Zygomycetes and Myxomycetes. The prevalent fungi observed are listed in Table 3.

Examination of studies of the fungal populations of the water column and sediments of southeastern Florida coastal waters *(21, 22, 26, 27)* reveals that the seedling fungus populations, although similar to the ambient fungus populations in some respects, do not merely reflect them. (It must be noted that these earlier investigations did not involve precisely the same isolation and observation techniques as the present study.) Genera of relatively high occurrence frequency in seedling tissues *(Pestalotia, Septonema, Zalerion, Cytoplea, Cytospora)* were not reported or were rare in the water column and sediment studies. Prevalent genera from the water and sediment studies, such as *Aureobasidium, Fusarium, Trichoderma, Dendryphiella, Nigrospora, Cirrenalia, Humicola,* and *Halosphaeria,* showed a very limited occurrence or did not occur in red mangrove seedlings during the present study.

Similarities between seedling populations and ambient populations included the co-occurrence of *Cladosporium, Alternaria, Penicillium, Aspergillus, Phoma,*

Table 2.   Species[a] of fungi recorded from red mangrove seedlings

Hyphomycetes	Sphaeropsidales
Alternaria sp.	*Cytoplea* sp.
Aspergillus niger V. Tiegh.	*Cytospora rhizophorae* Kohlmeyer
A. repens de Bary	*Cytosporina* sp.
Aspergillus sp.	*Dendrophoma* sp.
Aureobasidium pullulans (de Bary) Arnaud	*Dothiorella* sp.
Botryotrichum sp.[b]	*Micropera* sp.
Cephalosporium sp.	*Naemosphaera* sp.[d]
Cladosporium cladosporioides	*Phlyctaena* sp.
(Fres.) De Vries	*Phoma* sp.
Cladosporium sp.	*Phomopsis* sp.
Curvalaria sp.	*Rabenhorstia* sp.
Cylindrocarpon sp.	*Robillarda rhizophorae* Kohlmeyer
Dendryphiella salina (Suth.) Pugh et Nicot	Melanconiales
Fusarium sp.	*Colletotrichum* sp.[d]
Haplobasidion sp.	*Pestalotia* sp.[e]
Harposporium baculiforme Drechsler	Ascomycetes
Hendersonia sp.	*Lophiotrema littorale* Speg.[f]
Nigrospora sp.	*Lulworthia grandispora* Meyers
Paecilomyces sp.	*L. medusa* var. *biscaynia* Meyers
Penicillium steckii Zaleski	unidentified species[g]
Periconia prolifica Anastasiou	Oomycetes
Septonema sp.	*Pythiogeton utriforme* v. Minden
Sporothrix sp.[c]	*Thraustochytrium* sp.
Trichoderma viride Pers.	Zygomycetes
Zalerion varium Anastasiou	*Rhizopus stolonifer* (Ehrenb. ex Fr.) Vuill.
Zygosporium masonii Hughes	Myxomycetes
	Labyrinthula sp.

[a] See Table 3 for the species which were prevalent.

[b] This species lacks conspicuous setae, but has light brown thick-walled spores. It may be a *Humicola*.

[c] Sometimes tends toward a somewhat verticillate arrangement and so may be a *Calcarisporium*.

[d] Identified after submission of manuscript.

[e] Although Kohlmeyer (*18*) lists five species of *Pestalotia* occurring on *Rhizophora*, none of those listed corresponds to the present species.

[f] A tentative identification.

[g] Includes the perfect stage of the *Pestalotia* species, not readily identifiable, and a species with 7-celled brown ascospores without appendages. Not enough material of the latter was available for identification.

and *Lulworthia*. The first five are ubiquitous fungi which occur on a wide variety of substrates and are common in seawater environments (*12*). *Lulworthia*, an Ascomycete restricted to the marine environment, was reported by Meyers (*21*, sub *Halophiobolus*) to be one of the initial invaders of wooden panels submerged in southeastern Florida estuarine waters. *Lulworthia* is a genus of active degraders of cellulosic materials (*22*), and its spores can attach to substrates within twenty-four hours after submergence (*14*). It is not surprising, therefore, that this fungus is one of the prevalent invaders of the seedlings after they fall into the water.

As is the case with water and sediment of southeastern Florida, the seedling

Table 3. Prevalent[a] fungi from red mangrove seedlings, listed by major fungal group in order of decreasing frequency of overall occurrence

Fungal Species	Months Resident in Water					
	0	1	2	3	4	5
Hyphomycetes						
Cladosporium cladosporioides	60	57	16	22	26	7
Penicillium steckii	0	27	5	16	10	7
Septonema sp.	0	9	16	2	2	0
Alternaria sp.	15	9	14	4	5	0
Aspergillus repens	0	7	5	6	2	3
Cladosporium sp.	5	7	5	8	2	0
Zygosporium masonii	5	7	5	8	2	0
Zalerion varium	0	0	5	0	5	7
Sphaeropsidales						
Cytoplea sp.	0	3	5	2	7	7
Cytospora rhizophorae	0	1	0	2	7	7
Phoma sp.	0	5	7	4	3	3
Melanconiales						
Pestalotia sp.	45	53	39	30	21	29
species RZ 85	0	5	5	0	5	7
Ascomycetes						
Lulworthia grandispora	0	3	11	20	26	36
Oömycetes						
Thraustochytrium sp.	—	40	33	30	22	14

[a] Prevalent refers to fungi occurring with at least 5% frequency, either overall or in at least two sampling periods.

[b] Percentage frequency of occurrence is listed to the right of each species.

mycoflora is similar in only some respects to the mycoflora of mangrove muds reported by Swart *(30)* and Rai et al. *(25)* from East Africa and India, respectively. Swart, however, found *Pestalotia* to be common under *Rhizophora* stands, and he isolated *Cladosporium, Alternaria, Aspergillus, Penicillium, Phoma,* and *Septonema* from the muds. Ulken *(31)*, working with lower fungi, found *Schizochytrium* and *Rhizophydium* along with *Thraustochytrium* in mangrove sediments from Brazilian estuaries. Neither *Schizochytrium* nor *Rhizophydium* has been observed from red mangrove seedlings. Even the extensive studies of Kohlmeyer (summarized in references *18* and *19*) on fruiting marine fungi do not provide a predictor of the seedling fungus community through five months of seedling residence in the water. Only one (*Lulworthia grandispora*) of the several Ascomycetes found by Kohlmeyer was observed to be prevalent on seedlings; the only marine Basidiomycete (*Nia vibrissa*) reported was not observed in this study; and of the six imperfect fungi reported by Kohlmeyer, only two, *Cytospora rhizophorae* and *Phoma* sp., appeared to play a significant role in attack on the seedlings in the estuarine waters of southeastern Florida.

Some of the fungi observed with dominant frequency from seedlings exhibited a marked pattern of distribution with respect to station location. One of these was the Oömycete, *Thraustochytrium*, which was the only regularly isolated lower fungus. Its occurrence was limited to stations 2, 6, and 7, which were located farthest inshore, in direct proximity to red mangrove trees. It was consistently isolated from these stations, but never isolated from stations nearby (1, 3, and 4) which were subject to tidal flow from the inshore locales. Neither salinity nor temperature appeared to be the cause of this distribution pattern, for salinities were vastly different (Table 1) among stations 2, 6, and 7, and temperatures were not markedly different between stations 2, 6, and 7 and the remaining stations. Nothing from the knowledge of *Thraustochytrium* ecology gives a clear indication of the reason for the absence of the *Thraustochytrium* species on red mangrove seedlings at stations not directly adjacent to the mangrove stands. The Thraustochytriaceae, according to the literature (e.g., 28), exhibit distributions from inshore shallow to offshore deep (>100 m) water. They can be easily isolated from diverse substrates, and they show a strong tendency to grow on solid surfaces (3).

Other species showing marked distributional patterns were *Lulworthia grandispora* and *Zalerion varium*. *L. grandispora* occurred with highest frequency at stations closest to the shore (1, 2, and 7) and did not occur at all at station 5, outside the estuarine waters. *Z. varium* occurred only at station 7. *Zalerion* is a genus of marine and salt-lake restricted fungi, and so its occurrence at station 7, where salinities varied widely, both daily and seasonally, may have been due to its ability to compete under these conditions.

A succession of fungi on red mangrove seedlings became apparent after five months of seedling residence in the water (Fig. 3). The four major primary invader species, *Cladosporium cladosporioides*, *Alternaria* sp., *Zygosporium masonii*, and *Pestalotia* sp., were present in seedlings sampled directly from the trees. These maintained high frequency levels through the first month in the water, but fell to markedly lower levels by the end of the third month (Table 3). By the end of the fifth month, *Alternaria* sp. and *Z. masonii* had disappeared, *C. cladosporioides* had fallen to the low frequency of 7%, and *Pestalotia* sp. had leveled off at about 30% frequency.

The Hyphomycetes, in general, fell (Fig. 3) from a very high relative occurrence, 70% after one month, to about a 50% level between two and four months, and down to 34% after five months. The primary invader Hyphomycetes were largely replaced by secondary invader Hyphomycete species. Two Hyphomycetes which showed increases in frequency in the fourth and fifth months were *Z. varium* and *Botryotrichum* sp.

The Sphaeropsidales stayed at a level of about 10% relative occurrence from the first to the third month, but rose to 23% after the fifth month. The Ascomycetes went from 2% after the first month, to 16% after the third month, and rose to 23% after the fifth month.

The pattern of succession of prevalent fungi, then, through the fifth month of seedling residence in the water, appeared to be: primary invader Hyphomycetes → primary and secondary invader Hyphomycetes → Ascomycetes + Sphaeropsidales + secondary invader Hyphomycetes. The Melanconiales, represented for the most part by *Pestalotia* sp., were prevalent throughout. A rather steady

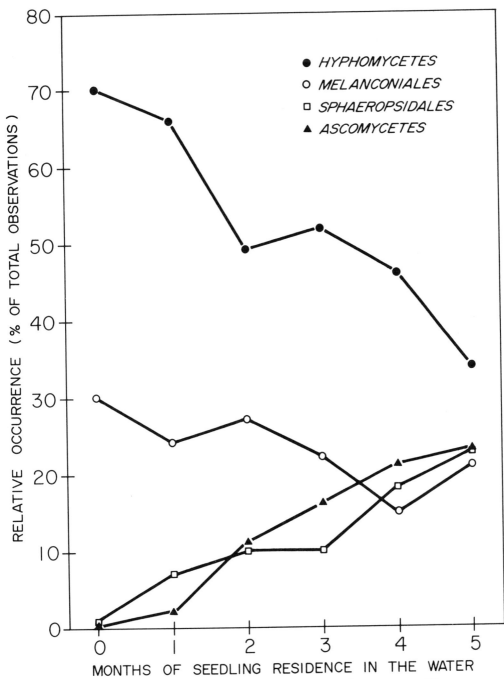

Fig. 3. Succession of fungi, by major fungal group, on red mangrove seedlings.

progressive shift in occurrence from Hyphomycetes to Sphaeropsidales and Ascomycetes took place after the first month (Fig. 3). The only prevalent Oömycete, *Thraustochytrium* sp., was isolated only by the baiting technique; it was uncertain whether its rhizoids penetrated seedling tissues and had an active part in seedling degradation.

This pattern of succession differs in one respect from the generalized pattern described by Garrett *(8)* for terrestrial materials decomposing on or in soil. Garrett's pattern is given as weak parasites → primary saprophytic or sugar fungi, especially Zygomycetes → cellulose decomposers + sugar fungi → lignin decomposers + associated fungi. The succession patterns are compared below.

The weak parasitic stage on seedlings was represented by the primary invaders of the seedlings, prominently *C. cladosporioides* and *Pestalotia* sp. Both of these fungi produced viable elements on the seedling·surface and within the seedling tissues while the seedlings were still on the tree, although no marked disease symptoms were apparent. *C. cladosporioides* was largely confined to the seedling surface, as evidenced by its lower overall frequency (18%) on disc plates than on wash plates (58%). The Zygomycetes played no observed role in the degradation of red mangrove seedlings. Instead, the opportunist or sugar fungi were represented by *Alternaria*, *Aspergillus*, *Penicillium*, *Septonema*, and *Zygosporium*. Of these, *Penicillium* produced more viable elements at the seedling surface than within seedling tissues. *Alternaria* and *Aspergillus* were more common within seedling tissues, and *Septonema* and *Zygosporium* were observed almost exclusively on damp chamber incubation (never on wash-water plates), perhaps indicating that they were surface colonizers which were resistant to washing. These sugar fungi largely gave way after five months to the Ascomycetes, Sphaeropsidales, and secondary invader Hyphomycetes, which perhaps represent the onset of the cellulose-decomposer stage. The actively cellulolytic Ascomycete, *Lulworthia grandispora*, reached its highest observed frequency in the fourth month. The Sphaeropsidales and Ascomycetes appeared only on disc plates and/or damp chamber incubation up to the third month, but they began to appear on wash plates after the third month.

Jones *(15, 16)* reported on the succession of marine fungi on wood submerged in marine waters. The succession involved initial attack by Hyphomycetes → two stages of Ascomycete attack → Ascomycetes + Sphaeropsidales. This is somewhat similar to the succession, to date, of fungi on mangrove seedlings, in that here also Hyphomycetes were initially prevalent, then began to give way to Sphaeropsidales and Ascomycetes. The succession on *Rhizophora* seedlings bears more similarity to the succession on marine-submerged wood than to the succession on *Rhizophora* leaves reported by Fell and Master (this volume). The leaf succession involved Sphaeropsidales, Ascomycetes, and Hyphomycetes (including *Cladosporium* and *Zygosporium*) as the weak parasites, and lower fungi (*Phytophthora* and *Rhizophydium*) as the primary saprophytes, while the Sphaeropsidales appeared to play a relatively small role in the later stages of the leaf succession.

The succession of fungi on wood submerged in southeastern Florida estuarine waters reported by Meyers *(21)* involved the occurrence of the marine Ascomycetes *Halosphaeria* (sub "Form 2") and *Corollospora* (sub *Peritrichospora*) following *Lulworthia* (sub *Halophiobolus*) attack. Although neither *Halosphaeria* nor *Corollo-*

spora has been observed on *Rhizophora* seedlings, they may appear later in the succession, for they were not isolated from wood panels until after seven months of submergence. Also, the absence of Zygomycetes in the primary saprophyte stage does not rule out their playing a part in degradation later in the succession, as observed by Kendrick and Burges *(17)* in the succession of fungi on pine needles and Frankland *(7)* on bracken fronds. In these terrestrial successions, the Zygomycetes were involved in the final stages of decomposition.

The pattern of succession of fungi on the seedlings did not differ markedly between the injured set of seedlings and the uninjured set, but the injured set exhibited more frequent occurrences of fungi (the ratio was 1.27:1, injured to uninjured) and a slightly larger species diversity (39 species as opposed to 33). Although uninjured seedlings lost a very small amount of dry weight, they did not rot extensively and were still viable (capable of rooting and plumule growth in the laboratory) at the five-month point. This was in contrast to the extensive rotting and deaths among injured seedlings.

However, the small dry weight loss exhibited by uninjured seedlings may have been due to a counterbalancing of degradative weight loss by the normal density gain of red mangrove seedlings after their fall into seawater *(4)*. Evidence from cross sections through the radicular end of the seedlings stained for fungal hyphae indicated that both injured and uninjured seedlings were subject to extensive fungal invasion of their tissues. It may be that the uninjured seedlings slowly lost weight through attack on the tissue and starch reserves external to the vascular bundles, for hyphae did not penetrate beyond the vascular tissues to a significant extent. This observation correlates with the very low observed frequency of occurrence of fungi in the seedling pith cores incubated on the disc plates. Mixed colonies of bacteria and protozoa, however, developed from the pith cores of injured seedlings, and the mean number of viable elements of bacteria from injured seedlings (as determined on wash-water plates) was five times that from uninjured seedlings. The bacterial-protozoan invasion caused rotting and softening of the pith tissues of the injured seedlings.

The fungi, then, attacked primarily the seedling tissues exterior to the vascular layer, i.e. the cortex and epidermis. Entry of the fungi to these tissues was probably gained through lenticel openings, injuries to the epidermis and cortex, and points in the radicle tip where new rootlets emerged. All of these types of locations exhibited dense development of mycelium in the damp chambers. The most successful entries appeared to occur at the level of the radicle, for the frequency of occurrence of fungi there (97%) was much higher than elsewhere (28% from mid-hypocotyl, 40% from upper hypocotyl).

In summary, the role of fungi in the degradation of red mangrove seedlings appeared to be as invaders of the protective external tissues. In injured seedlings, bacteria and protozoa entered the pith tissues via the wound and rotted them. Both injured and uninjured sets of seedlings exhibited similar fungal successions, involving a procession from Hyphomycetes to Sphaeropsidales, Ascomycetes, and Hyphomycetes. It seemed that the fungi might take the place of mechanical injury as the agents which would cause uninjured seedlings to be susceptible to bacterial-protozoan destruction of the starch-laden pith tissues. In any case, a unique community of fungi appeared to play an important part, in conjunction with bacteria, in the first stages of conversion of seedling biomass to microbial biomass, the first step in the passage of seedling energy into the estuarine food web.

Acknowledgments

The author was supported during the period of this study by a National Science Foundation Graduate Fellowship. Additional support was provided by the National Institutes of Health Grant No. 5 Rol FD–00031 and the Atomic Energy Commission Grant No. AT–(40–1) 3801. Thanks are due to Dr. J. W. Fell for his guidance and advice, to Mr. I. M. Master for his invaluable advice and assistance in fieldwork, and to Dr. J. Kohlmeyer for his stimulation of the author's interest in this study. Dr. W. E. Odum, University of Virginia, and Dr. E. J. Heald, Tela Railroad Company, kindly supplied prepublication information dealing with their studies of the importance of mangrove detritus to estuarine ecology.

Literature Cited

1. Argo, V. N. 1963. Root growth claims soil from sea. Mangrove spreads by unique adaption. *Nat. Hist.* **72:**52–55.
2. Bowman, H. H. M. 1917. Ecology and physiology of the red mangrove. *Proc. Amer. Philos. Soc.* **56:**589–672.
3. Clokie, J. 1970. Some substrate relationships of the family Thraustochytriaceae. *Veröff. Inst. Meeresforsch. Bremerh.* **12:**329–51.
4. Davis, J. H. 1940. The ecology and geologic role of mangroves in Florida. *Papers from Tortugas Lab.* **32:**307–412.
5. Egler, F. E. 1948. The dispersal and establishment of red mangrove, *Rhizophora,* in Florida. *Carib. Forest.* **9:**299–310.
6. Fell, J. W., and N. van Uden. 1963. Yeasts in marine environments. In *Symposium on marine microbiology,* ed. C. H. Oppenheimer, pp. 329–40. Springfield, Ill.: Charles C Thomas Co.
7. Frankland, J. C. 1966. Succession of fungi on decaying petioles of *Pteridium aquilinum. J. Ecol.* **54:**41–63.
8. Garrett, S. D. 1963. Soil fungi and soil fertility. Oxford: Pergamon Press.
9. Heald, E. J., and W. E. Odum. 1969. The contribution of mangrove swamps to Florida fisheries. *Proc. Gulf and Caribb. Fish. Inst.* **22:**130–35.
10. Heald, E. J., and W. E. Odum. 197–. The role of detritus in a South Florida estuary. II. Production and breakdown of vascular plant material. (In manuscript.)
11. Hering, T. F. 1965. Succession of fungi in the litter of a lake district oakwood. *Trans. Br. Mycol. Soc.* **48:**391–408.
12. Johnson, T. W., Jr. 1968. Saprobic marine fungi. In *The fungi. An advanced treatise. Volume III. The fungal population,* ed. G. C. Ainsworth and A. S. Sussman, pp. 95–104. New York: Academic Press Inc.
13. Johnson, T. W., Jr., H. A. Ferchau, and H. S. Gold. 1959. Isolation, culture, growth, and nutrition of some lignicolous marine fungi. *Phyton* **12:**65–80.
14. Johnson, T. W., Jr., and F. K. Sparrow. 1961. *Fungi in oceans and estuaries.* Weinheim, Germany: J. Cramer.
15. Jones, E. B. G. 1963. Observations on the fungal succession on wood test blocks submerged in the sea. *J. Inst. Wood Sci.* **11:**14–23.
16. Jones, E. B. G. 1968. The distribution of marine fungi on wood submerged in the sea. In A. H. Walter and J. J. Elphick (ed.), *Biodeterioration of materials. Micro-*

biology and allied aspects. Proceedings of the First International Biodeterioration Symposium, pp. 460–85. New York: Elsevier Publishing Co.

17. Kendrick, W. B., and A. Burges. 1962. Biological aspects of the decay of *Pinus sylvestris* leaf litter. *Nova Hedwigia* **4**:313–42.

18. Kohlmeyer, J. 1969. Ecological notes on fungi in mangrove forests. *Trans. Brit. Mycol. Soc.* **53**:237–50.

19. Kohlmeyer, J., and E. Kohlmeyer. (In press). Marine fungi from tropical America and Africa. *Mycologia.*

20. La Rue, C. D., and T. J. Muzik. 1954. Growth, regeneration, and precocious rooting in *Rhizophora mangle. Papers Michigan Acad. Sci., Arts and Letters* **39**:9–29.

21. Meyers, S. P. 1954. Marine fungi in Biscayne Bay, Florida. II. Further studies of occurrence and distribution. *Bull. Mar. Sci.* **3**:307–27.

22. Meyers, S. P. 1968. Degradative activities of filamentous marine fungi. In *Biodeterioration of materials. Microbiology and allied aspects. Proceedings of the First International Biodeterioration Symposium*, ed. A. H. Walter and J. J. Elphick, pp. 594–609. New York: Elsiver Publishing Co.

23. Meyers, S. P., and E. S. Reynolds. 1958. A wood incubation method for the study of lignicolous marine fungi. *Bull. Mar. Sci.* **8**:342–47.

24. Odum, W. E., and E. J. Heald. 197–. The role of detritus in a South Florida estuary. III. Pathways of energy flow. (In manuscript).

25. Rai, J. N., J. P. Tewari, and K. G. Mukerji. 1969. Mycoflora of mangrove mud. *Mycopath. Mycol. Appl.* **38**:17–31.

26. Roth, F. J., Jr., D. G. Ahearn, J. W. Fell, S. P. Meyers, and S. A. Meyer. 1962. Ecology and taxonomy of yeasts isolated from various marine substrates. *Limnol. Oceanogr.* **7**:178–85.

27. Roth, F. J., Jr., P. A. Orpurt, and D. G. Ahearn. 1964. Occurrence and distribution of fungi in a subtropical marine environment. *Can. J. Bot.* **42**:375–83.

28. Sparrow, F. K. 1969. Zoosporic marine fungi from the Pacific Northwest (U. S. A.). *Arch. Mikrobiol.* **66**:129–46.

29. Stephens, W. M. 1962. Tree that makes land. *Sea Frontiers* **8**:219–30.

30. Swart, H. J. 1958. An investigation of the mycoflora in the soil of some mangrove swamps. *Acta. Bot. Nederl.* **7**:741–68.

31. Ulken, A. 1970. Phycomyceten aus der Mangrove bei Canaéia (São Paulo, Brasilien). *Veröff. Inst. Meeresforsch. Bremerh.* **12**:313–19.

Comments

ZOBELL: Did you find that the fungi were active anaerobically?

NEWELL: None of my stations was located such that the seedlings were subject to anaerobic ambient conditions. I have not looked into anaerobic activity within the seedlings.

WATSON: There are a couple of questions. *Thraustochytrium* and *Labyrinthula* seem to occupy the same ecological niche on algae, normally. I wonder if you have looked for *Labyrinthula*, and is *Thraustochytrium* the only obligate marine fungus that is involved here?

NEWELL: I found *Labyrinthula* only twice, out of some 266 observations. *Zalerion* and *Lulworthia* are also obligate marine genera; they are found only in the marine or high-saline environment. The remaining prevalent genera on seedlings appear to be facultatively marine, for they are well known in terrestrial environments also.

WARNER: With respect to injured and uninjured seedlings, how much of the loss of firmness in the injured seedlings might have been caused by mechanical soaking of the water into the tissues and by effects of plasmolysis on the cells by an increase in salinity?

NEWELL: I do not know whether osmotic stress caused by mechanical soaking of water into the tissues of injured seedlings played any role in their degradation. Factors bearing on this question are these: rotting of seedlings occurs at different rates at different stations (most extensively at stations 5 and 7, least extensively at station 2); salinity characteristics do not correlate with extent of rotting; tissue fluids of *Rhizophora* are highly saline, although those of the seedling are less so when they are maturing; an increase in density to greater than that of seawater normally occurs after the fall from the parent tree, perhaps indicating a normal increase in salinity of the seedling tissue fluids. All these things considered, it does not seem likely that the tissues of the seedlings would be adversely affected by seawater directly.

Section Eight
MICROBE-ANIMAL INTERACTIONS

Nematological-Microbial Interrelationships and Estuarine Biodegradative Processes

Samuel P. Meyers and Bruce E. Hopper

Biological transformation of plant material to detritus constitutes an initial and complex step in the estuarine energy flow system *(1)*. A variety of taxonomic and physiological groups of microorganisms are involved in this process, both in detrital-microbial interrelationships and in association with the meiobenthic population *(4)*. Thus, the subject of animal interaction in this symposium is indeed in order, for the faunistic component of the meiobenthos significantly affects the environment, as well as our interpretation of its microbiological activities. The significance of the marine meiofauna in the benthos has been noted repeatedly by various workers *(3, 15)*. Gerlach, in a recent discussion of the importance of the marine meiofauna, calculated that a meiofaunal biomass representative of only 3% of the macrofaunal biomass, would contribute about 15% of the food consumed and of the biomass provided to the food chain *(5)*.

The omnipresence of nematodes in the estuarine environment, and their significant position of often being the dominant meiofaunal representative, similarly, has been repeatedly recognized *(15, 22, 23)*. The trophic levels of this diverse metazoan group range from types which feed selectively to nonselectively on sedimentary deposits, to epigrowth feeders, and finally to forms of predatory capability. However, general knowledge of the biology of marine nematodes is scanty. Certain groups regarded as being predatory on the basis of stomatal morphology (i.e., presence of teeth and/or powerful mandibles) may function adequately as omnivores, while other species may actually feed nonselectively on deposits. "Predatory species" probably derive their nutrition in the natural habitat by passage of sediment and/or detrital material through their gut.

Not intended as a nematological treatise, this paper rather illustrates some of the potentially significant relationships between nematodes and microorganisms, such as marine algae, bacteria, and fungi. In view of our investigation of fungal-induced cellulose degradation *(19)*, we have chosen to restrict our attention to the activities of nematodes at sites of plant and detrital degradation in the estuarine environment. The importance of algal and bacterial-nematode associations, is not, thereby, slighted but rather the paucity of quantitative information on this particular subject is stressed. Pertinent aspects discussed subsequently in this paper include studies on turtle grass, *Thalassia testudinum,* in Biscayne Bay (Miami), Florida *(8–12, 16–20)*, as well as some concern analyses of biodegradation of oyster grass *(Spartina alterniflora)* in marshlands of southeastern Louisiana *(13, 21)*. Both habitats show unspecific aggregations of nematological activity correlated with microbial conversion of plant cellulose in the ecosystem. Investigations have included analysis of nematode feeding on fungi and direct and indirect responses of these animals to fungal-infested cellulosic substrates in the laboratory and in the field. We will briefly examine these associations from the following four broad aspects: (1) growth of *Aphelenchoides marinus* on fungal mycelia, (2) response of *Metoncholaimus scissus* to fungal-cellulose substrates, (3) nematode populations and fungal-degraded cellulose, and (4) patterns of nematode distribution.

Growth of Aphelenchoides marinus on Fungal Mycelia

The feeding of nematodes on fungi is well recognized and of considerable biological importance in the terrestrial environment, but has been seldom reported in the sea *(16, 17)*. In the latter studies, growth and reproduction of *Aphelenchoides marinus,* a stylet-bearing nematode extracted from the rhizosphere of turtle grass, were supported on a range of viable mycelia from the same habitat. The response of the nematode to each fungus was studied on a comparative basis at various stages of fungal growth. Utilization ranged from <100 animals/mg mycelium of terrestrial deuteromycetous fungi to >5,000 animals/mg mycelium of the widespread marine deuteromycete, *Dendryphiella arenaria.* This species, along with the marine ascomycete *Halosphaeria mediosetigera,* was most effective in supporting animal growth. Species of *Dendryphiella* are involved in various degradative processes in the marine environment, including decomposition of cellulose and laminarin, as well as ecologically significant associations with nematode representatives of the meiofauna. In addition to the nutritional value of *Dendryphiella,* the fungus-cellulose degradation site serves as a locus for aggregation of *Aphelenchoides marinus.* Further evidence of the "attractive" effect of such sites to nematode colonization is noted in the following section.

Response of Metoncholaimus scissus to Fungus-Cellulose Substrates

Considerable attention has been devoted to bionomical investigations of the large oncholaimid, *Metoncholaimus scissus,* probably one of the most intensively studied marine nematodes to date *(8, 11, 12, 18, 20)*. Fungus-cellulose mats, especially those employing the marine fungus *Halosphaeria mediosetigera,* have proved effective in establishing alterations in animal concentrations. Significantly, the nematode fauna migrating to test substrates placed in the field, consisted almost entirely of gravid female *M. scissus* specimens. Often such accumulations attained extrapolated densities of nearly 3,000. Sediment samples taken from the immediate vicinity of the test site showed a nearly equal male:female sex ratio in the *M. scissus* population. The response of *M. scissus* was not a feeding reaction alone, but more likely a positive chemotactic response of the animal to biologically active substances released from and/or generated by the fungal degradative process on the cellulose. In this regard, the cellulolytic and proteolytic activities of marine fungi, typified by *Halosphaeria mediosetigera*, are of considerable ecological significance. The migration of large numbers of *M. scissus* to the site of fungus-cellulose interaction demonstrates the complexity of microbial relationships associated with degradation of submerged cellulose substrates. The associated microbiota, including bacteria, diatoms, ciliates, and protozoa that rapidly colonize the mats following submergence, may affect the microsite and favor the aggregation of *M. scissus*. These organisms themselves may readily serve as food for various nematodes, including *M. scissus*. A frequent "visitor" to the mats was the large red ciliate, *Keronopsis rubra*. Blooms of *K. rubra* on mats (in the field and in enrichment culture), were in all likelihood a response to prior bacterial colonization of the cellulose. The ability of *M. scissus* to ingest *K. rubra* is considered as additional evidence of the complexity of the microbe-nematode associations. We have observed the ability of *M. scissus* to ingest bacteria in tubes of enriched sediment, thus emphasizing the importance of including predacious activity when undertaking analyses of substrate utilization by microorganisms.

Meyers and Hopper *(18)* postulated that the fungus-cellulose disc may be effective in evaluating alterations of marine nematode population density and microsite dissimilarities that could have significant ecological implications. Patterns of activity of *M. scissus*, in numbers and in sex ratio, suggested that discrete loci of organic material (i.e., fungal-infested leaves and decaying plant tissue) notably affect biological activity of segments of the meiobenthos. In ancillary work *(9)*, the laboratory incubation of exposed mats and the cultivation of nematodes on mycelia have provided information on the life cycle of nematodes in subtropical waters. These procedures have provided considerable information on generation time, reproductive rates, and other ecologically relevant data.

Another instance of microbial-animal association was observed *(20)* in the striking concentration of *M. scissus* in mats of the large benthic diatom, *Pleurosigma balticum*. *M. scissus* occupied microsites in irregular assemblages and in numbers far exceeding those previously reported. Nematode concentrations per sample ranged from 1,000 to 7,000, with *M. scissus* comprising from 4% to 93% of the total nematode biomass. The maximum number of *M. scissus* isolated was $268/cm^2$ or $2.68 \times 10^6/m^2$. Nematode populations calculated to be as great as 10 million specimens/m^2 have been reported, with from 70 to 7,300 nematodes per 10 cm^2 area. Dense nematode colonization occurred in older portions of the diatom mats. Laboratory tests demonstrated the inability of *Metancholaimus* to utilize directly cells of *P. balticum*. The mat-forming effect of the diatom upon sediment particles apparently provided a favorable physicochemical environment for nematode colonization. Studies of chemical by-products from such communities, and their effect on the meiobenthos may reveal "attractive" stimuli comparable to those reported by Gray *(6, 7)* in colonization studies of the interstitial archiannelid, *Protodrilus symbioticus*.

Nematode Populations and Fungus-Degraded Cellulose

Further studies *(19)* have verified earlier observations on the association of fungi and nematodes with cellulosic substrates undergoing degradation. Cellulose filters removed from the seawater system at the Institute of Marine Sciences, University of Miami, showed striking stages of decomposition. Fibers from such filters were attacked by the marine fungus *Lulworthia floridana* and supported a numerically large and characteristically structured nematode population, including *Viscosia*, *Leptolaimus*, a chromadorid, and *Monhystera* sp. The animals represent three, possibly four, different trophic levels based on the stomatal characteristics of each genus: *Viscosia*—an omnivore; chromadorid—an epistratum feeder; *Leptolaimus* and *Monhystera*—either selective or nonselective deposit feeders. The pattern of nematode species succession observed was a reaction to the altering physicochemical properties of the fiber matrix as well as a response to a varying food source. These observations demonstrate the need for a more realistic consideration of the *composite* biology of the organisms prevalent at the biodegradation site.

Patterns of Nematode Distribution

Investigations of benthic and foliicolous nematodes *(10, 11)* have demonstrated the feasibility of using certain taxa as biological indicators. Populations of foliicolous species in turtle grass beds are affected by the type and concentration of associated algal epigrowth, hydrographic conditions, and the state of development of the total plant community. Patterns of seasonal variation of foliicolous nematodes differ widely among turtle grass localities. Senescence or death of the algal epigrowth often results in significant increases in total numbers of nematodes on the leaf surfaces. Often the state or condition of the epiphyte has a greater effect on nematode colonization than does the specificity of the alga. The omnipresent algal feeders, *Chromadora macrolamoides*, may serve as species indicative of the epiphytic biomass. Changes in bionomics of the foliicolous nematode population may be used to characterize relative concentrations of organic deposition in the immediate locality. The foliicolous nematode fauna differs markedly from that found in the immediately adjacent benthos.

The use of nematodes and their microbial feeding patterns as biological indicators is further complicated by the tremendous variability in colonization between sites within individual areas as well as at the same site at different sampling periods *(11)*. Examples of this variability are seen in studies of the four benthic species, *Theristus fistulatus*, *Metoncholaimus scissus*, *Spirinia parasitifera*, and *Gomphionema typica*, which comprised over 90% of the nematode population at one specific test site (Site C). Maximum peaks in population densities were noted and correlated with physiographic environmental alterations. The *M. scissus* population declined concurrently with changes in the sediment load deposited in the community. Species ratios, especially shifts in the dominant forms present, may be used to gauge biological and physical changes in a particular environment. These data support our observations, using fungal mats, on the significance of shifts in animal concentrations and faunal microzonicity characteristics.

Mechanisms affecting nematode colonization of a particular substrate are poorly understood. Information is lacking on time required for aggregation *in situ* as well as the effect of the activities of the animal on the microbiological characteristics of the particular site. We know that the nematode biota is sensitive to small changes in substrate composition and that loci of animal activity may occur within an apparently uniform and homogenous substrate. Such phenomena are highly significant in calculations of meiobenthos biomass, and must be considered in any proposed application of nematode species to reflect changes in the chemical-biological characteristics of the benthos.

Recently, a bacterial (and possibly detrital) feeder, *Diplolaimelloides bruciei*, from the *Spartina* biosphere was described *(13)*. This animal, the dominant nematode representative within the meiofauna, was associated with the fungal-infested *Spartina* tissue *(21)*, notably upon the culm and within the intercellular spaces around the periphery of the culm. Evidence to date indicates that the nematode population increases with decomposition of *Spartina*, and in all likelihood feeds on the developing microbial flora. Study of the terrestrial plant rhizosphere has revealed numerous nematode-microbial interrelationships of considerable ecological significance. Analyses of comparable relationships in the salt marsh and estuarine plant ecosystem should facilitate our understanding of detrital turnover.

Discussion

A number of questions arise from nematode-microbial investigations developed to date, and further studies are needed before definite answers are possible, particularly in view of the incomplete knowledge of marine nematode systematics at present. We lack information concerning the significance of nematological predation on the microbial biomass and material transformation into the ecosystem. Since biomass data often are used to estimate microbial processes and biological characteristics of the environment, the impact of predation cannot be ignored, especially since numerous representatives of the marine nematoda actively feed on the microbiota. Recently, Wilt and Smith *(24)* presented additional evidence on interactions of aquatic bacteria and nematodes involving active ingestion of specific bacteria as food, as well as migration of nematodes to loci containing viable bacteria. Comparable observations have been made in our laboratory. Obviously, more detailed information is needed on feeding habits and nutrition of marine nematodes. Other investigations *(14)* provide such data for *Rhabditis marina*, a cosmopolitan marine species commonly recovered from seaweeds stranded between tides in the upper littoral zone. Using tracer feeding and synxenic culture experiments, it has been demonstrated that the specific nutritional requirements of the animal can be satisfied by a relatively restricted group of microorganisms.

The feasibility of using selected nematodes as indicator organisms of pollution or ecosystem imbalances needs to be explored thoroughly. Nematological studies, done in terrestrial and aquatic localities, suggest that various species can serve as indicators of levels of pollution, while other species occupy an important developmental niche in the biological buildup, resulting in imbalances and pollution of water masses. Analyses of factors affecting attraction and aggregation of nematodes to sites of organic deposition and decay in estuarine areas have special relevance in understanding nutrient regeneration in marine plant communities.

The role of nematodes as vectors in transport of viruses, bacteria, and protozoa through aquatic systems, especially in transmission of disease-causing organisms, is an area of promising research. Chang and Kalber (2) noted the presence of coliforms within nematodes in aerobic treatment plant effluents in the range of 5 to 10% of the total bacteria count, thus demonstrating the potential of the animals as carriers of human enteric organisms. Our recent report of a high incidence of microsporidian infection among marine nematodes suggests their possible role in parasite transmission to various commercially important invertebrates (12).

Obviously, the meiofauna is the domain of the microbiologist as well as the zoologist, for in our elucidation of environmental processes at the microbial level we often find ourselves working with unusual and unfamiliar organisms. The bacterial feeder may be a better "tool" than the bacterium and may ultimately provide more definite answers on the role of the bacterium itself. Certainly, new methodology must be developed as we attempt to view the microbial estuarine system as a complex mechanism of interrelated factors. Modern approaches to analyses of population dynamics of marine nematodes, coupled with basic biological studies, have much to offer in this area.

Literature Cited

1. Burkholder, P. R., and G. H. Bornside. 1957. Decomposition of marsh grass by aerobic marine bacteria. *Bull. Torrey Bot. Club* **84**:366–83.
2. Chang, S. L., and P. W. Kalber. 1962. Free-living nematodes in aerobic treatment plant effluent. *J. Wat. Poll. Cont. Fed.* **34**:1256–61.
3. Fenchel, T. 1969. The ecology of marine microbenthos. IV. Structure and function of the benthic ecosystem, its chemical and physical factors and the microfauna communities with special reference to the ciliated protozoa. *Ophelia* **6**:1–182.
4. Fenchel, T. 1970. Studies on the decomposition of organic detritus derived from the turtle grass *Thalassia testudinum*. *Limnol. Oceanogr.* **15**:14–20.
5. Gerlach, S. A. 1971. On the importance of marine meiofauna for benthos communities. *Oecologia (Berl.)* **6**:176–90.
6. Gray, J. S. 1966. The attractive factor of intertidal sands to *Protodrilus symbioticus*. *J. Mar. Biol. Ass. U.K.* **46**:627–45.
7. Gray, J. S. 1967. Substrate selection by the archiannelid *Protodrilus rubropharyngeus*. *Helgol. Wiss. Meeresunters.* **15**:253–69.
8. Hopper, B. E. 1970. *Diplolaimelloides bruciei* n. sp. (Monhysteridae: Nematoda) prevalent in marsh grass, *Spartina alterniflora* Loisel. *Can. J. Zool.* **48**:573–75.
9. Hopper, B. E., and S. P. Meyers. 1966. Observations on the bionomics of the marine nematode, *Metoncholaimus* sp. *Nature (London)* **209**:899–900.
10. Hopper, B. E., and S. P. Meyers. 1966. Aspects of life cycle development of marine nematodes. *Helgol. Wiss. Meeresunters.* **13**:444–49.
11. Hopper, B. E., and S. P. Meyers. 1967. Foliicolous marine nematodes on turtle grass, *Thalassia testudinum* Konig, in Biscayne Bay, Florida. *Bull. Mar. Sci.* **17**:471–517.
12. Hopper, B. E., and S. P. Meyers. 1967. Population studies of benthic nematodes within a subtropical seagrass community. *Mar. Biol.* **1**:85–96.
13. Hopper, B. E., S. P. Meyers, and R. Cefalu. 1970. Microsporidian infection of a marine nematode, *Metoncholaimus scissus*. *J. Invert. Path.* **16**:371–77.

14. Lee, J. J., J. H. Tietjen, R. J. Stone, W. A. Muller, J. Rullman, and M. McEnery. 1970. The cultivation and physiological ecology of members of salt marsh epiphytic communities. *Helgol. Wiss. Meeresunters* **20**:136–56.

15. McIntyre, A. D. 1969. Ecology of marine meiobenthos. *Biol. Rev.* **44**:245–90.

16. Meyers, S. P., W. A. Feder, and K. M. Tsue. 1963. Nutritional relationships among certain filamentous fungi and a marine occurring nematode. *Science* **141**:520–22.

17. Meyers, S. P., W. A. Feder, and K. M. Tsue. 1964. Studies of relationships among nematodes and filamentous fungi in the marine environment. *Develop. Indust. Microbiol.* **5**:354–64.

18. Meyers, S. P., and B. E. Hopper. 1966. Attraction of the marine nematode, *Metoncholaimus* sp., to fungal substrates. *Bull. Mar. Sci.* **16**:143–50.

19. Meyers, S. P., and B. E. Hopper. 1967. Studies on marine fungal/nematode associations and plant degradation. *Helgol. Wiss. Meeresunters* **15**:270–81.

20. Meyers, S. P., B. E. Hopper, and R. Cefalu. 1970. Ecological investigations of the marine nematode, *Metoncholaimus scissus*. *Mar. Biol.* **6**:43–47.

21. Meyers, S. P., M. E. Nicholson, P. Miles, J. S. Rhee, and D. G. Ahearn. 1970. Mycological studies in Barataria Bay, Louisiana, and biodegradation of oyster grass, *Spartina alterniflora. Louisiana State Univ. Coastal Studies Bull.* **5**:111–24.

22. Tietjen, J. H. 1969. The ecology of shallow water meiofauna in two New England estuaries. *Oecologia (Berl.)* **2**:251–91.

23. Warwick, R. M., and J. B. Buchanan. 1970. The meiofauna off the coast of Northumberland. I. The structure of the nematode population. *J. Mar. Biol. Ass. U.K.* **50**:129–46.

24. Wilt, G. R., and R. E. Smith. 1970. *Studies on the interactions of aquatic bacteria and aquatic nematodes.* Auburn Univ. Bull. 701. Auburn, Ala.: Water Resources Research Institute.

Interrelationships between Bacteria and Protozoa

Robert D. Hamilton

The relationship between protozoa and their environment can be exceedingly complex, and if one thing could be said to be characteristic of communities containing protozoa, it would be that they possess the capacity for sudden change in dominant organisms. Personally, I believe that anyone interested in this field should gain an intimate appreciation of successional changes. To this end, I would like to suggest a very simplistic experiment. Place a little water from some standing source, such as a pond or ditch, in a flat-bottomed flask or even a petri plate. Add an equal volume of any general purpose phytoplankton medium and place the culture in the light for a few days. After the algae are well up, transfer the container to the stage of a dissecting or inverted microscope with as little physical disturbance as is possible and try to spend at least 10 min each day observing the culture. After a very short time, you will discover that you can quickly recognize dominant organisms without any need to investigate their taxonomy. Appearance, size, and swimming motion are valuable clues. This experiment also demonstrates the narrowness of certain niches in that you will quickly become aware that certain small, indeed microscopic, areas are likely spots to find the cell that looks like a beer barrel with spikes, while other equally small areas should harbor the one that looks like an animated shopping bag. Spending a short time each day with a culture such as this will quickly provide you, or your student, with an illustration of successional changes, their magnitude, their effect on the community, and above all, an appreciation of the speed with which they can occur. It is an experience that I, for one, could not gain from observing the gradual appearance of bacterial colonies on agar plates some days after taking the samples.

Successional changes in dominant species are a reflection of prior variations in parameters basic to community structure. It is generally accepted that one of the basic parameters governing the occurrence of protozoa is the food supply (9, and references therein). By choice, or by accident, many protozoa feed upon bacteria; yet the interrelationships between the two trophic levels are little understood.

For example, it is fairly common to find references in the literature to the fact that a pure culture of bacterium X will not support the growth of a certain protozoan. However, this is usually accompanied by either the observation that good growth is attained if bacterium Y is mixed with bacterium X or by the observation that all is well if bacterium X is grown on an ill-defined substrate such as hay infusion.

The adequacy of specific bacteria in protozoan diets extends beyond the bacteriovore trophic level. Burbanck and Eisen (2) reported that *Paramecium*, if fed certain pure cultures of bacteria, will not support the reproduction of the predatory protozoan *Didinium*. If the pure culture is replaced with a mixed culture, *Didinium* reproduces successfully.

I am sure we could all speculate as to the biochemical and physiological basis for these observations, but it would be pure speculation. The point I wish to stress is that few, if any, of these relationships have ever been analyzed with modern techniques. Many other examples of our ignorance of the causes of observed interrelationships might be chosen, but the following four may serve as adequate illustrations.

First, some bacteria are definitely toxic to certain protozoans. In some cases this toxicity has been suggested as being due to pigment produced by the bacteria (26). However, the cause is unknown in most recorded cases.

Second, other bacteria are beneficial in an ill-defined way. For example, some cause the encystment of soil protozoa. The active factor is unknown, although free amino acids have been implicated in the process (1).

Third, Singh (22) has reported that soil amoebae ingest certain bacteria much more readily than others. He notes that not only do some protozoans display distinct preferences for certain bacteria but also that these preferences can influence their behavior. Amoebae and ciliates, if placed in circumstances where they are free to migrate to loci of different bacteria, display remarkable consistency and speed in reaching the loci of choice. This behavior implies the presence of water-soluble behavior-influencing chemicals; yet I know of few serious attempts to define these chemicals in spite of significant ecological implications.

Finally, we should consider the bacteria for a moment. In the early 1900s, it was suggested that predation of soil bacteria by protozoans was detrimental to soil fertility (24). This was not experimentally verified by further studies. Indeed, it may be that exactly the opposite is the case. It now appears that a measure of predation results not only in enhanced multiplication rates but also in enhanced efficiency of substrate utilization that we would expect to be associated with elevated growth rate. For example, increased rates of oxygen uptake, nitrogen fixation, and ammonia production have been observed (6, 19, 20). Similar observations have been recorded recently during studies on the microbial communities of reservoirs (16, 27, 28). One may conclude that the trophodynamic relationship between protozoa and their bacterial prey might markedly affect the trophodynamic relationship between bacteria and their substrate.

Attempts to study the trophodynamics of microbial prey-predator systems and the successional changes associated with such systems date from the introduction of the microscope. Indeed, I believe that I would not be far wrong in suggesting that 95% of all such studies are based solely upon that instrument. Direct observation can produce a wealth of information (the excellent study of Picken, 21) and is probably essential to the most modern of investigations. However, this approach simply documents changes and does not provide the basic reasons behind them. These reasons might be forthcoming if the advanced techniques of other areas, such as molecular biology and bacteriology, were to be applied to this field.

That they have not been applied is obvious when one realizes that factors as basic to the trophodynamic pattern as the caloric equivalents of either prey or predator are little known *(4)*. Indeed, Drake et al. *(7)* have noted that while we possess a relative wealth of information regarding certain food chains, we know very little about the process by which constituents of prey are, to use their term, "upgraded" by the predator. While I question the precise meaning of the term "upgrade," I do agree that very little is known of the importance that is assumed by protozoans and other microscopic predators in the process of making molecular and microparticulate material, such as bacteria, available to higher trophic levels.

Protozoa are part of the nutrient regenerative process in that they degrade particulate biological materials, releasing soluble organic and inorganic compounds that can then be utilized by the primary producers *(17)*. However, they can also be considered, in some sense, as producers of, or at least, mediators in that they convert soluble organic compounds and microparticulate material into their own physical form. This process makes these materials available to the larger zooplankton. The importance of one function over the other is quite unexplored. Considering these gaps in our knowledge, it is not surprising to find that the biochemical changes associated with predation and the biochemical bases for the existence of one microbial prey-predator system over another are, literally, *terra incognito.*

In the past, trophodynamic experiments have been carried out, for the most part, on the basis of visual observations alone and usually in closed systems such as flasks or bottles. In such environments, the parameters contributing to the observed changes in population density must be in more rapid flux than the gross changes themselves. It is certain that the following trophodynamic parameters can change in a very short time: prey density, predator density, and by-product concentrations. While it is possible to obtain useful information from the study of closed systems, the timing of the observations can be very important. The selection of appropriate intervals is difficult, and these intervals can change during the course of the experiment. As the quality of one's observation is directly dependent upon the degree to which sampling intervals are properly selected, one can easily be faced with a collection of data in which the individual observations are variably accurate in their reflection of reality.

One must also be aware that, in closed systems, variables other than those mentioned above can be in flux. I refer here to such things as physiological state, changes in tactic responses, or subtle shifts in physical factors such as illumination, pH, and oxygen tension. It is even possible that yield may change (unit predator produced/unit prey consumed). If one can "freeze" these changing parameters long enough to spend one's time making more accurate or technically more sophisticated observations rather than deciding upon sampling intervals, it would appear that one would have a much better chance to tease apart the intricate relationships that govern the microbial prey-predator systems. Such a cessation of change implies steady state conditions. Steady state is characteristic of, and is most easily obtained under conditions of continuous culture in a chemostat. I have been investigating the use of chemostats in trophodynamic studies and propose to spend some time discussing the approach.

The chemostat is a powerful tool that is routinely used in microbiology but has been little employed in microbial trophodynamic studies. It is really quite a simple arrangement but, of course, can be as intricate as the individual investigator's budget permits. Particularly good discussions of chemostat design have been presented by Jannasch (14) and by Evans et al. (8). All that is required for a bacterial chemostat is equipment which will preserve the following parameters invariant: medium composition, medium flow rate, culture volume, pH, oxygen tension, and temperature. When investigating organisms other than bacteria, there are other factors to consider keeping constant (for example, light in the case of a photoautotroph or turbulence in the case of a protozoan). Indeed, this latter point introduces some difficulty, for basic to most chemostats is the accomplishment of near-perfect mixing. Near-perfect mixing requires a great deal of turbulence. Many protozoa seem to prefer a quiet life and express displeasure at turbulence by lysing. We have found that stirring by means of streams of bubbles provides adequate mixing yet does not appreciably affect the protozoan with which we are working. I have the distinct impression that we have been fortunate in our choice of organism. Some of the organisms that have been reported to withstand this sort of treatment are now reported to disintegrate.*

In the design of the chemostat for work with protozoans, one must plan for long-term reliability. We commonly run our chemostats for over 500 hr while others run theirs for over 800 hr. The admonitions regarding reliability which are contained in the paper of Evans et al. (8) are, therefore, to be taken quite seriously, even though certain aspects of their design (such as that of the impeller) are inappropriate to the culture of many protozoans. The monetary investment in a long-term culture makes the initial equipment investment, which guarantees reliability, very small indeed.

In a trophodynamic experiment which is to be carried out in a chemostat, I must admit that one is limited in his choice of organisms and must carefully select those with which he wishes to work. Examples of desirable characteristics are as follows: (1) ability to tolerate chemical and physical restraints of the system (e.g., turbulence, mutually acceptable pH, etc.); (2) absence of clumping or other grouping of bacterial cells which would render them unavailable to the protozoan in question; and (3) absence of, or ability to control, resting or sexual stages which would be washed out of a flowing system. One would also look for a situation in which bacterial metabolites do not inhibit the protozoan, or vice versa. However, this may be real life and therefore may be accepted for certain experimental purposes.

In addition, one must have a definitive sequence of material transfer. For example, in a four-stage chemostat wherein a carbon source produces bacteria which are consumed by a protozoan which are in turn consumed by a carnivore, one would want to ensure that neither protozoan nor carnivore was able to use

* Tsuchiya, personal communication.

the carbon source. Ideally, it must be solely available to the bacteria. In addition, one would want to be sure that the protozoan or the carnivore was neither stimulated nor repressed by bacterial metabolites or that these effects were defined. In addition, there must be no dependence of the bacterial population upon materials produced by the protozoan or predator. Finally, the carnivore must not be able to use the bacteria but must be solely dependent upon the protozoan itself. Of course, all these points for selection are really ideals and, in practice, one almost always has to sacrifice one or more of the ideals or spend time defining their variants, in order to carry out a desired experiment.

The equations that describe the population dynamics of a single heterotrophic organism under continuous culture have been developed by a number of workers. The presentation which I find most useful is that of Herbert, Elsworth, and Telling *(13)* or that of Tempest *(29)*. It is not immediately apparent that the growth of a phagotrophic organism should be limited by the concentration of its prey in the manner required by these equations. Caperon *(3)* has advanced a possible explanation as to why this should be so, should certain assumptions be correct. Hamilton and Preslan *(11)* paraphrased these assumptions to include protozoa-bacteria systems. Moreover, a number of workers have shown that the growth rate of certain protozoa satisfies the basic criteria by apparently obeying an equation which is identical in form to that describing saturation, or Michaelis-Menten enzyme kinetics *(5, 10, 22)*. It should be noted here that these studies were performed in closed systems but with strict attention being paid to appropriate sampling intervals and the prey concentration which occurred during those intervals. There have been very few attempts to treat the population dynamics of prey-predator systems under continuous culture. Perhaps the most useful and interesting treatment to date is that of Drake et al. *(7)*.

Before we began our initial attempts to study microbial prey-predator systems in a chemostat, we had become convinced of the complexity of those systems in which both prey and predator were growing. Therefore, we took what seemed to be a first logical step in exploring the field and eliminated the growth of the prey organism entirely. This was accomplished by growing, harvesting, and cleaning batch cultures of the prey bacteria which were then resuspended in a medium and environment that did not permit their growth. We had established that the protozoan we were investigating required certain physical standards in its handling. Once these were satisfied, we could then inoculate the protozoan into a chemostat and pump in the bacterial suspension in the same manner one would pump in a glucose solution to a continuous culture of *Aerobacter aerogenes*.

Study of this simplified system revealed a number of interesting observations. The maximum growth rate as estimated by the chemostat was essentially the same as that estimated in flask culture but the K value was depressed by about 50% (0.225 μg carbon /ml vs. 0.486 μg carbon/ml). It has been suggested that the K of a protozoan which is feeding upon bacteria may be related to the ratio between prey and predator. Such was not the case in these experiments as this ratio was varied continually and we can provide no convincing explanation for this result.

Yield as calculated by cell numbers proved to be a constant over a wide range of flow rates as well as a wide range of incoming food concentrations. However, we noted that protozoan cell volume changed in apparent response to protozoan cell density. We believed that some conditioning factor was involved, as has been reported in bacterial chemostats run under similar conditions *(15)*. The net effect of this change in cell volume was that yield, based upon cell carbon, proved to be a variable (5 to 16%). This is a much lower yield than has been reported for other protozoan systems *(5)*. Indeed, yields up to 78% have been reported in some cases *(22)*. This seems excessively high but may, perhaps, be due to the fact that protozoans have no nongrowing adult stage, as suggested by Heal *(12)*.

The other unexpected observation was that the protozoan population washed out at incoming food concentrations which theoretically should have produced steady state. This effect occurred at low protozoan concentrations and therefore at those concentrations where the protozoan cell volumes were small. Therefore, one explanation for this effect is that there is a critical cell volume below which cell division cannot occur and, of course, in a flowing system this would mean outwash of the population. The summation of these effects means that, although the protozoan appears to obey a saturation or Michaelis-Menten type function, that portion of the curve below K simply does not exist when the protozoan is in an open system.

We have begun investigating systems in which both prey and predator are growing. At this point, I can only say that steady state is attained and that our systems has not exhibited the fluctuations in population densities reported by other workers *(7, 18)*.

In conclusion, I submit that the trophodynamics of bacteria-protozoa systems deserves rather more attention than it has received in the past. It specifically needs the application of modern techniques. The use of the chemostat may not only provide interesting data when coupled with classical observations (cell carbon determinations are, after all, not that different from cell counts) but also its use could free the researcher from cultural details, provide stable conditions, and afford the time to apply other sophisticated techniques.

Literature Cited

1. Alexander, M. 1971. *Microbial ecology.* New York: Wiley.
2. Burbanck, W. D., and J. D. Eisen. 1960. The inadequacy of monobacterially fed *Paramecium aurelia* as food for *Didinium nasutum. J. Protozool.* **7:**201–206.
3. Caperon, J. 1967. Population growth in microorganisms limited by food supply. *Ecology* **48:**715–22.
4. Cummins, A. W., and J. C. Wuycheck. 1971. Caloric equivalents for investigations in ecological energetics. *Mitt. int. Ver. Limnol.* **18:**1–158.
5. Curds, C. R., and A. Cockburn. 1968. Studies on the growth and feeding of *Tetrahymena pyriformis* in axenic and monoxenic culture. *J. Gen. Microbiol.* **54:**343–58.
6. Cutler, D. W., and D. B. Bal. 1926. Influence of protozoa on the process of nitrogen fixation by *Azotobacter chroococcum. Ann. Appl. Biol.* **13:**513–34.
7. Drake, J. F., J. L. Jost, A. G. Frederickson, and H. M. Tsuchiya. 1966. *The food chain, bioregenerative systems.* NASA Spec. Sp. 165. Washington, D.C.
8. Evans, C. G. T., D. Herbert, and D. W. Tempest. 1970. The continuous culture of microorganisms. II. Construction of a chemostat. In J. R. Norris and D. W. Ribbons *Methods in microbiology,* pp. 277–327. New York: Academic Press Inc.

9. Fenchel, T. 1968. The ecology of marine microbenthos. II. The food of marine benthic ciliates. *Ophelia* **5**:73–121.
10. Hamilton, R. D., and J. E. Preslan. 1969. Cultural characteristics of a pelagic marine hymenostome ciliate, *Uronema* sp. *J. Exp. Mar. Biol. Ecol.* **4**:90–99.
11. Hamilton, R. D., and J. E. Preslan. 1970. Observations on the continuous culture of a planktonic phagotrophic protozoan. *J. Exp. Mar. Biol. Ecol.* **5**:94–104.
12. Heal, O. W. 1967. Methods of study of soil protozoa. In *Proceedings of the symposium on methods of study of soil ecology*, ed. J. Philipson, Paris: UNESCO.
13. Herbert, D., R. Elsworth, and R. C. Telling. 1956. The continuous culture of bacteria: A theoretical and experimental study. *J. Gen. Microbiol.* **14**:601–22.
14. Jannasch, H. W. 1965. Continuous culture in microbial ecology. *Lab. Pract.* **14**:1162–67.
15. Jannasch, H. W. 1967. Growth of marine bacteria at limiting concentrations of organic carbon in sea water. *Limnol. Oceanogr.* **12**:264–71.
16. Javornicky, P., and V. Prokesova. 1963. The influence of protozoa and bacteria upon the oxidation of organic substances in water. *Int. Rev. Ges. Hydrobiol.* **48**:335–50.
17. Johannes, R. E. 1965. Influence of marine protozoa on nutrient regeneration. *Limnol. Oceanogr.* **10**:434–42.
18. Koelling, H. R. 1966. Oscillation of prey-predator systems in continuous culture. M. S. thesis, Virginia Polytechnic Institute.
19. Meiklejohn, J. 1932. The effect of *Colpidium* on ammonia production by soil bacteria. *J. Morph. Physiol.* **47**:85–129.
20. Nasie, S. M. 1923. Some preliminary investigations on the relationship of protozoa to soil fertility with special reference to nitrogen fixation. *Ann. Appl. Biol.* **10**:122–33.
21. Picken, L. E. R. 1937. The structure of some protozoan communities. *J. Ecol.* **25**:368–84.
22. Proper, G., and J. C. Garver. 1966. Mass culture of *Colpoda steinii*. *Biotechnol. Bioeng.* **8**:287–96.
23. Purdy, W. D., and C. T. Butterfield. 1918. The effect of plankton animals upon the bacterial death rates. *Amer. J. Public Health* **8**:499.
24. Russell, E. J., and H. B. Hutchinson. 1909. The effect of partial sterilization of soil on the production of plant food. *J. Agric. Sci.* **3**:111–14.
25. Singh, B. N. 1941. Selectivity in bacterial food by soil amoebae in pure mixed culture and in sterilized soil. *Ann. Appl. Biol.* **28**:52–64.
26. Singh, B. N. 1945. The selection of bacterial food by soil amoebae, and the toxic effects of bacterial pigments and other products on soil protozoa. *Brit. J. Exp. Pathol.* **26**:316–25.
27. Straskrabova, V., and M. Legner. 1969. The qualitative relation of bacteria and ciliates to water pollution. In *Advances in water pollution research*, ed. S. H. Jenkins, pp. 57–74. New York: Pergamon Press.
28. Straskrabova, V., and M. Legner. 1966. Interrelations between bacteria and protozoa during glucose oxidation in water. *Int. Rev. Ges. Hydrobiol.* **5**:279–93.
29. Tempest, D. W. 1970. The continuous culture of microorganisms. I. Theory of the chemostat. In *Methods in microbiology*, ed. J. R. Norris and D. W. Ribbons, pp. 259–75. New York: Academic Press Inc.

Estuarine Meiofauna: A Review: Trophic Relationships and Microbial Interactions

Bruce C. Coull

The term "meiobenthos" (meiofauna) was first coined by Mare *(52)* to describe those benthic metazoans of intermediate size. These are the animals smaller than those traditionally called "macrobenthos," but larger than the "microbenthos," i.e., bacteria and protozoa. McIntyre asserts that there is no clear-cut distinction between the meio- and the macrobenthos *(50);* the former simply refers to those metazoans, which, because of their small size, can be most adequately sampled by techniques different from those used with the larger animals. The meiofauna is by no means a homogeneous group and is not, like the psammon (interstitial fauna), restricted to a particular habitat. Other authors *(50, 56, 76)* further break the meiofauna down into two groups: (1) the temporary meiobenthos, those which spend only their larval stages as part of the meiobenthos (usually larvae of the macrofauna), and (2) the permanent meiobenthos, including Rotifera, Gastrotricha, Nematoda, Archiannelida, Tardigrada, Copepoda, Ostracoda, Mystococarida, Turbellaria, Oligochaeta, Polychaeta, Acarina, Gnathastomulida, and some specialized members of the Hydrozoa, Nemertina, Bryozoa, Gastropoda, Soelenogastres, Holothuroidea, Tunicata, Priapulida, and Sipunculida, which are always part of this group. Despite the fluidness of the definition, it is generally accepted that meiofauna refers to those animals which pass through a 0.5 mm sieve and are retained on a sieve with mesh widths smaller than 0.1 mm.

In recent years, aroused interest in meiofauna has produced an ever increasing volume of literature. Examining this literature, however, one finds that little data are available on the quantitative significance of meiofauna in estuarine systems. Most estuarine benthic research has been centered on the macrofauna, probably because of their commercial importance. In fact, Carriker in his review of estuarine benthic ecology *(17)* devotes but one subtitle to meiofauna and states that the lack of data is even more "thwarting" now that benthic ecologists recognize more clearly the profound ecological significance of smaller benthic organisms. Even though most estuarine benthic research has neglected the meiofauna, it has been suggested *(30)* that they may be responsible for five times the food supply of the macrofauna at any given spot. Obviously, their importance must continue to be investigated.

Review

Traditionally, estuarine (salt marsh) meiofaunal research has followed three distinct and rarely overlapping pathways, i.e., (1) faunal analysis of distinct

499

taxonomic groups and their distribution in the system *(4, 5, 8, 10, 16, 22, 25, 58, 63, 66, 71, 73, 75, 81, 83, 86)*; (2) abiotic effects (e.g., salinity, O_2, sediment size, Eh, pH) on meiofaunal distribution, with or without quantitative assessments of total meiofaunal populations *(6, 7, 9, 14, 27, 42, 59)*; and (3) examination of the total meiofaunal assemblage (often with an emphasis on a particular group) in quantitative terms and its interaction with environmental pressures *(12, 56, 62, 65, 70, 77, 79)*. As is obvious from the list of references provided (which does not intend to be complete, but salient), quantitative analysis of estuarine meiofauna communities is still in need of increased research effort. Estimates of seasonal fluctuation, population density, biomass, species composition, organism-sediment interrelationships, and community structure are limited to a few geographically isolated examples. Of course, many of the relationships apparent in nonestuarine conditions are similar in the estuaries and these data (see *20, 50, 51, 52, 85*) can be widely utilized by estuarine meiofaunal ecologists.

Comparisons of total meiobenthic populations from estuarine and nonestuarine systems indicate once again the generally high productivity of estuaries (Table 1). Although quantitative estuarine meiofaunal studies are limited to the few which we have listed, estuarine population densities are, in general, higher than those of nonestuaries. If, as is generally assumed, estuaries are one of the most productive noncultivated systems in existence *(60)*; the high density/biomass of meiofauna is not at all·unusual. The energetic role of these omnipresent forms is as yet unknown. Very little is known of the importance of meiofauna in the nutrition of other organisms, especially the commercially important species, many of which are known to use the estuary as a nursery ground. Possibly the large biomass of meiofauna may supply more nutritional value than the usually accepted major source of secondary production, the zooplankton. In fact, Reid reports meiofaunal numbers of $63 \times 10^3/m^2$ and zooplankton numbers of $40 \times 10^3/m^2$ for the same area at the same time *(65)*. This somewhat fragmented piece of data again points to a need for a better understanding of the role of meiofauna in the estuarine system.

The data presented in Table 1 were chosen (out of many other possibilities) because most represent mean values of a yearly study. To attempt to quantify meiofauna population density on the basis of a few seasonally restricted samples invites a misconception of the importance of these forms in the systems. For example, I noted values as low as $12/10$ cm^2 and as high as $1330/10$ cm^2 on the Bermuda platform *(20)*. Had samples been taken only at the time of the ebb ($12/10$ cm^2) or the peak ($1,330/10$ cm^2), a totally different picture of meiofauna density would have appeared. Thus, some of the values in Table 1 may not represent true estimates of the population density *(52, 65, 73, 85)*.

Due to the known patchiness of distribution in meiofauna *(38, 80)*, Reid's values (Table 1) must be viewed with further caution *(65)*. Her sampling was not quantitative, since "at each site one sample was taken by scooping about 25 ml from the upper 1 cm of the contents of a 0.2 m^2 Van Veen bottom sampler" *(65)*. She then converted these values into numbers per 100 ml sediment which, in turn, I have converted to numbers per 10 cm^2 in order to standardize the units in Table 1. However, if the inherent errors in her sampling are real-

Table 1. Summary of quantitative meiofauna data from estuarine and some nonestuarine environments[a]

Habitat	Author	Locality	Dominant Organism	× No/10 cm²	Total Meiofauna × No/10 cm²
Estuary	Brickman (12)	New Jersey salt marsh			
		1. 4.2‰ salinity	Copepoda	667	776
		2. 11‰ salinity	Copepoda	238	586
		3. 16.5‰ salinity	Nematoda	585	885
	Capstick (16)	River Blyth, England (Intertidal)			
		1. middle-high salinity	Nematoda	808–1,848	—
		2. upper-low salinity	Nematoda	203– 799	—
	Fenchel (25)	Danish sands	Nematoda	?	500–2,000[b]
	Muus (57)	Danish mesohaline	Nematoda	658	968
	Reid (65)	Pamlico River, N.C. (Summer)	Nematoda	123	136
	Smidt (70)	Danish Waddensea	Nematoda	223	—
	Teal and Wieser (73)	Georgia salt marsh (March)	Nematoda	12,400	—
	Tietjen (79)	Rhode Island			
		1. *Zostrea* Flat	Nematoda	2,030	2,369
		2. Sands	Nematoda	2,132	2,686
	Warwick (unpublished)	River Exe, England (Intertidal)	Nematoda	503	—
Nonestuary	Coull (20)	Bermuda (3–13 M)	Nematoda	427	560
	Guille and Soyer (39)	Mediterranean (35 M)	Nematoda	79	150
	McIntyre (50)	North Sea (101 M)	Nematoda	853	1,068
	Mare (52)	English Channel (45 M) (March–May)	Nematoda	83	145
	Wieser (85)	Buzzards Bay, Mass. (13–30 M) (July–September)	Nematoda	364	375

[a] The estimates are based on at least one year of data, except where indicated.
[b] Not including Protozoa.

ized, her data are assumed to represent a summer minimum (65)*. Fall and spring increases in the macrofauna of the Pamlico River (74) are paralleled by meiofaunal increases.† Yet unpublished data‡ show five- to tenfold increases in the meiofauna numbers over the summer values she has reported.

Species distribution within the estuarine system is often delimited by the physical conditions appreciated by the organisms. Salinity fluctuation has generally been assumed to be the major limiting factor in the distribution of estuarine benthos. Sanders et al. (68), however, have indicated that interstitial salinity variations depend on the type of estuary studied. In fluctuating (tidal) estuaries, salinity changes are regular and of short duration (diurnal) and the sediment salinities, buffered within a narrow range, change very little during the tidal cycle. In gradient (nontidal) estuaries, however, salinity changes are unpredictable and of longer duration, usually occurring seasonally, and the sediment and overlying water salinities are generally the same. Thus, the measurement of overlying water column salinity in a fluctuating estuary is in no way a valid measurement of the conditions appreciated by the forms. However, one must not be misled. Those upper reaches of an estuary with less saline water (even though there may be diurnal fluctuations), will probably also have a less saline (nonfluctuating) interstitial water, while those areas of higher salinity water will then also have a higher interstitial salinity. Thus, there does appear to be a distinct relationship between salinity and the meiofaunal assemblage. The relationship is reflected in species composition, density, and species diversity. Bilio was able to distinguish four distinct groupings of meiofauna in relation to a salinity gradient (9). Although quantitative numbers are lacking, he did notice a decrease in the number of species proceeding from the "euryhaline Meerestiere" to the "holoeuryhaline Tiere." Reasonably enough, several authors (9, 12, 16, 65, 81) found the lower salinity samples with a strong influence of freshwater forms. Reid (65) and Brickman (12) found Darthythompsonia inopinata (a freshwater harpacticoid copepod) at their 5–10‰ stations; Bilio (9) found several freshwater genera of turbellarians, nematodes, and copepods (Horsiella, Itunella) at his low salinity brackish water station. Capstick (16) noted distinct changes in the nematode fauna between the middle and upper reaches of the River Blyth, as did Warwick (81) in the River Exe.

Associated with the switch in fauna, there is usually a decrease in the number of species proceeding to brackish water. Bilio (9) lists 60, 59, 2, and 4 meiofaunal species respectively, going from high salinity to fresh water. Brickman (12) also notes an increase in the number of species approaching the sea; however, he feels a more significant factor is the change from one to two dominant species in the low salinity areas to a more equal representation of four to five species in the higher salinity areas. Reid (65) found the number of benthic copepod species increased from one to seven as higher salinities were approached. An increase in the number of species and/or a more equitable distribution of the existing species reflects a more stable, predictable environment (69).

The reduction in number of species is most common in the gradient estuary

* K. Tenore, personal communication.
† K. Tenore, personal communication.
‡ Reid, personal communication.

(9, 12, 65) where great physiological stress is put on the organism in the areas of salinity change. An animal living in the tidal estuary with regular diurnal salinity changes may adapt in either of two ways. It may become euryhaline or it may become stenohaline, with behavioral mechanisms developed to respond to the tidal changes. The behavioral adaption of meiofauna in the tidal estuary has been a burrowing or interstitial mode of existence as they maintained themselves in the relatively constant salinity conditions of the sediment. If the salinity change is irregular and unpredictable, the only possible adaption is a broad physiological tolerance. Even so, animals adapted to living in this irregular type of environment are occasionally encountered with conditions beyond their broad tolerances and may be entirely eliminated *(69)*. Thus, the regularity and amplitude of the environmental fluctuations may exert as much stress as the change itself. Theoretically then, the gradient estuary should have much fewer species (although biomass and numbers might be greater) than the fluctuating estuary with its constant interstitial salinity.

I have attempted to compare the meiofaunal species diversity in these estuarine types, but it is futile. Sampling methods differ significantly, no author considers the same taxonomic group as another, and the studies treat different biotopes. Capstick *(16)* and Warwick *(81)* treat the fluctuating intertidal; Muus *(57)* and Tietjen *(77, 79)* treat the subtidal mesohaline and poly-euhaline areas exclusively; and Brickman *(12)* and Reid *(65)* treat subtidal meiofauna but in different systems (a salt marsh and a gradient estuary, respectively).

From the small amount of data available, there also appears to be a decrease in the number of animals per unit area as one proceeds up the estuary. However, other parameters distort this idealized picture. Warwick *(81)*, for example, recorded his lowest population densities at his most saline station, but attributed this to sediment granulometry and not the salinity conditions. In other studies of meiofauna distance from sea *(12, 16, 65)*, the number of animals did increase seaward, probably because euryhaline marine animals make up the greatest portion of estuarine faunas *(40)* and, therefore, most of the changes in faunal distribution are due to the failure of marine animals to penetrate salinities below their adaptive capacities. Truly euryhaline estuarine faunas are rare and euryhaline freshwater forms almost nonexistent. Thus, with the preponderance of marine forms responsible for the estuarine assemblage, the decreases in population parameters experienced are expected.

Omitting salinity, which is usually the first considered limiting factor in an estuary, and ignoring sediment granulometric effects which have been discussed many times before *(20, 51, 79, 81, 85* and others), the oxidation-reduction (redox) potential and its relationship to estuarine faunal distribution probably become most significant. Fenchel, in his most comprehensive study of benthic ciliates *(25)*, states that due to the generally high organic production of estuaries, the redox discontinuity layer is often so close to (or at) the surface that it often overlaps the photic zone. Previous authors *(51, 79)* have indicated the significance of the redox layer in limiting the vertical distribution of the fauna. Most metazoans cannot tolerate the reducing conditions and are thus restricted to the oxidized sediment zone above it. If, as Fenchel *(25)* and Jansson *(46)* assert, the redox layer is at or near the surface in estuaries, then the meiofauna is either completely wiped out, living on the surface, or restricted to those few species that can survive anaerobically. This then leads to the recently described sulfide community (28) where the dominant forms are ciliates and those few metazoans capable of existing in this reducing environment. These highly specialized and physiologically adapted forms become the most significant in the system. Many aerobic metazoans then find it necessary to employ an epipelic (on the surface) existence or become restricted to a very narrow (1 mm–1 cm) vertical range.

Trophic Interactions

Classical meiofauna food webs have been drawn (52, 61) and the general consensus has been that meiofauna are detrital feeders or indiscriminate feeders on benthic diatoms and bacteria. A somewhat more sophisticated approach has considered each meiofaunal group as a feeder on a particular entity, e.g., harpacticoids eat diatoms; gastrotrichs eat bacteria, etc.

Nothing could be further from the truth. Probably what is most striking is the complexity and interactions within the meiobenthic food web. Each major taxonomic category may be split into distinct and different feeding types, with each species (or even the same species at different times) feeding on as varied a diet as exists in the sediment. The classical predator-prey size relationship (big fish eats little fish, eats littler fish, etc.) is also not necessarily true in the meiobenthic world. There is no reason to believe that a classification based on size directly reflects trophic relationships. Relatively small turbellarians, for example, may be the top of the food chain in some systems (15, 71). As active predators they will attack larger forms and either swallow their prey whole or suck out the prey with strong jaws and a muscular pharynx, respectively. Many macrofauna elements completely skip the intermediate meiofauna link and feed directly on bacteria and protozoa. Thus, they occupy the same trophic level as their smaller meiobenthic counterparts and, in fact, rather than being predators on, they are competitors with the meiofauna.

Recent comprehensive studies on individual groups have demonstrated the great diversity of feeding and nutritional gathering within these taxa. Fenchel's food web diagram (25) dealing with only the ciliates (not truly the meiobenthos, as defined earlier) is reproduced here (Fig. 1) to demonstrate the wide range of food procured by these unicellular organisms. If the system is as complex at the protozoan level as Figure 1 indicates, one can only speculate as to the interactions at the metazoan level.

Food of meiofauna within the estuary can come from a variety of sources. Recently, several new possibilities have been proposed, including assimilation of dissolved organic matter directly (19), although there is some doubt as to whether small crustaceans can use this source (3). Particulate organic detritus has been proposed and suggested as the most important food source of the estuarine benthos (1, 23, 24), but as of yet there are no data to indicate its nutritional value to the meiofauna. I have further suggested that carbon-rich aggregate particles formed by a physicochemical process may further serve as a food source (20).

Contrary to Adam and Angelovic (1), Fenchel and Hargrave both indicate

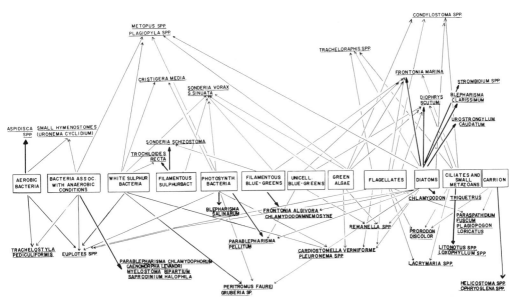

Fig. 1. The food relationships of a single group of Protozoans, the ciliates in Danish estuarine sediments. Note the complex interactions even within this "simple" group. (Redrawn from Fenchel *(25)* by permission of the Marine Biological Laboratory, Helsingor, Denmark.)

that it is not the detritus which supplies the energy, but rather the associated microbiota *(26, 44)*. Growth of the associated microbiota, however, is enhanced when the detrital feeder mechanically breaks down the particle and increases the surface to volume ratio, thus further encouraging microbial growth and decomposition. There is little doubt that in most estuaries particulate organic detritus is abundant and is consumed in great quantities by the meiofauna. Often the substrate itself may provide nutritional value, especially if cellulytic enzymes (assuming plant remains compose most of the detritus) occur in the gut as in some amphipods *(43)*. However, the critical feeding experiments on most animals remain to be done.

With the organic detritus problem still posing many questions, I very gingerly attempted to erect a new meiobenthic food web (Fig. 2). With the appearance of some new data and clarification of some old, this seems a justifiable task. Obviously, as more component fauna are studied from an autecological viewpoint and the feeding habits are more completely mapped, this schematic diagram will necessarily be altered. Only those food pathways known are drawn. (A reference citing the source of information occurs on each line of the figure.) Attempts to include food material out of the meiofaunal system proved futile and only served to make Figure 2 even more confusing. As generalizations, however, three pathways from meiofauna are apparent: (1) to macrofauna, e.g., shrimp *(31)* or nereid polychaetes *(20, 79)*; (2) to natant nektonic forms *(11, 13, 23, 55, 70)*; and (3) to nutrient regeneration. McIntyre *(51)* and Tietjen *(78)* indicate that dead meiofauna are rapidly broken down by bacterial action. Since there are relatively few meiofauna predators *(51)*, the main purpose of these forms may then be to assist in the recycling of nutrients at a low trophic level.

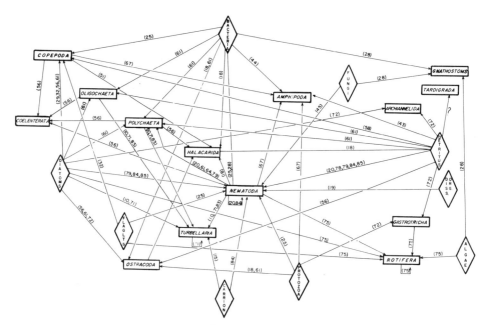

Fig. 2. Schematic diagram of the known food interrelationships of meiofauna. Detritus as used is an all-inclusive term to incorporate dead plant and animal remains, carbon-rich aggregated particles and fecal matter. The term "algae" refers primarily to blue-greens and filamentous greens, "Diss. org." to dissolved organics, and "Gnathostoms" to Gnathastomulida. The lines run from the food source to the consumer. The diamonds indicate a food source, the rectangles a consumer, except where predation is apparent. The numbers of each line refer to the data source.

Estimates of energy flow through the meiobenthos have been attempted only by Tietjen *(77)*, who concluded that benthic microflora growth was sufficient to support the meiobenthic population. One problem in accurately estimating meiobenthic metabolism has been the lack of adequate respiration data to estimate the caloric requirements of the component taxa. Recently, however, with a new impetus toward energy flow analysis, several works estimating the respiration of certain taxa have appeared *(21, 47, 48, 49, 86)* and should make the energetic role of meiofauna more easily assessed. This is an area which should continue to grow in development and significance and prove of great value in answering several questions regarding the trophic significance of these forms in the estuarine system.

Microbial Interactions

Besides the direct food benefit derived from microbes by meiofauna and the microbial degradation of dead meiofauna to regenerate nutrients, microbial entities, particularly bacteria and fungi (see Meyers, this volume) are important in governing the distribution of many forms within the system.

Preference experiments with marine meiofauna indicate that most forms will select sands having an organic coating. This coating is usually bacterial in origin and is not necessarily uniform on the sand grain, but occurs in crevices, hollows, or randomly interspersed on bare surfaces *(54)*. Each sand grain may have a distinct and characteristic microbial population and, on the surface of a given grain, there may be obvious differences in the structure of the microbial assemblage over short areas *(53)*. Localized variations in microbial coating will also vary with salinity, runoff, and wave action *(2)*. The microbial population may also alter the sediment granulometry. Bacterial films can change the properties of sand by altering the adhesiveness of the particles and by increasing their size and uniformity in shape, thereby modifying the interstitial lattice work *(82)*. If, as is generally assumed, meiofaunal populations are governed by a myriad of factors, the most important of which is sediment granulometry, then the bacterial alteration of sediment size may have a profound effect on determining species distribution and population density.

Many species of marine meiofauna can distinguish between sands containing different species of bacteria and are attracted to and inhabit sands where the preferred bacteria dominate *(33, 34, 35, 36, 37)*. Localized distribution cannot be explained simply by the fact that a bacterial film is or is not present. In fact, the species of bacteria and not their numbers per unit area is the factor controlling the attractiveness of a particular sediment.

Gray and Johnson *(37)* have found that certain coccoid bacteria, which occur more regularly than other bacterial morphotypes in marine sands *(54)*, are responsible for the attractiveness of certain sediments. The bacteria that were most attractive to their experimental organism (a gastrotrich, *Turbanella hyalina* Schultze) were those that were coccoid in shape, sensitive to EDTA and lysozyme treatment, but still attractive after the treatments, indicating that the attractive characters reside in the cell wall, not in the internal structure *(37)*.

If the attractive factor of bacterial films do reside in the cell wall, such recognition of the cells by meiofauna must occur by a "tactile chemical sense," i.e., direct contact with the bacteria. Gray has demonstrated, for example, that archiannelids *(Protodrilus* sp.) show avoidance of an unattractive (acid-washed) sand only after direct contact, thereby inferring a "tactile" rather than a chemotactic response which could be detected some distance from the substrate *(33, 34, 35)*. A similar tactile response is apparent in pelagic larvae of macrofauna *(76)*.

Many meiofaunal organisms are known to exhibit a patchiness of distribution *(38, 80)*. Meiofauna are sensitive to microhabitat changes and are able to select narrow areas of habitation within a relatively homogeneous (to the observer) substrate. The evidence seems to point to bacteria or bacterial films as the controlling factor within a substrate type. Where favorable (attractive) species exist, meiofauna population density may be high or dominated by those attracted species and vice versa. The observed bacterial species distribution differences may therefore account for the localized distribution of various meiofaunal species and may suggest reasons for the discrete substrate preferences of so many marine organisms. The relationship of specific microbes to meiofauna indicates the bacteria are an important ecological factor in the benthic biocenose and deserve much more attention.

Literature Cited

1. Adam, S. M., and J. W. Angelovic. 1970. Assimilation of detritus and its associated bacteria by three species of estuarine animals. *Chesapeake Sci.* **11**:249–54.
2. Anderson, J. G., and P. S. Meadows. 1969. Bacteria on intertidal sand grains. *Hydrobiol.* **33**:33–46.
3. Anderson, J. W., and G. C. Stephens. 1969. Uptake of organic material by aquatic invertebrates. VI. Role of epiflora in apparent uptake of glycine by marine crustaceans. *Mar. Biol.* **4**:243–49.
4. Barnett, P. R. O. 1968. Distribution and ecology of harpacticoid copepods of an intertidal mudflat. *Int. Rev. ges. Hydrobiol.* **53**:177–209.
5. Bilio, M. 1964. Die aquatische Bodenfauna von Salzwiesen der Nord-und Ostsee. I. Biotop und okologische Faunenanalyse: Turbellaria. *Int. Rev. ges. Hydrobiol.* **49**:509–62.
6. Bilio, M. 1965. Die Verteilung der aquatischen Bodenfauna und die Gliederung der Vegatation im Strandbereich der deutschen Nord-und Osteseekuste. *Botan. Gothoburgensia* **3**:25–42.
7. Bilio, M. 1966a. Die aquatische Bodenfauna von Salzwiesen der Nord-und Ostsee. II. Okologische Fauneanalyse: Hydrozoa, Nematodes, Rotatoria, Gastrotricha, Halacaridae, Ostracoda, Copepoda. *Int. Rev. ges. Hydrobiol.* **51**:147–95.
8. Bilio, M. 1966b. Characteristische Unterschiede in der Besiedlung finnischer, deutscher und hollandischer Kusten salzwiesen durch Turbellarien. *Veroff. Inst. Meeresforsch. Bremerhaven* **2**:305–18.
9. Bilio, M. 1967a. Die aquatische Bodenfauna von Salzwiesen der Nord-und Ostsee. III. Die Biotopeinflusse auf die Faunenverteilung. *Int. Rev. ges. Hydrobiol.* **52**:487–533.
10. Bilio, M. 1967b. Nahrungsbeziehungen der Turbellarien in Kustensalzwiesen. *Helgol. Wiss. Meeresunters.* **15**:602–21.
11. Bregnballe, F. 1961. Plaice and flounder as consumers of the microscopic bottom fauna. *Medd. Danm. Fisk-og. Havunders., N.S.* **3**:133–82.
12. Brickman, L. M. 1972. Base food chain relations in a coastal salt marsh ecosystem. Ph.D. dissertation, Lehigh University.
13. Bruun, A. 1949. The use of nematodes as food for larval fish. *J. du Conseil.* **16**:96–99.
14. Burbank, W. D., and G. P. Burbank. 1967. Parameters of interstitial water collected by a new sampler from the biotopes of *Cyathura polita* (Isopoda) in six southeastern states. *Chesapeake Sci.* **8**:14–27.
15. Bush, L. F. 1966. Distribution of sand fauna in beaches at Miami, Florida. *Bull. Mar. Sci.* **16**:58–75.
16. Capstick, C. K. 1959. The distribution of free living nematodes in relation to salinity in the upper and middle reaches of the river Blyth estuary. *J. Anim. Ecol.* **28**:189–210.
17. Carriker, M. R. 1967. Ecology of estuarine benthic invertebrates; A perspective. In *Estuaries*, ed. G. H. Lauff, pp. 442–87. Washington, D.C.: AAAS.
18. Chandraesekhara Rao, G. 1970. On some interstitial fauna in the marine sands on Indian Coast. *Curr. Sci. India.* **39**:504–507.
19. Chia, F. S., and R. W. Warwick. 1969. Assimilation of labelled glucose from seawater by marine nematodes. *Nature (London)* **224**:720–721.
20. Coull, B. C. 1970. Shallow water meiobenthos of the Bermuda platform. *Oecologia (Berl.)* **4**:325–57.
21. Coull, B. C., and W. B. Vernberg. 1970. Harpacticoid copepod respiration: *Enhydrosoma propinquum* and *Longipedia helgolandica*. *Mar. Biol.* **5**:341–44.
22. Crandell, G. F. 1967. Seasonal and spatial distribution of harpacticoid copepods in relation to salinity in Yaquina Bay, Oregon. Ph.D. dissertation, Oregon State University.
23. Darnell, R. M. 1961. Trophic structure of an estuarine community based on studies of Lake Pontchartrain, Louisiana. *Ecology* **42**:553–68.

24. Darnell, R. M. 1967. Organic detritus in relation to the estuarine system. In *Estuaries*, ed. G. H. Lauff, pp. 376–82. Washington, D.C.: AAAS.
25. Fenchel, T. 1969. The ecology of marine microbenthos. IV. Structure and function of the benthic ecosystem, its physical factors and the microfauna communities with special reference to the ciliated Protozoa. *Ophelia* **6:**1–182.
26. Fenchel, T. 1970. Studies on the decomposition of organic detritus derived from the turtle grass *Thalassia testudinum. Limnol. Oceanogr.* **15:**14–20.
27. Fenchel, T., and B. O. Jansson. 1966. On the vertical distribution of the microfauna in the sediments of a brackish water beach. *Ophelia* **3:**161–77.
28. Fenchel, T. M., and R. J. Riedl. 1970. The sulfide system: A new biotic community underneath the oxidized layer of marine sand bottoms. *Mar. Biol.* **7:**255–68.
29. Fraser, J. H. 1936. The occurrence, ecology and life history of *Tigriopus fulvus* (Fischer). *J. Mar. Biol. Ass. U.K.* **20:**523–36.
30. Gerlach, S. A. 1971. On the importance of marine meiofauna for benthos communities. *Oecologia (Berl.)* **6:**176–90.
31. Gerlach, S. A., and M. Schrage. 1969. Freilbende Nematoden als Nährung der Sandgarnele *Crangon crangon*. Experimentalle Untersuchungen über die Bedeutung der Meiofauna als Nahrung für das marine Makrobenthos. *Oecologia (Berl.)* **2:**362–75.
32. Gilat, E. 1967. On the feeding of a benthonic copepod, *Tigriopus brevicornis* O. F. Muller. *Sea Fish. Res. Sta. Hafia Bull.* **45:**79–94.
33. Gray, J. S. 1966. The attractive factor of intertidal sands to *Protodrilus symbioticus. J. Mar. Biol. Ass. U.K.* **46:**627–45.
34. Gray, J. S. 1967a. Substrate selection by the archiannelid *Protodrilus rubropharyngeus. Helgol. Wiss. Meeresunters.* **15:**253–69.
35. Gray, J. S. 1967b. Substrate selection by the archiannelid *Protodrilus hypoleucus* Armenante. *J. Exp. Mar. Biol. Ecol.* **1:**47–54.
36. Gray, J. S. 1968. An experimental approach to the ecology of the harpacticoid *Leptastacus constrictus* Lang. *J. Exp. Mar. Biol. Ecol.* **2:**278–92.
37. Gray, J. S., and R. M. Johnson. 1970. The bacteria of a sandy beach as an ecological factor affecting the interstitial gastrotrich *Turbanella hyalina* Schultze. *J. Exp. Mar. Biol. Ecol.* **4:**119–33.
38. Gray, J. S., and R. M. Rieger. 1971. A quantitative study of the meiofauna of an exposed sandy beach, at Robin's Hood Bay, Yorkshire. *J. Mar. Biol. Ass. U.K.* **51:**1–20.
39. Guille, A., and J. Soyer. 1968. La faune benthique des substrates meubles de Banyuls-sur-Mer. Premieres donnees qualitatives et quantitatives. *Vie et Milieu* **19:**323–60.
40. Gunter, G. 1961. Some relations of estuarine organisms to salinity. *Limnol. Oceanogr.* **6:**182–90.
41. Hagerman, L. 1969a. Respiration, anerobic survival and diel locomotory periodicity in *Hirschmannia viridis* Muller (Ostracoda). *Oikos* **20:**384–91.
42. Hagerman, L. 1969b. Environmental factors affecting *Hirschmannia viridis* (O. F. Muller) (Ostracoda) in shallow brackish water. *Ophelia* **7:**79–99.
43. Halcrow, K. 1971. Cellulase activity in *Gammarus oceanicus* Segerstrale (Amphipoda). *Crustaceana* **20:**121–24.
44. Hargrave, B. T. 1970. The utilization of benthic microflora by *Hyallela azteca* (Amphipoda). *J. Anim. Ecol.* **39:**427–37.
45. Hopper, B. E., and S. P. Meyers. 1966. Aspects of the life cycle of marine nematodes. *Helgol. Wiss. Meeresunters.* **13:**444–49.
46. Jansson, B. O. 1969. Factors and fauna of a Baltic mud bottom. *Limnologica (Berl.)* **7:**47–52.
47. Lasker, R., J. B. J. Wells, and A. D. McIntyre. 1970. Growth, reproduction, respiration and carbon utilization of the sand-dwelling harpacticoid copepod, *Asellopsis intermedia. J. Mar. Biol. Ass. U.K.* **50:**147–60.
48. Lasserre, P. 1969. Relations energetiques entre le metabolism respiratoire et la

regulation ionique chez Annelide oligochete euryhaline *Marionina achaeta* (Hagen). *C. R. Acad. Sci. Paris* **268**:1541–44.

49. Lasserre, M. P., and J. Renaud-Mornant. 1971. Consommation d'oxygene chez un Crustace meiobenthic-interstitielle de lasons—classe des Mystococarides. *C. R. Acad. Sci. Paris* **272**:1011–15.

50. McIntyre, A. D. 1964. Meiobenthos of sublittoral muds. *J. Mar. Biol. Ass. U.K.* **44**:665–74.

51. McIntyre, A. D. 1969. Ecology of marine meiobenthos. *Biol. Rev.* **44**:245–90.

52. Mare, M. F. 1942. A study of the marine benthic community with special reference to the micro-organisms. *J. Mar. Biol. Ass. U.K.* **25**:517–54.

53. Meadows, P. S., and J. G. Anderson. 1966. Micro-organisms attached to marine and freshwater sand grains. *Nature (London)* **212**:1059–60.

54. Meadows, P. S., and J. G. Anderson. 1968. Micro-organisms attached to marine sands. *J. Mar. Biol. Ass. U.K.* **48**:161–75.

55. Mulkana, M. S. 1964. The growth and feeding habits of juvenile fishes in two Rhode Island estuaries. M. S. thesis, University of Rhode Island.

56. Muus, K. 1966. A quantitative 3-year survey on the meiofauna of known macrofauna communities in the Oresund. *Ver. des Instituts Meeresforsch.* **2**:289–92.

57. Muus, B. J. 1967. The fauna of Danish estuaries and lagoons. Distribution and ecology of dominating species in the shallow reaches of the mesohaline zone. Medd. fra Danmarks Fisk.-og. Havunders., N.S. **5**:3–316.

58. Noodt, W. 1957. Zur Okologie der Harpacticoidea (Crust. Cop.) des Eulitorals der deutschen Meereskuste und der nagrenzenden Brackewasser. *Z. Morph. u. Okol. Tiere.* **46**:149–242.

59. Noodt, W. 1969. Substratspezifitat bei Brackwasser-Copepoden. *Limnologica (Berl.)* **7**:139–45.

60. Odum, E. P. 1959. *Fundamentals of ecology.* Philadelphia: W. B. Saunders.

61. Perkins, E. J. 1958. The food relationships of the microbenthos with particular reference to that found at Whitestable, Kent. *Ann. Mag. Nat. Hist.* (ser. 13) **1**:64–77.

62. Purasjoki, K. J. 1945. Quantitative Untersuchungen über die Mikrofauna des Meeresbondens in der Umgebung der zoologischen Station Tvarminne an der Sudkuste Finnlands. *Soc. Scient. Fennica. Comment. Biol.* **9**:14.

63. Raibaut, A. 1967. Recherches ecologiques sur les copepodes harpacticoides des etangs cotiers et des eaux saumatres temporaires dur Languedoc et de Camargue. Ph.D. dissertation, Univ. de Montpellier.

64. Rees, C. B. 1940. A preliminary study of the ecology of a mud flat. *J. Mar. Biol. Ass. U.K.* **24**:195–99.

65. Reid, J. W. 1970. The summer meiobenthos of the Pamlico River Estuary, North Carolina, with particular reference to the harpacticoid copepods. M. S. thesis, North Carolina State University.

66. Riemann, F. 1966. Die interstitielle Fauna im Elbe estuar. *Arch. Hydrobiol.* **31**:1–279.

67. Sameoto, D. D. 1969. Comparative ecology, life histories, and behaviour of intertidal sand-burrowing amphipods (Crustacea: Haustoriidae) at Cape Cod. *J. Fish. Res. Bd. Can.* **26**:361–88.

68. Sanders, H. L., P. C. Mangeldorf, Jr., and G. E. Hampson. 1965. Salinity and faunal distribution in the Pocasset River, Massachusetts. *Limnol. Oceanogr.* **10**:216–29.

69. Slobodkin, L. R., and H. L. Sanders. 1969. On the contribution of environmental predictability to species diversity. In *Diversity and stability in ecological systems.* Brookhaven Symp. Biol. **22**:82–95.

70. Smidt, E. L. B. 1951. Animal production in the Danish Waddensea. *Medd. Fra. Kommiss. for Danmarks Fisk.-og Havundersoger-ser. Ser:Fiskeri* **11**:151.

71. Straarup, B. J. 1970. On the ecology of turbellarians in a sheltered brackish shallow water bay. *Ophelia* **7**:185–216.

72. Swedmark, B. 1964. The interstitial fauna of marine sand. *Biol. Rev.* **39**:1–42.

73. Teal, J., and W. Wieser. 1966. The distribution and ecology of nematodes in a Georgia salt marsh. *Limnol. Oceanogr.* **11**:217–22.
74. Tenore, K. 1970. *The macrobenthos of the Pamlico River Estuary, North Carolina.* Chapel Hill: Water Res. Inst., Univ. of North Carolina, *Report* 40.
75. Thane-Fenchel, A. 1968. Distribution and ecology of nonplanktonic brackish water rotifers from Scandinavian waters. *Ophelia* **5**:273–99.
76. Thorson, G. 1966. Some factors influencing the recruitment and establishment of marine benthic communities. *Netherl. J. Sea Res.* **3**:267–93.
77. Tietjen, J. H. 1966. The ecology of estuarine meiofauna with particular reference to the class Nematoda. Ph.D. dissertation, University of Rhode Island.
78. Tietjen, J. H. 1967. Observations on the ecology of the marine nematode *Monohystera filicaudata* Allgen 1929. *Trans. Amer. Microscop. Soc.* **86**:304–306.
79. Tietjen, J. H. 1969. The ecology of shallow water meiofauna in two New England estuaries. *Oecologia* **2**:251–91.
80. Vitiello, P. 1968. Variations de la densite du microbenthos sur une aire restreinte. *Rec. Trav. St. Mar. d'Endoume Bull.* **43**:261–70.
81. Warwick, R. M. 1971. Nematode associations in the Exe Estuary. *J. Mar. Biol. Ass. U.K.* **51**:439–54.
82. Webb, J. E. 1969. Biologically significant properties of submerged beach sands. *Proc. Roy. Soc. London, B* **174**:355–402.
83. Webb, M. G. 1956. An ecological study of brackish water ciliates. *J. Anim. Ecol.* **25**:149–75.
84. Wieser, W. 1953. Die Beziehung zwischen Mundhohlengestalt, Ernährungsweise und Vorkommen bei freilebenden Marinen Nematoden. Eine okologisch-morphologische Studie. *Ark. Zool.*, Ser. II. **4**:439–84.
85. Wieser, W. 1960. Benthic studies in Buzzards Bay II. The Meiofauna. *Limnol. Oceanogr.* **5**:121–37.
86. Wieser, W., and J. Kanwisher. 1961. Ecological and physiological studies on marine nematodes from a small salt marsh near Woods Hole, Mass. *Limnol. Oceanogr.* **6**:262–70.

Comments

WATSON: Could you say a little more about what percentage of the nematodes are aerobic versus anaerobic? Do you have any wild guess? I always assumed that most of these in the sediments can live anaerobically, and I was somewhat surprised to hear you say that some are obligate aerobes.

COULL: I do not know. I would not even take a guess as to what percentage are aerobic or anaerobic. I think Wieser and Kanwisher *(86)* have been able to show that some will shift their metabolic pattern and that they will become anaerobic or aerobic depending on the situation that they find themselves in.

WATSON: Is there any evidence that anything eats nematodes?

COULL: Yes, according to my chart, Copepods, Halicarids, and Polychaetes. In fact, I have found a direct relationship and have seen nematode parts in the gut of nereid polychaetes. There is also an indication in the literature that larval fish eat them. In a recent paper, Gerlach and Schrage *(31)* report that they were able to raise shrimp on nematodes. I was also going to talk about what happened to this after it leaves the meiofaunal system, but I did not get a chance to. There is evidence that nematodes are actively being eaten by a myriad of things.

WATSON: What is the greatest number of nematodes you found per 10 cm^2?

COULL: I think the greatest number in the literature is probably 12,000 per 10 cm^2.

WATSON: Would this be in a contaminated area?

COULL: No, this was in a Georgia salt marsh, an area relatively free from pollution.

WATSON: How does this compare to offshore sediments?

COULL: Do you mean on the shelf or in the deep sea?

WATSON: On the shelf.

COULL: On the shelf, the values are ranging between 200 and 1,000 per 10 cm^2.

WATSON: And would you estimate most estuaries in the thousands per 10 cm^2?

COULL: Yes, as a generalization.

WATSON: Are nematodes a good indicator for pollution studies?

COULL: No, I say that they are not, because they are horrible to work with taxonomically.

WATSON: I mean as sheer numbers.

COULL: I do not think we know enough about them to say. I can not even comment on it.

Bacteria as Potential Nutritional Resources for Three Sympatric Species of Tubificid Oligochaetes

K. E. Chua and R. O. Brinkhurst

Studies on the biology of tubificid oligochaetes have indicated that the nature of the food available to the worms may be a primary factor in determining the distribution and abundance of species *(1, 3)*. In a preliminary study of the food available to three tubificid species obtained in Toronto Harbour, Brinkhurst and Chua *(2)* demonstrated qualitative differences in the bacteria surviving passage through the gut under experimental conditions. Further investigation was carried out by Wavre and Brinkhurst *(9)* to demonstrate the degree of destruction of bacteria on passage through the gut of two tubificid species. Since the methods were described in detail by Brinkhurst and Chua *(2)* and again by Wavre and Brinkhurst *(9)*, they are mentioned here only briefly. By culturing the sediments, the gut contents of worms killed in the field and of worms kept for one week in worm saline, it was possible to show the reduction in diversity of bacteria after passage of mud through the gut of the worms. By culturing the sediments and the fecal pellets collected using the Alsterberg technique (Fig. 1) or by using sterile beads above the sediments under the "normal orientation" technique *(9)*, it was possible to measure the degree of destruction of bacteria on passage through the worm gut. Most of the culture studies were done using nutrient agar incubated at 25 C.

The difference in the microflora after one week was obvious (Table 1), as were the consistent differences among the worm species. Only one bacterial species survived passage through the gut, a different species in each instance. *Aeromonas* sp., possibly ingested only by *Peloscolex multisetosus*, was the only bacterium to survive in that worm species. *Micrococcus* sp. and *Pseudomonas* sp. were well represented in the ingestion samples (32 out of 60, and 51 out of 60, respectively, and about the same frequency for each tubificid species), but *Micrococcus* survived in *Limnodrilus hoffmeisteri* alone and *Pseudomonas* in *Tubifex tubifex*.

A marked reduction in the number of bacteria in feces as compared with the number in the original mud was demonstrated by both the Alsterberg (inverted) and the "normal orientation" techniques (Table 2–3). The reduction of bacterial species, even under such artificial conditions, and the degree of destruction of bacteria on passage through the gut, indicate that these bacteria are a primary food source for the worms, as they are for other benthic sediment-feeders *(4, 5, 6, 7, 8, 10, 11)*.

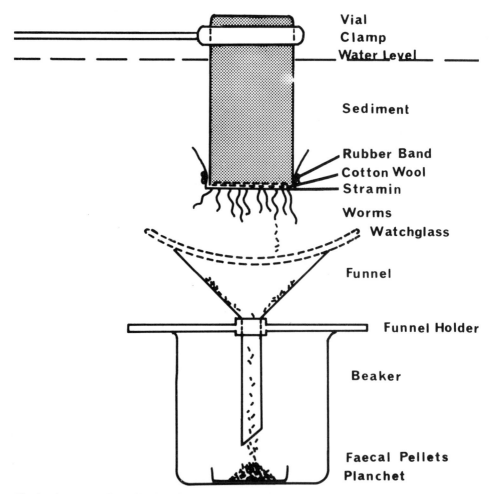

Fig. 1. Apparatus for collecting feces from inverted worm culture, the whole apparatus being immersed in a small thermostatically controlled water bath (Reprinted from Wavre and Brinkhurst (9) by permission of the Fisheries Research Board of Canada).

The differences between the worm species observed to date are consistent with the view of Brinkhurst and Chua (2) that partitioning of the nutritional resources may provide a mechanism by which three or more unspecialized detritus feeders are able to coexist despite continuous sharing of the same sediment. The reduction of diversity of bacteria in the worm after one week and the differences between the relative abundance of bacteria in mud and in feces deposited by the worms suggest that the worm population affects the composition of the microflora in the sediment as well as being affected by it.

Acknowledgment

The work was supported by grants to the junior author from the National Research Council, the Department of Energy, Mines and Resources, the Fisheries Research Board of Canada, and the U.S. Federal Water Quality Administration (16010 ECQ).

Table 1. Bacteria present in mud, in samples of worms killed in the field (after some digestion), and in worms starved for a week (complete digestion)[a]

Bacteria	No. Mud Samples[bc]	Samples from Worms Killed in the Field			Samples from Feces and Gut after 1 Wk		
		T.t[d]	L.h[d]	P.m[d]	T.t	L.h	P.m
Flavobacterium sp.	20	8	5	16	0	0	0
Pseudomonas fluorescens	10	0	2	5	0	0	0
Bacillus mycoides	19	2	1	1	0	0	0
Aeromonas sp.	20	0	0	14	0	0	20
Micrococcus sp.	16	10	12	10	0	15	0
Pseudomonas sp.	20	15	16	20	15	0	0
Bacillus cereus	20	2	4	10	0	0	0
No. of species present	7	5	6	7	1	1	1
Expected[e]	140	140	140	140	140	140	140
Observed[e]	125	37	40	76	15	15	20

[a] Reprinted from Brinkhurst and Chua (2) by permission of the Fisheries Research Board of Canada.

[b] Samples were collected from Toronto Harbour on four dates in May–June 1967.

[c] Numbers represent number of plates showing positive growth.

[d] T.t = Tubifex tubifex, L.h = Limnodrilus hoffmeisteri, P.m = Peloscolex multisetosus.

[e] "Expected" based on 5 plates per run, 4 runs, 7 species recognized equals 140 possible positive records. "Observed" equals number of plates on which positive growth was observed.

Table 2. Number of bacterial colonies per mg dry weight of sediment and of feces of Tubifex tubifex, cultured by Alsterberg technique and by normal orientation[a]

Expt.	Mud	Feces	Percentage Difference
Alsterberg technique			
1a	12,571	2,428	81
1b		1,080	91
2a	11,750	2,667	77
2b		3,615	69
3a	10,111	3,352	67
3b		4,000	60
Normal orientation			
1a	12,267	3,760	69
1b		3,836	69
2a	8,739	2,822	68
2b		2,601	70
3a	10,000	3,776	62
3b		3,791	62

[a] Reprinted from Wavre and Brinkhurst (9) by permission of the Fisheries Research Board of Canada.

Table 3. Number of bacterial colonies per mg dry weight of sediment and of feces of *Limnodrilus hoffmeisteri*, cultured by Alsterberg technique and by normal orientation[a]

Expt.	Mud	Feces	Percentage Difference
Alsterberg technique			
1a	11,444	2,000	83
1b	14,667	2,262	85
2a	10,785	2,250	79
2b	11,235	1,560	86
3a	12,500	2,740	78
3b	11,384	2,625	77
Normal orientation			
1a	12,091	3,707	69
1b		2,860	76
2a	10,917	3,732	66
2b		3,064	72
3a	11,909	4,857	59
3b		3,614	70

[a] Reprinted from Wavre and Brinkhurst *(9)* by permission of the Fisheries Research Board of Canada.

Literature Cited

1. Brinkhurst, R. O. 1970. Distribution and abundance of tubificid (Oligocheta) species in Toronto Harbour, Lake Ontario. *J. Fish. Res. Bd. Can.* **27:**1961–69.
2. Brinkhurst, R. O., and K. E. Chua. 1969. Preliminary investigation of the exploitation of some potential nutritional resources by three sympatric tubificid oligochaetes. *J. Fish. Res. Bd. Can.* **26:**2659–68.
3. Brinkhurst, R. O., and B. G. Jamieson. 1971. *The aquatic Oligochaeta of the world.* Edinburgh: Oliver and Boyd.
4. Hargrave, B. T. 1970. The utilization of benthic microflora by *Hyalella azteca* (Amphipoda). *J. Anim. Ecol.* **39:**427–37.
5. Marzolf, G. R. 1964. The trophic position of bacteria and their relation to the distribution of invertebrates. In *Organism-substrate relationships in streams,* ed. K. W. Cummins, C. A. Tryon, and R. T. Hartman, pp. 131–135. Pymatuning Laboratory of Ecology, Spec. Publ. 4. Pittsburgh: Univ. of Pittsburgh.
6. Newell, R. 1965. The role of detritus in the nutrition of two marine deposit feeders, the prosobranch *Hydrobia ulvae* and the bivalve *Macoma balthica. Proc. Zool. Soc. London* **144:**25–45.
7. Rodina, A. G. 1963. Microbiology of detritus of lakes. *Limnol. Oceanogr.* **8:**388–93.
8. Satchell, J. E. 1967. Lumbricidae. In *Soil biology,* ed. A. Burges and F. Raws, pp. 259–322. New York: Academic Press Inc.
9. Wavre, M., and R. O. Brinkhurst. 1971. Interactions between some tubificid oligochaetes and bacteria found in the sediments of Toronto Harbour, Ontario. *J. Fish. Res. Bd. Can.* **28:**335–41.

10. Zhukova, I. A. 1963. On the quantitative significance of microorganisms in nutrition of aquatic invertebrates. In *A symposium of marine microbiology*, ed. C. H. Oppenheimer, pp. 699–710. Springfield: C. C Thomas, Publisher.
11. ZoBell, C. E., and C. B. Feltham, 1938. Bacteria as a food source for certain marine invertebrates. *J. Mar. Res.* **1**:312–27.

Section Nine
SUMMATIONS

Summations

Rita R. Colwell

Rather than make a point-by-point summary of the discussions that have taken place over the past three days, I shall make a statement to provoke some thought and, hopefully, some action, or at least some future cooperative projects arising from the exchange of ideas and information occurring during this conference. First, it is sobering to reflect on the fact that those of us present, those of us brought together by the Belle W. Baruch Coastal Research Institute, comprise the working, practicing, estuarine microbiologists in the United States today, with the exception, of course, of about a half a dozen of our colleagues who are ill, out of the country, or on previously scheduled cruises. Practically speaking, I suspect our Canadian colleagues present today represent nearly the total population of estuarine and marine microbiologists in Canada. In terms of the tremendous amount of research to be done on microbiological problems of the estuaries, we are few in numbers and have been drawn, in many cases, from other disciplines. This is something to contemplate. We may need to consider recruitment efforts to increase our ranks.

As has been made obvious by the presentations we have heard, each of us is keenly interested in one or several aspects of the estuary and the estuarine biota, i.e., bacteria, yeast, and/or viruses. The number of estuarine and marine virologists is extremely small. Only two are here today and each is just beginning his research.

It is clear that we are very much aware that we are looking at only a fraction of the biota present in the estuary or open ocean. Some of us try to estimate total viable populations, for example, of bacteria, viruses, etc., keeping in mind that there are many kinds of other forms also operative and equally important in the biological cycle. The estuary system is very complex; none of us can do all that needs to be done. Thus, cooperation is going to be vital if we are to develop a more comprehensive approach to the study of estuarine and marine microbial ecology.

Cooperation can be effected in a variety of ways. A direct means of cooperative research is by setting up cruises with biologists and microbiologists, including participants such as a bacteriologist, an algologist, a mycologist, and a virologist. Each would measure the populations of living organisms and the data would be put together. It may be possible, also, to set up model ecosystems for analyses.

521

An urgent need we all recognize is to improve the present sampling devices and to invent new ones. We are at the moment practicing a most rudimentary kind of estuarine microbiology. As we have learned at this meeting, even the Niskin sampler does not help us significantly because the populations are inhibited by components leaching from the plastic sampling bag.

It may be necessary to establish, as is so commonly done in Great Britain, a working group to meet regularly, with a national society, as, for example, The American Society of Limnology and Oceanography. The working group can focus on an aspect of a developing area of science.

Frankly, it is quite obvious that the desperation we sense in the popular press of the need to "save our estuaries" does have validity. Furthermore, it is we who ultimately will do the job. This has been amply demonstrated by the papers read at this conference on effects of sewage effluent, plant wastes, and excess nutrients on the streams and estuaries feeding into the oceans. Fish kills are one example of the effects. So it seems we need to communicate with each other and with the public and we must share scientific expertise and the technology we now possess. New methods, instruments, and ideas are badly needed. As Dr. Coull indicated, in his search for literature on meiofaunal studies, few data were available. No seasonal analyses or other long-term studies had been done. Thus, it is not yet possible to correlate in a preliminary fashion our knowledge of the meiofauna. Even less so are we able to make extrapolations for microorganisms. The job ahead looms large and the challenge is great. The rewards are even greater, fortunately, for it is the survival of our estuaries and oceans that is at stake.

Claude E. ZoBell

I would like to take advantage of this opportunity to make a few postconference observations and appraisals. First, I want to express my appreciation to members of the organizing committee and the Institute for the invitation to speak. I found the Symposium to be very exciting. Every paper has been interesting to me. I felt that the quality of the context and presentation was a couple of levels higher than at most meetings of scientific societies. The illustrations were exceptionally good. Indeed, those of Dr. Foerster were good enough to merit an Emmy Award.

I particularly want to express my appreciation for the volume being dedicated to me. You will excuse me if I become just a bit emotional about this. At the risk of being thought overly sentimental for a grown man, I appreciate this more than words can tell. Whether deserved or not, I am very happy to accept this honor. I do it largely in the name of the forty or fifty graduate and postdoctoral students who have worked with me during the last forty years in aquatic microbiology. There is only one thing that could make it mean more to me, and that is if one piece of paper could be prepared with the signatures of all of you on it. This would not mean that you are signing a petition saying that I deserve it. But I would appreciate having your autographs in my copy of the book.

There are two closely related criticisms that I would offer concerning the proceedings. First, hardly enough time was allowed for discussion. It would have been nice to have more time for comments, questions, and suggestions for further work. The second criticism concerns my keynote address, in which I attempted to put in proper perspective all aspects of aquatic microbiology. In so doing, I fear that I was looking back instead of looking ahead. If we look back, we only see where we were. We have to look ahead to see where we are going.

In her summing up, Dr. Colwell made some very cogent remarks about looking ahead. Where are we going? What additional problems are we going to solve? Speaking retrospectively, perhaps instead of devoting most of the time discussing trivial details after each paper presented during the last three days, more attention should have been given to the discussion of general principles and unsolved problems.

What are the most pressing problems confronting aquatic microbiologists? As the major body of personnel that is responsible for advances in marine microbiology, what are we going to do about it? How can we use our limited personnel to maximum advantage? In order to see a significant impact of aquatic microbiologists on problems and processes in estuarine ecology in our lifetime, it is going to take much careful planning, perseverance, and concerted effort.

As I approach the status of professor emeritus, it is disappointing to look back and see how little has been achieved. A few things have been accomplished, but too few. I would like to be starting anew now. More rapid advances can be made because of better laboratory facilities, more sophisticated analytical apparatus, and more allies in the field. Currently, cooperation is available from several kinds of aquatic microbiologists having diverse backgrounds, interests, and expertise and also cooperation from chemists, physicists, biologists of all kinds, ecologists, and geologists. All such scientists are very important allies, because we are not a world unto ourselves. We have to cooperate with these scientists specializing in other disciplines.

One reason there are so few aquatic microbiologists is because aquatic microbiology is a multidisciplinary science. Besides having a familiarity with the fundamental principles of general microbiology, the aquatic microbiologist should also have preparation in biochemistry, physics, submarine geology, and ecology. Most of you and the new recruits either have had or are getting such multidisciplinary training. I think we are now in a much better position to meet the challenges than we ever were before.

As it looks to me, these challenges consist of obtaining meaningful information on the occurrence, abundance, characteristics, activities, and importance of bacteria and allied microorganisms in aquatic environments. Most urgently needed is information on how microbial activities influence (a) the productivity of natural waters, (b) the well-being of higher organisms, (c) geochemical conditions, (d) man-made structures, and (e) the pollution problem.

In looking forward, I would say that we should devote more attention to getting quantitative information rather than qualitative information. We devote too much time to counting various kinds of cells rather than getting information on how much biomass is present or produced per unit of time. For example, Holm-Hansen pointed out that when examining blue-green algae, it is better to report the results in terms of biomass rather than in terms of the number of cells. This applies even more to fungi, in which there is so much difference in cell or mycelium size. I believe this is true of most kinds of microorganisms.

Whenever possible, it seems desirable to express results in terms of biomass rather than merely the number of cells. Equally important in appraising the impact of various kinds of microorganisms is information on generation times or growth rates.

Speaking of various kinds of microorganisms reminds me of a major group of organisms which were not discussed during this conference. I have reference to the microflagellates, which I believe are quite important. From a point of view of numbers, biomass, and primary production, the microflagellates are very important. By microflagellates, I mean minute, chlorophyll-bearing organisms which are mostly planktonic. Owing to their small size, most microflagellates escape ordinary plankton nets. Consequently they are not counted along with the diatoms, dinoflagellates, zooplankton, and the like. Most microflagellates are not heterotrophic so they do not form colonies on conventional nutrient medium. They have to be collected and treated very carefully because they are fragile. They are highly labile. They soon deteriorate and disintegrate. They change in captivity when subjected to sudden change in climate such as temperature, salinity, pH, etc. Shrunken or partially disintegrated cells are not recognizable. One must rely largely on direct microscopic procedures to detect, identify, and enumerate marine microflagellates. I believe much more attention has been given to the microflagellates in France, Russia, and Great Britain than in the United States or Canada.

To jump abruptly to another topic, I would like to stress the importance of considering all significant environmental parameters. In many of the presentations during the last three days, most of the environmental conditions which could influence the results were specified. But in some cases, I wanted to ask, "What was the oxygen tension? What was the redox potential, the pH, salinity, or osmotic pressure? What was the surface tension?" Only a few speakers mentioned surface tension in discussing field conditions. At air–water interfaces and cell–water interfaces, surface or interfacial tension (which may range from 40 to 70 dynes per centimeter) may be a limiting or controlling factor. In making field observations as in laboratory experiments, I believe in monitoring and reporting all significant environmental conditions. In the deep sea, hydrostatic pressure is an important environmental condition.

I would like to make a remark about numerical taxonomy. This impresses me as being one of the great advances in systematic biology during the last fifty years. It seems so much better to compare groups of organisms according to similarities in their characteristics, the way Dr. Colwell, Dr. James Shewan, and others have done so nicely.[1] This is a big step forward. Now I would like to see more attention devoted to biochemical systematics. This system relies more on the chemical and biochemical composition of the organisms than on their physiological activities.

To what extent does the occurrence of chemical components in different

[1] Colwell, R. R. and J. Liston. 1961. Taxonomic analysis with the electronic computer of some *Xanthomonas* and *Pseudomonas* species. *J. Bacteriol.,* **82:**913–19.
  Colwell, R. R. 1969. Numerical taxonomy of the flexibacter. *J. Gen. Microbiol.,* **58:**207–15.
  Hodgkiss, W., and J. M. Shewan. 1968. Problems and modern principles in the taxonomy of marine bacteria. In *Advances in microbiology of the sea,* ed. M. R. Droop and E. J. F. Wood, pp. 127–66. New York: Academic Press Inc.

microbial species indicate the degree of their phylogenetic relationships? Such specific substances as pigments, unique proteins, nucleic acids, amino acids, fatty acids, polysaccharides, alkaloids, quinones, and hydrocarbons, for example, help to differentiate certain bacteria and higher plants.[2] My plea is for more such chemical analytical tests (e.g., spectrographic or chromatographic) which can be made in a matter of minutes or a few hours instead of relying so much on physiological and nutritional characterization which usually requires more time and is subject to more biological variability. The principles of biochemical taxonomy are well known, but not practiced as much as is now warranted by the advanced state of the art. Biochemical taxonomy is not a substitute for numerical analysis. On the contrary, biochemical taxonomy is a useful adjunct to numerical taxonomy. The point that I have been trying to make all week is that microbiologists, who are concerned with systematics, should be making more effort to discover unique molecules in different microbial species or varieties. This is largely a problem in analytical chemistry as opposed to microbial physiology. For example, instead of testing a culture for its ability to liquefy gelatin (a test which generally takes much time and is influenced by the chemical composition of the medium, temperature, pH, biological variability, etc.), would it not be much better if gelatinase could be detected by its optical properties? My plea is to obtain more significant and less variable characteristics of microorganisms. Computerized taxonomy as an indicator of the phylogenetic relationships of organisms can be no better than the data (characteristics of organisms) fed to the computer. More effort should be made to find distinctive, readily detectable molecules to serve as microbial "fingerprints."

Finally, I would like to make a few remarks about a "fun and games" approach to new problems, the solution of which requires new information. If the desired information involves the properties or physiological activities of microorganisms, I say ask the organism what you want to know. What do I mean by this? To begin with, the semantics of microorganisms are fairly simple. Microbes use only sign language. Therefore, we must ask them questions that they can answer in sign language. For example, a few years ago Holm-Hansen became interested in whether deep-sea organisms were alive or dead. So he asked the organisms such questions as, "Do you respire? Do you reproduce?" He did not receive satisfactory answers from the signs that the organisms gave. Then he asked, "Do you contain ATP?" Asking this question in such a way that it could be answered in sign language involved a lot of work. Finally, the question was asked in such a way that every living cell answered that it contained ATP. The answers came when the cells were treated with luciferin and luciferase. Then they lighted up to say, "Yes, yes." This is now a pretty good test. This is an example of asking the organism.

As a theoretical example, suppose you want to know how much ammonium is liberated from proteinaceous material in a given area in an estuary. Ordinarily, one would first collect a water sample for examination in the laboratory. Organisms in the samples may be upset by such treatment. Some of them may die. Many things could change during sampling, transition, and storage. Usually one

---

[2]R. E. Alston and B. L. Turner. 1963. *Biochemical systematics.* Englewood Cliffs, N.J.: Prentice-Hall, Inc.

starts by getting the organisms in pure culture. Then one may determine whether up to 10^9 or 10^{10} cells per ml of medium enriched with aspargine produce ammonia under conditions which may differ from natural conditions in the estuary. Would it not be much better to ask the organisms in the estuary whether they produce ammonia?

To illustrate this point of asking the organisms *in situ* whether they produce ammonia from asparagine, suppose one prepares a large number of rubber–bulb type water samples from ordinary 2 ml medicine droppers. Such samplers are also to serve as *in situ* cultures tubes. The tip of each is sealed in a flame after pulling it out to a fine capillary. With the desired quantity of asparagine solution (or other substrate) in a vial within the sampler, the assembly is autoclave–sterilized with the bulb collapsed. Now the samplers are ready to be submerged to the desired depth to collect water samples. Breaking off the capillary tip permits water to be drawn into the sampler. Methods of resealing the tip and of securing the samplers at the desired depth are details to be discussed elsewhere. After different periods of incubation (maybe hours or days), subsamples can be examined to determine whether or how much ammonia has been produced. The procedure is so simple and the required equipment is so inexpensive that dozens or hundreds of *in situ* tests can be made at different depths, at different stations, during different seasons, with different substrates, etc. Incidentally, we have under advisement placing on the deep-sea floor an automated carrier of a few hundred "packages" designed to ask the organisms some questions about their physiological reaction rates at deep-sea pressures.

I am not so "hung up" on *in situ* experiments as to believe that all problems of microbial processes in estuaries can be solved in this way. Far from it. But I believe that some very important problems can be solved by asking the organisms *in situ*. We should ask the organisms what they can do and how they do it under conditions which closely approximate conditions in nature.

I thank you all for your attention, for this opportunity to meet with you, and especially for your cordiality and friendship.

Osmund Holm-Hansen

I normally do not like to indulge in mutual admiration, but I feel rather inclined at this time to say that Dr. ZoBell's eloquence and humility are very inspiring to all of us at this conference.

I would like to comment on a couple of things of significance that you mentioned, Dr. ZoBell. One is the subject of microflagellates. Some of us are spending considerable time with these organisms. In our floristic analyses, which we do by settling techniques, we do count everything down to about one micron. We therefore do enumerate and measure the microflagellates. In all our productivity measurements, we also measure the effects of any biological activity carried out by these microflagellates as they are retained by the microfine glass fiber filters which we use.

Also, I talked quite a bit about biomass. Determination of biomass is one step in the long progression ahead of us in trying to understand the recycling of carbon and energy in the world's aquatic environments. The next step, of course, is the rate of transfer, the kinetics, the biological activity. The next step, there-fore, must be one of activity measurements or estimates based on biomass.

One procedure that did not receive any attention during this conference is the use of microradioautographic techniques. I think these techniques will be a strong tool combining microscopic techniques (including electron microscopy) with the heterotrophic uptake studies outlined these past few days. I am very hopeful that these will be informative in both estuarine and deep ocean studies.

Galen E. Jones

I would like to make several comments about some of the challenges that face marine microbiologists today. The federal government, state agencies, and private industry are coming to us asking for solutions to certain practical prob-lems that are of great importance to society. Among these problems are heavy metal, oil, herbicide, pesticide, sewage and thermal pollution, corrosion of metals and other materials in the marine environment, and the harvesting of mineral deposits in the ocean. The scope of the problem in these areas is tremendous. The answers to many of these problems are extremely complex and yet these agencies want rapid, accurate, and reasonably simple answers.

It is a great temptation to direct one's work toward the answers to some of these problems because of the interest and financial support in these areas. I would like to make a plea for those who still have an interest in basic individual research to pursue their interests and not lose the impact of the individual effort in the temptation to get interdisciplinary teams of people to cooperate to solve these very complex practical problems. In the end it is the individual who usually makes the truly important discovery. Nevertheless, microbiologists should work together with other biologists in the estuaries in terms of collecting and doing synoptic sampling which will make the importance of each individual contribu-tion more meaningful. The data of each scientist are important to other scien-tists in unraveling complex ecological and economic phenomena. The example of the Food Chain Group in the Institute of Marine Resources at the Scripps Institution of Oceanography in La Jolla, California, has been an outstanding model of cooperation without loss of the individual excellence of each investiga-tor. A number of human problems are involved in scientific cooperation as well as elucidating the natural phenomena. These problems are well worth the time and the trouble to overcome.

One of the practical aspects that marine biologists will be asked to solve in the coming years is aquaculture. I noticed at this conference that there were no pa-pers on pathogenic microorganisms for marine plants and animals. For reason-able aquaculture in this country, the knowledge of pathogens in the marine en-vironment is essential.

I would like to return to Dr. ZoBell's remark about trying to work closely with the conditions in the natural environment. We should do that. Sometimes in-vestigators want to make a quick survey of a phenomenon or a suspected phe-nomenon to determine whether it is a feasible area to work on. This is the type of investigation which Ed Gonye and I presented at this conference. As the feasibility and enthusiasm for the phenomenon increase, one should develop techniques where it is directly applicable to the environment rather than resorting to the more conventional techniques which we used here, offering a great deal of data in a short period of time. The problem is that there are not many people working in marine microbiology but there is a tremendous number of problems of interest. I hope and trust that marine microbiologists will continue to concern themselves with basic and applied problems in the estuaries and oceans and that no one approach will become so popular that other approaches are forgotten or overlooked.

F. John Vernberg

I was very heartened by Dr. Colwell's comments about the need for creating some type of continuing group of marine microbiologists that would meet on a regular basis. I hope that this group would not lose sight of the total ecosystem and the role of microorganisms in it. There must be an interchange of data and ideas between investigators concerned with other segments of the ecosystem and microbiologists. Meetings such as this should take place again in the future. I want to thank you all for coming to the University of South Carolina.

INDEX

Estuarine Microbial Ecology

Composition, photoengraving, offset printing, and binding by Kingsport Press, Inc., Kingsport, Tennessee. The primary typeface is Linofilm Baskerville, with selected lines in Linofilm Trade Gothic, and the paper Warren's University Text watermarked with the emblem of the University of South Carolina Press.

Estuarine Microbial Ecology

Composition, photoengraving, offset printing, and binding by Kingsport Press, Inc., Kingsport, Tennessee. The primary typeface is Linofilm Baskerville, with selected lines in Linofilm Trade Gothic, and the paper Warren's University Text watermarked with the emblem of the University of South Carolina Press.